Subcellular Biochemistry

Volume 26

myo-Inositol Phosphates, Phosphoinositides, and Signal Transduction

SUBCELLULAR BIOCHEMISTRY

Recent Volumes in This Series:

Volume 17 **Plant Genetic Engineering**
Edited by B. B. Biswas and J. R. Harris

Volume 18 **Intracellular Parasites**
Edited by J. L. Avila and J. R. Harris

Volume 19 **Endocytic Components: Identification and Characterization**
Edited by J. J. M. Bergeron and J. R. Harris

Volume 20 **Mycoplasma Cell Membranes**
Edited by S. Rottem and I. Kahane

Volume 21 **Endoplasmic Reticulum**
Edited by N. Borgese and J. R. Harris

Volume 22 **Membrane Biogenesis**
Edited by A. H. Maddy and J. R. Harris

Volume 23 **Physicochemical Methods in the Study of Biomembranes**
Edited by Herwig J. Hilderson and Gregory B. Ralston

Volume 24 **Proteins: Structure, Function, and Engineering**
Edited by B. B. Biswas and Siddhartha Roy

Volume 25 **Ascorbic Acid: Biochemistry and Biomedical Cell Biology**
Edited by J. Robin Harris

Volume 26 ***myo*-Inositol Phosphates, Phosphoinositides, and Signal Transduction**
Edited by B. B. Biswas and Susweta Biswas

A Continuation Order Plan is available for this series. A continuation order will bring delivery of each new volume immediately upon publication. Volumes are billed only upon actual shipment. For further information please contact the publisher.

Subcellular Biochemistry

Volume 26

myo-Inositol Phosphates, Phosphoinositides, and Signal Transduction

Edited by

B. B. Biswas
University of Calcutta
Calcutta, India

and

Susweta Biswas
Bose Institute
Calcutta, India

PLENUM PRESS • NEW YORK AND LONDON

The Library of Congress cataloged the first volume of this title as follows:

Sub-cellular biochemistry.
London, New York, Plenum Press.
v. illus. 23 cm. quarterly.
Began with Sept. 1971 issue. Cf. New serial titles.
1. Cytochemistry—Periodicals. 2. Cell organelles—Periodicals.
QH611.S84 574.8'76 73-643479

ISBN-13: 978-1-4613-8007-8 e-ISBN-13: 978-1-4613-0343-5
DOI: 10.1007/ 978-1-4613-0343-5

This series is a continuation of the journal *Sub-Cellular Biochemistry*,
Volumes 1 to 4 of which were published quarterly from 1972 to 1975

Softcover reprint of the hardcover 1st edition 1996
A Division of Plenum Publishing Corporation
233 Spring Street, New York, N.Y. 10013

10 9 8 7 6 5 4 3 2 1

INTERNATIONAL ADVISORY EDITORIAL BOARD

Contributors

Sailen Barik Department of Biochemistry and Molecular Biology, College of Medicine, University of South Alabama, Mobile, Alabama 36688-0002

B. B. Biswas Department of Biophysics, Molecular Biology and Genetics, University of Calcutta, Calcutta 700 009, India

Susweta Biswas Department of Biochemistry, Bose Institute, Calcutta 700 054, India

Martin D. Bootman The Babraham Institute Laboratory of Molecular Signaling, Department of Zoology, Cambridge University, Cambridge CB2 3EJ, United Kingdom

Anne Burmester Institute for Organic Chemistry, Laboratory of Bioorganic Chemistry, University of Bremen, 28359 Bremen, Germany

Rik Casteels Laboratory for Physiology, Catholic University of Leuven, Campus Gasthuisberg, B-3000 Leuven, Belgium

Meena R. Chandok School of Life Sciences, Jawaharlal Nehru University, New Delhi 110 067, India

Jie Chen Bernard Laboratory, Division of Cardiology, Department of Medicine and Department of Cell Biology and Anatomy, The Johns Hopkins University School of Medicine, Baltimore, Maryland 21287

Gary G. Coté Department of Molecular and Cell Biology, The University of Connecticut, Storrs, Connecticut 06269-3125; *present address:* Department of Biology, Millikin University, Decatur, Illinois 62522

Richard C. Crain Department of Molecular and Cell Biology, The University of Connecticut, Storrs, Connecticut 06269-3125

Humbert De Smedt Laboratory for Physiology, Catholic University of Leuven, Campus Gasthuisberg, B-3000 Leuven, Belgium

Paolo Gerbasi USDA-ARS Range Weeds and Cereals Research Unit, Montana State University, Bozeman, Montana 59717

Pascal J. Goldschmidt-Clermont Bernard Laboratory, Division of Cardiology, Department of Medicine and Department of Cell Biology and Anatomy, The Johns Hopkins University School of Medicine, Baltimore, Maryland 21287

John N. Hawthorne Department of Biochemistry, Medical School, Queen's Medical Centre, Nottingham NG7 2UH, United Kingdom; *present address:* International Medical College, 46050 Petaling Jaya, Selangor D.H., Malaysia

Lowell E. Hokin Department of Pharmacology, University of Wisconsin Medical School, Madison, Wisconsin 53706

David E. Kandzari Bernard Laboratory, Division of Cardiology, Department of Medicine and Department of Cell Biology and Anatomy, The Johns Hopkins University School of Medicine, Baltimore, Maryland 21287

Ludwig Missiaen Laboratory for Physiology, Catholic University of Leuven, Campus Gasthuisberg, B-3000 Leuven, Belgium

Pushpalatha P. N. Murthy Chemistry Department, Michigan Technological University, Houghton, Michigan 49931

Jan B. Parys Laboratory for Physiology, Catholic University of Leuven, Campus Gasthuisberg, B-3000 Leuven, Belgium

Victor Raboy USDA-ARS Range Weeds and Cereals Research Unit, Montana State University, Bozeman, Montana 59717

Alan R. Saltiel Department of Signal Transduction, Parke-Davis Pharmaceutical Research, Division of Warner-Lambert Co., Ann Arbor, Michigan 48105

Carsten Schultz Institute for Organic Chemistry, Laboratory of Bioorganic Chemistry, University of Bremen, 28359 Bremen, Germany

Stephen B. Shears Inositol Lipid Section, Laboratory of Cellular and Molecular Pharmacology, National Institute of Environmental Health Sciences, National Institutes of Health, Research Triangle Park, North Carolina 27709

Ilse Sienaert Laboratory for Physiology, Catholic University of Leuven, Campus Gasthuisberg, B-3000 Leuven, Belgium

Sudhir K. Sopory School of Life Sciences, Jawaharlal Nehru University, New Delhi 110 067, India

Christoph Stadler Institute for Organic Chemistry, Laboratory of Bioorganic Chemistry, University of Bremen, 28359 Bremen, Germany

Yir G. Yueh Department of Molecular and Cell Biology, The University of Connecticut, Storrs, Connecticut 06269-3125; *present address:* Department of Biochemistry and Molecular Biology, Mayo Clinic Scottsdale, Scottsdale, Arizona 85259

Preface

The biochemical and molecular basis of signal transduction across the cell membrane has long been a subject of great interest. And this interest has indeed made an impact in answering several questions raised on the perception, amplification, and responses elicited by different cells exposed to different agents. The principal architecture of the major signaling systems is not known. Different signaling pathways consist of a series of proteins, such as specific receptors, GTP-binding proteins, second-messenger-generating enzymes, protein kinases, and phosphatases, including other regulatory and effector proteins. The technique of cloning and sequencing of genes has revealed that all these proteins exhibit enormous heterogeneity, differential tissue expression, and specific intracellular localization. However, the biological significance of this heterogeneity has not been understood.

There are also diverse interactions between different signaling systems. These interactions include potentiation, cooperation, antagonism, and synergism in cellular responses. The present volume is designed to include *myo*-inositol phosphate signaling systems in both animal and plant responses.

There has recently been rapid progress in understanding receptors that generate intracellular signals from inositol lipids. One of these lipids, phosphatidylinositol bisphosphate, is hydrolyzed to diacylglycerol (DAG) and inositol trisphosphate [Ins(1,4,5)P_3] as part of a signal transduction mechanism for controlling a variety of cellular processes, including secretion, metabolism, phototransduction, cell proliferation, and neural and contractile processes. DAG operates within the plane of the membrane to activate protein kinase C, whereas Ins(1,4,5)P_3 is released into the cytosol to function as a second messenger for mobilizing intracellular calcium. It has been known for a long time that Ca^{2+} plays a vital role in regulating a variety of cellular activities. Moreover, it is now known that Ca^{2+} homeostasis is maintained by complex interactions among different signaling systems. In the first chapter the history of phosphoinositide research and discovery of *myo*-inositol trisphosphate as a second messenger is presented by Hokin. He depicts the story of how the concept of inositol trisphosphate unexpectedly emerged as a universal second messenger, systematically and sequentially, out of the work in his and other laboratories. Concomitantly, Hawthorne, in Chapter 2, reviews the recent work on phosphoinositides and synaptic transmission. He discusses metabolism of phosphoinositides, including phosphatidylinositol glycan neuromodulation and plasticity, along with the long-term potentiation of synaptic transmission involving glutamate receptors, influx of Ca^{2+} ions, and protein kinase C activation.

In the third chapter Missiaen and colleagues discuss the control of Ca^{2+} release induced by *myo*-inositol trisphosphate and its implication in signal transduction. They dwell on the complexities of inositol trisphosphate receptor modulation by various cellular components. The regulation by Ca^{2+} can modify the release in the presence of a constant concentration of $InsP_3$ and could therefore be the basis for generating Ca^{2+} oscillations and waves. The inositol phospholipid signaling pathway has been found to be regulated by cytoskeletal structure, and, in particular, some actin-binding proteins have been implicated in this process. Kandzari, Chen, and Goldschmidt-Clermont very succinctly discuss this aspect in Chapter 4.

A protein kinase cascade is now a well-known concept for amplifying a single signal to regulate a large number of cellular functions. The regulation of gene transcription and cell cycle involves a network of such protein kinase cascades. Besides the protein kinases, protein phosphatases are also implicated in this network. The involvement of some protooncogenes, such as *ras, raf, erk, rsk, rap, sos,* and *GAP,* in this cascade is now gradually being unraveled. This has been very ably reviewed by Barik in Chapter 5. He further touches on the direct and indirect effects of nitric oxide on the cascades of phosphorylation and dephosphorylation of proteins involved in signal transduction.

Chapter 6 contains an interesting review presented by Saltiel on the state of the art of structural and functional roles of a particular class of phosphoinositides, i.e., glycosylphosphoinositides (GPIs), linked to proteins for anchorage of proteins to membranes. It is further pointed out how a single amino acid at the GPI attachment site determines which membrane anchor is to be utilized. Recently the role of higher homologues of inositol phosphates in certain cellular functions has been sought. Shears, who is one of the proponents of the involvement of higher homologues of inositol phosphates on regulation of cellular activities, describes the novel effects of inositol pentakis- and hexakisphosphates on ion channels and vesicle traffic in Chapter 7.

The next five chapters have primarily been devoted to reviewing the state of the art of biochemistry and genetics of *myo*-inositol phosphates and phosphoinositide accumulation and degradation and induction of signal transduction, including the effect of light in plant systems. Plant systems are less explored than animal systems as far as the phosphoinositide–inositol trisphosphate signaling pathway is concerned. Therefore, much emphasis has been given to focus in this volume on different aspects of signaling mediated through *myo*-inositol phosphates involving Ca^{2+} as well as through light and other physical, chemical, and mechanical agents.

Murthy briefly describes the metabolic network of *myo*-inositol phosphates in plants in Chapter 8. The genetics of *myo*-inositol phosphate synthesis and accumulation in plant and animal systems have been ably presented by Raboy and Gerbasi in Chapter 9, though very little work has so far been done on the genetics of accumulation of different inositol phosphates. Biswas and Biswas present a new concept of *myo*-inositol hexakisphosphate–phytase signaling in plants while discussing an overall picture of the implications of *myo*-inositol phosphates in metabolism and calcium mobilization in plants. This is included in Chapter 10. Coté, Yueh, and Crain effectively deal with the phosphoinositide turnover in certain plant responses, assigning a definite role of inositol trisphosphate in signal transduction; this is included in Chapter 11. Light-induced signal transduction pathways involving inositol trisphosphate are discussed by Sopory and Chandok in Chapter 12. The involvement of the phytochrome system involved in this pathway is an interesting aspect to follow. Finally, but not least, a chapter has been earmarked for the synthesis, separation, and identification of different inositol phosphates. This assignment has been very ably fulfilled by Schultz, Burmester, and Stadler. Thus, Chapter 13 is a valuable addition to this volume for the benefit of all workers in this field of inositol phosphates.

From the progress of research in this area as depicted in this volume, it is now clear that cross talk among different signaling systems and spatiotemporal aspects of biochemical or molecular events occurring within cells, be they animal or plant, is a prerequisite for an understanding of physiological responses in normal and abnormal growth, including diseases and development of an organism. Finally, we hope that this volume will serve a variety of interests of workers in this field.

B. B. Biswas
Susweta Biswas

Calcutta, India

Contents

Chapter 1
History of Phosphoinositide Research
Lowell E. Hokin

1. Historical Background 1
   1.1. Discovery of the "Phosphoinositide Effect" 1
   1.2. Discovery of the Phosphatidylinositol Cycle 5
   1.3. Cellular Site of the Phosphatidylinositol Effect 7
   1.4. Early Studies Suggesting an Involvement of Inositol Lipids in Calcium-Dependent Responses 8
   1.5. Early Studies on Polyphosphoinositide Turnover 9
   1.6. Formation of Inositol Phosphates 9
2. Modern Era 10
   2.1. Stimulation of Polyphosphoinositide Metabolism 10
   2.2. Role of GTP-Binding Proteins in Receptor Coupling to Phospholipase C 18
   2.3. Protein Kinase C 20
   2.4. Phosphoinositides, Cell Proliferation, and Oncogenes 25
   2.5. Pharmacological Action of Lithium 27
   2.6. Fertilization 29
3. References 30

Chapter 2
Phosphoinositides and Synaptic Transmission
John N. Hawthorne

1. The Brain Phosphoinositides 43
   1.1. Phosphoinositide Metabolism 43
   1.2. Receptor-Linked Phosphoinositide Hydrolysis 44

2. Further Metabolism of Inositol Trisphosphate ... 45
3. Phosphoinositide-Specific Phospholipase C ... 45
4. G Proteins and Phosphoinositide-Linked Receptors ... 46
5. Phosphoinositide-Linked Receptors in Nervous Tissue ... 47
6. The Inositol Trisphosphate Receptor and Calcium Release ... 48
7. Inositol-1,3,4,5-Tetrakisphosphate ... 49
8. Lithium Salts ... 49
9. Protein Kinase C ... 50
10. Neuromodulation and Plasticity ... 51
11. Phosphatidylinositol 3-Kinase ... 52
12. Phosphatidylinositol Glycans ... 52
13. The Future ... 53
14. References ... 54

Chapter 3

Control of the Ca^{2+} Release Induced by *myo*-Inositol Trisphosphate and the Implication in Signal Transduction

Ludwig Missiaen, Jan B. Parys, Humbert De Smedt, Ilse Sienaert, Martin D. Bootman, and Rik Casteels

1. Introduction ... 59
2. Functional Regulation ... 60
   2.1. Control by $InsP_3$ ... 60
   2.2. Control by Ca^{2+} ... 61
   2.3. Control by Phosphorylation ... 67
   2.4. Control by Sulfhydryl Oxidation ... 68
   2.5. The Adenine Nucleotide-Binding Site ... 69
   2.6. Interaction with the Cytoskeleton ... 70
3. Heterogeneity of $InsP_3R$ ... 71
   3.1. Molecular Diversity ... 71
   3.2. Functional Comparison ... 73
   3.3. Cell-Specific Expression ... 75
4. Partial Ca^{2+} Release in Response to Submaximal $InsP_3R$ Stimulation ... 77
   4.1. All-or-None Ca^{2+} Release ... 81
   4.2. Steady-State Ca^{2+} Release ... 81
   4.3. $InsP_3R$ Inactivation ... 82
5. Conclusions ... 83
6. References ... 83

Chapter 4

Regulation of the Actin Cytoskeleton by Inositol Phospholipid Pathways

David E. Kandzari, Jie Chen, and Pascal J. Goldschmidt-Clermont

1. Introduction 97
2. Effect of Membranes on Actin and Satellite Proteins 99
3. ras and the Cytoskeleton 101
4. Phospholipase A_2 and Arachidonate Metabolism 103
5. Phospholipase C and the Cytoskeletal Response 104
6. Phospholipid Kinases and Cytoskeletal Reorganization 107
7. Effects of Other Phospholipids on the Actin Structure 109
8. Conclusions 109
9. References 109

Chapter 5

Protein Phosphorylation and Signal Transduction

Sailen Barik

1. Introduction 115
2. Primary Signal Transduction Mechanisms in Eukaryotes 116
   2.1. Receptor Tyrosine Kinases 117
   2.2. Nonreceptor Tyrosine Kinases 122
   2.3. Protein Serine/Threonine Kinases 125
   2.4. Protein Kinase C and Inositol Phosphates 133
   2.5. Soluble Guanylate Cyclase and Nitric Oxide 136
   2.6. Phosphorylation of Transcription Factors 137
   2.7. Protein Phosphatases 143
3. Signal Transduction Systems in Prokaryotes 148
   3.1. Bacterial Chemotaxis 149
   3.2. Nitrogen Assimilation 150
   3.3. Outer Membrane Protein Synthesis 151
4. Malfunction and Disease 153
5. Future Directions 154
6. References 156

Chapter 6

Structural and Functional Roles of Glycosylphosphoinositides

Alan R. Saltiel

1. Introduction 165
2. Structure and Distribution of GPI Molecules 166
3. Biosynthesis and Attachment of GPI 168
   3.1. Biosynthesis of GPI 168
   3.2. Attachment of GPI Anchors 169
4. Glycosylphosphatidylinositol in Insulin Action 171
5. Functional Roles of GPI Anchors 174
   5.1. Protein Distribution and Targeting 175
   5.2. Regulation of Lateral Mobility 176
   5.3. Regulated Release at the Cell Surface 177
6. Conclusions 178
7. References 178

Chapter 7

Inositol Pentakis- and Hexakisphosphate Metabolism Adds Versatility to the Actions of Inositol Polyphosphates: Novel Effects on Ion Channels and Protein Traffic

Stephen B. Shears

1. Introduction 187
2. Synthesis of Ins(1,3,4,5,6)P_5 188
3. Ins(3,4,5,6)P_4 and Ins(1,4,5,6)P_4 193
   3.1. Identification 193
   3.2. Synthesis and Metabolism of Ins(1,4,5,6)P_4 194
   3.3. Ins(1,4,5,6)P_4 and Pleckstrin Homology Domains 197
   3.4. Synthesis and Metabolism of Ins(3,4,5,6)P_4 198
   3.5. Evidence for Physiological Effects of Ins(3,4,5,6)P_4 200
4. Metabolism and Functions of InsP_6 205
   4.1. Early Thoughts on InsP_6 205
   4.2. Synthesis of InsP_6 207
   4.3. Dephosphorylation of InsP_6 208
   4.4. Synthesis and Metabolism of Diphosphoinositol Polyphosphates 209
   4.5. Proteins that Bind InsP_6 with High Affinity: A Functionally Related Family? 211
5. Concluding Statement 216
6. References 216

Chapter 8

Inositol Phosphates and Their Metabolism in Plants

Pushpalatha P. N. Murthy

1. Introduction 227
2. Nomenclature of *myo*-Inositol and Its Phosphates 228
3. Central Role of *myo*-Inositol in Plant Metabolism 231
4. Metabolism of *myo*-Inositol Phosphates 234
   4.1. Inositol Monophosphates 234
   4.2. Inositol Bisphosphates 236
   4.3. Inositol Trisphosphates 236
   4.4. Inositol Tetrakisphosphates 238
   4.5. Inositol Pentakisphosphates 238
   4.6. Inositol Hexakisphosphate (Phytic Acid) 238
5. Synthetic Analogues of Inositol Phosphates 243
6. Inositol Pyrophosphates 246
7. Metabolic Processes Involving Inositol Phosphates 249
   7.1. Signal Transduction 249
   7.2. Stress-Related Processes 249
8. Conclusions 250
9. References 251

Chapter 9

Genetics of *myo*-Inositol Phosphate Synthesis and Accumulation

Victor Raboy and Paolo Gerbasi

1. Introduction 257
2. *myo*-Inositol and L-*myo*-Inositol 1-Phosphate: Synthesis/Regulation 259
3. Phosphatidylinositol Synthesis and Transport 261
4. Phosphatidylinositol Kinases and Phosphate Kinases 262
5. Phosphatidylinositol-Specific Phospholipase C and 1,2-Cyclic Inositol Phosphates 264
6. Inositol Phosphate Receptors and Kinases 267
7. Inositol Polyphosphate Phosphatases other than Phytases 269
8. Phytases 271
9. Phytic Acid 272
10. Summary: Current Status of Inositol and Phosphatidylinositol Phosphate Genetics 276
11. References 277

Chapter 10

Metabolism of *myo*-Inositol Phosphates and the Alternative Pathway in Generation of *myo*-Inositol Trisphosphate Involved in Calcium Mobilization in Plants

Susweta Biswas and B. B. Biswas

1. Introduction 287
2. Metabolism of Inositol Phosphates in Plants 288
   2.1. Biosynthetic Pathways of *myo*-Inositol Phosphates 289
   2.2. Dephosphorylation of *myo*-Inositol Polyphosphates 291
   2.3. Metabolic Cycle Involving Glucose-6-Phosphate and *myo*-Inositol Phosphates in Plants 292
3. Calcium Mobilization and Its Role in Plant Responses 293
   3.1. Calcium Channels in Plants 294
   3.2. Calmodulin: The Calcium Receptor 294
   3.3. Calcium- and Calmodulin-Dependent Protein Phosphorylations 295
   3.4. Calcium- and Calmodulin-Independent Protein Kinases 295
   3.5. Inositol-Trisphosphate-Activated Ca^{2+} Channels in Plants 295
   3.6. Abscisic-Acid-Activated Ca^{2+} Channels in Plants 296
   3.7. Voltage-Dependent Ca^{2+} Channels in Plants 297
   3.8. Calcium-Activated Ca^{2+} Channels 297
4. Phosphoinositides in Plants 297
   4.1. Phospholipase C in Plants 299
   4.2. GTP-Binding Proteins in Plants 299
   4.3. The *myo*-Inositol Trisphosphate Receptor in Plants 300
   4.4. Cascades of Protein Kinases and Phosphatases in Plants 301
5. Phosphoinositide and Other *myo*-Inositol Phosphate Signaling Systems in Plants 303
6. Signaling in Plants by Wounding 304
7. Signaling in Leaf Movement 305
8. Concluding Remarks 306
9. References 307

Chapter 11

Phosphoinositide Turnover and Its Role in Plant Signal Transduction

Gary G. Coté, Yir G. Yueh, and Richard C. Crain

1. Introduction 317
   1.1. Plants' Perception of Their Environment 317

1.2. Evidence for a Phosphoinositide Signaling System in Plants . . . 319
2. Stress Responses in Unicellular Green Algae . . . 320
2.1. Osmotic Stress . . . 320
2.2. Deflagellation of *Chlamydomonas reinhardtii* . . . 322
3. Pathogen Responses in Higher Plants . . . 324
4. Plant Movements Mediated by Osmotic Changes . . . 326
4.1. Hydraulic Movement . . . 326
4.2. Regulation of Stomatal Aperture . . . 327
4.3. Regulation of Leaf Movement . . . 331
4.4. Plant Hydraulic Movement: A Common Signaling Mechanism? . . . 333
5. Future Prospects . . . 334
5.1. Conclusive Demonstration of Phosphoinositide Signaling in Plants . . . 334
5.2. Other Possible Roles of Phosphoinositides . . . 334
6. References . . . 335

Chapter 12

Light-Induced Signal Transduction Pathway Involving Inositol Phosphates

Sudhir K. Sopory and Meena R. Chandok

1. Introduction . . . 345
2. Inositol Phospholipid Turnover as Signal Tranduction in Plants . . . 346
3. Light-Stimulated Inositol Turnover . . . 349
3.1. Changes in Inositol Lipids and Inositol Phosphates . . . 350
3.2. Changes in Kinases/Phosphatases and Phospholipase C . . . 351
3.3. Changes in $InsP_3$ and Calcium Level as a Consequence of Polyphosphoinositide Turnover . . . 353
3.4. Changes in DAG Level and Role of Protein Kinases . . . 354
3.5. Is Polyphosphoinositide Turnover in Plants Coupled to G Protein? . . . 356
4. Relationship between Inositol Phospholipid Turnover and Physiological Responses . . . 357
4.1. *Samanea* Leaf . . . 357
4.2. *Ceratodon purpureus* . . . 358
4.3. Maize (*Zea mays*) . . . 359
4.4. Tissue and Protoplast Culture . . . 361
4.5. Bud Formation in *Funaria* . . . 362
5. Alternate Roles for Inositol Phospholipids in Signal Transduction . . . 362

6. Conclusions 363
7. References 364

Chapter 13

Synthesis, Separation, and Identification of Different Inositol Phosphates

Carsten Schultz, Anne Burmester, and Christoph Stadler

1. Introduction 371
1.1. Inositol Phosphate Turnover 372
1.2. The Structure of *myo*-Inositol 374
2. Synthetic Inositol Phosphates 375
2.1. Synthesis of Natural and Unnatural Inositol Phosphates 378
2.2. Synthesis of Inositol Phosphate Analogues 384
2.3. Inositol Monophosphate Analogues as Inhibitors of the Inositol Monophosphatase 387
2.4. Ins(1,4,5)P_3 Analogues and Some of Their Biological Properties 388
2.5. Membrane-Permeant Derivatives of Inositol Phosphates 394
3. Separation 396
3.1. Extraction 397
3.2. Anion-Exchange Chromatography 398
3.3. High-Performance Liquid Chromatography 398
3.4. Gas Chromatography 399
3.5. Detection 399
4. Identification 401
4.1. Nuclear Magnetic Resonance Spectroscopy 401
4.2. Chemical Analysis 401
4.3. Discovery of Novel Inositol Phosphates 402
5. References 403

Index 415

Chapter 1

History of Phosphoinositide Research

Lowell E. Hokin

1. HISTORICAL BACKGROUND

1.1. Discovery of the "Phosphoinositide Effect"

The discovery of the "phosphoinositide effect" arose out of an accidental observation over 40 years ago. The details of this discovery have been reviewed (Hokin, 1987; Hokin and Hokin-Neaverson, 1989). This chapter deals with studies dating up to about 1990.

This story dates back to 1949, when I arrived at H. A. Krebs's laboratory to study for a Ph.D. in Biochemistry. Krebs's policy was to have a student formulate his own problem, pursue it independently with a minimum of supervision, and publish by himself. With some background in gastroenterology, which I had received in the laboratory of Warren S. Rehm while attending medical school, I looked into whether I could use pancreatic tissue as a model to study the synthesis and secretion of proteins, using amylase as a measure of these two processes.

Pigeon pancreas slices proved to be an excellent system for this purpose. Near the end of my doctoral studies, I had become interested in the possible involvement of RNA in protein synthesis. This was based on some cytological studies in the late 1940s by Caspersson and Brachet showing that RNA levels in a variety of dividing and nondividing cells were correlated with protein synthetic activity. RNA levels were particularly high in the pancreas. At about that time,

Lowell E. Hokin Department of Pharmacology, University of Wisconsin Medical School, Madison, Wisconsin 53706.

Subcellular Biochemistry, Volume 26: myo-Inositol Phosphates, Phosphoinositides, and Signal Transduction, edited by B. B. Biswas and Susweta Biswas. Plenum Press, New York, 1996.

Kenneth Burton, who was then a Lecturer in Krebs's department, drew my attention to an abstract by two Russian workers, M. A. Guberniev and L. I. Il'ina, who showed that on cholinergic stimulation of secretion in pancreas, stomach, and salivary glands in anesthetized dogs (shades of Pavlov), there were marked increases in the incorporation of ^{32}P into "nucleoproteins" isolated by the method Schmidt and Thannhauser (1945). This method was really designed to isolate phosphorus fractions, including DNA and RNA (separated by alkaline hydrolysis), for chemical analysis.

The use of ^{32}P as a tracer had been adopted by Krebs and his associates a year or two earlier. I found that on stimulation of enzyme secretion by cholinergic agents in pancreas slices, there was about a 100% increase in the incorporation of [^{32}P]orthophosphate into the RNA fraction, which confirmed the observations of Guberniev and Il'ina on "nucleoproteins." Near the end of this work, I began to suspect that perhaps an alkaline hydrolytic product of lipids was contaminating the RNA fraction and that this might be responsible for the stimulation of ^{32}P incorporation into RNA.

This idea came to me shortly after I had completed my requirements for the Ph.D. and was about to set sail with Mabel Hokin for Halifax. I did manage to do one experiment, but I did not have time to go through the necessary procedures to rid the lipid fraction of contaminating ^{32}P and to count the radioactivity. We took with us on the boat a rack of about a dozen rather large test tubes containing the ethanol-ether extracts containing the ^{32}P-labeled lipids and contaminating ^{32}P. (I don't know whether I would attempt this these days.) When we arrived at J. H. Quastel's laboratory at McGill, we counted the purified total lipid fractions and found an enormous increase (tenfold) in the specific radioactivity in the lipids from pancreas slices that had been stimulated with carbachol. When better methods for isolation of nucleic acids from small samples of tissue were developed by others in 1952, we were able to show that there was no stimulation of ^{32}P incorporation into RNA. The lipid work was published in the *Journal of Biological Chemistry* in 1953 (M. Hokin and Hokin, 1953). It is of interest to point out that the Watson–Crick double-helix structure of DNA was published that same year; nucleic acids had a much brighter immediate future than phospholipids. In fact, for many years nobody paid much attention to these unique observations.

The 1953 *Journal of Biological Chemistry* paper arising from this work is often incorrectly quoted as demonstrating the stimulated turnover of phosphatidylinositol (PI). In the initial studies, only the incorporation of [^{32}P]Pi into the total phospholipid fraction was followed, because methods were not available for separation of individual phospholipids from small samples of tissue. The stimulated ^{32}P incorporation appeared to be confined to the phospholipids because there was no stimulation of ^{32}P incorporation into other phosphorus-containing fractions (M. Hokin and Hokin, 1953). The stimulation was also blocked by the muscarinic cholinergic antagonist atropine. In a subsequent study (M. Hokin and

Hokin, 1954), it was found that enzyme secretion and ^{32}P incorporation into phospholipids did not parallel each other as the acetylcholine (ACh) concentration was increased. For example, enzyme secretion was 80% of maximal at 10^{-7} M ACh, but incorporation of ^{32}P into phospholipids was only 15% of maximal. At 10^{-6} M ACh, enzyme secretion was maximal, but incorporation of ^{32}P into phospholipids was only 65% of maximal. These observations indicated that enzyme secretion and the incorporation of ^{32}P into phospholipids were not tightly coupled. The stimulated incorporation of ^{32}P into phospholipids was not a unique effect of cholinergic drugs unrelated to stimulated enzyme secretion because cholecytokinin–pancreozymin (CCK/PZ), which is also a physiological secretagogue for enzyme secretion but binds to a separate receptor, gave identical results to those with ACh (L. Hokin and Hokin, 1956), except that its actions were not blocked by atropine. In the intervening years, a similar dissociation between phosphoinositide effects and physiological responses at increasing agonist concentrations has been observed (L. Hokin, 1968; Michell *et al.*, 1977). It appears that the phospholipid effect corresponds more closely to agonist binding to receptors and that there are spare receptors in some tissues (Michell *et al.*, 1977).

When methods became available for separation of the deacylated products of the individual phospholipids (Dawson, 1954) and somewhat later the intact phospholipids (Marinetti and Stotz, 1956) from small samples of tissue, it was possible to determine which phospholipids showed the phospholipid response. In pigeon pancreas slices, the incorporation of ^{32}P into PI was increased about 15-fold, and that into phosphatidic acid (PA) about threefold (L. Hokin and Hokin, 1955b, 1958a). The incorporation of [^{3}H]*myo*-inositol into PI was also stimulated in pancreas (L. Hokin and Hokin, 1958a) and brain (L. Hokin and Hokin, 1958b). Figure 1 shows an autoradiogram of lipids from unstimulated and stimulated pigeon pancreas slices, separated on silicic-acid-impregnated paper. The major stimulated spot is PI. Below PI is lysophosphatidylinositol, which was artifactually derived from PI at that time and was eliminated if the tissues were immediately frozen. At the top of the chromatogram is PA, which also showed good stimulation.

The incorporation of ^{32}P and the respective head groups was also increased in phosphatidylcholine (PC) and phosphatidylethanolamine (PE) (L. Hokin and Hokin, 1955a), but to a considerably lesser extent than that in PI and PA. These smaller effects in PC and PE, which comprise the major fraction of the total phospholipid phosphorus but usually turn over much more slowly than PI and PA, have not received much attention until relatively recently. Data have accumulated that suggest that in some tissues agonists activate phosphodiesteratic cleavage of PC and PE, releasing diacylglycerol (DAG) (Besterman *et al.*, 1986; see also below), and in fact, in some tissues the phosphodiesteratic cleavage of PC seems to be the main source of DAG on agonist stimulation.

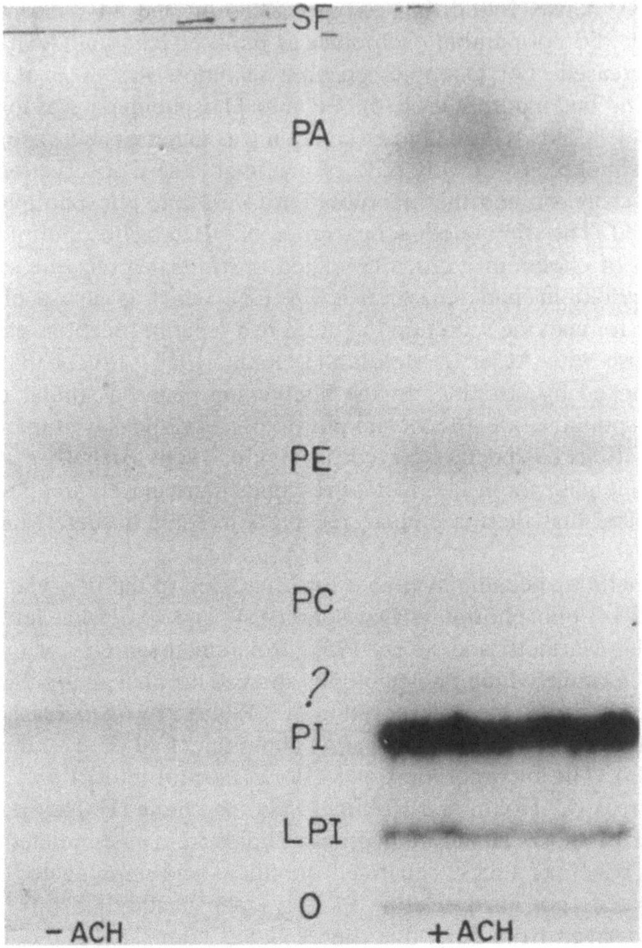

FIGURE 1. The effects of acetylcholine on the incorporation of ^{32}P into phospholipids in pancreas slices. Left lane, no ACh; right lane, ACh. O, origin; LPI, lysophosphatidylinositol; PI, phosphatidylinositol; PC, phosphatidylcholine; PE, phosphatidylethanolamine; PA, phosphatidic acid. This autoradiogram is similar to Figure 1 in L. Hokin and Hokin (1958b).

In the case of the pancreas, some of the stimulated incorporation of ^{32}P and $[^{3}H]$*myo*-inositol into PI and of ^{32}P into PA was a result of increased total turnover of these phosphatides. This was indicated by the observation that ACh increased the incorporation of $[^{14}C]$glycerol into PI fivefold and into PA threefold (L. Hokin and Hokin, 1958a,b). Stimulated *de novo* synthesis of PI in exocrine pancreas has been confirmed and extended (Calderon *et al.*, 1979). These observations are in line with more recent studies that show a substantial release of free fatty acids and glycerol from PI on stimulation of enzyme secretion in mouse pancreatic minilobules (Dixon and Hokin, 1984). These moieties can reenter lipids only by the sequential action of glycerol kinase and *sn*-glycero-3-phosphate transacylase to form PA, which is the branch point for the synthesis of other lipids (see Kennedy, 1986).

At the time that the phospholipid effect in pancreas was first reported (M. Hokin and Hokin, 1953), ACh was also shown to stimulate the incorporation of ^{32}P into phospholipids in guinea pig brain cortex slices, albeit to a much smaller extent than that in pancreas. There was no increased incorporation of [^{14}C]glycerol into PI or PA in guinea pig brain cortex slices on stimulation with ACh, suggesting that in this tissue the DAG moiety is largely conserved (L. Hokin and Hokin, 1958a,b), and there was independent turnover of the inositol monophosphate moiety.

Since these early studies, a phosphoinositide effect in brain slices and in cell-free preparations containing synaptosomes has been studied extensively (see Abdel-Latif, 1983; Agranoff, 1983). Very high concentrations of ACh were required to achieve maximal effects in brain cortex slices (L. Hokin and Hokin, 1955a, 1958b). This was shown to be at least in part a result of very poor permeation of ACh into the slices, because maximal effects in cell-free preparations of brain were seen with concentrations of ACh three orders of magnitude lower (L. Hokin, 1969; Redman and Hokin, 1964). Synaptosomes, which are really "minicells," are the only "cell-free" system in which stimulated incorporation of precursors into PI and PA has been reported, in contrast to PI breakdown, which occurs in many cell-free preparations fortified with GTP or a nonmetabolizable GTP analogue (see below).

In addition to pancreas and brain cortex already discussed, phospholipid effects were seen in a wide variety of tissues stimulated with appropriate neurotransmitters or hormones. Thus, it was realized early that the phosphoinositide effect was a broadly based mechanism thrown into play on activation of a variety of receptors (L. Hokin, 1968).

1.2. Discovery of the Phosphatidylinositol Cycle

Acetylcholine is a muscarinic agonist for stimulated NaCl secretion into the avian salt gland (Schmidt-Nielson, 1960). When albatross or seagull salt gland slices were incubated with [^{32}P]Pi in the absence and presence of ACh, the incorporation of ^{32}Pi into PA and PI was increased 15- and threefold, respectively, by ACh (L. Hokin and Hokin, 1960). Detailed kinetic studies of the turnover of PA and PI under various conditions revealed several interesting features. If ACh was added to salt gland slices preincubated with [^{32}P]Pi, a fraction of PA amounting to no more than 20% of the total PA became rapidly labeled and continued to turn over (M. Hokin and Hokin, 1964). Under these conditions, in contrast to the addition of ACh and [^{32}P]Pi at zero time, there was only a small increase in radioactivity in PI caused by ACh, presumably as a result of simultaneous breakdown of prelabeled PI and its increased turnover (see below). If the action of ACh was blocked by atropine, the renewing fraction of PA rapidly disappeared and was accompanied by a rapid spurt in radioactivity in PI. The rise in radioactivity in PI was quite close to the loss in radioactivity in PA. If the blocking action of atropine was overridden by washing the slices and addition of a concentration of ACh 33 times higher than the initial ACh concentration, the labeled PI that had been formed on adding atropine disappeared, and PA gained the label lost from PI. Similar results were obtained with [^{3}H]inositol prelabeling; i.e., PI gained [^{3}H]inositol on atropine blockade and lost this fraction of [^{3}H]inositol on restimulation with ACh.

On the basis of the observations in the avian salt gland, a scheme was proposed, called the "phosphatidylinositol–phosphatidic acid" (PI–PA) cycle (M. Hokin and Hokin, 1964; Figure 2 in Rana and Hokin, 1990) in which, on stimulation with ACh, PI breaks down to DAG and inositol-1-phosphate, catalyzed by PI phosphodiesterase [phospholipase C (PLC) or phosphoinositidase C], and DAG is phosphorylated by ATP to form PA. On removal of ACh, PA is converted back to PI by the sequential actions of CMP-PA cytidyltransferase and PI synthase. From the data presented earlier, it is obvious that all steps in the cycle are occurring under constant stimulation.

Another interesting conclusion from these data was that there was a discrete compartment of PI and PA that turned over, while the remaining fraction of these lipids did not turn over, in response to ACh. This has been confirmed and extended (see below). Interconversions between PA and PI similar to those in salt gland were later shown for the exocrine mouse pancreas (see Hokin-Neaverson, 1977), where the secretory process and products are quite different from those in the avian salt gland; however, in both systems secretion is coupled to receptor occupancy, leading to mobilization of intracellular Ca^{2+} (Rubin, 1982).

If albatross salt gland slices were incubated without and with ACh, the average PI phosphorus in the stimulated tissue was 40% lower than that in the control (L. Hokin and Hokin, 1960), if the stimulated value was expressed as percentage of its control from each animal. Hokin-Neaverson (1974) observed a decrease in mass of PI and a roughly equivalent increase in mass of PA on stimulation of mouse pancreas with ACh or CCK/PZ, and Jones and Michell (1974) noted a decrease in mass of PI on stimulation of the parotid with ACh or epinephrine. This confirmed and extended the studies on the kinetics of ^{32}P and [^{3}H]*myo*-inositol labeling as well as the chemical measurements in the avian salt gland made more than a decade earlier. As indicated above, if brain cortex slices were incubated with and without ACh in the presence of [^{14}C]glycerol, there was no increased incorporation of radioactivity into PI or PA (L. Hokin and Hokin, 1958a,b). This made it highly likely that DAG kinase was the enzyme responsible for the stimulated incorporation of ^{32}P into PA and PI, rather than *sn*-glycero-3-phosphate transacylase.

1.3. Cellular Site of the Phosphatidylinositol Effect

If pigeon pancreas slices were incubated without and with ACh, followed by homogenization and differential centrifugation, about 80% of the ACh-induced increment in ^{32}P incorporation or [^{14}C]glycerol incorporation in PI was in the microsomal fraction (Redman and Hokin, 1959). In the pancreas, this fraction is derived from the smooth-surfaced and rough-surfaced endoplasmic reticulum (ER) (Palade *et al.*, 1962) and presumably the plasma membrane. The remaining increment was distributed as a contaminant in the various fractions. The stimulated ^{32}P incorporation in the avian salt gland was also microsomal, derived primarily from invaginated plasma membrane (L. Hokin and Hokin, 1960).

An autoradiographic technique for measuring the incorporation of [^{3}H]*myo*-inositol into PI was developed (L. Hokin, 1965; L. Hokin and Huebner, 1967), and grain counts showed the stimulation of [^{3}H]*myo*-inositol incorporation into PI in pancreas to be about equally distributed between the basophilic cytoplasm (site of rough-surfaced ER) and nonbasophilic cytoplasm (site of smooth-surfaced ER and Golgi apparatus). Increased graining over the Nissl bodies (site of the ER) was also seen in cat superior cervical ganglion. There was no increased graining around the plasma membrane on stimulation, excluding this structure as the primary site of PI turnover. Because cell fractionation had already identified the microsomal fraction as the site of stimulation, the combined autoradiographic and differential centrifugation studies strongly suggested that the site of stimulated [^{3}H]*myo*-inositol incorporation into PI was in the ER and excluded the plasma membrane as a major site of stimulation. Similar autoradiographic results were later obtained with photoreceptor cells of the retina in response to light (Anderson *et al.*, 1983; Schmidt, 1983). The data indicate that the resynthesis of PI, and presumably the majority of its breakdown, occur in the ER. In fact, the magnitude of PI loss in pancreas is far greater than what could be accounted for by total plasmalemmal PI breakdown (Dixon and Hokin, 1984). In pancreas, nearly 70% of the total cellular PI undergoes breakdown on maximal stimulation, and the plasma membrane contains only about 5–10% of total PI. Flux studies of phosphoinositides in platelets (Wilson *et al.*, 1985) and GH_3 cells (Imai and Gershengorn, 1986) also suggest that a large portion of PI may be degraded directly during thrombin stimulation. A kinetic analysis of the formation and degradation of inositol cyclic phosphates (Ins cP) indicates that approximately half of Ins(c1:2,4,5)P_3 is formed by direct phosphodiesteratic cleavage of PI (Dixon and Hokin, 1989).

It now appears that there are two separate phosphoinositide cycles, one in the plasma membrane and one in the ER. There is now good evidence that polyphosphoinositides and the kinases that form them are in the ER as well as the plasma membrane (Helms *et al.*, 1991). The polyphosphoinositide cycle functions in the plasma membrane to generate inositol-1,4,5-trisphosphate [Ins(1,4,5)P_3] and some DAG. A function for the phosphoinositide cycle in the ER would be to provide a steady large supply of DAG and arachidonic acid on prolonged stimulation (derived primarily from PI), which in fact is observed in mouse pancreas (Dixon and Hokin, 1984). The fate and function of polyphosphoinositides in the ER remain unanswered. They may have local functions as yet unexplained, or they may be transferred to the plasma membrane. A PI transfer protein has been isolated and is present in all mammalian cells (Snoek *et al.*, 1993), and this may be involved in movements of PI between the plasma membrane and the ER. In any event, the PI cycle in the ER must be turned on by some as yet unidentified signal unless there is direct continuity between the plasma membrane and the ER at numerous points.

1.4. Early Studies Suggesting an Involvement of Inositol Lipids in Calcium-Dependent Responses

It had been known since the time of Ringer over a century ago (Ringer, 1883) that Ca^{2+} is essential for contraction of the heart. Beginning with the classical work of Katz on ACh release from nerve terminals and of Douglas on stimulus–secretion coupling (see Douglas, 1974; Katz, 1969), it was becoming apparent that Ca^{2+} was an essential link in many agonist-evoked responses. That is, the physiological response required extracellular Ca^{2+}, and increases in intracellular Ca^{2+} derived from intracellular stores or from extracellular Ca^{2+} were seen.

By presoaking pigeon pancreas slices in EDTA, followed by incubation without and with Ca^{2+}, it was found that with the low levels of Ca^{2+} achieved, amylase secretion in response to ACh was inhibited 98%, but the stimulated ^{32}P incorporation into PI and PA was inhibited only 38% and 41%, respectively (L. Hokin, 1966). These results again showed loose coupling between the phospholipid effect and enzyme secretion and suggested that the phospholipid effect was concerned with some step in the "overall process of excitation and secretion other than exocytotic enzyme secretion *per se*." These studies also showed that the phospholipid effect was partially independent of Ca^{2+}. These observations were part of the underpinnings of the Ca^{2+} gating hypothesis proposed by Michell in his seminal review a decade later (Michell, 1975). That is to say, when Michell (1975) surveyed a large number of tissues showing the PI response, he noted: (1) the stimulus–response mechanism required extracellular Ca^{2+}, and rises in intracellular Ca^{2+} occurred on agonist stimulation; (2) the PI effect was independent or only partially dependent on Ca^{2+} in many tissues; and (3) in some tissues, Ca^{2+} ionophores stimulated the physiological response but not PI breakdown. Taken together, these observations favored PI breakdown as antecedent to elevated cell Ca^{2+}, and it was suggested that PI breakdown opened "Ca^{2+} gates" in the plasma membrane. More direct support for the "Ca^{2+}-gating" hypothesis was obtained by Berridge and Fain in the blowfly salivary gland (Berridge and Fain, 1979; Fain and Berridge, 1979a,b), where stimulation with 5-HT caused PI loss and increased exchange of Ca^{2+} in the epithelial cells. Supramaximal stimulation with 5-HT caused a loss of the small pool of responsive PI and a fall in Ca^{2+} transport (desensitization). Incubation of washed glands with *myo*-inositol restored both PI sensitivity to 5-HT and Ca^{2+} transport. Although there was considerable controversy in those early years over the Michell hypothesis (see L. Hokin, 1985), the concept proved to be correct, albeit not in its details.

1.5. Early Studies on Polyphosphoinositide Turnover

As early as 1962, the polyphosphoinositides were shown to turn over rapidly in brain, and it was suggested that they may play an important physiological role (Brockerhoff and Ballou, 1962). They were present on radiochromatograms of lipid extracts from brain particulate fractions undergoing oxidative phosphorylation in the presence of [^{32}P]Pi (L. Hokin and Hokin, 1958a,b), and the radioactivity in "spot C" (later identified as PIP) (Santiago-Calvo *et al.*, 1964) was 30% lower when the fractions were incubated for 1 hr with ACh. Later, it was shown that there was substantial radioactivity in polyphosphoinositides in a wide variety of tissues incubated with [^{32}P]Pi (Lee and Huggins, 1968; Santiago-Calvo *et al.*, 1964). If sea gull salt gland slices were prelabeled with [^{32}P]Pi, followed by addition of ACh, there was a statistically significant drop in radioactivity in PIP and PIP_2 to 68% and 77% of that of control slices, respectively (Santiago-Calvo *et al.*, 1964). Unfortunately, no time courses were done at that time, which undoubtedly would have shown an earlier breakdown of the polyphosphoinositides than of PI.

Phosphatidylinositol kinase and PIP kinase were also discovered at that time by incubating erythrocyte membranes with [γ-^{32}P]ATP, which led to the labeling of PIP and PIP_2 in their monoesterified phosphates (M. Hokin and Hokin, 1964). If exogenous PI was added to erythrocyte membranes, the incorporation of ^{32}P from [γ-^{32}P]ATP into PIP was stimulated, providing further evidence for PI kinase. Phosphatidylinositol kinase was demonstrated shortly afterwards in brain (Colodzin and Kennedy, 1965). Several years later, PI kinase was reported to be present in plasma membranes other than that of the erythrocyte (Harwood and Hawthorne, 1969; Michell and Hawthorne, 1965).

1.6. Formation of Inositol Phosphates

In the late 1960s, Durell and associates (Durell *et al.*, 1969) reported the formation of inositol mono- and bisphosphates in synaptosomes in response to ACh, and they suggested that the primary reaction in response to ACh might be a phosphodiesteric cleavage of PIP and PIP_2. In the late 1970s, Abdel-Latif *et al.* (1977) showed a rapid breakdown of PIP_2 in iris smooth muscle after muscarinic or α-adrenergic stimulation. Subsequently, they showed that the breakdown of PIP_2 was accompanied by the formation of Ins(1)P, inositol bisphosphate ($InsP_2$), and inositol trisphosphate ($InsP_3$) (Akhtar and Abdel-Latif, 1980). However, they did not believe that the release of inositol phosphates was involved in Ca^{2+} mobilization because this response was partially dependent on Ca^{2+} and could be mimicked by a Ca^{2+} ionophore. Stimulation of PIP_2 breakdown by Ca^{2+} ionophore in this tissue has now been shown to be a secondary effect attributed to the release of Ca^{2+}-mobilizing neurotransmitters by the ionophore (Akhtar and Abdel-Latif, 1984).

Further developments that have led to the recognition that polyphosphoinositides play a central role in transduction of signals from Ca^{2+}-mobilizing receptors are discussed below.

2. MODERN ERA

2.1. Stimulation of Polyphosphoinositide Metabolism

Even though polyphosphoinositide breakdown in response to agonists had been described earlier (see above), little attention was paid to it until the early 1980s, when reports describing the rapid breakdown of polyphosphoinositides were published (Creba *et al.*, 1983; Thomas *et al.*, 1983). This led to a revival of interest in polyphosphoinositides because, unlike the effect in iris smooth muscle reported earlier (Abdel-Latif *et al.*, 1977), the vasopressin-stimulated breakdown of polyphosphoinositides in liver was independent of Ca^{2+} (Creba *et al.*, 1983).

Studies of the effects of receptor activation on phosphoinositide metabolism in various tissues focused on three aspects: (1) effects of agonist stimulation on hydrolysis of polyphosphoinositides; (2) characterization of products formed as a result of this hydrolysis; and (3) determination of physiological functions of these products.

2.1.1. Breakdown of Polyphosphoinositides

2.1.1a. PIP_2 Hydrolysis Is Rapid and Precedes that of PIP and PI. The rapid breakdown of polyphosphoinositides following receptor activation was reported in numerous tissues in response to stimuli such as neurotransmitters, growth factors, hormones, and light. Several reviewers have tabulated agonists and tissues where a polyphosphoinositide response had been demonstrated (see Abdel-Latif, 1986; Berridge, 1986; Sekar and Hokin, 1986). A key feature of the studies summarized in these reviews is that breakdown of PIP_2 and PIP is rapid, usually observed within seconds, and precedes that of PI.

2.1.1b. PIP_2 Hydrolysis Is Independent of Calcium Mobilization. The Ca^{2+} requirement for the breakdown of PIP_2 was found to be somewhat variable. In some tissues, such as parotid gland, platelets, superior cervical sympathetic ganglia, hepatocytes, and human fibroblasts, PIP_2 breakdown was found to be independent of Ca^{2+} (see Rana and Hokin, 1990). On the other hand, in some cases, such as brain synaptosomes, neutrophils, pancreatic islets, pancreas, cerebral cortical slices, and iris smooth muscle, polyphosphoinositide breakdown was partially dependent on Ca^{2+} and in many cases could be activated by a Ca^{2+} ionophore.

Based on the Ca^{2+} requirements for the phosphoinositide effect in various tissues, it is now generally accepted that in some tissues under physiological conditions of ionic strength and pH, the Ca^{2+} concentration required for the breakdown of at least PIP_2 is at or below the resting cytosolic Ca^{2+} concentration (0.1 μM). In an evaluation of the Ca^{2+} requirement for PLC activity in liver cells (Renard *et al.*, 1987), where the $[Ca^{2+}]_i$ was clamped at 29 nM, which is below resting $[Ca^{2+}]$, vasopressin caused a twofold stimulation of Ins(1,4,5)P_3 formation at the clamped $[Ca^{2+}]_i$, suggesting that Ins(1,4,5)P_3 formation was not dependent on a rise in $[Ca^{2+}]_i$. Furthermore, Ins(1,4,5)P_3 formation was little affected by clamping $[Ca^{2+}]_i$ at the resting level of 193 nM and up to 1130 nM. The PLC reaction at the low $[Ca^{2+}]_i$ concentrations can occur in the presence of G proteins and GTP. It is clear that in some cells, such as hepatocytes, the Ca^{2+} requirement for PLC is very low, and its resting intracellular concentration is not rate limiting. On the other hand, in tissues such as cerebral cortex slices, resting Ca^{2+} is rate limiting for PLC.

2.1.1c. Only a Select Pool of Phosphoinositides Is Recognized to Undergo Hydrolysis in Response to Calcium-Mobilizing Agonists. In some tissues, it appeared that there is a unique pool of phosphoinositides that are hydrolyzed in response to agonists. This was first shown in the avian salt gland, where a discrete pool of PI broke down on ACh stimulation, forming a discrete pool of PA. On atropine blockade, this PA pool reverted to the original pool of PI (M. Hokin and Hokin, 1964; see above). Further support for the existence of an agonist-responsive pool of phosphoinositides was obtained by studying the breakdown of phosphoinositides after labeling of cells under stimulatory and nonstimulatory conditions (Monaco and Woods, 1983) and by the demonstration that Mn^{2+} stimulates incorporation of inositol into an agonist-unresponsive pool of phosphoinositides (Gonzales and Crews, 1985). A discrete pool of polyphosphoinositides in the red cell membrane was shown by the Michell group (King *et al.*, 1987).

2.1.2. Products of Polyphosphoinositide Hydrolysis

Early on, it was recognized that phosphodiesteratic cleavage of PIP_2 by PLC enzyme preparations leads to the formation of DAG and Ins(1,4,5)P_3 (Kemp *et al.*, 1961a,b; Thompson and Dawson, 1964). The reaction can be measured by determining the loss of PIP_2 (reviewed above), the formation of DAG (reviewed in Section 2.3 below), or the formation of Ins(1,4,5)P_3 (discussed here).

It was shown that the inositol trisphosphate produced on hydrolysis of PIP_2 can be either inositol-1,4,5-trisphosphate [Ins(1,4,5)P_3] or inositol-1,2-cyclic-4,5-trisphosphate (cInsP_3), depending on the source of hydroxyl group utilized in the reaction. The entry of the hydroxyl group from water produces Ins(1,4,5)P_3, whereas the use of the resident hydroxyl at the 2-position of the inositol ring leads to production of cInsP_3.

2.1.2a. Formation of Ins(1,4,5)P_3. Rapid agonist-induced formation of Ins(1,4,5)P_3 (usually reaching a peak within a few seconds) was reported in numerous tissues. Detailed lists of agonists and tissues where formation of Ins(1,4,5)P_3 has been documented can be found in several earlier reviews (Abdel-Latif, 1986; Berridge, 1986; Downes and Michell, 1985; Sekar and Hokin, 1986). More recent demonstrations of agonist-stimulated formation of Ins(1,4,5)P_3 are reviewed by Rana and Hokin (1990). Of the various inositol phosphates, Ins(1,4,5)P_3 appears first on agonist stimulation, and this is then followed by the appearance of Ins(1,3,4,5)P_4, Ins(1,3,4)P_3, and Ins(1,4)P_2, in that order (Wollheim and Biden, 1986).

2.1.2b. Formation of cInsP_3. In the late 1950s, Dawson (1959) showed that incubation of PI with PLC released, in addition to inositol monophosphate, inositol-1,2-cyclic phosphate (cInsP). However, the formation of cInsP on agonist-stimulated PI breakdown in cells was not shown until relatively recently (Dixon and Hokin, 1985; Graham *et al.*, 1987). It was suggested (Dixon and Hokin, 1985) that the formation of cInsP was evidence for the direct breakdown of PI because this compound is a product of PLC action on PI. This interpretation became more complicated when Wilson *et al.* (1985) showed that PLC action on PIP and PIP_2 produced inositol-1,2-cyclic 4-bisphosphate (cInsP_2) and cInsP_3, and Connolly *et al.* (1986b) showed that no enzyme could be found for decyclizing cInsP_2 and cInsP_3, in contrast to cInsP. Thus, agonist-stimulated formation of cInsP could result from PLC action on PI and/or sequential dephosphorylation of cInsP_3 (see below). In lieu of these enzymatic studies, the possibility that cInsP_3 and cInsP_2 might also be formed on stimulation of intact cells with an agonist was investigated. Substantial amounts of both of these compounds were shown to be formed in pancreatic minilobules on stimulation with cerulein (Sekar *et al.*, 1987) or carbachol in pancreas or parotid (Dixon and Hokin, 1987a,b). The amount of cInsP_3 formed equaled the amount of Ins(1,4,5)P_3 formed at 30 min. A few months earlier, Ishii *et al.* (1986) showed a small amount of cInsP_3 formed on thrombin stimulation of platelets. The conclusion that cInsP_3 was formed on agonist stimulation was initially not accepted by many of the British workers, largely because Hawkins *et al.* (1987) published a paper after the above two positive reports in which they claimed no formation of

$cInsP_3$ on cholinergic stimulation of rat parotid acinar cells. These experiments were repeated using rat parotid minilobules, and substantial amounts of $cInsP_3$ were formed on carbachol stimulation (Dixon and Hokin, 1987a). The results were quite similar to those in pancreatic mililobules. Hughes *et al.* (1988) confirmed these results in rat parotid. The controversy over the issue of $cInsP_3$ formation was finally resolved by the demonstration that $cInsP_3$ forms very slowly but accumulates over time in both pancreas (Dixon and Hokin, 1987b) and parotid (Dixon and Hokin, 1987a; Hughes *et al.*, 1988). The only HPLC scans that Hawkins *et al.* (1987) showed were from parotid stimulated for 5 sec with carbachol. Results similar to those of Hawkins *et al.* (1987) were recently reported in vasopressin-stimulated WRK1 cells (Wong *et al.*, 1988), but this study, like that of Hawkins *et al.* (1987), focused only on early changes in inositol phosphates.

Majerus and his co-workers reported the formation of both $Ins(1,4,5)P_3$ and $cInsP_3$ on incubation of PIP_2 with a purified phosphoinositidase C (Wilson *et al.*, 1985). They reported that the relative rates of formation of $Ins(1,4,5)P_3$ and $cInsP_3$ on incubation with the purified seminal vesicle enzyme were 10:1. Assuming that this ratio does not change during the course of agonist stimulation, the relative abundance of these compounds at any given time would depend on the rates of their further metabolism. Two enzymes are involved in the removal of $Ins(1,4,5)P_3$, i.e., a 3′-kinase and a 5′-phosphatase (see below). Cyclic $InsP_3$ is not phosphorylated by the 3′-kinase, and the phosphatase is only about one-tenth as active toward $cInsP_3$ as $Ins(1,4,5)P_3$ (Connolly *et al.*, 1987). This can explain the different kinetic behavior of the cyclic compounds.

2.1.2c. Formation of $Ins(1,3,4)P_3$. This $InsP_3$ isomer was first identified in parotid gland (Irvine *et al.*, 1984) and has subsequently been seen in many types of cells. The formation of $Ins(1,3,4)P_3$ involves phosphorylation of $Ins(1,4,5)P_3$ to $Ins(1,3,4,5)P_4$, which is then dephosphorylated to $Ins(1,3,4)P_3$ (Hansen *et al.*, 1986; Hawkins *et al.*, 1986). Earlier studies suggested that the enzyme that removes the 5-phosphate from $Ins(1,3,4,5)P_4$ is similar or identical to $Ins(1,4,5)P_3$-phosphatase, which acts on $Ins(1,4,5)P_3$, but Hansen *et al.* (1987) purified two types of soluble inositol phosphate 5-phosphomonoesterases from rat brain that differ in molecular size and kinetic constants toward $Ins(1,4,5)P_3$ and $Ins(1,3,4,5)P_4$, although there is overlap in their activities toward both inositol polyphosphates.

Compared to $Ins(1,4,5)P_3$, which has a half-life of about 4 sec (Storey *et al.*, 1984), $Ins(1,3,4)P_3$ appeared to turn over much more slowly (Hawkins *et al.*, 1986). Consequently, when cells were stimulated by agonists, the proportion of $Ins(1,3,4)P_3$ relative to $Ins(1,4,5)P_3$ continued to rise after $Ins(1,4,5)P_3$ dropped to a lower steady-state level or to baseline. Lithium enhances accumulation of $Ins(1,3,4)P_3$ (Burgess *et al.*, 1985). Lithium has been a useful tool because it amplifies the inositol phosphate response, but because it causes selective accumulation of some but not all inositol phosphates, it distorts their quantitative proportions.

2.1.3. Inositol Phosphates as Second Messengers

2.1.3a. Ins(1,4,5)P_3. Cytosolic Ca^{2+} can be mobilized either from intracellular stores or from the extracellular compartment or both. In most cells, activation by Ca^{2+}-mobilizing agonists appears to lead to two distinct phases of Ca^{2+} mobilization. Initially, the Ca^{2+} is mobilized from intracellular stores (Putney, 1978), and this is followed by an influx of Ca^{2+} from the extracellular compartment (Rasmussen and Barrett, 1984; Schulz and Stolze, 1980). Some of the cytosolic Ca^{2+} derived from intracellular stores is subsequently released from the cell (Altin and Bygrave, 1987), although some of the Ins(1,4,5)P_3-releasable Ca^{2+} is taken up by mitochondria (Biden *et al.*, 1986) and pumped back into the ER. Elevated intracellular Ca^{2+} is sustained at later times by an influx of extracellular Ca^{2+} (Kojima *et al.*, 1985), which maintains the response and also tends to replenish the depleted intracellular store. Based on the hypothesis first enunciated by Michell (1975) and on studies in the insect salivary gland by Berridge and Fain (Berridge and Fain, 1979; Fain and Berridge, 1979a,b), it was anticipated that a second messenger generated during the breakdown of polyphosphoinositides in the plasma membrane may be involved in the mobilization of intracellular Ca^{2+}.

Overwhelming evidence has accumulated that implicates Ins(1,4,5)P_3 as the molecule that links receptor-activated phosphoinositide breakdown to Ca^{2+} mobilization from intracellular stores. These studies were initiated in 1983 with the demonstration by Streb *et al.* (1983) that the addition of Ins(1,4,5)P_3 to permeabilized pancreatic acinar cells resulted in the release of Ca^{2+} from nonmitochondrial internal stores. Subsequently, literally hundreds of reports have appeared confirming this property of Ins(1,4,5)P_3 in permeabilized cells or microsomal fractions representing a wide variety of tissues (for a list of tissues and references, see Berridge, 1987b). The release of Ca^{2+} in permeabilized cells was rapid and occurred at less than micromolar concentrations of Ins(1,4,5)P_3. This release by Ins(1,4,5)P_3 was transient because of its rapid metabolism and was accompanied by Ca^{2+} reuptake into the ER (Streb *et al.*, 1983).

Although the demonstration that Ins(1,4,5)P_3 releases Ca^{2+} from permeabilized cells was crucial in establishing the link between Ins(1,4,5)P_3 formation and Ca^{2+} mobilization, experiments involving intracellular injection of Ins(1,4,5)P_3 in intact cells were important in confirming this function of Ins(1,4,5)P_3. Thus, intracellular injection of Ins(1,4,5)P_3 was shown to mimic the muscarinic-induced depolarizing chloride current in *Xenopus* oocytes (Oron *et al.*, 1985) and stimulation by light of *Limulus* photoreceptor cells (Brown *et al.*, 1984; Fein *et al.*, 1984). Inositol-1,4,5-trisphosphate was also found to produce a reversible hyperpolarization of salamander rod membranes (Waloga and Anderson, 1985) and to mimic norepinephrine activation of K^+ and Cl^- conductances in hepatocytes (Capiod *et al.*, 1987). Furthermore, intracellular injection of Ins(1,4,5)P_3 into mature eggs of several species appeared to initiate responses that closely resembled those normally produced after fusion with spermatozoa (see below).

Consistent with a second messenger role for Ins(1,4,5)P_3 was the observation that in pancreas the time course for Ins(1,4,5)P_3 formation paralleled secretagogue-induced Ca^{2+} release, and blockade of Ins(1,4,5)P_3 production by neomycin inhibited release of Ca^{2+} by the agonist (Streb *et al.*, 1985). The formation of Ins(1,4,5)P_3 and mobilization of intracellular Ca^{2+} are both very rapid events, and because Ins(1,4,5)P_3 is rapidly metabolized to either Ins(1,4)P_2 or Ins(1,3,4,5)P_4, it is not easy to see a clear separation of these two events. Using a rapid mixing device, Tashjian *et al.* (1987) attempted to determine subsecond changes in inositol polyphosphates in GH_4C1 cells. They demonstrated that changes in inositol polyphosphates and cytosolic Ca^{2+} could be detected as early as 400 to 500 msec.

On the basis of cell fractionation experiments and the use of metabolic inhibitors, it was shown that Ins(1,4,5)P_3 releases Ca^{2+} from a membrane fraction that is a component of the ER (Streb *et al.*, 1984). Depending on the tissue under investigation, there seems to be considerable variability in the proportion of total ER Ca^{2+} released in response to Ins(1,4,5)P_3. The amount of ER Ca^{2+} released in various permeabilized cells in response to Ins(1,4,5)P_3 and that released in their intact counterparts in response to an agonist appear to parallel each other (Burgess *et al.*, 1984). Quantitative aspects of Ins(1,4,5)P_3-responsive Ca^{2+} in the ER have been analyzed. There appear to be distinct Ins(1,4,5)P_3-sensitive and -insensitive compartments because in GH_3 cells additional Ca^{2+} is released when the ATPase inhibitor vanadate is added subsequent to the release of Ins(1,4,5)P_3-sensitive Ca^{2+} (Bond *et al.*, 1984). An electron microprobe X-ray analysis of smooth muscle (Bond *et al.*, 1984) and intracellular injection of Ins(1,4,5)P_3 at various depths in *Xenopus* oocytes (Busa *et al.*, 1985) suggested localization of the Ins(1,4,5)P_3-sensitive pool of Ca^{2+} in a region of ER that is close to the cell surface, although more recent studies indicate no localization of this pool in the ER (M. Hansen, personal communication).

Recently, the ER as the sole site of the Ins(1,4,5)P_3-responsive pool of Ca^{2+} has been questioned. The Ins(1,4,5)P_3 binding sites appear to be relatively abundant in a liver fraction enriched in plasma membrane (Guillemette *et al.*, 1988), and the release of Ca^{2+} from this fraction in response to Ins(1,4,5)P_3 appears to be fourfold higher than that from the ER. These studies suggest that the Ins(1,4,5)P_3-responsive site may lie either in the plasma membrane or in a fraction currently inseparable from it. Mikoshiba's laboratory has recently demonstrated, by antibody staining, Ins(1,4,5)P_3 receptors in the plasma membrane (Mikoshiba *et al.*, 1994).

2.1.3b. Specificity of Inositol Phosphates. The specificity of the 1,4,5-isomer of InsP_3 for Ca^{2+} release was demonstrated in pancreas (Streb *et al.*, 1983) and Swiss mouse 3T3 cells (Irvine *et al.*, 1986b). Inositol-1-phosphate and Ins(1,4)P_2, which are products of Ins(1,4,5)P_3 hydrolysis and can also be derived from direct breakdown of PI and PIP by PLC, were unable to release Ca^{2+}; cInsP_3 appeared to be slightly active (see below). It appears that the phosphates on the 4- and 5-position of the inositol ring are necessary for Ca^{2+} release because Ins(4,5)P_2 and Ins(2,4,5)P_3 are somewhat active, although their potencies are much lower than that of Ins(1,4,5)P_3 (Burgess *et al.*, 1984).

2.1.3c. Identification of the Ins(1,4,5)P_3 Receptor. The interaction of Ins(1,4,5)P_3 with the ER appears to involve specific binding sites. Identification of these specific binding sites in numerous cells (see Rana and Hokin, 1990) and in microsomes from adrenal cortex (Baukal *et al.*, 1985) suggested that Ins(1,4,5)P_3 stimulates Ca^{2+} release from the ER via interaction with a receptor. Mitochondrial membranes appear to lack the Ins(1,4,5)P_3 binding sites. Further support for the existence of a receptor derived from specific photoaffinity labeling of the receptor (Hirata *et al.*, 1985) as well as the recent purification of the Ins(1,4,5)P_3 receptor from rat cerebellum by Snyder and his associates (Supattapone *et al.*, 1988). The binding studies with the Ins(1,4,5)P_3 receptor show a high degree of specificity toward Ins(1,4,5)P_3. Recently, Ins(1,4,5)P_3 receptor isoforms have been cloned.

The molecular mechanism by which Ins(1,4,5)P_3 releases Ca^{2+} from the ER remains to be elucidated, although evidence indicates that binding of Ins(1,4,5)P_3 to the ER leads to opening of a Ca^{2+} channel (Joseph and Williamson, 1986; Muallem *et al.*, 1985). Evidence indicates that Ca^{2+} continuously cycles across the ER and that Ins(1,4,5)P_3 stimulates release by enhancing the passive efflux of Ca^{2+} from the ER (Prentki *et al.*, 1984) without affecting the ATP-driven influx (Delfert *et al.*, 1986). It has been shown that K^+ is required (Muallem *et al.*, 1985) and acts as a counterion in Ins(1,4,5)P_3-induced Ca^{2+} release (Joseph and Williamson, 1986). This is supported by the ability of potassium channel blockers to inhibit Ins(1,4,5)P_3-induced Ca^{2+} release (Shah and Pant, 1988).

In addition to its well-established role in Ca^{2+} release from intracellular stores, Ins(1,4,5)P_3 may also have a potential role in stimulation of extracellular Ca^{2+} influx (Parker and Miledi, 1987). This is now made more likely by the demonstration of Ins(1,4,5)P_3 receptors in the plasma membrane (Mikoshiba *et al.*, 1994).

2.1.3d. cInsP_3. In studies from our laboratory in which we chemically synthesized cIP_3 from Ins(1,4,5)P_3, and the product contained 0.7% Ins(1,4,5)P_3

as a contaminant, we found $cInsP_3$ to be about one-20th as potent as Ins(1,4,5)P_3 in releasing Ca^{2+} from permeabilized pancreatic acinar cells (Lee and Hokin, 1989). Meyer *et al.* (1988) found that 20 nM Ins(1,4,5)P_3 released all of the Ins(1,4,5)P_3-mobilizable Ca^{2+} within 1 min in a rat basophilic leukemia cell line, but there was no effect of $cInsP_3$ at this concentration. Also, Crossley *et al.* (1988) showed that $cInsP_3$ was only one-tenth as potent as Ins(1,4,5)P_3 in activating fertilization in sea urchin eggs. These studies are in general agreement that $cInsP_3$ is only 5–10% as potent as Ins(1,4,5)P_3 in releasing Ca^{2+}.

In any case, it seems unlikely that the accumulation of large amounts of $cInsP_3$ on agonist stimulation in some tissues over time is of no biological significance. Recently, J. F. Dixon and L. E. Hokin (unpublished observations) found no Ins(1,4,5)P_3 in resting pancreatic minilobules incubated in the absence of agonist but found appreciable amounts of $cInsP_3$. It is possible that $cInsP_3$ is involved in the maintenace of the basal Ca^{2+} concentrations in the resting cell.

2.1.3e. Ins(1,3,4,5)P_4 Formation and Its Function. It was shown that the Ins(1,4,5)P_3 formed on agonist stimulation is subject to further metabolism by either a kinase specific for the 3-position of inositol or a phosphomonoesterase specific for the 5-position. The former produces Ins(1,3,4,5)P_4, and the latter produces Ins(1,4)P_2. The Ins(1,4,5)P_3 3-kinase is a soluble enzyme, as first shown in brain (Irvine *et al.*, 1986a) and later shown in a variety of tissues. The enzyme activity was shown to be stimulated by Ca^{2+} (Biden and Wollheim, 1986). Therefore, it appears likely that Ins(1,4,5)P_3-induced intracellular Ca^{2+} mobilization enhances the formation of Ins(1,3,4,5)P_4 and is a mechanism contributing to fine-tuning of Ins(1,4,5)P_3 levels.

There was suggestive evidence that Ins(1,3,4,5)P_4 may be a second messenger in Ca^{2+} influx (Irvine and Moor, 1986, 1987), but more recent studies have not confirmed this. Highly specific binding sites for this molecule have been detected in adrenal cortex (Enyedi and Williams, 1988), and an Ins(1,3,4,5)P_4 binding protein has been purified. Very recently, the binding protein has been identified as a member of the GAP1 family (Cullen *et al.*, 1995).

2.1.3f. IP_5 and IP_6. Although IP_5 and IP_6 were recognized as natural metabolites of inositol as early as 1969 (Cosgrove, 1969), their detection in mammalian cell types has been relatively recent (Heslop *et al.*, 1985; Stephens *et al.*, 1988a,b). Their relative distribution in various regions of the brain following [^{3}H]*myo*-inositol labeling has been studied (Vallejo *et al.*, 1987). The IP_5 was tentatively determined to be Ins(1,3,4,5,6)pentakisphosphate. Both labeled IP_5 and IP_6 were found in approximately equal proportions in the midbrain and the hypothalamus; IP_5 was more abundant in the medulla oblongata, and IP_6 was more abundant in the hippocampus, whereas the corpus striatum lacked IP_5. The biosynthetic pathways for the synthesis of IP_5 and IP_6 have not been adequately characterized. Stephens *et al.* (1988b) reported that Ins(1,3,4,5,6)P_5 is synthesized by the phosphorylation of Ins(1,4,5,6)P_4 by a kinase that shows specificity for the 3-OH. Presumably, IP_6 is produced by a 2-kinase acting on Ins(1,3,4,5,6)P_5.

So far, there is no evidence to implicate IP_5 and IP_6 in intracellular signaling, as their levels do not show the transient alterations characteristic of inositol phosphate second messengers. Further work is required to clarify the biosynthesis, degradation, and function of these two inositol polyphosphates.

2.2. Role of GTP-Binding Proteins in Receptor Coupling to Phospholipase C

2.2.1. Evidence Supporting the Involvement of G Proteins

The subject of G proteins and their involvement in transduction of agonist signals has been reviewed (see Gilman, 1987; Litosch and Fain, 1986). A role for GTP in coupling receptor activation to phosphoinositide breakdown was first suggested from studies on the effects of guanine nucleotides on the binding of agonists to their receptors. It was observed that guanine nucleotides lowered the agonist affinities but not the antagonist affinities in several tissues and receptors involving phosphoinositide metabolism, such as vasopressin, muscarinic cholinergic agonists, α_1-adrenergic agonists, and f-Met-Leu-Phe (Goodhardt *et al.*, 1982; see Fain, 1987). These observations were consistent with the popular model for the coupling of adenylate cyclase to G proteins, a concept initiated by Rodbell *et al.* (1971). According to this model, the interaction of agonist with the receptor leads to the formation of a high-affinity receptor–agonist–G-protein complex that, on interaction with GTP, dissociates to a GTP-liganded G protein and a low-affinity agonist–receptor complex (see Gilman, 1987).

Studies using permeabilized cells or isolated membranes have proven invaluable in establishing the involvement of GTP-binding proteins in receptor-mediated activation of PLC. In many such preparations, agonist-induced responses, although somewhat attenuated, can be mimicked by the addition of Ca^{2+} and potentiated on the addition of GTP or GTP analogues. Using such an approach, Gomperts (1983) demonstrated that guanine nucleotides stimulated a Ca^{2+}-dependent secretion of histamine from permeabilized mast cells. He proposed that a guanine nucleotide binding protein was involved in Ca^{2+}-dependent secretory events. The successful demonstration of the potentiating effects of GTP analogues in permeabilized mast cells led to similar studies in numerous cell types. Nonhydrolyzable analogues of GTP, GTPγS and Gpp(NH)p, were shown to stimulate phosphoinositide-linked serotonin release and DAG production in permeabilized platelets (Haslam and Davidson, 1984). It was concluded that the stimulation of secretory responses by GTP or its analogues occurred through the stimulation of phosphoinositide breakdown.

Potentiation of agonist-induced phosphoinositide breakdown by GTP,

GTPγS, or Gpp(NH)p has been widely reported in numerous cell and membrane preparations (see references in Rana and Hokin, 1990). In addition to the potentiation of agonist-induced responses, the ability of GTPγS or Gpp(NH)p alone to stimulate the breakdown of endogenously labeled phosphoinositides has been demonstrated in many cell types. In many of these studies, the stimulatory effect of GTP analogues was blocked in the presence of GDPβS, indicating the involvement of a G protein, as GDPβS arrests G proteins in their inactive state.

Further evidence favoring the involvement of a guanine nucleotide binding protein in coupling receptor occupation to PLC activation derived from studies on the effect of AlF_4 on phosphoinositide breakdown or phosphoinositide-mediated events (reviewed by Rana and Hokin, 1990).

2.2.2. Identity of G Proteins Involved in the Activation of Phospholipase C

The investigations of the identity of the putative G proteins that may be involved in the receptor-mediated activation of PLC relied earlier on the use of ADP-ribosylation as a tool (for review, see Ui, 1986). Two bacterial toxins, cholera toxin and pertussis toxin, catalyze an NAD^+-dependent ribosylation that leads to inactivation of the protein.

Following the successful use of ADP-ribosylation in mast cells, this approach was extensively followed in establishing the identities of G proteins that might mediate the activation of PLC in many tissues and cell types. The inhibition by pertussis toxin of phosphoinositide breakdown in response to Ca^{2+}-mobilizing agonists was reported in many cell types (reviewed by Rana and Hokin, 1990). In some cases, Ins(1,4,5)P_3 formation also appeared to be sensitive to cholera toxin.

Inasmuch as three of the well-characterized G proteins (G_i, G_o, and transducin) can be ribosylated by pertussis toxin, and G_s can be ribosylated by cholera toxin, a simple demonstration that a receptor-mediated event is inhibited by pertussis toxin or cholera toxin is insufficient to identify the G protein involved. The failure of some agonist-induced phosphoinositide responses to be inhibited by either pertussis toxin or cholera toxin in several cell types led to the suggestion that PLC coupling to receptors may be mediated by a unique G protein, which has been named G_q (Litosch and Fain, 1986). A family of G_q proteins, which are specific for coupling receptors to PLC, have now been purified and cloned.

2.2.3. G Proteins Enhance Ca^{2+} Sensitivity of Phospholipase C

It was widely reported that membranes derived from [^{32}P]- or [^{3}H]*myo*-inositol-labeled cells undergo phosphoinositide hydrolysis when incubated in the presence of Ca^{2+}. However, the concentration of Ca^{2+} required for this stimulation was very high. In the presence of GTP or nonhydrolyzable analogues of GTP, the Ca^{2+} requirement was reduced to the physiological range for resting cells. For example, the Ca^{2+} requirement for platelet cytosolic PLC was reduced 100-fold in the presence of GTPγS (Deckmyn *et al.*, 1986). These observations and others suggested that GTP-binding proteins increase the sensitivity of PLC to Ca^{2+}, although the mechanism for this enhancement is not as yet defined. Modulation of Ca^{2+} sensitivity of PLC by G proteins is distinct from the role of these proteins in the receptor-mediated coupling discussed above.

2.3. Protein Kinase C

The other product of PLC-mediated breakdown of phosphoinositides is DAG, which acts as a second messenger by activating protein kinase C (PKC) and, in addition to influencing many cellular processes, is also involved in the attenuation of the phosphoinositide response itself. I briefly review developments pertaining to this enzyme.

2.3.1. Discovery and Mechanism of Activation

Protein kinase C was discovered by Nishizuka and associates in 1977 (Inoue *et al.*, 1977) and subsequently shown to require Ca^{2+} and phospholipid for activity (Takai *et al.*, 1979a,b). Early on, it was thought that DAG arising from the PI cycle was the exclusive activator of PKC. An enrichment of stearic acid and arachidonate (AA) in the incremental fraction of DAG on stimulation with agonists had suggested that it must be derived from inositol phospholipids (Bocckino *et al.*, 1985; Hughes *et al.*, 1984). Later, Kennerly (1987) showed that in stimulated mast cells, half of this incremental fraction in DAG is derived from non-phosphoinositide lipids. As mentioned above, it is now known that DAG derived from phosphodiesteratic cleavage of PC is also a major source of DAG. Thus, DAG formation can no longer be regarded as belonging exclusively to the PI cycle.

Two key observations heightened interest in PKC: (1) DAG was found to stimulate the enzyme by lowering its Ca^{2+} requirement into a physiological range (0.1 to 1 μM) (Kirk *et al.*, 1981; Takai *et al.*, 1979a,b); and (2) the potent tumor promoter phorbol myristate acetate (PMA) stimulated the enzyme. This latter property quickly caught the attention of many workers in the field of chemical carcinogenesis.

Fully active PKC is believed to be a quaternary complex consisting of phospholipid, Ca^{2+}, DAG, and the enzyme. The ternary complex lacking DAG displays activity, but only at a 100-fold higher concentration of Ca^{2+}. Phosphatidylserine was found to be the most effective phospholipid in reconstituting enzyme activity (Takai *et al.*, 1979a,b).

2.3.2. Acceptance of DAG as a Second Messenger

Although the activation of PKC by DAG clearly indicated the possibility of a second-messenger function for DAG, it was important to show that the cellular mass of DAG increased after the application of a cell surface stimulus and that the exogenous addition of cell-permeable DAGs induced responses identical to those of growth factors, hormones, and neurotransmitters. These two aspects are summarized below. A rapid net increase in the mass of cellular DAG on stimulation was reported in a number of tissues and cells (reviewed by Rana and Hokin, 1990). Nonphosphoinositide sources of DAG were also reported in several systems. Several cell-permeable DAGs were synthesized and evaluated for their ability to mimic the responses induced by cell surface stimuli. Cell-permeable DAGs were proven to be valuable tools for the study of PKC-mediated responses, and studies with these agents provided additional support to the second-messenger status of DAG. Early studies with synthetic DAGs were largely unsuccessful because they were carried out with long-chain DAGs, which did not interact with cells because of phase separation in an aqueous environment. Nishizuka and colleagues (reviewed by Nishizuka, 1984a,b, 1986) were able to overcome this problem by using 1-oleoyl-2-acetylglycerol (OAG). Initial studies with this compound in platelets, neutrophils, and mast cells indicated that OAG could function as a second messenger for secretion. The initial success with OAG encouraged Bell and his associates to synthesize and test DAGs with varying chain lengths. Optimal biological activity in many intact cell systems was seen with DAGs containing saturated acyl chain lengths of six to nine carbon atoms (Conn *et al.*, 1985; Davis *et al.*, 1985). In platelets, a 40-kDa protein (a PKC substrate) was phosphorylated either by physiological stimuli or by OAG (Friesen and Gerrard, 1985), dihexanoyl glycerol (diC6), and dioctanoyl glycerol (diC8) (Lapetina *et al.*, 1985). Cell-permeant DAGs were shown to stimulate superoxide generation in phagocytes and neutrophils (Fujita *et al.*, 1984).

2.3.3. Identification of PKC as the Phorbol Ester Receptor

The report by Castanga *et al.* (1982) that PMA could substitute for DAG and activate PKC *in vitro* was a major breakthrough in understanding the mechanism of phorbol ester action. The dose–response curve for PMA activation of PKC is similar to the saturation curve for PMA binding to the receptor, as are the structure–activity relationships for enzyme activation when compared with receptor binding, suggesting that PKC may act as a phorbol receptor. Furthermore, when the enzyme was maximally stimulated by either PMA or DAG alone, the addition of the other activator did not lead to additional stimulation, indicating that both of these agents acted by a common mechanism. Soon after the report by Castanga *et al.* (1982), PMA was shown to immediately activate PKC in intact cells (Sano *et al.*, 1983). Additional support for the idea that PKC was a phorbol ester receptor came from studies on receptor purification. It was invariably observed that both the binding activity and the Ca^{2+}/phospholipid-dependent enzyme activity copurified in the same fraction. Several laboratories reported a cytosolic form of the receptor that required PS and copurified with PKC and, in general, after reconstitution, exhibited the same order of affinities for phorbol ester analogues as seen with the receptor in cells or subcellular fractions or by biological assays (see references in Rana and Hokin, 1990).

In general, there were remarkable qualitative similarities between the pattern of responses induced by DAGs and phorbol esters in many cells and tissues. However, the generalization that all effects of phorbol esters or DAGs are mediated by PKC activation may be an oversimplification.

2.3.4. Translocation of PKC

The term translocation, in the context of PKC, refers to the change in the intracellular site occupied by the enzyme on exposure of certain cell types to phorbol esters and, in some cases, natural agonists. The phenomenon of PKC translocation was first observed by Kraft *et al.* (1982; Kraft and Anderson, 1983) in a study of the subcellular distribution of PKC activity in EL4 mouse thymoma cells following phorbol ester exposure. Since then, numerous reports have appeared that demonstrate that phorbol esters can induce a rapid association of PKC with the cellular particulate fraction in a variety of cell types.

2.3.5. PKC Substrates

Niedel and Blackshear (1986) compiled a list of some of the proteins believed to be phosphorylated by PKC either *in vitro* or in intact cells. At that time, this list included eight receptors and 18 endogenous proteins or enzymes.

Of the large number of proteins that appear to be good candidates as PKC substrates, the 40K protein and guanine nucleotide binding regulatory proteins are particularly interesting. Majerus and his associates (Connolly *et al.*, 1986a,b) obtained evidence that the 40K protein in platelets might be an Ins(1,4,5)P_3 5′-phosphomonoesterase. The identification of the protein was based on comigration in gels, similar peptide maps, and phosphorylation studies. It was reported that the phosphorylation of this protein by PKC enhanced its phosphatase activity. The PKC-induced phosphorylation of this protein was thought to be important in the regulation of inositol phosphate levels. Because guanine nucleotide binding proteins (discussed earlier) are thought to couple receptor activation to PLC stimulation, phosphorylation of these proteins may be important in the modulation of phosphoinositide breakdown (see below). Of course, in recent years the MARCKS protein with a molecular weight of 80 kD has been shown to be a major PKC substrate.

Investigations in the late 1980s revealed that PKC exists as a family of multiple subspecies with subtle individual characteristics. In brain tissue, structures of the subspecies have been deduced from the analysis of their cDNA sequences. The structures of the subspecies (α, βI, βII, and γ) are highly homologous and reveal four conserved and five variable regions (for review, see Nishizuka, 1988). Two of the variable regions are within the regulatory domain. Therefore, subtle differences in the regulation of various subspecies may be expected. Type I PKC (encoded by the γ sequence) is less sensitive to DAG but is significantly activated by relatively low concentrations of free arachidonate. Type II PKC (derived from splicing of βI and βII sequences) exhibits substantial activity at basal Ca^{2+} levels and responds well to DAG and, to some extent, arachidonate. Type III PKC (encoded by the α sequence) is most sensitive to 1-stearoyl-2-arachidonylglycerol (the major species of DAG derived from inositol phospholipids). This subspecies of PKC, at elevated Ca^{2+}, can be activated by high concentrations of arachidonate in the absence of phospholipid.

2.3.6. Modulatory Functions of PKC in Phosphoinositide-Mediated Signal Transduction

The phosphoinositide signal transduction mechanism is unique in the sense that receptor activation leads to production of two messengers, Ins(1,4,5)P_3 and DAG, each of which in turn, either independently or together, leads to production or mobilization of other messengers and modulators. Together, the two branches of this pathway [Ins(1,4,5)P_3 and DAG] represent a highly versatile mechanism to control a host of cellular responses. The two branches of the signal pathway exert their influence by phosphorylating distinct sets of target proteins, DAG acting via PKC and Ca^{2+} acting via calmodulin-dependent kinases. It became increasingly clear that the two branches of the phosphoinositide messenger system not only interact with each other but also interact in rather complex ways with other regulatory molecules.

2.3.6a. Modulatory Interactions Related to Ca^{2+} Signaling. These interactions seem to affect Ca^{2+} signaling, either directly via modulation of influx and efflux or indirectly via modulation of polyphosphoinositide formation or degradation.

Feedback Inhibition of Polyphosphoinositide Turnover. It appears that one of the important functions of the DAG–PKC pathway may be to inhibit Ca^{2+} signaling. This action of PKC may prevent the rise of cytosolic Ca^{2+} to unphysiological levels (Berridge, 1986). In a number of cell types, elevation of intracellular free Ca^{2+} by Ca^{2+}-mobilizing agonists known to act by receptor-mediated stimulation of inositol phospholipid turnover was shown to be inhibited by phorbol esters. This was demonstrated in a number of cells (see Rana and Hokin, 1990, for references). The phorbol-ester-induced inhibition of the agonist-stimulated rise in intracellular free Ca^{2+} is believed to be a consequence of the inhibition of inositol phospholipid turnover because in many instances the inhibition of responsiveness was correlated with the inhibition of formation of inositol phosphates. The synthetic cell-permeant DAG, OAG, was also found to inhibit agonist-dependent inositol phosphate generation (Watson and Lapetina, 1985). As discussed previously, the major mode of action of phorbol esters and DAG is PKC activation, and the most likely interpretation of these findings is that PKC plays a negative modulatory role in phosphoinositide turnover.

The most widely used experimental approach in establishing the involvement of PKC was to use the down-regulation of PKC as a tool. With this approach, it was shown in 3T3 cells that down-regulation of PKC by prolonged treatment with phorbol dibutyrate (PDBu) abolished the PDBu or OAG inhibition of vasopressin-induced Ca^{2+} mobilization (Mendoza *et al.*, 1986). Similarly, TPA inhibition of bombesin-induced Ca^{2+} release and inositol phosphate generation was abolished by down-regulating PKC with prolonged pretreatment of 3T3 cells with TPA (Brown *et al.*, 1987).

Although it has generally been accepted that PKC activation, either by endogenous DAGs or by phorbol esters, can modulate phosphoinositide responses by feedback inhibition, the mechanisms underlying this inhibition appear quite diverse. There are numerous instances where phorbol ester treatment of cells modifies agonist responses via PKC-mediated phosphorylation of the receptor (see, for example, Safran *et al.*, 1987). In the case of the α_1-adrenergic receptor, a decrease in the affinity and number of receptors was reported (Cotecchia *et al.*, 1985). The generally held view is that receptor phosphorylation reduces the affinity for agonists or interferes with agonist binding by promoting internalization of the receptor.

In many cases, phorbol ester treatment inhibited phosphoinositide responses but without affecting the affinity or the number of receptors examined (see references in Rana and Hokin, 1990). In these cases, it was thought by some that the coupling of receptors to PLC was interrupted, possibly as a result of PKC-induced phosphorylation of guanine nucleotide binding regulatory protein(s) (Katada *et al.*, 1985). This, in fact, was confirmed (Sagi-Eisenberg, 1989).

Acceleration of Ins(1,4,5)P_3 Degradation. The rate of Ins(1,4,5)P_3 degradation may be an important factor in modulating Ins(1,4,5)P_3-mediated intracellular Ca^{2+} mobilization. In this context, the report by Connolly *et al.* (1986b), which suggested that PKC, by phosphorylating inositol 5′-trisphosphatase, may enhance the dephosphorylation of Ins(1,4,5)P_3, could be important but it has not been proven. However, in human platelets, phorbol-12,13-dibutyrate and OAG stimulated inositol triphosphate dephosphorylation (Molina Y Vedia and Lapetina, 1986).

2.3.6b. Synergism between Ca^{2+} and PKC. One of the earliest questions concerning the roles of the two messengers, Ca^{2+} and DAG, was whether either one or both of these messengers was required for sustained cellular responses. In many tissues, it was observed that although activation of PKC alone by phorbol esters or elevation of cytosolic Ca^{2+} alone by ionophores induced small responses, activation of both branches was required for induction of sustained maximal responses equivalent to those induced by natural agonists. This synergism between the two branches was reported in many tissues and cell types (for references, see Rana and Hokin, 1990).

An interesting feature of PKC modulation is that although it affects in a positive manner the events that lie distal to the formation of Ins(1,4,5)P_3, it negatively modulates the formation of Ins(1,4,5)P_3.

2.4. Phosphoinositides, Cell Proliferation, and Oncogenes

Several reviews on this subject appeared before 1990 (Berridge, 1987a,b; Weinstein, 1987; Whitman *et al.*, 1986). This topic has expanded explosively in recent years.

2.4.1. Mitogens and Phosphoinositide Turnover

Historically, the first studies suggesting a possible link between growth stimulation and phosphoinositide turnover were studies in T lymphocytes by Fisher and Mueller (1971). They observed that phytohemagglutinin specifically stimulated incorporation of [^{32}P]orthophosphate into PI and PA at early time points, prior to any changes in radioactivity in other phospholipids. Later, it was shown that stimulation of T cells with concanavalin A or anti-T cell receptor antibody (Imboden and Stobo, 1985; Taylor *et al.*, 1984, 1988) led to rapid production of Ins(1,4,5)P_3, DAG, and release of Ca^{2+} from intracellular stores, leading to increased cytosolic Ca^{2+} levels.

Several mitogens, such as platelet-derived growth factor (PDGF), bombesin, thrombin, vasopressin, or bradykinin, and $PGF_{2\alpha}$ (see references in Rana and Hokin, 1990), stimulated rapid PI turnover when added to quiescent 3T3 fibroblasts. In addition to increased PI turnover following exposure to mitogens, there was an increased production of Ins(1,4,5)P_3 (Berridge *et al.*, 1984; Heslop *et al.*, 1986), increases in DAG (Habernicht *et al.*, 1981), activation of PKC (Rodriguez-Pena and Rozengurt, 1985), and intracellular Ca^{2+} mobilization (McNeil *et al.*, 1985).

That the turnover of polyphosphoinositides constituted a crucial step in the signaling pathway for some mitogens was supported by the observation that the microinjection of an antibody to PIP_2 led to complete abolition of nuclear labeling with [^{3}H]thymidine in NIH 3T3 cells in response to PDGF or bombesin (Matuoka *et al.*, 1988). A similar conclusion was reached from an earlier study that showed that neomycin, an inhibitor of PIP_2 phosphodiesteratic cleavage, inhibited thrombin-induced proliferation in hamster fibroblasts (Carney *et al.*,

1985). This inhibition was observed only at doses of neomycin that were sufficient to block thrombin-induced phosphoinositide turnover completely. It should be borne in mind that neomycin inhibits PIP_2 breakdown by forming a salt complex with PIP_2, and this is unlikely to be a highly specific effect.

2.4.2. Phosphoinositide Turnover in Transformed Cells

An association between changes in PI turnover and cellular transformation was first suggested by Diringer and Friis (1977). They observed that under conditions that limited normal but not transformed cell growth, such as serum omission or high cell density, there was a decline in the PI turnover rate in normal quail cells but not in the Rous-sarcoma-transformed cells. These studies suggested that PI turnover in transformed cells might be constitutively activated. In serum-starved RSV-transformed chicken embryo fibroblasts containing a temperature-sensitive pp60v-*src* transforming protein, increased polyphosphoinositide turnover was reported at the permissive but not at the nonpermissive temperature or in uninfected cells (Sugimoto *et al.*, 1984), suggesting a causal relationship between the *src* gene product and phosphoinositide turnover. Similar observations were reported with the tyrosine kinase-transforming protein, pp68v-*ros*, of the avian sarcoma virus UR2 (Macara *et al.*, 1985). Evidence for constitutively activated phosphoinositide turnover was also reported in NIH 3T3 cells transformed with the Abelson murine leukemia virus (Fry *et al.*, 1985). This was also seen in NIH 3T3 cells infected with the wild-type polyoma virus but not the mutant polyoma virus lacking transforming ability (Kaplan *et al.*, 1986).

2.4.3. Phosphatatidylinositol-3-Kinase

Studies indicated that there were two types of PI kinases (Whitman *et al.*, 1987) and that the tyrosine kinase-associated PI kinase specifically phosphorylated the D-3 position of the inositol in PI to generate PI-3-phosphate [PI(3)P] (Whitman *et al.*, 1988). The detection of PI kinase activity with specificity for the D-3 position in immunoprecipitates of the protein product of polyoma middle T/pp60c-*src* competent for transformation or of the ligand-activated PDGF receptor indicated that PI(3)P itself could be a critical mediator of mitogenic signals. It is possible that PLC may be regulated by tyrosine kinase-catalyzed phosphorylation because GTP-dependent PLC activity is stimulated in mink cells transformed with either v-*fes* or v-*fms*, both of which code for tyrosine kinase (see Hunter, 1986, for review). Also, antityrosine immunoprecipitates of cell extracts prepared from A431 cells treated with EGF exhibited a tenfold increase in the activity of PLC (Wahl *et al.*, 1988). The biochemical mechanism for this is not understood. Current efforts to unravel the relationship between oncogene-derived tyrosine kinases and PLC have focused on determination of amino acid sequences for these enzymes. Investigations (Mayer *et al.*, 1988; Stahl *et al.*, 1988) that compared the cDNA-deduced amino acid sequence of PLC with those of three oncogene product tyrosine kinases (*fps, src, gag-crk*) reported that PLC contained several sequences that were homologous to those present in the regulatory domains of the tyrosine kinases. However, there are at least three PLC isozymes (Carter and Smith, 1987; Ryu *et al.*, 1987), and two of these (PLC I and PLC III) do not contain appreciable structural homology with known tyrosine kinases (Suh *et al.*, 1988).

Besides the plasma membrane, the nuclear membrane also appears to contain PI kinases. Polyphosphoinositides also appear to be associated with chromatin and undergo marked increases at the time of differentiation.

2.5. Pharmacological Action of Lithium

Lithium has been successfully used in the treatment of both the manic and depressive phases of bipolar disorders. Possible mechanisms of action of lithium have recently been reviewed by Jope and Williams (1994). Advances in deciphering the role of inositol lipids in signal transduction and the observation that lithium perturbs inositol lipid metabolism had suggested that there is a causal link between the effects of Li^+ on phosphoinositide metabolism and its pharmacological action.

Inhibition of inositol monophosphatase by lithium was reported by Allison and Stewart (1971) over 25 years ago. This action of lithium appeared to interfere with the inositol lipid cycle in a selective manner, leading to elevation of specific inositol phosphates in brain of animals given lithium *in vivo* and in incubated brain slices (Sherman *et al.*, 1986). The best-known hypothesis for lithium's therapeutic action has been that put forward by Berridge *et al.* (1989). This is based on the fact that in agonist-stimulated brain preparations, lithium traps considerable inositol in the form of inositol monophosphates but also $Ins(1,3,4)P_3$ and $Ins(1,4)P_2$. The accumulation of inositol phosphates was suggested to ultimately reduce the supply of phosphoinositides and thus attenuate agonist-induced $Ins(1,4,5)P_3$ formation and intracellular Ca^{2+} mobilization (Berridge *et al.*, 1989). In support of the above hypothesis were the findings that Li^+ produced small inhibitions (13–25%) in accumulation of $Ins(1,4,5)P_3$ and a somewhat larger inhibition (50%) in accumulation of $Ins(1,3,4,5)P_4$ in rat brain cortex slices (Kennedy *et al.*, 1989) and inhibitions in $Ins(1,3,4,5)P_4$, but not $InsP_3$ (see above), in mouse brain cortex slices (Whitworth and Kendall, 1988). These inhibitions were seen only when brain cortex slices were stimulated with high concentrations of cholinergic agents (0.1–1 mM), which produce massive breakdown of phosphoinositides, a condition that, in the presence of Li^+, is more likely to trap considerable inositol as inositol phosphates.

It should be pointed out that Downes and Stone (1986) found that in parotid glands lithium, in combination with carbachol, greatly reduced the labeling of PI and PIP, but the labeling of PIP_2 was not appreciably affected. More recent studies have also failed to show a decrease in PIP_2 in the brains of rats given short-term chronic administration of lithium (Navidi *et al.*, 1991). Recently, we confirmed that in mouse and rat brain cortex slices lithium decreased Ins(1,4,5)P_3 and Ins(1,3,4,5)P_4, but we could see increases in these inositol phosphates if the incubation medium was supplemented with inositol (Lee *et al.*, 1992). In monkey cerebral cortex slices, there was no effect of 1–10 mM inositol on the percentage stimulation of Ins(1,4,5)P_3 accumulation by Li^+ in the presence of 0.1 and 1.0 mM ACh and in the presence of 1.0, 10, and 25 mM Li^+ (Dixon *et al.*, 1992). Apparently, the "washout" of inositol that occurs in monkey brain cortex slices is insufficient to reduce the substrate to the point that makes it rate limiting, so as to reduce or reverse the Li^+ effect. The inhibitions of Ins(1,4,5)P_3 and Ins(1,3,4,5)P_4 by Li^+ in rat and mouse cerebral cortex slices appear to be the result of artifactual *in vitro* "washout" of inositol, which is already at a low level in the rodents as compared to guinea pig, for example, and this is supported by our restoration of Li^+ stimulations in these inositol polyphosphates in rat and mouse by supplementation with inositol (Lee *et al.*, 1992). Washout does not occur *in vivo*, and existing *in vivo* studies do not show inhibitions of resting Ins(1,4,5)P_3 by Li^+ (Jope and Williams, 1994).

Because addition of agonist was not required to demonstrate the Li^+ effect, in monkey cerebral cortex slices it was not unreasonable to assume that the Li^+ effect was under the control of an endogenous agonist or agonists, as brain cortex slices contain many intact neurons and synapses. To test this possibility, Dixon *et al.* (1994) tested a variety of antagonists to various receptors in monkey cerebral cortex slices. Ketanserin, phentolamine, chlorpheniramine, and atropine, antagonists at $5HT_2$-serotoninergic, α_1-noradrenergic, H_1-histaminergic, and muscarinic receptors, respectively, had no significant effect on Ins(1,4,5)P_3 in the presence of Li^+, even though these receptors are coupled to phosphoinositidase C. On the other hand, antagonists to the NMDA receptor/channel complex, i.e., (±)3-(2-carboxypiperazine-4-yl)-propyl-1-phosphonic acid (CPP), ketamine, and MK801 (dizoclipine), abolished or reduced considerably the Li^+ effect, suggesting that the Li^+ effect in the absence of added agonist is rather specifically dependent on glutamatergic transmission via the NMDA receptor, presumably by regulation of intracellular Ca^{2+}. Antagonists to other glutamate receptors did not inhibit the Li^+ effect. Very recently, Dixon *et al.* (1994) found that the primary effect of lithium in cerebral cortex slices was to release the excitatory neurotransmitter glutamate. This in turn activates the NMDA receptor, permitting more Ca^{2+} to enter the postsynaptic neuron, leading to activation of PLC and release of Ins(1,4,5)P_3.

2.6. Fertilization

Calcium ions play a central role in fertilization. The possibility that a transient increase in intracellular free Ca^{2+} is a pivotal event during fertilization was suggested as early as 1974, when treatment of sea urchin eggs with the Ca^{2+} ionophore A23187 was shown to initiate changes in protein and DNA synthesis, mimicking the initiating effects of sperm penetration (Steinhardt and Epel, 1974). Depending on the type of egg, the Ca^{2+} source can be either intracellular or extracellular.

Evidence implicating the phosphoinositide messenger system in fertilization was obtained by studying (1) the effect of intracellular injection of Ins(1,4,5)P_3, (2) the effect of pretreatment of eggs with phorbol esters or synthetic cell-permeable DAGs, and (3) the effect of nonhydrolyzable analogues of GTP. Most of the experiments using these approaches have been carried out with sea urchin eggs and *Xenopus* oocytes. We briefly summarize these studies here.

In unfertilized sea urchin eggs (deuterosomes), microinjection of Ins(1,4,5)P_3 produced many of the changes normally associated with fertilization. The transient increase in $[Ca^{2+}]_i$ elicited by Ins(1,4,5)P_3 had a time course similar to the Ca^{2+} transient following fertilization and was not affected by removing external Ca^{2+}. Furthermore, the eggs contained substantial amounts of phosphoinositides, which begin to turn over rapidly at fertilization, and there were increases in the levels of Ins(1,4,5)P_3 immediately after insemination. The Ins(1,4,5)P_3 activated voltage-dependent Ca^{2+} influx in permeabilized *Xenopus* oocytes, and a rise in intracellular free Ca^{2+} was correlated with nuclear envelope breakdown in sea urchin embryos. The Ins(1,4,5)P_3 also appeared to stimulate chromatin condensation immediately after fertilization. Thus, it appeared highly likely that Ins(1,4,5)P_3 was the crucial second messenger for fertilization.

A wave of Ca^{2+} occurs at fertilization. The $[Ca^{2+}]_i$ first increases at the point of sperm entry and spreads throughout the cytoplasm. This Ca^{2+} wave in sea urchin eggs can be initiated by a very localized microinjection of Ins(1,4,5)P_3. Somewhat similar conclusions have also been reached from studies with *Xenopus* oocytes.

Diacylglycerol, the other second messenger produced by hydrolysis of PIP_2, is also involved in the transduction of a fertilization signal, although the mechanism for this is not as well defined as with Ins(1,4,5)P_3. Diacylglycerol is produced, along with Ins(1,4,5)P_3, after fertilization of sea urchin eggs. Also, treatment of sea urchin eggs with dioctanoylglycerol induces an increase in cytoplasmic pH, which is likely to be a cytoplasmic signal responsible for stimulating protein and DNA synthesis.

Treatment of unfertilized mouse eggs with PMA was shown to cause oscillations in $[Ca^{2+}]_i$ similar to those that occur at fertilization (Cuthbertson and

Cobbold, 1985) and also triggered a zona reaction, consistent with exocytosis of the egg's cortical granules (Endo and Schultz, 1987), by mechanisms not yet established.

ACKNOWLEDGMENTS. The author wishes to thank Karen Wipperfurth and Teresa Schiffer for their dedication and skill in preparation of the manuscript. This review was written during the tenure of NIH Grant HL16318 and an Established Investigatorship from the National Alliance for Research in Schizophrenia and Depression (NARSAD).

3. REFERENCES

Abdel-Latif, A. A., 1983, Metabolism of phosphoinositides, in *Handbook of Neurochemistry* (A. Lajtha, ed.), pp. 91–131, Plenum Press, New York.

Abdel-Latif, A. A., 1986, Calcium-mobilizing receptors, polyphosphoinositides, and the generation of second messengers, *Pharmacol. Rev.* **38:**227–272.

Abdel-Latif, A. A., Akhtar, R. A., and Hawthorne, J. N., 1977, Acetylcholine increases the breakdown of trisphosphoinositide in rabbit iris muscle prelabelled with [^{32}P]phosphate, *Biochem. J.* **162:**61–73.

Agranoff, B. W., 1983, Biochemical mechanisms in the phosphatidylinositol effect, *Life Sci.* **32:**2047–2054.

Akhtar, R. A., and Abdel-Latif, A. A., 1980, Requirement for calcium ions in acetylcholine-stimulated phosphodiesteratic cleavage of phosphatidyl-*myo*-inositol 4,5-bisphosphate in rabbit iris smooth muscle, *Biochem. J.* **142:**599–604.

Akhtar, R. A., and Abdel-Latif, A. A., 1984, Carbachol causes rapid phosphodiesteratic cleavage of phosphatidylinositol 4,5-bisphosphate and accumulation of inositol phosphates in rabbit iris smooth muscle; prazosin inhibits noradrenaline and ionophore A23187-stimulated accumulation of inositol phosphates, *Biochem. J.* **224:**291–300.

Allison, J. H., and Stewart, M. A., 1971, Reduced brain inositol in lithium-treated rats, *Nature* **233:**267–268.

Altin, J. G., and Byrave, F. L., 1987, The influx of Ca^{2+} administration of glucagon and Ca^{2+} mobilizing agents to the perfused liver could involve at least two separate pathways, *Biochem. J.* **242:**43–50.

Anderson, R. E., Maude, M. B., Kelleher, P. A., Rayborn, M. E., and Hollyfield, J. G., 1983, Phosphoinositide metabolism in the retina. Localization to horizontal cells and regulation by light and divalent cations, *J. Neurochem.* **41:**764–771.

Baukal, A. J., Guillemette, G., Rubin, R., Spat, A., and Catt, K. J., 1985, Binding sites for inositol trisphosphate in the bovine adrenal cortex, *Biochem. Biophys. Res. Commun.* **133:**532–538.

Berridge, M. J., 1986, Inositol phosphates as second messengers, in *Receptor Biochemistry and Methodology. Phosphoinositides and Receptor Mechanisms* (J. W. Putney, Jr., ed.), pp. 25–45, Alan R. Liss, New York.

Berridge, M. J., 1987a, Inositol lipids and cell proliferation, *Biochim. Biophys. Acta* **907:**33–45.

Berridge, M. J., 1987b, Inositol trisphosphate and diacylglycerol: Two interacting second messengers, *Annu. Rev. Biochem.* **56:**159–193.

Berridge, M. J., and Fain, J. N., 1979, Inhibition of phosphatidylinositol synthesis and the inactivation of calcium entry after prolonged exposure of the blowfly salivary gland to 5-hydroxytryptamine, *Biochem. J.* **178:**59–69.

Berridge, M. J., Heslop, J. P., Irvine, R. F., and Brown, K. D., 1984, Inositol trisphosphate formation and calcium mobilization in Swiss 3T3 cells in response to platelet-derived growth factor, *Biochem. J.* **222:**195–201.

Berridge, M. J., Downes, C. P., and Hanley, M. R., 1989, Neural and developmental actions of lithium: a unifying hypothesis, *Cell* **53:**411–419.

Besterman, J. M., Duranio, V., and Cuatrecasas, P., 1986, Rapid formation of diacylglycerol from phosphatidylcholine: A pathway for generation of a second messenger, *Proc. Natl. Acad Sci. U.S.A.* **83:**6785–6789.

Biden, T. J., and Wollheim, C. B., 1986, Ca^{2+} regulates the inositol tris/tetrakisphosphate pathway in intact and broken preparations of insulin-secreting RINm5F cells, *J. Biol. Chem.* **261:**11931–11934.

Biden, T. J., Wollheim, C. B., and Schlegel, W., 1986, Inositol 1,4,5-trisphosphate and intracellular Ca^{2+} homeostasis in clonal pituitary cells (GH_3). Translocation of Ca^{2+} into mitochondria from a functionally discrete portion of the nonmitochondrial store, *J. Biol. Chem.* **261:**7223–7229.

Bocckino, S. B., Blackmore, P. F., and Exton, J. H., 1985, Stimulation of 1,2-diacylglycerol accumulation in hepatocytes by vasopressin, epinephrine, and angiotensin II, *J. Biol. Chem.* **260:**14201–14207.

Bond, M., Kitazawa, T., Somlyo, A. P., and Somlyo, A. V., 1984, Release and recycling of calcium by the sarcoplasmic reticulum in guinea-pig portal vein smooth muscle, *J. Physiol.* **355:**677–695.

Brockerhoff, H., and Ballou, C. E., 1962, Phosphate incorporation in brain phosphoinositides, *J. Biol. Chem.* **237:**49–52.

Brown, J. E., Rubin, L. J., Ghalayini, A. J., Tarver, A. P., Irvine, R. F., Berridge, M. J., and Anderson, R. E., 1984, Myoinositol polyphosphate may be a messenger for visual excitation in limulus photoreceptors, *Nature* **311:**160–163.

Brown, K. D., Blakeley, D. M., Harmon, M. H., Laurie, M. S., and Corps, A. N., 1987, Protein kinase C-mediated negative-feedback inhibition of unstimulated and bombesin-stimulated polyphosphoinositide hydrolysis in Swiss-mouse 3T3 cells, *Biochem. J.* **245:**631–639.

Burgess, G. M., Irvine, R. F., Berridge, M. J., Mckinney, J. S., and Putney, J. W., Jr., 1984, Actions of inositol phosphates on Ca^{2+} pools in guinea-pig hepatocytes, *Biochem. J.* **224:**741–746.

Burgess, G. M., McKinney, J. S., Irvine, R. F., and Putney, J. W., Jr., 1985, Inositol 1,4,5-trisphosphate and inositol 1,3,4-trisphosphate formation in Ca^{2+}-mobilizing-hormone-activated cells, *Biochem. J.* **232:**237–243.

Busa, W. B., Ferguson, J. Z., Joseph, S. K., Williamson, J. R., and Nuccitelli, R., 1985, Activation of frog (*Xenopus laevis*) eggs by inositol trisphosphate, *J. Cell Biol.* **101:**677–682.

Calderon, P., Furnelle, J., and Cristophe, J., 1979, *In vitro* lipid metabolism in the rat pancreas. II. Effects of secretagogues on fatty acid metabolism, net lipolysis and ATP levels, *Biochim. Biophys. Acta* **574:**391–403.

Capiod, T., Field, A. C., Ogden, D. C., and Sandford, C., 1987, Internal perfusion of guinea-pig hepatocytes with buffered Ca^{2+} or inositol 1,4,5-trisphosphate mimics noradrenaline activation of K^+ and CI conductances, *FEBS Lett.* **217:**247–252.

Carney, D. H., Scott, D. L., Gordon, E. A., and Labelle, E. F., 1985, Phosphoinositides in mitogenesis: Neomycin inhibits thrombin-stimulated phosphoinositide turnover and initiation of cell proliferation. *Cell* **42:**488–497.

Carter, H. R., and Smith, A. D., 1987, Resolution of the phosphoinositide-specific phospholipase C isolated from porcine lymphocytes into multiple species, *Biochem. J.* **244:**639–645.

Castanga, M., Takai, Y., Kaibuchi, K., Sano, K., Kikkawa, U., and Nishizuka, Y., 1982, Direct activation of calcium-activated, phospholipid-dependent protein kinase by tumor-promoting phorbol esters, *J. Biol. Chem.* **257**:7847–7851.

Celtin, J. H., and Bygrave, F. L., 1985, The Ca^{2+}-mobilizing actions of vasopressin and angiotensin differ from those of the α-adrenergic agonist phenylephrine in the α-adrenergic agonist phenylephrine in the perfused rat liver, *Biochem. J.* **242**:43–50.

Colodzin, M., and Kennedy, E. P., 1965, Biosynthesis of diphosphoinositide in brain, *J. Biol. Chem.* **240**:3771–3780.

Conn, P. M., Ganong, B. R., Ebeling, J., Staley, D., Neidel, J. E., and Bell, R. M., 1985, Diacylglycerols release LH: Structure–activity relations reveal a role for protein kinase C, *Biochem. Biophys. Res. Commun.* **126**:532–539.

Connolly, T. M., Lawing, W. J., Jr., and Majerus, P. W., 1986a, Protein kinase C phosphorylates human platelet inositol trisphosphate 5′-phosphomonoesterase, increasing the phosphatase activity, *Cell* **46**:951–958.

Connolly, T. M., Wilson, D. B., Bross, T. E., and Majerus, P. W., 1986b, Isolation and characterization of the inositol cyclic phosphate products of phosphoinositide cleavage by phospholipase C. Metabolism in cell-free extracts, *J. Biol. Chem.* **261**:122–126.

Connolly, T. M., Bansal, V. S., Bross, T. E., Irvine, R. F., and Majerus, P. W., 1987, The metabolism of tris- and tetraphosphates of inositol by 5-phosphomonoesterase and 3-kinase enzymes, *J. Biol. Chem.* **262**:2146–2149.

Cosgrove, D. J., 1969, Ion-exchange chromatography of inositol polyphosphates, *Ann. N.Y. Acad. Sci.* **165**:677–686.

Cotecchia, S., Leeb-Lundberg, L. M. F., Hagen, P. O., Lefkowitz, R. J., and Caron, M. G., 1985, Phorbol ester effects on α_1-adrenoceptor binding and phosphatidylinositol metabolism in cultured vascular smooth muscle cells, *Life Sci.* **37**:2389–2398.

Creba, J. A., Downes, C. P., Hawkins, P. T., Brewster, G., Michell, R. H., and Kirk, C. J., 1983, Rapid breakdown of phosphatidylinositol 4-phosphate and phosphatidylinositol 4,5-bisphosphate in rat hepatocytes stimulated by vasopressin and other Ca^{2+}-mobilizing hormones, *Biochem. J.* **212**:733–747.

Crossley, I., Swann, K., Chambers, E., and Whitaker, M., 1988, Activation of sea urchin eggs by inositol phosphates is independent of external calcium, *Biochem. J.* **252**:257–262.

Cullen, P. J., Hsuan, J. J., Truong, O., Letcher, A. J., Jackson, T. R., Dawson, A. P., and Irvine, R. F., 1995, Identification of a specific Ins(1,3,4,5)P_4-binding protein as a member of the GAP 1 family, *Nature* **376**:527–530.

Cuthbertson, K. S. R., and Cobbold, P. H., 1985, Phorbol ester and sperm activate mouse oocytes by inducing sustained oscillations in cell Ca^{2+}, *Nature* **316**:541–542.

Davis, R. J., Ganong, B. R., Bell, R. M., and Czech, M. P., 1985, Structural requirements for diacylglycerols to mimic tumor-promoting phorbol diester action on the epidermal growth factor receptor, *J. Biol. Chem.* **260**:5315–5322.

Dawson, R. M. C., 1954, The measurement of ^{32}P labelling of individual kephalins and lecithin in a small sample of tissue, *Biochim. Biophys. Acta* **14**:374–379.

Dawson, R. M. C., 1959, Studies on the enzymic hydrolysis of monophosphoinositide by phospholipase preparations from *P. notatum* and ox pancreas, *Biochim. Biophys. Acta* **33**:68–77.

Deckmyn, H., Tu, S. M., and Majerus, P. W., 1986, Guanine nucleotides stimulate soluble phosphoinositide-specific phospholipase C in the absence of membranes, *J. Biol. Chem.* **261**:16553–16558.

Delfert, D. M., Hill, S., Pershadsingh, H. A., Sherman, W. R., and McDonald, J. M., 1986, *myo*-Inositol 1,4,5-trisphosphate mobilizes Ca^{2+} from isolated adipocyte endoplasmic reticulum but not from plasma membranes, *Biochem. J.* **236**:37–44.

Diringer, H., and Friis, R. R., 1977, Changes in phosphatidylinositol metabolism correlated with growth state of normal and Rous sarcoma virus-transformed Japanese quail cells, *Cancer Res.* **37:**2978–2984.

Dixon, J. F., and Hokin, L. E., 1984, Secretagogue-stimulated phosphatidylinositol breakdown in the exocrine pancreas liberates arachidonic acid, stearic acid, and glycerol by sequential actions of phospholipase C and diacylglycerol lipase, *J. Biol. Chem.* **259:**14418–14425.

Dixon, J. F., and Hokin, L. E., 1985, The formation of inositol 1,2-cyclic phosphate on agonist stimulation of phosphoinositide breakdown in mouse pancreatic minilobules. *J. Biol. Chem.* **260:**16068–16071.

Dixon, J. F., and Hokin, L. E., 1987a, Inositol 1,2-cyclic 4,5-trisphosphate is formed in the rat parotid gland on muscarinic stimulation, *Biochem. Biophys. Res. Commun.* **149:**1208–1213.

Dixon, J. F., and Hokin, L. E., 1987b, Inositol 1,2-cyclic 4,5-trisphosphate: Concentration relative to inositol 1,4,5-trisphosphate in pancreatic minilobules on stimulation with carbamylcholine in the absence of lithium. Possible role as a second messenger in long- but not short-term responses, *J. Biol. Chem.* **262:**13892–13895.

Dixon, J. F., and Hokin, L. E., 1989, Kinetic analysis of the formation of inositol 1:2 cyclic phosphate in carbachol-stimulated pancreatic minilobules. Half is formed by direct phosphodiestratic cleavage of phosphatidylinositol, *J. Biol. Chem.* **264:**11721–11724.

Dixon, J. F., Los, G. V., and Hokin, L. E., 1992, Lithium enhances accumulation of [^{3}H] inositol radioactivity and mass of second messenger inositol 1,4,5-trisphosphate in monkey cerebral cortex slices, *J. Neurochem.*, **5:**2332–2335.

Dixon, J. F., Los, G. V., and Hokin, L. E., 1994, Lithium stimulates glutamate release and inositol 1,4,5-trisphosphate accumulation via activation of the N-methyl-D-aspartate receptor in monkey and mouse cerebral cortex slices, *Proc. Natl. Acad. Sci. U.S.A.* **91:**8358–8362.

Douglas, W. W., 1974, Involvement of calcium in exocytosis and the exocytosis–vesiculation sequence. *Biochem. Soc. Symp.* **39:**1–28.

Downes, C. P., and Michell, R. H., 1985, Inositol phospholipid breakdown as a receptor-controlled generator of second messengers, in *Molecular Mechanisms of Transmembrane Signalling* (P. Cohen and M. D. Houslay, eds.), pp. 4–56, Elsevier, Amsterdam.

Downes, C. P., and Stone, M. A., 1986, Lithium-induced reduction in intracellular inositol supply in cholinergically stimulated parotid gland, *Biochem. J.* **234:**199–204.

Durell, J., Garland, J. T., and Friedel, R. O., 1969, Acetylcholine action: Biochemical aspects, *Science* **165:**862–866.

Endo, Y., and Schultz, R. M., 1987, Effects of phorbol esters and a diacylglycerol on mouse eggs: Inhibition of the fertilization and modification of the zona pellucida, *Dev. Biol.* **119:**199–209.

Enyedi, P., and Williams, G. H., 1988, Heterogeneous inositol tetrakisphosphate binding sites in the adrenal cortex, *J. Biol. Chem.* **263:**7940–7942.

Fain, J. N., 1987, Activation of phosphoinositide specific phospholipase C by ligands in the presence of guanine nucleotides, in *Mechanisms of Signal Transduction by Hormones and Growth Factors* (M. C. Cabot, and W. L. McKeehan, eds.), pp. 133–147, Alan R. Liss, New York.

Fain, J. N., and Berridge, M. J., 1979a, Relationship between hormonal activation of phosphatidylinositol hydrolysis, fluid secretion and calcium flux in the blowfly salivary gland, *Biochem. J.* **178:**45–58.

Fain, J. N., and Berridge, M. J., 1979b, Relationship between phosphatidylinositol synthesis and recovery of 5-hydroxytryptamine-responsive Ca^{2+} flux in blowfly salivary glands, *Biochem. J.* **180:**655–661.

Fein, A., Payne, R., Corson, D. W., Berridge, M. J., and Irvine, R. F., 1984, Photoreceptor excitation and adaptation by inositol 1,4,5-trisphosphate, *Nature* **311:**157–160.

Fisher, D. B., and Mueller, G. C., 1971, Studies of the mechanism by which PHA stimulates phospholipid metabolism of human lymphocytes, *Biochim. Biophys. Acta* **248:**434–448.

Friesen, L. L., and Gerrard, J. M., 1985, The effects of I-oleoyl-2-acetylglycerol on platelet protein phosphorylation and platelet ultrastructure, *Am. J. Pathol.* **121:**79–87.

Fry, M. J., Gebhardt, A., Parker, P. J., and Foljlkes, G., 1985, Phosphatidylinositol turnover and transformation of cells by Abelson murine lukemia virus, *EMBO J.* **4:**3173–3178.

Fujita, I., Irita, K., Takeshiga, K., and Minakami, S., 1984, Diacylglycerol, I-oleoyl-2-acetylglycerol, stimulates superoxide generation from human neutrophils, *Biochem. Biophys. Res. Commun.* **120:**318–324.

Gilman, A. G., 1987, G proteins: Transducers of receptor-generated signals, *Annu. Rev. Biochem.* **56:**615–649.

Gomperts, B. D., 1983, Involvement of guanine nucleotide-binding protein in the gating of Ca^{2+} by receptors, *Nature* **306:**64–66.

Gonzales, R. A., and Crews, F. T., 1985, Cholinergic- and adrenergic-stimulated inositide hydrolysis in brain: Interaction, regional distribution and coupling mechanisms, *J. Neurochem.* **45:**1076–1084.

Goodhardt, M., Ferry, N., Geynet, P., and Hanoune, J., 1982, Hepatic α_1-adrenergic receptors show agonist-specific regulation by guanine nucleotides, *J. Biol. Chem.* **257:**11577–11583.

Graham, R. A., Meyer, R. A., Szwergold, B. S., and Brown, T. R., 1987, Observation of *myo*-inositol 1,2-(cyclic) phosphate in a Morris hepatoma by 31p NMR, *J. Biol. Chem.* **262:**35–37.

Guillemette, G., Balla, T., Baukal, A. J., and Catt, K. J., 1988, Characterization of inositol 1,4,5-trisphosphate receptors and calcium mobilization in a hepatic plasma membrane fraction, *J. Biol. Chem.* **263:**4541–4548.

Habernicht, A. J. R., Glomset, J. A., King, W. C., Nist, C., Mitchell, C. D., and Ross, R., 1981, Early changes in phosphatidylinositol and arachidonic acid metabolism in quiescent Swiss 3T3 cells stimulated to divide by PDGF, *J. Biol. Chem.* **256:**12329–12335.

Hansen, C. A., Mah, S., and Williamson, J. R., 1986, Formation and metabolism of inositol 1,3,4,5-tetrakisphosphate in liver, *J. Biol. Chem.* **261:**8100–8103.

Hansen, C. A., Johansen, R. A., Williamson, M. T., and Williamson, J. R., 1987, Purification and characterization of two types of soluble inositol phosphate 5-phosphomonoesterases from rat brain, *J. Biol. Chem.* **262:**17319–17326.

Harwood, J. L., and Hawthorne, J. N., 1969, The properties and subcellular distribution of phosphatidylinositol kinase in mammalian tissues, *Biochim. Biophys. Acta* **171:**75–88.

Haslam, R. J., and Davidson, M. M. L., 1984, Guanine nucleotides decrease the free Ca^{2+} required for secretion of serotonin from permeabilized platelets. Evidence of a role of GTP-binding protein in platelet activation, *FEBS Lett.* **174:**90–95.

Hawkins, P. T., Stephens, L., and Downes, C. P., 1986, Rapid formation of inositol 1,3,4,5-tetrakisphosphate and inositol 1,3,4-trisphosphate in rat parotid glands may both result indirectly from receptor-stimulated release of inositol 1,4,5-trisphosphate from phosphatidylinositol 4,5-bisphosphate, *Biochem. J.* **238:**507–516.

Hawkins, P. T., Berrie, C. P., Morris, A. J., and Downes, C. P., 1987, Inositol 1,2-cyclic 4,5-trisphosphate is not a product of muscarinic receptor-stimulated phosphatidylinositol 4,5-bisphosphate hydrolysis in rat parotid glands, *Biochem. J.* **243:**211–218.

Helms, J. B., deVries, K.-J., and Wirtz, K. W. A., 1991, Synthesis of phosphatidylinositol 4,5-bisphosphate in the endoplasmic reticulum of Chinese hamster ovary cells, *J. Biol. Chem.* **266:**21368–21374.

Heslop, J. P., Irvine, R. F., Tashjian, A. H., Jr., and Berridge, M. J., 1985, Inositol tetrakis- and pentakisphosphates in GH4 cells, *J. Exp. Biol.* **119:**395–401.

Heslop, J. P., Blakeley, D. M., Brown, K. D., Irvine, R. F., and Berridge, M. J., 1986, Effects of bombesin and insulin on inositol (1,4,5)trisphosphate and inositol (1,3,4)trisphosphate formation in Swiss 3T3 cells, *Cell* **47:**703–709.

Hirata, M., Sasaguri, T., Hamachi, T., Hashimoto, T., Kukita, M., and Koga, T., 1985, Irreversible inhibition of Ca^{2+} release in saponin-treated macrophages by the photoaffinity derivative of inositol 1,4,5-trisphosphate, *Nature* **317**:723–725.

Hokin, L. E., 1965, 'Autoradiographic localization of the acetylcholine-stimulated synthesis of phosphatidylinositol in the superior cervical ganglion, *Proc. Natl. Acad. Sci. U.S.A.* **53**:1369–1376.

Hokin, L. E., 1966, Effects of calcium omission on acetylcholine-stimulated amylase secretion and phospholipid synthesis in pigeon pancreas slices, *Biochim. Biophys. Acta* **115**:219–221.

Hokin, L. E., 1968, Dynamic aspects of phospholipids during protein secretion, *Int. Rev. Cytol.* **23**:187–208.

Hokin, L. E., 1969, Phospholipid metabolism and functional activity of nerve cells, in *The Structure and Function of Nervous Tissue* (G. H. Bourne, ed.), pp. 161–184, Academic Press, New York.

Hokin, L. E., 1985, Receptors and phosphoinositide-generated second messengers, *Annu. Rev. Biochem.* **54**:205–235.

Hokin, L. E., 1987, The road to the phosphoinositide-generated second messengers, *Trends Pharmacol. Sci.* **8**:53–56.

Hokin, L. E., and Hokin, M. R., 1955a, Effects of acetylcholine on phosphate turnover in phospholipids of brain cortex *in vitro*, *Biochim. Biophys. Acta* **16**:229–237.

Hokin, L. E., and Hokin, M. R., 1955b, Effects of acetylcholine on the turnover of phosphoryl units in individual phospholipids of pancreas slices and brain cortex slices, *Biochim. Biophys. Acta* **18**:102–110.

Hokin, L. E., and Hokin, M. R., 1956, The actions of pancreozymin in pancreas slices and the role of phospholipids in enzyme secretion, *J. Physiol.* **132**:442–453.

Hokin, L. E., and Hokin, M. R., 1958a, Phosphoinositides and protein secretion in pancreas slices, *J. Biol. Chem.* **233**:805–810.

Hokin, L. E., and Hokin, M. R., 1958b, Acetylcholine and the exchange of inositol and phosphate in brain phosphoinositide, *J. Biol. Chem.* **233**:818–821.

Hokin, L. E., and Hokin, M. R., 1960, Studies on the carrier function of phosphatidic acid in sodium transport. I. The turnover of phosphatidic acid and phosphoinositide in the avial salt gland on stimulation of secretion, *J. Gen. Physiol.* **44**:61–85.

Hokin, L. E., and Hokin-Neaverson, M. R., 1989, Commentary on effects of acetylcholine on the turnover of phosphoryl unites in individual phospholipids of pancreas slices" by L. E. Hokin, and M. R. Hokin, Biochim. Biophys. Acta 18 (1955) 102–110, *Biochim. Biophys. Acta* **1000**:465–469.

Hokin, L. E., and Huebner, D., 1967, Radioautographic localization of the increased synthesis of phosphatidylinositol in response to pancreozymin or acetylcholine in guinea pig pancreas slices, *J. Cell Biol.* **33**:521–530.

Hokin, M. R., 1968, Studies on chemical mechanisms of the action of neurotransmitters and hormones: Relationship between hormone-stimulated ^{32}P incorporated into phosphatidic acid and into phosphatidylinositol in pigeon pancreas slices, *Arch. Biochem. Biophys.* **124**:271–279.

Hokin, M. R., and Hokin, L. E., 1953, Enzyme secretion and the incorporation of ^{32}P into the phospholipids of pancreas slices, *J. Biol. Chem.* **203**:967–977.

Hokin, M. R., and Hokin, L. E., 1954, Effects of acetylcholine on phospholipids in the pancreas, *J. Biol. Chem.* **209**:549–558.

Hokin, M. R., and Hokin, L. E., 1964, Interconversions of phosphatidylinositol and phosphatidic acid involved in the response to acetylcholine in the salt gland, in *Metabolism and Physiological Significance of Lipids* (R. M. C. Dawson, and D. N. Rhodes, eds.), pp. 423–434, John Wiley & Sons, New York.

Hokin-Neaverson, M., 1974, Acetylcholine causes a net decrease in phosphatidylinositol and a net increase in phosphatidic acid in mouse pancreas, *Biochem. Biophys. Res. Commun.* **58**:763–768.

Hokin-Neaverson, M., 1977, Metabolism and role of phosphatidylinositol in acetylcholine-stimulated membrane function, *Adv. Exp. Biol. Med.* **83:**429–446.

Hughes, A. R., Takemura, H., and Putney, J. W., Jr., 1988, Kinetics of inositol 1,4,5-trisphosphate and inositol cyclic 1:2,4,5-trisphosphate metabolism in intact rat parotid acinar cells: Relationship to calcium signalling, *J. Biol. Chem.* **263:**10314–10319.

Hughes, B. P., Rye, K. A., Pickford, L. B., Barritt, G. J., and Chalmers, A. H., 1984, A transient increase in diacylglycerols is associated with the action of vasopressin on hepatocytes, *Biochem. J.* **222:**535–540.

Hunter, T., 1986, Cell growth control mechanisms, *Nature* **322:**14–16.

Imai, A., and Gershengorn, M. C., 1986, Phosphatidylinositol 4,5-bisphosphate turnover is transient while phosphatidylinositol turnover is persistent in thyrotropin-releasing hormone-stimulated rat pituitary cells, *Proc. Natl. Acad. Sci. U.S.A.* **83:**8540–8544.

Imboden, J. B., and Stobo, J. D., 1985, Transmembrane signalling by T cell antigen receptor. Perturbation of the T3-antigen complex generates inositol phosphates and releases calcium ion intracellular stores, *J. Exp. Med.* **161:**446–456.

Inoue, M., Kishimoto, A., Takai, Y., and Nishizuka, Y., 1977, Studies on cyclic nucleotide-independent protein kinase and its proenzyme in mammalian tissues, *J. Biol. Chem.* **252:**7610–7616.

Irvine, R. F., and Moor, R. M., 1986, Micro-injection of inositol 1,3,4,5-tetrakisphosphate activates sea urchin eggs by a mechanism dependent on external Ca^{2+}, *Biochem. J.* **240:**917–920.

Irvine, R. F., and Moor, R. M., 1987, Inositol (1,3,4,5)tetrakisphosphate-induced activation of sea urchin eggs requires the presence of inositol trisphosphate, *Biochem. Biophys. Res. Commun.* **146:**284–290.

Irvine, R. F., Letcher, A. J., Lander, D. J., and Downes, C. P., 1984, Inositol trisphosphates in carbachol-stimulated rat parotid glands, *Biochem. J.* **223:**237–243.

Irvine, R. F., Letcher, A. J., Heslop, J. P., and Berridge, M. J., 1986a, The inositol tris/tetrakisphosphate pathway—demonstration of Ins(1,4,5)P_3 3-kinase activity in animal tissues, *Nature* **320:**631–634.

Irvine, R. F., Letcher, A. J., Lander, D. J., and Berridge, M. J., 1986b, Specificity of inositol phosphate-stimulated Ca^{2+} mobilization from Swiss-mouse 3T3 cells, *Biochem. J.* **240:**301–304.

Ishii, H., Connolly, T. M., Bross, T. E., and Majerus, P. W., 1986, Inositol cyclic trisphosphate [inositol 1,2-(cyclic)-4,5-trisphosphate] is formed upon thrombin stimulation of human platelets, *Proc. Natl. Acad. Sci. U.S.A.* **83:**6397–6401.

Jones, L. M., and Michell, R. H., 1974, Breakdown of phosphatidylinositol provoked by muscarinic cholinergic stimulation in rat parotid-gland fragments, *Biochem. J.* **142:**583–590.

Jope, R. S., and Williams, M. B., 1994, Lithium and brain signal transduction systems, *Biochem. Pharmacol.* **47:**429–441.

Joseph, S. K., and Williamson, J. R., 1986, Characteristics of inositol trisphosphate-mediated Ca^{2+} release from permeabilized hepatocytes, *J. Biol. Chem.* **261:**14658–14664.

Kaplan, D. R., Whitman, M., Schaffhausen, B., Raptis, L., Garcea, R. L., Pallas, D., Roberts, T. M., and Cantley, L., 1986, Phosphatidylinositol metabolism and polyoma-mediated transformation, *Proc. Natl. Acad. Sci. U.S.A.* **83:**3624–3628.

Katada, T., Gilman, A. G., Watanabe, Y., Bauer, S., and Jakobs, K. H., 1985, Protein kinase C phosphorylates the inhibitory guanine-nucleotide-binding regulatory component and apparently suppresses its function in hormonal inhibition of adenylate cyclase, *Eur. J. Biochem.* **151:**431–437.

Katz, B., 1969, *The Release of Neural Transmitter Substances*, p. 1–60, Charles C. Thomas, Springfield, IL.

Kemp, P., Hubscher, G., and Hawthorne, J. N., 1961a, A liver phospholipase hydrolysing phosphoinositides, *Biochim. Biophys. Acta* **31:**585–586.

Kemp, P., Hubscher, G., and Hawthorne, J. N., 1961b, Phosphoinositides. 3. Enzymic hydrolysis of inositol-containing phospholipids, *Biochem. J.* **79:**193–200.

Kennedy, E. D., Challis, R. A. J., and Nahonski, S. R., 1989, Lithium reduces the accumulation of inositol polyphosphate second messengers following cholinergic stimulation of cerebral cortex slices, *J. Neurochem.* **53:**1652–1655.

Kennedy, E. P., 1986, The biosynthesis of phospholipids, in *Lipids and Membranes: Past, Present and Future* (J. A. F. Op den Kemp, B. Roelofsen, and K. W. A. Wirtz, eds.), pp. 171–206, Elsevier, Amsterdam.

Kennerly, D. A., 1987, Diacylglycerol metabolism in mast cells, *J. Biol. Chem.* **262:**16305–16313.

King, C. E., Stephens, L. R., Hawkins, P. T., Guy, G. R., and Michell, R. H., 1987, Multiple metabolic pools of phosphoinositides and phosphatidate in human erythrocytes incubated in a medium that permits rapid transmembrane exchange of phosphate, *Biochem. J.* **244:**209–217.

Kirk, C. J., Creba, J. A., Downes, C. P., and Michell, R. H., 1981, Hormone-stimulated metabolism of inositol lipids and its relationship to hepatic receptor function, *Biochem. Soc. Trans.* **9:**377–379.

Kojima, I., Kojima, K., and Rasmussen, H., 1985, Role of calcium fluxes in the sustained phase of angiotensin II-mediated aldosterone secretion from adrenal glomerulosa cells, *J. Biol. Chem.* **260:**9177–9184.

Kraft, A. S., and Anderson, W. B., 1983, Phorbol esters increase the amount of Ca^{2+}, phospholipid-dependent protein kinase associated with plasma membrane, *Nature* **301:**621–623.

Kraft, A. S., Anderson, W. B., Cooper, H. L., and Sando, J. J., 1982, Decrease in cytosolic calcium phospholipid-dependent protein kinase activity following phorbol ester treatment of EL4 thymoma cells, *J. Biol. Chem.* **257:**13193–13196.

Lapetina, E. G., Reep, R., Ganong, B. R., and Bell, R. M., 1985, Exogenous *sn*-1,2-diacylglycerols containing saturated fatty acids function as bioregulators of protein kinase C in human platelets, *J. Biol. Chem.* **260:**1358–1361.

Lee, C. H., and Hokin, L. E., 1989, Inositol 1,2-cyclic 4,5-trisphosphate is an order of magnitude less potent than inositol 1,4,5-trisphosphate in mobilizing intracellular stores of calcium in mouse pancreatic acinar cells, *Biochem. Biophys. Res. Commun.* **159:**561–565.

Lee, C. H., Dixon, J. F., Reichman, M., Moummi, C., Los, G., and Hokin, L. E., 1992, Li^+ increases accumulation of inositol 1,4,5-trisphosphate and inositol 1,3,4,5-tetrakisphosphate in cholinergically stimulated brain cortex slices in guinea pig, mouse, and rat. The increases require inositol supplementation in mouse and rat but not in guinea pig, *Biochem. J.* **282:**377–385.

Lee, T. C., and Huggins, C. G., 1968, Triphosphoinositide phosphomonoesterase in rat kidney cortex. I. General properties and subcellular localization, *Arch. Biochem. Biophys.* **126:**206–213.

Litosch, I., and Fain, J. N., 1986, Minireview. Regulation of phosphoinositide breakdown by guanine nucleotides, *Life Sci.* **39:**187–194.

Macara, I. G., Marinetti, G. V., Livingston, J. N., and Balduzzi, P. C., 1985, Lipid phosphorylating activities and tyrosine kinases: A possible role for phosphatidylinositol turnover in transformation, in *Growth Factors and Transformation* (F. Feramisco, ed.), pp. 365–368, Cold Spring Harbor Laboratory, New York.

Marinetti, G. V., and Stotz, E., 1956, Chromatography of phosphatides on silicic acid impregnated paper, *Biochim. Biophys. Acta* **21:**168–170.

Matuoka, K., Fukami, K., Nakanishi, O., Kawai, S., and Takenawa, T., 1988, Mitogenesis in response to PDGF and bombesin abolished by microinjection of antibody to PIP_2, *Science* **239:**640–643.

Mayer, B. J., Hamagijchi, M., and Hanafusa, H., 1988, A novel viral oncogene with structural similarity to phospholipase C, *Nature* **332:**272–275.

McNeil, P. L., Mckenna, M. P., and Taylor, D. L., 1985, A transient rise in cytosolic calcium follows stimulation of quiescent cells with growth factors and is inhibitable with phorbol myristate acetate, *J. Cell Biol.* **101**:372–379.

Mendoza, S. A., Lopez-Rivas, A., Sinnett-Smith, J. W., and Rozengurt, E., 1986, Phorbol esters and diacylglycerol inhibit vasopressin-induced increases in cytoplasmic-free Ca^{2+} and $^{45}Ca^{2+}$ efflux in Swiss 3T3 cells, *Exp. Cell. Res.* **164**:536–545.

Meyer, T., Holowka, D., and Stryer, L., 1988, Highly cooperative opening of calcium channels by inositol 1,4,5-trisphosphate, *Science* **240**:653–656.

Michell, R. H., 1975, Inositol phospholipids and cell surface receptor function, *Biochim. Biophys. Acta* **415**:81–147.

Michell, R. H., and Hawthorne, J. N., 1965, The site of diphosphoinositide synthesis in rat liver, *Biochem. Biophys. Res. Commun.* **21**:333–338.

Michell, R. H., Jafferji, S. S., and Jones, L. M., 1977, The possible involvement of phosphatidylinositol breakdown in the mechanism of stimulus-response coupling at receptors which control cell-surface calcium gates, in *Function and Biosynthesis of Lipid* (N. G. Bazan, B. B. Brenner, and N. M. Buisto, eds.), pp. 447–464, Plenum Press, New York.

Mikoshiba, K., Furuichi, T., and Miyawaki, A., 1994, $InsP_3$ receptor and its regulation in the nervous system, in *Proceeding of the XVIth International Congress of Biochemistry and Molecular Biology,* Vol. I, p. 103, Publications & Information Directorate, Dr. K. S. Krishan Mara, New Delhi, India.

Molina Y Vedia, L. M., and Lapetina, E. G., 1986, Phorbol 12,13-dibutyrate and I-oleoyl-2-acetyldiacylglycerol stimulate inositol trisphosphate dephosphorylation in human platelets, *J. Biol. Chem.* **261**:10493–10495.

Monaco, M. E., and Woods, D., 1983, Characterization of the hormone-sensitive phosphatidylinositol pool in WRK-I cells, *J. Biol. Chem.* **258**:15125–15129.

Muallem, S., Schoeffield, M., Pandol, S., and Sachs, G., 1985, Inositol trisphosphate modification of ion transport in rough endoplasmic reticulum, *Proc. Natl. Acad. Sci. U.S.A.* **82**:4433–4437

Navidi, M., Yoa, F. G., and Sun, G. Y., 1991, Brief chronic effects of lithium administration on rat brain phosphoinositides and phospholipids, *J. Neurosci. Res.* **28**:428–433.

Niedel, J. E., and Blackshear, P. J., 1986, Protein kinase C, in *Receptor Biochemistry and Methodology. Phosphoinositides and Receptor Mechanisms* (J. W. Putney, Jr., ed.), pp. 47–88, Alan R. Liss, New York.

Nishizuka, Y., 1984a, Turnover of inositol lipids and signal transduction, *Science* **225**:1365–1370.

Nishizuka, Y., 1984b, The role of protein kinase C in cell surface signal transduction and tumour promotion, *Nature* **308**:693–698.

Nishizuka, Y., 1986, Studies and perspectives of protein kinase C, *Science* **233**:305–312.

Nishizuka, Y., 1988, The heterogeneity and differential expression of multiple species of the protein kinase C family, *Biofactors* **1**:17–20.

Oron, Y., Dascal, N., Nadler, E., and Lupu, M., 1985, Inositol 1,4,5-trisphosphate mimics muscarinic response in *Xenopus* oocytes, *Nature* **313**:141–143.

Palade, G. E., Siekevitz, P., and Caro, L. G., 1962, Structure, chemistry and function of the pancreatic exocrine cell, in *Ciba Foundation Symposium on Endocrine Pancreas* (A. V. S. de Reuck, and M. P. Cameron, eds.), pp. 23–55, Chruchill, London.

Parker, I., and Miledi, R., 1987, Inositol trisphosphate activates a voltage-dependent calcium influx in *Xenopus* oocytes, *Proc. R. Soc.* **231**:27–36.

Prentki, M., Janjic, D., Biden, T. J., Blondel, B., and Wollheim, C. B., 1984, Regulation of Ca^{2+} transport by isolated organelles of a rat insulinoma, *J. Biol. Chem.* **259**:10118–10123.

Putney, J. W., Jr., 1978, Stimulus-permeability coupling: role of calcium in the receptor regulation of membrane permeability, *Pharmacol. Rev.* **30**:209–245.

Rana, R. S., and Hokin, L. E., 1990, Role of phosphoinositides in transmembrane signaling, *Physiol Rev.* **70**:115–164.

Rasmussen, H., and Barrett, P. Q., 1984, Calcium messenger system: An integrated view, *Physiol. Rev.* **64:**938–984.

Redman, C. M., and Hokin, L. E., 1959, Phospholipide turnover in microsomal membranes of the pancreas during enzyme secretion, *J. Biophys. Biochem. Cytol.* **6:**207–214.

Redman, C. M., and Hokin, L. E., 1964, Stimulation of the metabolism of phosphatidylinositol and phosphatidic acid in brain cytoplasmic fractions by low concentrations of cholinergic agents, *J. Neurochem.* **11:**155–163.

Renard, D., Poggioli, J., Berthon, B., and Claret, M., 1987, How far does phospholipase C activity depend on the cell calcium concentration? A study of intact cells, *Biochem. J.* **243:**391–398.

Ringer, S., 1883, A further contribution regarding the influence of different constituents of the blood on the contraction of the heart, *J. Physiol.* **4:**29–42.

Rodbell, M., Birnbaumer, L., Pohl, S. L., and Krans, H. M., 1971, The glucagon-sensitive adenyl cyclase system in plasma membranes of rat liver. V. An obligatory role of guanyl nucleotides in glucagon action, *J. Biol. Chem.* **246:**1877–1882.

Rodriguez-Pena, A., and Rozengurt, E., 1985, Serum, like phorbol esters, rapidly activates protein kinase C in intact quiescent fibroblasts, *EMBO J.* **4:**71–76.

Rubin, R. P., 1982, *Calcium and Cellular Secretion,* Plenum Press, New York.

Ryu, S. H., Suh, P., Cho, K. S., Lee, K., and Rhee, S. G., 1987, Bovine brain cytosol contains three immunologically distinct forms of inositol phospholipid-specific phospholipase C, *Proc. Natl. Acad. Sci. U.S.A.* **84:**6649–6653.

Safran, A., Sagi-Eisenberg, R., Neumann, D., and Fuchs, S., 1987, Phosphorylation of the acetylcholine receptor by protein kinase C and identification of the phosphorylation site within the receptor I subunit, *J. Biol. Chem.* **262:**10506–10510.

Sagi-Eisenberg, R., 1989, GTP-binding proteins as possible targets for PKC action, *Trends Biochem. Sci.* **14:**355–357.

Sano, K., Takai, Y., Yamanishi, J., and Nishizuka, Y., 1983, A role of calcium-activated phospholipid-dependent protein kinase in human platelet activation. Comparison of thrombin and collagen actions, *J. Biol. Chem.* **258:**2010–2013.

Santiago-Calvo, E., Mula, S., Redman, C. M., Hokin, M. R., and Hokin, L. E., 1964, The chromatographic separation of polyphosphoinositides and studies on their turnover in various tissues, *Biochim. Biophys. Acta* **84:**550–562.

Schmidt, G., and Thannhauser, S. J., 1945, A method for the determination of desoxyribonucleic acid, ribonucleic, acid, and phosphoproteins in animal tissues, *J. Biol. Chem.* **161:**83–89.

Schmidt, S. Y., 1983, Light-dependent and cytidine-dependent phosphatidylinositol synthesis in photoreceptor cells of the rat, *J. Cell Biol.* **97:**832–837.

Schmidt-Nielson, K., 1960, The salt-secreting gland of marine birds, *Circulation* **21:**955–967.

Schulz, I., and Stolze, H. H., 1980, The exocrine pancreas: The role of secretagogues, cyclic nucleotides, and calcium in enzyme secretion, *Annu. Rev. Physiol.* **42:**127–156.

Sekar, M. C., and Hokin, L. E., 1986, The role of phosphoinositides in signal transduction, *J. Membr. Biol.* **89:**193–210.

Sekar, M. C., Dixon, J. F., and Hokin, L. E., 1987, The formation of inositol 1,2-cyclic 4,5-trisphosphate and inositol 1,2-cyclic 4-bisphosphate on stimulation of mouse pancreatic minilobules with carbamylcholine, *J. Biol. Chem.* **262:**340–344.

Shah, J., and Pant, H. C., 1988, Potassium-channel blockers inhibit inositol trisphosphate-induced calcium release in the microsomal fractions isolated from the rat brain, *Biochem. J.* **250:**617–620.

Sherman, W. R., Gish, B. G., Honchar, M. P., and Munsell, L. Y., 1986, Effects of lithium on phosphoinositide metabolism *in vivo, Fed. Proc.* **45:**2639–2646.

Snoek, G., deWit, J. S., and Wirtz, K. W., 1993, Properties and intracellular localization of phosphatidylinositol transfer protein in Swiss mouse 3T3 cells, *Biochem. Soc. Trans.* **21:**244–247.

Stahl, M. L., Ferenz, C. R., Kelleher, K. L., Kriz, R. W., and Knopf, J. L., 1988, Sequence similarity of phospholipase C with the non-catalytic region of *src*, *Nature* **332:**269–272.

Steinhardt, R. A., and Epel, D., 1974, Activation of sea urchin eggs by a calcium ionophore, *Proc. Natl. Acad. Sci. U.S.A.* **71:**1915–1919.

Stephens, L. R., Hawkins, P. T., Barker, C. J., and Downes, P. C., 1988a, Synthesis of *myo*-inositol 1,3,4,5,6-pentakisphosphate from inositol phosphates generated by receptor activation, *Biochem. J.* **253:**721–733.

Stephens, L. R., Hawkins, P. T., Morris, A. J., and Downes, P. C., 1988b, L-*Myo*-inositol 1,4,5,6-tetrakisphosphate (3-hydroxy)kinase, *Biochem. J.* **249:**283–292.

Storey, D. J., Shears, S. B., Kirk, C. J., and Michell, R. H., 1984, Stepwise enzymatic dephosphorylation of inositol 1,4,5-trisphopshate to inositol in liver, *Nature* **312:**374–376.

Streb, H., Irvine, R. F., Berridge, M. J., and Schulz, I., 1983, Release of Ca^{2+} from a non-mitochondrial intracellular store in pancreatic acinar cells by inositol 1,4,5-trisphosphate, *Nature* **306:**67–69.

Streb, H., Bayerdorffer, E., Haase, W., Irvine, R. F., and Schultz, I., 1984, Effect of inositol-1,4,5-trisphosphate on isolated subcellular fractions of rat pancreas, *J. Membr. Biol.* **81:**241–253.

Streb, H., Heslop, J. P., Irvine, R. F., and Berridge, M. J., 1985, Relationship between secretagogue-induced Ca^{2+} release and inositol phosphate production in permeabilized pancreatic acinar cells, *J. Biol. Chem.* **260:**7309–7315.

Sugimoto, Y., Whitman, M., Cantley, L. C., and Erikson, R. L., 1984, Evidence that the Rous sarcoma transformed gene product phosphorylates phosphatidylinositol and diacylglycerol, *Proc. Natl. Acad. Sci. U.S.A.* **81:**2117–2121.

Suh, P. G., Ruy, S. H., Moon, K. H., Suh, H. W., and Rhee, S. G., 1988, Cloning and sequence of multple forms of phospholipase C, *Cell* **54:**161–169.

Supattapone, S., Worley, P. F., Baraban, J. M., and Snyder, S. H., 1988, Solubilization, purification, and characterization of an inositol trisphosphate receptor, *J. Biol. Chem.* **263:**1530–1534.

Takai, Y., Kishimoto, A., Iwasa, Y., Kawahara, Y., Mori, T., and Nishizuka, Y., 1979a, Calcium-dependent activation of a multifunctional protein kinase by membrane phospholipids, *J. Biol. Chem.* **254:**3692–3695.

Takai, Y., Kishimoto, A., Kikkawa, U., Mori, T., and Nishizuka, Y., 1979b, Unsaturated diacylglycerol as a possible messenger for the activation of calcium-activated, phospholipid-dependent protein kinase system, *Biochem. Biophys. Res. Commun.* **91:**1218–1224.

Tashjian, A. H., Jr., Heslop, J. P., and Berridge, M. J., 1987, Subsecond and second changes in inositol polyphosphates in GH4CI cells induced by thyrotropin-releasing hormone, *Biochem. J.* **243:**305–308.

Taylor, M. V., Metcalfe, J. C., Hesketh, T. R., Smith, G. A., and Moore, J. P., 1984, Mitogens increase phosphorylation of phosphoinositides in tymocytes, *Nature* **312:**463–466.

Taylor, M. V., Hesketh, T. R., and Metcalfe, J. C., 1988, Phosphoinositide metabolism and the calcium response to concanavalin A in S49 T-lymphoma cells, *Biochem. J.* **249:**847–855.

Thomas, A. P., Marks, J. S., Coll, K. E., and Williamson, J. R., 1983, Quantitation and early kinetics of inositol lipid changes induced by vasopressin in isolated and cultured hepatocytes, *J. Biol. Chem.* **258:**5716–5725.

Thompson, W., and Dawson, R. M. C., 1964, The triphosphoinositide phosphodiesterase of brain tissue, *Biochem. J.* **91:**237–243.

Ui, M., 1986, Pertussis toxin as a probe of receptor coupling to inositol lipid metabolism, in *Receptor Biochemistry and Methodology. Phosphoinositides and Receptor Mechanisms* (J. W. Putney, Jr., ed.), pp. 163–195, Alan R. Liss, New York.

Vallejo, M., Jackson, T., Lightman, S., and Hanley, M. R., 1987, Occurrence and extracellular actions of inositol pentakis- and hexakisphosphate in mammalian brain, *Nature* **330:**656–658.

Wahl, M. I., Daniel, T. O., and Carpenter, G., 1988, Antiphosphotyrosine recovery of phospholipase C activity after EGF treatment of A-431 cells, *Science* **241:**968–970.

Waloga, G., and Anderson, R. E., 1985, Effects of inositol-1,4,5-trisphosphate injections into salamander rods, *Biochem. Biophys. Res. Commun.* **126:**59–62.

Watson, S. P., and Lapetina, E. G., 1985, 1,2-Diacylglycerol and phorbol ester inhibit agonist-induced formation of inositol phosphates in human platelets: Possible implications for negative feedback regulation of inositol phospholipid hydrolysis, *Proc. Natl. Acad. Sci. U.S.A.* **82:**2623–2626.

Weinstein, I. B., 1987, Growth factors, oncogenes, and multistage carcinogenesis, *J. Cell Biochem.* **33:**213–224.

Whitman, M., Fleischman, L., Chahwala, S. B., Cantley, L., and Rosoff, P., 1986, Phosphoinositides, mitogenesis, and oncogenesis, in *Receptor Biochemistry and Methodology. Phosphoinositides and Receptor Mechanisms* (J. W. Putney, Jr., ed.), pp. 197–217, Alan R. Liss, New York.

Whitman, M., Kaplan, D. R., Roberts, T. M., and Cantley, L., 1987, Evidence for two distinct phosphatidylinositol kinases in fibroblasts, *Biochem. J.* **247:**165–174.

Whitman, M., Downes, C. P., Keeler, M., Keller, T., and Cantley, L., 1988, Type I phosphatidylinositol kinase makes a novel inositol phospholipid, phosphatidylinositol-3-phosphate, *Nature* **332:**644–646.

Whitworth, P., and Kendall, D. A., 1988, Lithium selectively inhibits muscarinic receptor-stimulated inositol tetrakisphosphate accumulation in mouse cerebral cortex slices, *J. Neurochem.* **51:**258–265.

Wilson, D. B., Bross, T. E., Sherman, W. R., Berger, R. A., and Majerus, P. W., 1985, Inositol cyclic phosphates are produced by cleavage of phosphatidylphosphoinositols (polyphosphoinositides) with purified sheep seminal vesicle phospholipase C enzymes, *Proc. Natl. Acad. Sci. U.S.A.* **84:**4013–4017.

Wollheim, C. B., and Biden, T. J., 1986, Second messenger function of inositol 1,4,5-trisphosphate. Early changes in inositol phosphates, cytosolic Ca^{2+}, and insulin release in carbamylcholine-stimulated RIN5F cells, *J. Biol. Chem.* **261:**8314–8319.

Wong, N. S., Barker, C. J., Shears, S. B., Kirk, C. J., and Michell, R. H., 1988, Inositol 1:2(cyclic)4,5-trisphopshate is not a major product of inositol phospholipid metabolism in vasopressin-stimulated WRKI cells, *Biochem. J.* **252:**1–5.

Chapter 2

Phosphoinositides and Synaptic Transmission

John N. Hawthorne

1. THE BRAIN PHOSPHOINOSITIDES

The isolation of a "diphosphoinositide" fraction from ox brain by Folch (1949) and the observation by Dawson (1954) that radioactive phosphate was rapidly incorporated into its lipids laid the foundations of our present knowledge of the brain phosphoinositides. The structures of the major phosphoinositides, phosphatidylinositol (I), phosphatidylinositol-4 phosphate (II) and phosphatidylinositol-4,5-biphosphate (III) are given in Figure 1. The generally accepted abbreviations used in this chapter are PtdIns (I), PtdIns 4-P (II), and PtdIns 4,5-P_2 (III).

1.1. Phosphoinositide Metabolism

Phosphatidylinositol is formed by a microsomal enzyme from free *myo*-inositol and CDP-diacylglycerol. Its phosphorylation by two ATP-requiring kinases produces, in turn, PtdIns 4-P and PtdIns 4,5-P_2. Though removal of fatty acids by phospholipase A is known, the major catabolic route involves a phos-

John N. Hawthorne Department of Biochemistry, Medical School, Queen's Medical Centre, Nottingham NG7 2UH, United Kingdom; *present address:* International Medical College, 46050 Petaling Jaya, Selangor D.H., Malaysia.

Subcellular Biochemistry, Volume 26: myo-Inositol Phosphates, Phosphoinositides, and Signal Transduction, edited by B. B. Biswas and Susweta Biswas. Plenum Press, New York, 1996.

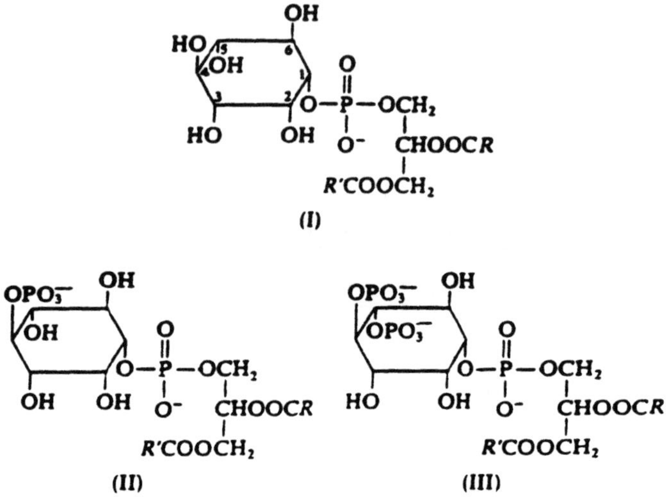

FIGURE 1. The major brain phosphoinositides: (**I**) phosphatidylinositol, (**II**) phosphatidyl-inositol-4-phosphate, (**III**) phosphatidylinositol-4,5-bisphosphate. The *R*CO- groups represent long-chain fatty acid residues. In brain, *R*CO- is mainly from arachidonic acid, and *R*'CO- is mainly from stearic acid.

pholipase C specific for inositol lipids, which we first prepared from rat liver (Kemp *et al.*, 1959). This Ca^{2+}-requiring enzyme hydrolyzes the phosphoinositides to diacylglycerol and an inositol phosphate. It plays a key role in the receptor-linked breakdown of PtdIns 4,5-P_2 described in the next section.

1.2. Receptor-Linked Phosphoinositide Hydrolysis

The pioneer work of Hokin and Hokin (1958) showed that acetylcholine stimulated the incorporation of ^{32}P into phosphatidic acid and phosphatidylinositol of brain slices, but the elucidation of this effect began only with the review of Michell (1975), who suggested that phosphoinositide-linked receptors raised the concentration of Ca^{2+} within cells. We had suggested (Kai *et al.*, 1968) that PtdIns 4,5-P_2 resembled a transmitter, and Abdel-Latif *et al.* (1977) showed that acetylcholine could specifically stimulate the hydrolysis of this lipid. Because the phosphoinositidase C involved was known to be a Ca^{2+}-dependent enzyme, there was confusion for some time. Was the phosphoinositide breakdown a cause or a consequence of Ca^{2+} mobilization? Creba *et al.* (1983) answered this question by showing that in hepatocytes, vasopressin caused PtdIns 4,5-P_2 loss but that this could not be imitated by ionophore-mediated influx of Ca^{2+}. The phosphoinositide hydrolysis was causing mobilization of Ca^{2+}. Berridge (1983) then made the key suggestion that this hydrolysis produced a second messenger, inositol-1,4,5-trisphosphate, which interacts with intracellular receptors to release Ca^{2+} from stores associated with the endoplasmic reticulum. The other product of this hydrolysis, diacylglycerol, can also act as a second messenger by activating protein kinase C (Nishizuka, 1988).

2. FURTHER METABOLISM OF INOSITOL TRISPHOSPHATE

All the inositol phosphates discussed here are numbered as D-*myo*-inositol derivatives. The nomenclature is outlined in the review by Berridge and Irvine (1989).

Inositol-1,4,5-trisphosphate is inactivated as a Ca^{2+}-releasing messenger by a 5-phosphatase, which is widely distributed. It occurs in rat brain in two soluble forms and also membrane-bound (Shears, 1989). The product, Ins 1,4-P_2, is hydrolyzed in brain by a soluble 1-phosphatase, and the resulting Ins 4-P is hydrolyzed by an inositol phosphatase that has been purified to homogeneity from brain by Gee *et al.* (1988). This enzyme can hydrolyze the 1-, 3-, 4-, 5-, and 6-phosphates of inositol, but it is inhibited by Ins 2-P. It is important in making free inositol available for the resynthesis of the phosphoinositides. It will also release inositol from the Ins 3-P produced by cyclization of glucose-6-phosphate.

The discovery of Ins 1,3,4,5-P_4 (Batty *et al.*, 1985) led to the detection of Ins 1,4,5-P_3 3-kinase in brain (Irvine *et al.*, 1986). The kinase has been purified to homogeneity from rat brain by Johanson *et al.* (1988) and shown to consist of two 53-kDa monomers, each bound to calmodulin. Increases in Ca^{2+} concentration stimulate the kinase, and the resulting Ins 1,3,4,5-P_4 promotes the Ca^{2+}-mobilizing function of Ins 1,4,5-P_3.

The 5-phosphatase that hydrolyzes Ins 1,4,5-P_3 will also hydrolyze Ins 1,3,4,5-P_4, accounting for the presence of Ins 1,3,4-P_3 in brain and other tissues.

3. PHOSPHOINOSITIDE-SPECIFIC PHOSPHOLIPASE C

The key enzyme in receptor-linked phosphoinositide metabolism is a phospholipase C that attacks PtdIns, PtdIns 4-P, and PtdIns 4,5-P_2 but not other phospholipids. The isolation of inositol monophosphate from liver (Hubscher and Hawthorne, 1957) suggested that a phospholipase C was present, and this was confirmed for pancreas (Dawson, 1959) and liver (Kemp *et al.*, 1959). A first attempt at purification was made by Kemp *et al.* (1961).

Recent studies, reviewed by Rhee *et al.* (1989), have shown that there are several isozymes of phosphoinositidase C, designated α, β, γ, and δ by these authors. All except the α-isozyme have been detected in brain. Immunohistochemical studies from Rhee's laboratory showed that the γ-enzyme was present in all neurons, whereas the β-isozyme has a more restricted neuronal distribution. The δ-enzyme was associated only with astroglia.

Most of the isozymes have been sequenced by deduction from the cloned cDNAs. The overall sequence homology among isozymes β, γ, and δ is low, but all three contain two regions, designated X and Y by Rhee *et al.* (1989), of significant similarity in all three enzymes. The X domain has about 150 amino acids, and the Y domain has about 120. The α-isozyme does not have these regions. Deletion of either X or Y domains caused complete loss of enzyme activity.

Yagisawa *et al.* (1994) expressed the rat brain δ_1-isozyme in *E. coli* and purified it from the bacterial lysate. The amino-terminal domain was shown to bind both Ins 1,4,5-P_3 and PtdIns 4,5-P_2. The authors suggested that binding of Ins 1,4,5-P_3 may provide feedback inhibition of this phosphoinositidase C.

Growth factors EGF and PDGF activate the widely distributed γ_1 isozyme by serine and tyrosine phosphorylation. The β- and δ-isozymes are unaffected (reviewed by Rhee, 1991).

Though these phosphoinositidases can hydrolyze all three phosphoinositides, PtdIns is not attacked at the low concentrations of Ca^{2+} found intracellularly. It seems likely that regulating proteins are present because these active enzymes would rapidly destroy all the PtdIns 4,5-P_2 present in a few seconds. The linkage of some phosphoinositidase C isozymes to the activating receptors involves G proteins, a topic discussed in the next section.

4. G PROTEINS AND PHOSPHOINOSITIDE-LINKED RECEPTORS

The phosphoinositidase C_γ enzymes are activated by tyrosine phosphorylation, and G proteins are not involved. Regulation of the α- and δ-isozymes is not understood, but G-protein regulation of the β-forms is well established (Rhee and Choi, 1992). On the basis of sequence homology, five G proteins have been identified in the G_q subfamily that all activate phosphoinositidase C_β. These are $G\alpha_q$, $G\alpha_{11}$, $G\alpha_{14}$, $G\alpha_{15}$, and $G\alpha_{16}$, none of which has a site for pertussis toxin modification. All of them stimulate phosphoinositidase $C_{\beta1}$, with $G\alpha_q$ and $G\alpha_{11}$ being most effective. The $G\alpha_{16}$ protein is the best activator of phosphoinositidase $C_{\beta2}$, the others being much less effective. A further isozyme, phosphoinositidase $C_{\beta3}$, is also activated by G_q proteins. It was cloned, purified, and sequenced from a rat brain particulate fraction by Jhon *et al.* (1993).

A phosphoinositidase C related to this $\beta3$ isozyme has been isolated from bovine brain cytosol by Blank *et al.* (1993). It has a molecular mass of 110 kDa and is unusual in being markedly activated by the $\beta\gamma$-subunits of G_q but not by the α-subunit.

Berstein *et al.* (1992) showed that phosphoinositidase $C_{\beta 1}$ is a GTPase-activating protein for G_{q11}. Purified M_1 muscarinic cholinergic receptor and G_{q11} were reconstituted together in lipid vesicles. In the presence of carbachol, added phosphoinositidase $C_{\beta 1}$ stimulated the GTPase activity of G_{q11}, thus providing a mechanism for the deactivation of G-protein-mediated signaling.

5. PHOSPHOINOSITIDE-LINKED RECEPTORS IN NERVOUS TISSUE

Many receptors in neural tissue are linked to phosphoinositide metabolism. The listing by Fisher *et al.* (1992) is reproduced in Table I, and the original publication should be consulted for references. The authors divide the receptors into two groups: category I receptors produce a robust phosphoinositide response when stimulated, whereas only a modest response is obtained with receptors of category II.

Dopamine is not listed in Table I because there are conflicting reports about its effects (Fisher *et al.*, 1992). It now seems clear that brain dopamine D_1

Table I
Receptor Activation of Phosphoinositide Turnover in Neural Tissues[a]

Category I	Category II
Muscarinic cholinergic (M_1, M_3, and M_5)	Purinergic (P_2)
Adrenergic (α_{1A} and α_{1B})	Thromboxane (A_2)
Histaminergic (H_1)	Nerve growth factor
Serotonergic (5-HT_2 and 5-HT_{1C})	Prostaglandin (E_2)
Glutamatergic (metabotropic)	Bradykinin (B_2)
Endothelin	Vasopressin (V_1)
	Cholecystokinin
	Neuropeptide Y
	Neurotensin
	Gastrin-releasing peptide
	Bombesin
	Substance P
	Neurokinin
	Oxytocin
	Eledoisin
	Vasointestinal peptide
	Angiotensin
	Gonadotropin-releasing hormone
	Platelet-activating factor

[a]Reproduced from Fisher *et al.* (1992) with the permission of the authors and publishers.

receptors are linked to phosphoinositidase C. There is also evidence that dopamine inhibits the activation of this enzyme by carbachol (muscarinic M_3 receptors) in rat brain cortical membranes (Salles *et al.*, 1993).

Substance P and neurokinins A and B are peptides known collectively as the tachykinins. In brain, substance P and neurokinin A can stimulate phosphoinositide turnover.

The phosphoinositide response was first demonstrated with acetylcholine (Hokin and Hokin, 1958) and shown to be associated with muscarinic but not nicotinic cholinergic receptors. Activation of nicotinic receptors depolarizes membranes such as those of skeletal muscle by allowing influx of sodium ions. The situation with the muscarinic receptor is more complex. Cloning studies have shown that there are five different muscarinic receptors, of which types M_1, M_3, and M_5 are linked to phosphoinositide metabolism. Types M_2 and M_4 cause inhibition of adenylate cyclase. The M_1 receptors are distinguished from the others by their sensitivity to the inhibitor pirenzepine.

The metabotropic glutamate receptor is linked to phosphoinositidase C via a G protein sensitive to pertussis toxin, unlike most other neuronal phosphoinositide-linked receptors. These glutamate receptors may be important in controlling neuronal plasticity.

6. THE INOSITOL TRISPHOSPHATE RECEPTOR AND CALCIUM RELEASE

Inositol-1,4,5-trisphosphate releases Ca^{2+} from stores associated with the smooth endoplasmic reticulum by binding to a receptor in this membrane. This receptor is particularly abundant in the Purkinje cells of the cerebellum, from which it has been isolated and sequenced (Mignery *et al.*, 1990). It appears to be a tetramer of four identical 300-kDa subunits, each of which binds a molecule of Ins 1,4,5-P_3. When three or four molecules are bound, a conformational change in the receptor occurs, and a calcium channel is opened. Further details may be found in the review by Fisher *et al.* (1992). There is a considerable sequence homology with the ryanodine receptor of skeletal and cardiac muscle, which responds to small concentrations of Ca^{2+} by opening a calcium channel (Berridge, 1993).

Several protein kinases are able to phosphorylate the Ins 1,4,5-P_3 receptor from rat cerebellum, and it is also capable of autophosphorylation on a serine residue (Ferris *et al.*, 1992). It is known that ATP enhances Ins 1,4,5-P_3-induced calcium flux, but the effect of autophosphorylation is not yet known.

In single cells Ins 1,4,5-P_3 causes oscillations in calcium ion concentration. These may involve positive feedback in which calcium amplifies its own release (Berridge, 1993). Calcium at low concentration can also potentiate inositol trisphosphate release. In permeabilized neuroblastoma cells, Ca^{2+} in the range 10–100 nM markedly increased muscarinic- and GTPγS-activated Ins 1,4,5-P_3 release (Fisher *et al.*, 1989). In cerebellar granule cells, depolarization by 20 mM KCl raised cytoplasmic Ca^{2+} by 50–60 nM and increased the release of the trisphosphate by carbachol threefold (del Rio *et al.*, 1994).

Aluminum ions inhibit muscarinic inositol trisphosphate release in human neuroblastoma cells (Wood *et al.*, 1994). Calcium mobilization by inositol trisphosphate itself was unaffected, suggesting that the aluminum acted at the stage of PtdIns 4,5-P_2 hydrolysis.

In saponin-permeabilized human neuroblastoma cells, Ins 1,3,4,5-P_4 is a weak but full agonist at the Ins 1,4,5-P_3 receptor (Wilcox *et al.*, 1993).

Finally, Wojcikiewicz *et al.* (1994) have shown that stimulation of human neuroblastoma cells with carbachol down-regulates the inositol trisphosphate receptor by 90% after 6 hr because the persistent discharge of calcium ions accelerates its degradation.

7. INOSITOL-1,3,4,5-TETRAKISPHOSPHATE

Batty *et al.* (1985) showed that muscarinic stimulation of rat brain cortical slices produced both inositol-1,4,5-trisphosphate and inositol-1,3,4,5-tetrakisphosphate. As outlined in Section 2, the latter is produced by a 3-kinase acting on Ins 1,4,5-P_3.

A specific high-affinity receptor for Ins 1,3,4,5-P_4 has been isolated from pig cerebellar membranes by Donie and Reiser (1991), suggesting that this polyphosphate may have a messenger function. Irvine (see review by Berridge and Irvine, 1989) has suggested that this may be the control of Ca^{2+} transfer between different intracellular pools and across the plasma membrane. Support for this concept comes from the finding that Ins 1,3,4,5-P_4 activates a calcium channel in endothelial cells of the aorta (Luckhoff and Clapham, 1992), but little is known about such functions in nerve cells.

8. LITHIUM SALTS

Lithium carbonate is effective in the treatment of manic–depressive illness, and lithium chloride reduces the concentration of free inositol in rat brain with a simultaneous increase in that of inositol phosphate (Allison *et al.*, 1976). These biochemical effects result from the noncompetitive inhibition of inositol phosphatase by lithium ions, and they may explain the therapeutic action of lithium. A reduction in the supply of inositol by lithium will in turn reduce the resynthesis of the phosphoinositides and so dampen down the signaling system. Because the inhibition is noncompetitive, the free phosphatase will be unaffected, with lithium binding only to the inositol phosphate–enzyme complex. Consequently, neurons with highly active receptors and thus high levels of inositol phosphates will be selectively inhibited.

Jope and Williams (1994) have criticized this hypothesis on the grounds that brain inositol is reduced only with toxic doses of lithium (6–14 mmol/kg) and then by a mere 35%. Furthermore, chronic administration of lithium does not reduce phosphoinositide or inositol trisphosphate concentrations. It is admitted, though, that inositol depletion could occur in selected brain regions, perhaps where cholinergic neurons predominate, so that the depletion would not be detected in whole brain. Jope and Williams suggested that lithium could act by weakening the G-protein/phosphoinositidase coupling, but as Berridge and Irvine (1989) point out, this would not explain the selectivity of the lithium effect or why it has no significant action on people who are not manic–depressives. Jope and Williams list other effects of lithium, including changes in mRNA for various enzymes and other proteins, and inhibition of protein kinases.

Gani *et al.* (1993) have also reviewed this field, discussing in detail the mechanism by which lithium inhibits inositol phosphatase. They also point out that neurons able to obtain inositol readily by transport from extracellular sources will be less susceptible to inhibition by lithium than those that depend on hydrolysis of inositol phosphates for their supply of inositol. These phosphates originate not only from the receptor-linked hydrolysis of phosphoinsitides but also from the biosynthetic pathway that produces inositol-3-phosphate from glucose-6-phosphate. Support for this concept comes from the recent study of Batty and Downes (1994) using astrocytoma cells in culture. In an inositol-deficient medium, these cells showed a marked reduction in the phosphoinositidase C response to carbachol in the presence of lithium ions. The response was measured as release of [^{3}H]inositol phosphates from cells that had been labeled with [^{3}H]inositol. Because brain cholinergic neurons seem particularly susceptible to lithium inhibition, this work suggests that transport of free inositol into such cells is not very effective.

9. PROTEIN KINASE C

There is evidence that the diacylglycerol released from PtdIns 4,5-P_2 by the receptor-linked activation of phosphoinositidase C stimulates protein kinase C so that it is active even at low concentrations of cytosolic calcium ions. This stimulation also involves transfer of soluble kinase C to membranes. As would be expected, however, the functions of this kinase in brain are complex.

There are at least ten isozymes of protein kinase C, of which the γ-form is found only in brain (Asaoka *et al.*, 1992). These authors indicate that after stimulation of phosphoinositide-linked receptors, the initial release of diacylglycerol corresponds with inositol trisphosphate formation but that there is a subsequent sustained rise in diacylglycerol concentration. This second release of diacylglycerol comes from the action of phospholipase D on phosphatidylcholine and hydrolysis of the resulting phosphatidate by a phosphatase. Protein kinase C may also be activated by various unsaturated fatty acids and by lysophosphatidylcholine, all of which result from the action of phospholipase A_2.

Fisher *et al.* (1992) refer to reports that kinase C activation leads to the inhibition of receptor-linked phosphoinositide hydrolysis, providing a mechanism for feedback inhibition. This may involve phosphorylation of phosphoinositidase C so that its interaction with G protein is impaired. Another important function of kinase C is enhancement of neurotransmitter release. Phorbol ester activation of the kinase in synaptosomes from rat cerebrum enhanced norepinephrine release initially, but the effect was lost with the subsequent downregulation of the kinase (Oda *et al.*, 1991). Further references to enhancement of release are given by these authors.

Shearman *et al.* (1991) showed that synaptosomal levels of the γ-form of protein kinase C from rat hippocampus or cerebellar cortex peaked at 14 days after birth and then steadily declined until this kinase was barely detectable in the adult. It was suggested that this isozyme is important in synaptogenesis.

Though most of his examples are from nonneural cells, Houslay (1991) showed that kinase C is important in modulating relationships between different signal transduction pathways (crosstalk). Gusovsky *et al.* (1989) provide examples of this, using the marine dinoflagellate toxin maitotoxin, which activates protein kinase C, causing it to move from cytosol to membrane. In pheochromocytoma cells (PC12), this caused a rise in cyclic AMP levels, whereas in neuroblastoma hybrid cells (NCB20) there was a reduction.

Another function of protein kinase C (reviewed by Shearman *et al.*, 1989) is the modulation of ion channel activity. Depending on the cell type, phosphorylation by the kinase can lead to increases or decreases of conduction through K^+ and Ca^{2+} channels. The calcium current is inhibited in sensory neurons and PC12 cells.

These varied functions of kinase C may account for the large number of isozymes.

10. NEUROMODULATION AND PLASTICITY

Berridge (1993) has reviewed recent studies indicating that phosphoinositide signaling plays a part in modulating neural activity. The hippocampus is important in memory formation, and its neurons have large numbers of inositol trisphosphate receptors. Long-term potentiation of synaptic transmission, which may be important in memory formation, can be demonstrated in hippocampal slices from rat brain. It involves activation of the N-methyl-D-aspartate type of glutamate receptors and the resulting influx of calcium ions. There is a concomitant increase in protein kinase C activity, and long-term potentiation is blocked by inhibitors of the kinase (Klann *et al.*, 1993). Potentiation is enhanced by activation of the metabotropic glutamate receptor, which is phosphoinositide linked (references in Berridge, 1993). The cholinergic innervation of the hippocampus may also be important in long-term potentiation and memory formation. This involves phosphoinositide-linked muscarinic receptors, and potentiation of N-methyl-D-aspartate responses by acetylcholine in hippocampal neurons can be reproduced by flash photolysis of caged Ins 1,4,5-P_3 (references in Berridge, 1993). It is also of interest that infusion of carbachol increases the extracellular concentration of the trisphosphate in the ventral hippocampus of the anesthetized rat (Minisclou *et al.*, 1994).

The memory loss associated with aging may be related to changes in cholinergic neurons. Free and total lipid inositol were shown to decrease with age in human brain (Stokes *et al.*, 1983). In Alzheimer's disease, the phosphoinositide concentration in the anterior temporal cortex was significantly lower than that of age-matched controls (Stokes and Hawthorne, 1987). Loss of phosphoinositides could damage muscarinic cholinergic systems important in memory recall, and such changes may be related to the pathology of Alzheimer's disease, as discussed in the review of Bothmer and Jolles (1994).

11. PHOSPHATIDYLINOSITOL 3-KINASE

Activation of various growth factor receptors leads to phosphorylation of tyrosine residues on their cytoplasmic surface. This in turn causes various proteins to bind to this surface, one of which is PtdIns 3-kinase, which can phosphorylate PtdIns, PtdIns 4-P, and PtdIns 4,5-P_2. This kinase plays a part in mitogenesis (Varticovski *et al.*, 1994) and has been purified from brain (Fry *et al.*, 1992). It has regulating (85-kDa) and catalytic (110-kDa) subunits, but its precise function in brain tissue is not known.

12. PHOSPHATIDYLINOSITOL GLYCANS

Lipid–protein interactions in biological membranes usually involve noncovalent bonding, but it is now known that many proteins are anchored into membranes by covalent linkage to a complex glycosylphosphatidylinositol. The earlier research is reviewed by Low (1987), whose work showing that a bacterial phosphoinositidase C could release plasma membrane enzymes opened up this field.

In a brief but remarkable paper many years before this, Klenk and Hendricks (1961) described the isolation from human brain of an inositol phosphatide containing three molecules of mannose, N-acetylglucosamine, ethanolamine, batyl alcohol, and smaller quantities of glucose and galactose. All the PtdIns glycans are now known to have this backbone: mannose 1-2 mannose 1-6 mannose 1-4 N-acetylglucosamine 1-6 *myo*-inositol. Mammalian forms are linked to the C terminus of a protein through an ethanolamine phosphate attached to the first mannose. An ether form of phosphatidic acid with batyl alcohol in the 1-position is linked to the D-1 position of the inositol.

Phosphatidylinositol glycans of differing structure are found in parasitic protozoa (McConville and Ferguson, 1993), and all the glycans are able to provide a stable anchor for proteins in the outer leaflet of the plasma membrane. Among the brain proteins anchored in this way are Thy-1, the neural cell adhesion molecule, neuronal cell recogition molecules F3 and F11, and the scrapie prion protein. In addition to plasma membrane proteins, two chromaffin granule proteins from the adrenal medulla are known to be anchored thus. Most of the plasma membrane proteins that are bound to PtdIns glycans have an extracellular function such as interaction with other cells or hydrolysis of external substrates, e.g., by cholinesterase, 5′-nucleotidase, and alkaline phosphatase. The proteoglycan glypican is anchored to glycosylated PtdIns in the surface of projection neurons in the rat brain, mainly in limbic structures such as the hippocampus and thalamus (Litwack *et al.*, 1994).

Phospholipases that could release the anchored proteins from brain membranes have been described. Bovine brain has a soluble PtdIns glycan-specific phospholipase D activated by calcium ions (Hoener *et al.*, 1990). Membranes of mouse brain have two specific phospholipases C, both particulate. One is lysosomal and acts at an acid pH; the other, active at neutral pH, is found in plasma membranes and myelin (Fouchier *et al.*, 1990).

Finally, a related lipid, though not known to be an anchor for membrane proteins, has been described by Mansson *et al.* (1991). It was isolated from human peripheral nerve and has the structure inositol-phosphate-galactosylceramide.

13. THE FUTURE

Predicting the direction of future research is usually foolhardy, but a few areas seem particularly worthy of attention.

Control of the responses to phosphoinositide-linked receptors is likely to be one area of interest. We do not know, for instance, why there are so many isozymes of phosphoinositidase C or whether particular isozymes are linked to certain receptor classes. Similarly, the detailed function of the kinase C isozymes requires study, including the way in which they are related to the hydrolysis of the phosphoinositides and other lipids.

Therapeutic intervention using inhibitors of phosphoinositidase C or analogues of Ins 1,4,5-P_3 is of interest to the drug companies, but it is likely that general inhibitors will be too toxic, as the phosphoinositide-linked receptors have so many functions. However, the usefulness of lithium salts in psychiatry provides an encouraging example of inhibition.

Further work on the phosphatidylinositol glycan of brain could throw light on the development of the nervous system because the glycan is important in cell recognition. Similarly, studies of phosphoinositide 3-kinase will help to elucidate responses to growth factors and could improve our understanding of the origin of brain tumors.

Finally, the part played by phosphoinositides in modulating neural activity and in plasticity will be an active area of research. It should help in our understanding of memory and of its derangement in Alzheimer's disease. Phosphoinositide research in the nervous system is likely to be active for some time yet.

14. REFERENCES

Abdel-Latif, A. A., Akhtar, R. A., and Hawthorne, J. N., 1977, Acetylcholine increases the breakdown of triphosphoinositide of rabbit iris muscle prelabelled with ^{32}P phosphate, *Biochem. J.* **162:**61–73.

Allison, J.H., Blisner, M. E., Holland, W. H., Hipps, P. P., and Sherman, W. R., 1976, Increased brain *myo*-inositol 1-phosphate in lithium-treated rats, *Biochem. Biophys. Res. Commun.* **71:**664–670.

Asaoka, Y., Nakamura, S., Yoshida, K., and Nishizuka, Y., 1992, Protein kinase C, calcium and phospholipid degredation, *Trends Biochem. Sci.* **17:**414–417.

Batty, I. H., and Downes, C. P., 1994, The inhibition of phosphoinositide synthesis and muscarinic-receptor-mediatied phospholipase C activity by Li^+ as secondary, selective consequences of inositol depletion in 1321N1 cells, *Biochem. J.* **297:**529–537.

Batty, I. R., Nahorski, S. R., and Irvine, R. F., 1985, Rapid formation of inositol 1,3,4,5-tetrakisphosphate following muscarinic receptor stimulation of rat cerebral cortical slices, *Biochem. J.* **232:**211–215.

Berridge, M. J., 1983, Rapid accumulation of inositol trisphosphate reveals that agonists hydrolyse polyphosphoinositides instead of phosphatidylinositol, *Biochem. J.* **212:**849–858.

Berridge, M. J., 1993, Inositol trisphosphate and calcium signalling, *Nature* **361:**315–325.

Berridge, M. J., and Irvine, R. F., 1989, Inositol phosphates and cell signalling, *Nature* **341:**197–205.

Berstein, G., Blank, J. L., Jhon, D. Y., Exton, J. H., Rhee, S. G., and Ross, E. M., 1992, Phospholipase $C_{\beta1}$ is a GTPase-activating protein for G_{q11}, its physiologic regulator, *Cell* **70:**411–418.

Blank, J. L., Shaw, K., Ross, A. H., and Exton, J. H., 1993, Purification of a 110-kDa phosphoinositide phospholipase C that is activated by G protein β-subunits, *J. Biol. Chem.* **268:**25184–25191.

Bothmer, J., and Jolles, J., 1994, Phosphoinositide metabolism, aging and Alzheimer's disease, *Biochim. Biophys. Acta* **1225:**111–124.

Creba, J. A., Downes, C. P., Hawkins, P. T., Brewster, G., Michell, R. H., and Kirk, C. J., 1983, Rapid breakdown of phosphatidylinositol 4-phosphate and phosphatidylinositol 4,5-bisphosphate in rat hepatocytes stimulated by vasopressin and other Ca^{2+}-mobilizing hormones, *Biochem. J.* **212**:733–747.

Dawson, R. M. C., 1954, The measurement of ^{32}P labelling of individual kephalins and lecithin in a small sample of tissue, *Biochim. Biophys. Acta* **14**:374–379.

Dawson, R. M. C., 1959, Studies on the enzymatic hydrolysis of monophosphoinositide by phospholipase preparations from *P. notatum* and ox pancreas, *Biochim. Biophys. Acta* **33**:68–77.

del Rio, E., Nicholls, D. G., and Downes, C. P., 1994, Involvement of calcium influx in muscarinic cholinergic regulation of phospholipase C in cerebellar granule cells, *J. Neurochem.* **63**:535–543.

Donie, F., and Reiser, G., 1991, Purification of a high-affinity inositol 1,3,4,5-tetrakisphosphate receptor from brain, *Biochem. J.* **275**:453–457.

Ferris, C. D., Cameron, A. M., Bredt, D. S., Huganir, R. L., and Snyder, S. H., 1992, Autophosphorylation of inositol 1,4,5-P_3 receptors, *J. Biol. Chem.* **267**:7036–7041.

Fisher, S. K., Domask, L. M., and Roland, R. M., 1989, Muscarinic receptor regulation of cytoplasmic Ca^{2+} concentrations in human SK-N-SH neuroblastoma cells: Ca^{2+} requirements for phospholipase C activation, *Mol. Pharmacol.* **35**:195–204.

Fisher, S. K., Heacock, A. M., and Agranoff, B. W., 1992, Inositol lipids and signal transduction in the nervous system: An update, *J. Neurochem.* **58**:18–38.

Folch, J., 1949, Complete fractionation of brain cephalin: Isolation from it of phosphatidyl serine, phosphatidyl ethanolamine and diphosphoinositide, *J. Biol. Chem.* **177**:497–504.

Fouchier, F., Baltz, T., and Rougon, G., 1990, Identification of glycosylphosphatidylinositol-specific phospholipases C in mouse brain membranes, *Biochem. J.* **269**:321–327.

Fry, M. J., Panayotou, G., Dhand, R., Ruiz-Larrea, F., Gout, I., Nguyen, O., Courtneidge, S. A., and Waterfield, M. D., 1992, Purification and characterization of a phosphatidylinositol 3-kinase complex from bovine brain by using phosphopeptide affinity columns, *Biochem. J.* **288**:383–393.

Gani, D., Downes, C. P., Batty, I., and Bramham, J., 1993, Lithium and *myo*-inositol homeostasis, *Biochim. Biophys. Acta* **1177**:253–269.

Gee, N. S., Ragan, C. I., Watling, K. J., Aspley, S., Jackson, R. G., Reid, G. G., Gani, D., and Shute, J. K., 1988, The purification and properties of *myo*-inositol monophosphatase from bovine brain, *Biochem. J.* **249**:883–889.

Gusovsky, F., Yasumoto, T., and Daly, J. W., 1989, Calcium-dependent effects of maitotoxin on phosphoinositide breakdown and on cyclic AMP accumulation in PC12 and NCB20 cells, *Mol. Pharmacol.* **36**:44–53.

Hoener, M. C., Stieger, S., and Brodbeck, U., 1990, Isolation and characterization of a phosphatidylinositol glycan-anchor-specific phospholipase D from bovine brain, *Eur. J. Biochem.* **190**:593–601.

Hokin, L. E., and Hokin, M. R., 1958, Acetylcholine and the exchange of inositol and phosphate in brain phosphoinositide, *J. Biol. Chem.* **233**:818–821.

Houslay, M. D., 1991, Crosstalk: A pivotal role for protein kinase C in modulating relationships between signal transduction pathways, *Eur. J. Biochem.* **195**:9–27.

Hubscher, G., and Hawthorne, J. N., 1957, The isolation of inositol monophosphate from liver, *Biochem. J.* **67**:523–527.

Irvine, R. F., Letcher, A. J., Heslop, J. P., and Berridge, M. J., 1986, The inositol tris/tetrakisphosphate pathway—demonstration of Ins 1,4,5-P_3 3-kinase activity in animal tissues, *Nature* **320**:631–634.

Jhon, D.-Y., Lee, H.-H., Park, D., Lee, C.-W., Lee, K.-H., Yoo, O. J., and Rhee, S. G., 1993, Cloning, sequencing, purification of G_q-dependent activation of phospholipase $C_{\beta 3}$, *J. Biol. Chem.* **268**:6654–6661.

Johanson, R. A., Hansen, C. A., and Williamson, J. R., 1988, Purification of D-*myo*-inositol 1,4,5-trisphosphate 3-kinase from rat brain, *J. Biiol. Chem.* **263:**7465–7471.

Jope, R. S., and Williams, M. B., 1994, Lithium and brain signal transduction systems, *Biochem. Pharmacol.* **47:**429–441.

Kai, M., Salway, J. G., and Hawthorne, J. N., 1968, The diphosphoinositide kinase of rat brain, *Biochem. J.* **106:**791–801.

Kemp, P., Hubscher, G., and Hawthorne, J. N., 1959, A liver phospholipase hydrolysing phosphoinositides, *Biochim. Biophys. Acta* **31:**585–586.

Kemp, P., Hubscher, G., and Hawthorne, J. N., 1961, Enzymic hydrolysis of inositol-containing phospholipids, *Biochem. J.* **79:**193–200.

Klann, E., Chen, S., and Sweatt, J. D., 1993, Mechanism of protein kinase C activation during the induction and maintenance of long-term potentiation probed using a selective peptide substrate, *Proc. Natl. Acad. Sci. U.S.A.* **90:**8337–8341.

Klenk, E., and Hendricks, U. W., 1961, An inositol phosphatide containing carbohydrate, isolated from human brain, *Biochim. Biophys. Acta* **50:**602–603.

Litwack, E. D., Stipp, C. S., Kumbasar, A., and Lander, A. D., 1994, Neuronal expression of glypican, a cell-surface glycosylphosphatidylinositol-anchored heparan sulphate proteoglycan, in the adult rat nervous system, *J. Neurosci.* **14:**3713–3724.

Low, M. G., 1987, Biochemistry of the glycosylphosphatidylinositol membrane protein anchors, *Biochem. J.* **224:**1–13.

Luckhoff, A., and Clapham, D. E., 1992, Inositol 1,3,4,5-tetrakisphosphate activates an endothelial Ca^{2+}-permeable channel, *Nature* **355:**356–358.

Mansson, J.-E., Rynmark, B.-M., and Svennerholm, L., 1991, A novel inositol-containing glycosphingolipid isolated from human peripheral nerve, *FEBS Lett.* **280:**251–253.

McConville, M. J., and Ferguson, M. A. J., 1993, The structure, biosynthesis and function of glycosylated phosphatidylinositols in the parasitic protozoa and higher eukaryotes, *Biochem. J.* **294:**305–324.

Michell, R. H., 1975, Inositol phospholipids and cell surface receptor function, *Biochim. Biophys. Acta* **415:**81–147.

Mignery, G. A., Newton, C. L., Archer, B. T. III, and Sudhof, T. C., 1990, Structure and expression of the rat inositol 1,4,5-trisphosphate receptor, *J. Biol. Chem.* **265:**12679–12685.

Minisclou, C., Rouquier, L., Benavides, J., Scatton, B., and Claustre, Y., 1994, Muscarinic receptor-mediated increases in extracellular inositol 1,4,5-trisphosphate levels in the rat hippocampus: An *in vivo* microdialysis study, *J. Neurochem.* **62:**557–562.

Nishizuka, Y., 1988, The molecular heterogeneity of protein kinase C and its implications for cellular regulation, *Nature* **334:**661–665.

Oda, T., Shearman, M. S., and Nishizuka, Y., 1991, Synaptosomal protein kinase C subspecies: Down-regulation promoted by phorbol ester and its effect on evoked norepinephrine release, *J. Neurochem.* **56:**1263–1269.

Rhee, S. G., 1991, Inositol phospholipid-specific phospholipase C: Interaction of the γ_1 isoform with tyrosine kinase, *Trends Biochem. Sci.* **16:**297–301.

Rhee, S. G., and Choi, K. D., 1992, Regulation of inositol phospholipid-specific phospholipase C isozymes, *J. Biol. Chem.* **267:**12393–12396.

Rhee, S. G., Suh, P.-G., Ryu, S.-H., and Lee, S. Y., 1989, Studies of inositol phospholipid-specific phospholipase C, *Science* **244:**546–550.

Salles, J., Wallace, M. A., and Fain, J. N., 1993, Modulation of the phospholipase C activity in rat brain cortical membranes by simultaneous activation of distinct monoaminergic and cholinergic muscarinic receptors, *Mol. Brain Res.* **20:**111–117.

Shearman, M. S., Sekiguchi, K., and Nishizuka, Y., 1989, Modulation of ion channel activity: A key function of the protein kinase C enzyme family, *Pharmacol. Rev.* **41:**211–237.

Shearman, M. S., Shinomura, T., Oda, T., and Nishizuka, Y., 1991, Synaptosomal protein kinase C subspecies: Dynamic changes in the hippocampus and cerebellar cortex concomitant with synaptogenesis, *J. Neurochem.* **56:**1255–1262.

Shears, S. B., 1989, Metabolism of the inositol phosphates produced upon receptor activation, *Biochem. J.* **260:**313–324.

Stokes, C. E., and Hawthorne, J. N., 1987, Reduced phosphoinositide concentrations in anterior temporal cortex of Alzheimer-diseased brains, *J. Neurochem.* **48:**1018–1021.

Stokes, C. E., Gillon, K. R. W., and Hawthorne, J. N., 1983, Free and total lipid *myo*-inositol concentrations decrease with age in human brain, *Biochim. Biophys. Acta* **753:**136–138.

Varticovski, L., Harrison-Findik, D., Keeler, M. L., and Susa, M., 1994, Role of PI 3-kinase in mitogenesis, *Biochim. Biophys. Acta* **1226:**1–11.

Wilcox, R. A., Whitham, E. M., Liu, C., Potter, B. V. L., and Nahorski, S. R., 1993, *myo*-Inositol 1,3,4,5-tetrakisphosphate can independently mobilize intracellular calcium, via the inositol 1,4,5-trisphosphate receptor: Studies with *myo*-inositol 1,4,5-trisphosphate 3-phosphorothioate and *myo*-inositol hexakisphosphate, *FEBS Lett.* **336:**267–271.

Wojcikiewicz, R. J. H., Furnichi, T., Nakada, S., Mikoshiba, K., and Nahorski, S. R., 1994, Muscarinic receptor activation down-regulates the type I inositol 1,4,5-trisphosphate receptor by accelerating its degradation, *J. Biol. Chem.* **269:**7963–7969.

Wood, P. C., Wojcikiewicz, R. J. H., Burgess, J., Castledeu, C. M., and Nahorski, S. R., 1994, Aluminium inhibits muscarinic agonist-induced inositol 1,4,5-trisphosphate production and calcium mobilization in permeabilized SH-SY5Y human neuroblastoma cells, *J. Neurochem.* **62:**2219–2223.

Yagisawa, H., Hirata, M., Kanematsu, T., Watanabe, Y., Ozaki, S., Sakuma, K., Tanaka, H., Yabuta, N., Kamata, H., Hirata, H., and Nojima, H., 1994, Expression and characterization of an inositol 1,4,5-triphosphate binding domain of phosphatidylinositol-specific phospholipase C-δ_1, *J. Biol. Chem.* **269:**20179–20188.

Chapter 3

Control of the Ca^{2+} Release Induced by *myo*-Inositol Trisphosphate and the Implication in Signal Transduction

Ludwig Missiaen, Jan B. Parys, Humbert De Smedt, Ilse Sienaert, Martin D. Bootman, and Rik Casteels

1. INTRODUCTION

Inositol-1,4,5-trisphosphate ($InsP_3$) is a diffusible messenger formed within the cell in response to external stimuli. It mobilizes Ca^{2+} from those nonmitochondrial Ca^{2+} pools that express the $InsP_3$ receptor ($InsP_3R$), a specific Ca^{2+}-release channel (Berridge and Irvine, 1989; Berridge, 1993). The nonmitochondrial pools were originally classified as $InsP_3$-sensitive and $InsP_3$-insensitive. Recent evidence suggests that the $InsP_3$-sensitive Ca^{2+} pool is much larger than hitherto expected (Bird *et al.*, 1992) and that $InsP_3$-insensitive Ca^{2+} pools can artifactually be formed during the permeabilization procedure (Hajnóczky *et al.*, 1994). Under conditions of very mild permeabilization, 95% of the nonmitochondrial Ca^{2+} pools can be $InsP_3$-sensitive, e.g., in A7r5 smooth muscle cells (Missiaen *et al.*, 1992b).

Ludwig Missiaen, Jan B. Parys, Humbert De Smedt, Ilse Sienaert, and Rik Casteels Laboratory for Physiology, Catholic University of Leuven, Campus Gasthuisberg, B-3000 Leuven, Belgium. **Martin D. Bootman** The Babraham Institute Laboratory of Molecular Signaling, Department of Zoology, Cambridge University, Cambridge CB2 3EJ, United Kingdom.
Subcellular Biochemistry, Volume 26: myo-Inositol Phosphates, Phosphoinositides, and Signal Transduction, edited by B. B. Biswas and Susweta Biswas. Plenum Press, New York, 1996.

The $InsP_3R$ is modulated in a complex way by various cellular components (Berridge, 1993). This regulation can appreciably modify the release in the presence of a constant [$InsP_3$] and could therefore be the basis for generating Ca^{2+} oscillations (Woods *et al.*, 1986) and various types of Ca^{2+} waves (Lechleiter *et al.*, 1991). These complex intracellular Ca^{2+} signals may even spread to neighboring cells through gap junctions (Boitano *et al.*, 1992).

This chapter focuses on some important regulatory mechanisms of $InsP_3$-induced Ca^{2+} release. Recent developments are highlighted. These findings are correlated with recent information on the molecular properties of the receptor. The physiological role and relevance of the regulation for the generation of complex Ca^{2+} signals in intact cells are discussed in their respective sections.

2. FUNCTIONAL REGULATION

2.1. Control by $InsP_3$

Deletion studies revealed that binding of $InsP_3$ occurs in the N-terminal fourth of the $InsP_3R$ (Mignery and Südhof, 1990; Miyawaki *et al.*, 1991) and that the region around splice domain SI could be important for binding (Miyawaki *et al.*, 1991). Photoaffinity labeling with an $InsP_3$ analogue localized the $InsP_3$ binding site between amino acids 476 and 501 (Mourey *et al.*, 1993), but the exact structure of the binding site is not yet known.

The high-affinity $InsP_3$ binding site recognizes not only $InsP_3$ and a number of related inositol phosphates (see below) but also the inhibitor heparin (Ghosh *et al.*, 1988) and the extremely potent agonists adenophostin A and B (Takahashi *et al.*, 1994).

Because of differences in binding assays (Taylor and Richardson, 1991; Missiaen *et al.*, 1994c) and the existence of different affinity states in a single receptor (Mauger *et al.*, 1994), it is difficult to compare the $InsP_3$ affinities of the various $InsP_3R$ isoforms. Published K_d values range from 2 to 300 nM (Taylor and Richardson, 1991). However, from recent data the following picture starts to emerge. The $InsP_3Rs$ of the type II family (i.e., II, IV, and V, see Section 3.1) have the highest affinity for $InsP_3$ (Südhof *et al.*, 1991; Parys *et al.*, 1995). Type I receptors have a lower affinity, and differences are observed between the nonneuronal and the neuronal splice variant, the latter having a lower affinity (Nakagawa *et al.*, 1991a; Parys *et al.*, 1995). Finally, the plasma membrane $InsP_3R$ (Khan *et al.*, 1992b) has the lowest affinity for $InsP_3$ described to now (Khan *et al.*, 1992a). For $InsP_3R$-III, a low (Maranto, 1994) as well as a high (Yamamoto-Hino *et al.*, 1994) affinity has been described.

A complete overview of the biological activity of all natural and synthetic inositol phosphates is beyond the scope of this review. Therefore, we limit ourselves to a description of the most potent compounds.

The selectivity of the $InsP_3Rs$ for different inositol phosphates could vary among isoforms, but displacement experiments on purified receptors (Supattapone *et al.*, 1988; Mourey *et al.*, 1990; Maeda *et al.*, 1991; Parys *et al.*, 1992) or on cloned receptors of different types (Furuichi *et al.*, 1989; Miyawaki *et al.*, 1990; Südhof *et al.*, 1991; Yoshikawa *et al.*, 1992; Kume *et al.*, 1993; Maranto, 1994; Yamamoto-Hino *et al.*, 1994) gave an essentially similar rank order of affinities with $InsP_3$ > Ins(2,4,5)P_3 > Ins(1,3,4,5)P_4 >> other inositol phosphates. The only exception to this rank order could be the plasma membrane $InsP_3R$. This protein has an almost similar but quite low affinity for $InsP_3$ and Ins(1,3,4,5)P_4 and a much lower affinity for heparin than the other $InsP_3Rs$ (Khan *et al.*, 1992a).

Functional assays, although more difficult to perform, have the advantage of measuring not only the binding characteristics but also the ability of the protein to induce Ca^{2+} release. In a number of studies, evidence was presented that a number of naturally occurring or synthetic inositol phosphates, such as Ins(2,4,5)P_3 (Irvine *et al.*, 1984), Ins(1,3,4,5)P_4 (e.g., Wilcox *et al.*, 1993), or Ins(1,2,4,5)P_4 (Hirata *et al.*, 1994), can also induce Ca^{2+} release. In the most exhaustive comparative study so far, the efficiencies of 34 different inositol phosphates to induce Ca^{2+} release in *Xenopus* oocytes were compared (De Lisle *et al.*, 1994). The most effective compounds were, in the following rank order, $InsP_3$ > Ins(2,4,5)P_3 > Ins(1,2,4,5)P_4, Ins(4,5)P_2 > Ins(1,4,6)P_3, Ins(1,3,4,6)P_4 >> other inositol phosphates. This rank order, which interestingly is somewhat different from the rank order observed in binding studies (see above), allows the drawing of a few structural rules. The important characteristics needed to induce Ca^{2+} release in *Xenopus* oocytes were the presence of two *trans*-equatorial phosphates and one equatorial hydroxyl on the ring carbons D-4, D-5, and D-6, respectively (or on the structurally equivalent D-1, D-6, and D-5) together with an equatorial hydroxyl on D-3 and/or an equatorial phosphate on D-1 (or on the structurally equivalent D-4) (De Lisle *et al.*, 1994). Interestingly, substituents at the D-2 position do not much affect the potency of the compound (Ozaki *et al.*, 1992; De Lisle *et al.*, 1994; Wilcox *et al.*, 1994). Adenophostins (Takahashi *et al.*, 1994) also contain two *trans*-equatorial phosphate groups, one equatorial hydroxyl group, and two other substitutions in equatorial positions on structurally equivalent carbon atom positions, which seems to confirm the validity of the proposed rule.

2.2. Control by Ca^{2+}

The control of the $InsP_3R$ by Ca^{2+} is physiologically important because the opening of the channel is associated with pronounced [Ca^{2+}] changes on both sides of the channel. We first focus on the control by Ca^{2+} at the cytosolic side of the receptor and then on that from the luminal side.

2.2.1. Activation by Cytosolic Ca^{2+}

Concentrations of cytosolic Ca^{2+} in the submicromolar range increase the effectiveness of $InsP_3$ in releasing Ca^{2+} (Iino, 1990; Bezprozvanny *et al.*, 1991; Finch *et al.*, 1991; Parys *et al.*, 1992; Iino and Endo, 1992; Yao and Parker, 1992; Missiaen *et al.*, 1992c, 1994a; Tshipamba *et al.*, 1993; Marshall and Taylor, 1993a). A rise in cytosolic [Ca^{2+}] in the presence of a constant [$InsP_3$] can therefore be a stimulus for releasing Ca^{2+}. The $InsP_3$R can, therefore, from a functional point of view, be considered a Ca^{2+}-induced Ca^{2+}-release channel because the latter channel can also be activated by a rise in cytosolic [Ca^{2+}].

This positive feedback by Ca^{2+} is exerted instantaneously (Iino and Endo, 1992) and can lead to a very rapid emptying of the internal stores. Such seemingly all-or-none responses ascribed to positive feedback of Ca^{2+} have already been observed with confocal imaging in intact cells (Parker and Ivorra, 1990b; Parker and Yao, 1991). The activation also increases with time, indicating that a time-dependent component contributes to the activation process (Missiaen *et al.*, 1994a; De Smedt *et al.*, 1994).

The levels of cytosolic Ca^{2+} needed to activate the release seem to vary among the various reports. For example, the $InsP_3$R in canine cerebellar membranes is activated by [Ca^{2+}] in the range 10–250 nM and inhibited by higher [Ca^{2+}] (Bezprozvanny *et al.*, 1991). However, Ca^{2+} release from rat brain synaptosomes is activated by Ca^{2+} with an EC_{50} of 660 nM (Finch *et al.*, 1991). Maximal stimulation in A7r5 cells occurs at 1 μM Ca^{2+} (Missiaen *et al.*, 1994a). These differences may be explained in part by differences in experimental conditions, e.g., the pH of the assay medium. Alkalinization has been found to increase the sensitivity of $InsP_3$-mediated Ca^{2+} release (Brass and Joseph, 1985; Guillemette and Segui, 1988; Tsukioka *et al.*, 1994), an effect that is usually ascribed to an increase in the affinity for $InsP_3$ (Worley *et al.*, 1987; Guillemette and Segui, 1988; Varney *et al.*, 1990; White *et al.*, 1991). Alkalinization also sensitizes $InsP_3$-mediated Ca^{2+} release to cytosolic Ca^{2+}, suggesting that protons also interact with the Ca^{2+} binding site of the $InsP_3$R (Tsukioka *et al.*, 1994; Bootman *et al.*, 1995). The Ca^{2+}–activation curve is also affected by the level of Mg^{2+}. Elevating the [Mg^{2+}] to 2 mM induces a rightward shift in the Ca^{2+}–activation curve (Bootman *et al.*, 1995).

The extent to which Ca^{2+} potentiates $InsP_3$-mediated Ca^{2+} release is variable. One possible reason may be related to the level of store loading (see Section 2.2.3). The stimulatory effect of cytosolic Ca^{2+} becomes more important when stores contain less Ca^{2+} (Missiaen *et al.*, 1992c, 1994a). The functional implication of this finding is that Ca^{2+} release from initially full stores is mainly determined by the [$InsP_3$] and is much less dependent on changes in the cytosolic [Ca^{2+}]. In contrast, Ca^{2+} release from less filled stores is absolutely dependent on the rise in cytosolic [Ca^{2+}]: even extremely high doses of $InsP_3$ are unable to deplete the stores significantly when the cytosolic [Ca^{2+}] is close to basal levels (100 nM). This interaction between activation by cytosolic and luminal Ca^{2+} also occurs with the ryanodine receptor (Sitsapesan and Williams, 1994).

A last possible explanation for the various degrees of potentiation is isoform variability in the dependency on cytosolic Ca^{2+}. The fact that the stimulatory effect of cytosolic Ca^{2+} is absent in some systems (Meyer *et al.*, 1990; Combettes *et al.*, 1993) would support this possibility.

Most of the experiments addressing the effects of cytosolic Ca^{2+} on $InsP_3R$ activity have relied on Ca^{2+} chelators such as EGTA and BAPTA to buffer the free [Ca^{2+}] in the submicromolar range. Richardson and Taylor (1993) demonstrated that these Ca^{2+} chelators, in their Ca^{2+}-free form, competitively antagonized the binding of $InsP_3$ to its receptor and the subsequent $InsP_3$-stimulated Ca^{2+} release in hepatocytes. The EGTA inhibited $InsP_3$ binding to the rat cerebellar $InsP_3R$ with an $IC_{50} > 20$ mM, and BAPTA was approximately 60-fold more potent. These studies were recently extended by Combettes *et al.* (1994a), who showed that the activation of $InsP_3$-mediated Ca^{2+} release by cytosolic Ca^{2+} in cerebellum was very weak when the concentration of Ca^{2+}-free chelator was fixed. These studies suggest that the decrease in Ca^{2+}-free chelator concentration that inevitably results from the addition of Ca^{2+} to a fixed concentration of chelator induces a decreased inhibition of the $InsP_3R$. It is therefore necessary to reassess the extent to which the Ca^{2+} activation of the $InsP_3R$ really represents a direct Ca^{2+} effect on the $InsP_3R$ or instead represents an indirect effect of decreasing the free chelator concentration. BAPTA and EGTA, even at concentrations up to 10 mM, have hardly any effect on the $InsP_3Rs$ expressed in A7r5 cells (Bootman *et al.*, 1995). Cytosolic Ca^{2+} was found to potentiate Ca^{2+} release either in the complete absence of Ca^{2+} chelator or at a constant Ca^{2+}-free chelator concentration. In order to reduce the free [Ca^{2+}] in the test media without adding EGTA, sarcoplasmic reticulum vesicles from skeletal muscle were used as an active Ca^{2+}-buffering system. A similar activation of the $InsP_3R$ in the absence of Ca^{2+} chelators was also observed in *Xenopus laevis* oocytes (Callamaras and Parker, 1994). Finch and Goldin (1994) also presented evidence that artifactual effects of Ca^{2+} chelators were not responsible for the Ca^{2+}-dependent activation of the $InsP_3R$ in brain-derived microsomes.

Cytosolic Ca^{2+} stimulates $InsP_3$ binding to permeabilized guinea pig hepatocytes (Spät *et al.*, 1986) and to rat basophilic leukemia cell microsomes (Mohr *et al.*, 1993). The increased binding in the latter study results not from a shift in affinity but from an increase in number of binding sites. In rat hepatocytes, increasing the free Ca^{2+} (or Cd^{2+} or Sr^{2+}) concentration in the medium to micromolar levels catalyzes the interconversion of a rapidly dissociating low-affinity state of the receptor to a slowly dissociating high-affinity state (Piétri *et al.*, 1990). The low-affinity and the high-affinity states display the same specificity for inositol phosphates and for heparin (Piétri Rouxel *et al.*, 1992). In the low-affinity state, which presumably represents an open channel state (Mauger *et al.*, 1989), two separate kinetic components can be identified (Piétri Rouxel *et al.*, 1992; Hilly *et al.*, 1993). The high-affinity state was first proposed to correspond to a desensitized form of the $InsP_3R$ (Piétri *et al.*, 1990), but it is also possible that the high-affinity state is present in both an active and an inactive form (Marshall and Taylor, 1993a; Poitras *et al.*, 1993; Hilly *et al.*, 1993). Calmodulin, which binds to the $InsP_3R$, does not modify $InsP_3$ binding (Maeda *et al.*, 1991) and is not believed to mediate the interconversion phenomenon (Piétri *et al.*, 1990). Part of the activation of $InsP_3$-induced Ca^{2+} release by Ca^{2+} can, however, be caused by activation of Ca^{2+}/calmodulin-dependent protein kinase II (Zhang *et al.*, 1993; see Section 2.3.3). The latter mechanism would be compatible with the observed time dependence of part of the activation (see above).

Finally, a putative Ca^{2+}-binding site in the region just preceding the transmembrane sequences of $InsP_3R$-I and -II was detected by $^{45}Ca^{2+}$ overlay on fusion proteins (Mignery *et al.*, 1992). No binding was observed to a recombinant protein representing the N-terminal 324 amino acids of $InsP_3R$-I.

2.2.2. Inhibition by Cytosolic Ca^{2+}

In addition to stimulation of Ca^{2+} release, Ca^{2+} also inhibits the $InsP_3R$. The concentrations of Ca^{2+} needed to inhibit the receptor are usually higher than those that stimulate the release (Hirata *et al.*, 1984; Iino, 1990; Bezprozvanny *et al.*, 1991; Missiaen *et al.*, 1992c, 1994a; Parys *et al.*, 1992; Tshipamba *et al.*, 1993). The inhibition by Ca^{2+} also develops more slowly than the activation by Ca^{2+} (Finch *et al.*, 1991; Iino and Endo, 1992; Marshall and Taylor, 1993a; Missiaen *et al.*, 1994a). The inhibition is reversible (Finch *et al.*, 1991; Bootman *et al.*, 1995), at least when the $[Ca^{2+}]$ is not raised too high (Van Delden *et al.*, 1993; Benevolensky *et al.*, 1994).

Calcium ions (0.1–1 μM) inhibit $InsP_3$ binding to brain microsomes (Worley *et al.*, 1987) but not to the purified brain $InsP_3R$ (Supattapone *et al.*, 1988) or to microsomes derived from a number of peripheral tissues (Piétri *et al.*, 1990; Mourey *et al.*, 1990). Part of the inhibitory effect of Ca^{2+} is probably related to the presence in certain tissues of a highly active Ca^{2+}-activated phospholipase C, producing unlabeled $InsP_3$, which then displaces $[^3H]InsP_3$ from binding sites (Mignery *et al.*, 1992). However, Ca^{2+} can inhibit $InsP_3$ binding in conditions where phospholipase C is not yet activated (Van Delden *et al.*, 1993; Benevolensky *et al.*, 1994).

The site of Ca^{2+} inhibition on the $InsP_3R$ has not been identified yet, although it might be different from the stimulatory site (Marshall and Taylor, 1994). It furthermore remains possible that Ca^{2+} is not interacting directly with the $InsP_3R$. There have indeed been suggestions that the inhibition could be exerted indirectly through binding of Ca^{2+} to the inositol phosphate (Luttrell, 1993), through activation of proteases (Magnussen *et al.*, 1993), through activation of phosphatase 2B (Zhang *et al.*, 1993), or through binding to an associated factor (Benevolensky *et al.*, 1994).

The onset of the inhibition is fast enough to account for the transience of the release response (Magnussen *et al.*, 1993). Negative feedback by Ca^{2+} can therefore be a useful safety valve and limit excess Ca^{2+} release (Iino, 1990), thereby preventing the intracellular [Ca^{2+}] from becoming too high during agonist stimulation. Calcium-dependent inactivation of the $InsP_3R$ may terminate Ca^{2+} release during a Ca^{2+} spike and is therefore a predominant feature in some models for Ca^{2+} oscillations in nonexcitable cells (Parker and Ivorra, 1990a; Zhang and Muallem, 1992).

The inhibition by cytosolic Ca^{2+} becomes less prominent at higher [$InsP_3$] and disappears at supramaximal $InsP_3R$ activation (Combettes *et al.*, 1994a; Bootman *et al.*, 1995). The fact that high $InsP_3$ concentrations prevent Ca^{2+}-dependent inactivation of the $InsP_3R$ raises the possibility that in discrete areas with high $InsP_3$ concentrations, the $InsP_3R$ may be continually activated despite the global changes in [Ca^{2+}]. Such discrete areas are likely to be located close to the $InsP_3$-generating enzymes near the plasma membrane (Zhao *et al.*, 1990a). In this location, the $InsP_3Rs$ may also be able to generate a signal for the entry of Ca^{2+} across the plasma membrane, and they could therefore be responsible for the continuous entry of Ca^{2+} that is required to sustain Ca^{2+} spiking.

2.2.3. Control by Luminal Ca^{2+}

The loading of the stores controls the $InsP_3R$ in rat hepatocytes (Missiaen *et al.*, 1991, 1992a; Nunn and Taylor, 1992), smooth muscle (Missiaen *et al.*, 1992b; Iino and Endo, 1992), and fibroblasts (De Smedt *et al.*, 1994). This control is such that stores with a low Ca^{2+} content are less sensitive to $InsP_3$ than more filled stores. Stores usually have to be appreciably depleted (usually below 30% of their steady-state loading) before the effects of luminal Ca^{2+} become apparent (Nunn and Taylor, 1992; Marshall and Taylor, 1993a; Parys *et al.*, 1993a). The latter finding may explain why luminal Ca^{2+} effects are usually not observed when stores are only marginally depleted (Shuttleworth, 1992; Combettes *et al.*, 1992, 1993; Sayers *et al.*, 1993). Luminal Ca^{2+} effects are also more pronounced when the cytosolic [Ca^{2+}] is low (Missiaen *et al.*, 1992c, 1994a).

Although the region of the $InsP_3R$ located in the lumen of the Ca^{2+} store is distant from the $InsP_3$ binding site, luminal Ca^{2+} can increase the affinity of the high-affinity $InsP_3$ binding site (Oldershaw and Taylor, 1993; Parys *et al.*, 1993a).

Little is known about a putative intraluminal Ca^{2+} sensor domain in the $InsP_3R$. The question must first be addressed whether or not the $InsP_3R$ has a large luminal loop. In the past, eight (Mignery *et al.*, 1990; Südhof *et al.*, 1991;

Blondel *et al.*, 1993) as well as six (Yoshikawa *et al.*, 1992; Kume *et al.*, 1993; Maranto *et al.*, 1994; Yamamoto-Hino *et al.*, 1994) transmembrane regions have been proposed. Recently, compelling evidence has been provided for the presence of six transmembrane regions (Michikawa *et al.*, 1994), in which case there is a large luminal loop between the fifth and the sixth transmembrane domain (Mikoshiba, 1993).

It is not known whether luminal Ca^{2+} ions interact with the $InsP_3R$ itself or interact with an associated protein. Recent evidence indicates that the luminal loop of the $InsP_3R$ from chromaffin cells can interact with chromogranin A (Yoo, 1994; Yoo and Lewis, 1994).

The question must finally be addressed to what extent the stimulatory effect of luminal Ca^{2+} on the release is actually exerted at the cytosolic side of the receptor (Iino and Endo, 1992). Stores with a high Ca^{2+} content have a higher passive Ca^{2+} leak and could thereby create a microenvironment of a locally elevated [Ca^{2+}], which would then activate the receptor from the cytosolic side. The loading dependence of the Ca^{2+} release in permeabilized A7r5 cells still occurs in the presence of 10 mM BAPTA (Parys *et al.*, 1993a; Missiaen *et al.*, 1994a), i.e., in the presence of a high concentration of a fast Ca^{2+} buffer, as well as in the presence of 10 μM cytosolic Ca^{2+}, a concentration sufficient to saturate all cytosolic Ca^{2+} sites (Missiaen *et al.*, 1994a). The possible existence of a shell of a localized elevated [Ca^{2+}] cannot therefore entirely explain the stimulatory effects of luminal Ca^{2+}.

2.2.4. Physiological Relevance

Receptor-mediated increases in $[Ca^{2+}]_i$ in intact cells result from mobilization of Ca^{2+} from intracellular stores as well as from transmembrane Ca^{2+} influx (Berridge, 1993). The Ca^{2+} release from the internal stores is often oscillatory (Woods *et al.*, 1986) and may be initiated at multiple focal sites (Lechleiter *et al.*, 1991). The oscillatory release can occur through the $InsP_3R$ (Petersen, 1992; Berridge, 1993), the ryanodine receptor (Lipscombe *et al.*, 1988), or both (Galione *et al.*, 1993; Lee *et al.*, 1993). There are two main types of oscillations involving the $InsP_3$-sensitive store: sinusoidal oscillations and baseline spiking. Baseline spiking, in contrast with sinusoidal oscillations (Bird *et al.*, 1993a), does not depend on protein kinase C activation (Zhao *et al.*, 1990b; Harootunian *et al.*, 1991) and can occur in the presence of nonoscillating [$InsP_3$] (Wakui *et al.*, 1989). Factors other than the [$InsP_3$] must therefore control the periodic opening of the $InsP_3R$.

Individual Ca^{2+} spikes in a baseline oscillating cell are preceded by a phase of Ca^{2+} accumulation into the stores (Berridge and Irvine, 1989), resulting in a gradual $[Ca^{2+}]_i$ rise (Jacob *et al.*, 1988) once the pools become overloaded. The $InsP_3R$, which is controlled by both cytosolic and luminal Ca^{2+}, is therefore exposed to a [Ca^{2+}] rise at its cytosolic and luminal side during the process of spike initiation. Most of the models for explaining this regenerative activity rely on the cyclic nature of stimulation and inhibition of the $InsP_3R$ by cytosolic Ca^{2+} (Berridge, 1993). An increasing Ca^{2+} load is, however, as effective a stimulus as cytosolic Ca^{2+} for initiating Ca^{2+} release in the presence of a constant [$InsP_3$] (Missiaen *et al.*, 1994a). It is therefore possible that the combined action of the buildup of luminal Ca^{2+} and the pacemaker $[Ca^{2+}]_i$ rise during the interspike interval finally sets the stage for spike initiation. Stimulation by cytosolic Ca^{2+} increases with the time of incubation (Missiaen *et al.*, 1994a), and this phenomenon may further contribute to the rather explosive $[Ca^{2+}]_i$ rise after the latency. Once the spike is initiated, the increase in cytosolic [Ca^{2+}] will further stimulate the $InsP_3R$ to produce the relatively fast upstroke phase of the spike.

2.3. Control by Phosphorylation

2.3.1. Cyclic-AMP-Dependent Protein Kinase (Protein Kinase A)

The $InsP_3R$-I can be phosphorylated at two different serine residues, one proximal and one distal to the SII splice domain. The inclusion or exclusion of SII influences the phosphorylation reaction (Danoff *et al.*, 1991): the peripheral isoform, which lacks the SII site, is more readily phosphorylated than the neuronal isoform. The proximal site is preferentially phosphorylated in the peripheral isoform, whereas the distal site is preferentially phosphorylated in the neuronal isoform.

The functional role of protein kinase A (PKA) phosphorylation of the $InsP_3R$ is still not understood, because multiple, and sometimes contradictory, effects have been observed between tissues or between species. The differences between these studies may result from differences in experimental conditions (e.g., the presence of Ca^{2+}, Joseph and Ryan, 1993), the presence of regulating factors, or $InsP_3R$ heterogeneity. In that respect, it has to be mentioned that, in most cases, no certainty exists on the isoform studied (a notable exception is Nakade *et al.*, 1994). In the latter study, it was shown that Ca^{2+} flux through immunoaffinity-purified cerebellar $InsP_3R$-I was enhanced by PKA phosphorylation. This result is similar to findings obtained in liver (Burgess *et al.*, 1991; Hajnóczky *et al.*, 1993; Joseph and Ryan, 1993) but is more difficult to reconcile with previous results obtained in cerebellum (Supattapone *et al.*, 1988; Volpe and Alderson-Lang, 1990).

2.3.2. Cyclic-GMP-Dependent Protein Kinase (Protein Kinase G)

Stoichiometric phosphorylation of the cerebellar $InsP_3R$ by protein kinase G (PKG) was recently demonstrated (Komalavilas and Lincoln, 1994). Phosphopeptide mapping indicated that the phosphorylation site corresponds to the distal PKA phosphorylation site. There is evidence that phosphorylation by PKG can be of functional importance in vascular smooth muscle cells because an increase in cyclic GMP level by treatment with atrial natriuretic peptide or sodium nitroprusside resulted in an increased phosphorylation of the $InsP_3R$. In addition, Koga *et al.* (1994) presented evidence that a type of $InsP_3R$ is the PKG substrate, stimulating the plasma membrane Ca^{2+} pump in vascular smooth muscle. In the latter study, no statement on the isoform or the subcellular localization was made.

2.3.3. Ca^{2+}/Calmodulin-Dependent Protein Kinase II

Ferris *et al.* (1991) demonstrated that the cerebellar $InsP_3R$ could be stoichiometrically phosphorylated by Ca^{2+}/calmodulin-dependent protein kinase II. The localization of the phosphorylation site is not yet known. Activation by Ca^{2+}/calmodulin-dependent protein kinase II and subsequent dephosphorylation by protein phosphatase 2B was proposed by Zhang *et al.* (1993) as part of the mechanism responsible for Ca^{2+}-modulated activation and inhibition of $InsP_3$-induced Ca^{2+} release.

2.3.4. Protein Kinase C

The $InsP_3R$ is also a substrate for protein kinase C. Here again, stoichiometric phosphorylation of a serine residue was observed (Ferris *et al.*, 1991). Protein kinase A, Ca^{2+}/calmodulin-dependent protein kinase II, and protein kinase C phosphorylation are additive, indicating that three different phosphorylation sites are involved. In isolated rat liver nuclei, phosphorylation of the $InsP_3R$ by protein kinase C resulted in an accelerated and enhanced $InsP_3$-induced Ca^{2+} release (Matter *et al.*, 1993).

2.3.5. Autophosphorylation

Purified cerbellar $InsP_3Rs$ can slowly autophosphorylate in the presence of Mg^{2+} (Ferris *et al.*, 1992a). Two different serines are autophosphorylated, one of them overlapping with the PKC phosphorylation site. The functional significance of this reaction is not known.

2.4. Control by Sulfhydryl Oxidation

The sulfhydryl reagent thimerosal induces Ca^{2+} oscillations in oocytes (Swann, 1991; Carroll and Swann, 1992; Miyazaki *et al.*, 1992; Cheek *et al.*, 1993; Kline and Kline, 1994), HeLa cells (Bootman *et al.*, 1992b), endothelial cells (Gericke *et al.*, 1993), and pancreatic acinar cells (Thorn *et al.*, 1992). This effect is not associated with a rise in intracellular [$InsP_3$] (Bootman *et al.*, 1992b). Small differences have been observed between the thimerosal-induced Ca^{2+} signals and the sperm-induced $[Ca^{2+}]_i$ rise in mouse oocytes: the Ca^{2+} gradients during a thimerosal-induced Ca^{2+} wave are less steep than those in sperm-induced Ca^{2+} wave (Cheek *et al.*, 1993). Some cell types, such as mouse lacrimal cells, seem to be unresponsive to thimerosal (Bird *et al.*, 1993b).

It is now established that thimerosal sensitizes the $InsP_3R$ to $InsP_3$ (Missiaen *et al.*, 1991, 1992a; Bootman *et al.*, 1992b; Parys *et al.*, 1993b; Sayers *et al.*, 1993; Poitras *et al.*, 1993; Hilly *et al.*, 1993; Tanaka and Tashjian, 1994). The ability of antibodies against the $InsP_3R$ to inhibit the thimerosal-induced Ca^{2+} oscillation in hamster eggs indicates that this stimulatory effect on the $InsP_3R$ is indeed involved in the induction of the oscillation (Miyazaki *et al.*, 1992). Higher thimerosal concentrations can inhibit the release in some experimental systems (Parys *et al.*, 1993b; Sayers *et al.*, 1993).

Thimerosal is not specific for the $InsP_3R$ because also other Ca^{2+}-transport processes of the internal stores are affected. However, the stimulation of the $InsP_3R$ usually occurs at lower doses than those required to inhibit the endoplasmic reticulum Ca^{2+} pump (Missiaen *et al.*, 1992a; Parys *et al.*, 1993b; Poitras *et al.*, 1993). Still higher doses nonspecifically increase the passive Ca^{2+} leak from the stores (Parys *et al.*, 1993b). Thimerosal is therefore a pharmacological tool that, only at low concentrations, presents some specificity for the $InsP_3R$.

Pharmacological agents that increase the cellular level of oxidized glutathione induce Ca^{2+} oscillations in hepatocytes (Rooney *et al.*, 1991). The sulfhydryl reagent oxidized glutathione also sensitizes the $InsP_3R$ in hepatocytes (Missiaen *et al.*, 1991, 1992a; Renard *et al.*, 1992; Hilly *et al.*, 1993). Effects of oxidized glutathione on $InsP_3Rs$ from other tissues have not been reported.

2.5. The Adenine Nucleotide-Binding Site

2.5.1. Effects of Adenine Nucleotides

Up to three different ATP-binding sites are present on the transducing domain of the $InsP_3R$ (Ferris and Snyder, 1992; Mikoshiba, 1993; Marshall and Taylor, 1993b), one of them at least being functionally active (Maeda *et al.*, 1991). ATP, ADP, and AMP, but not GTP, competitively inhibit [α-^{32}P]ATP binding to these sites, indicating that ATP probably interacts with the receptor through its adenine moiety (Maeda *et al.*, 1991).

Submillimolar ATP concentrations increase the open probability of the channel in the presence of $InsP_3$ but do not open the channel in the absence of $InsP_3$ (Ehrlich and Watras, 1988; Mayrleitner *et al.*, 1991; Bezprozvanny and

Ehrlich, 1993). This effect does not require phosphorylation of the receptor because nonhydrolyzable ATP analogues are also effective (Ehrlich and Watras, 1988).

High [ATP] inhibits the release (Iino, 1991; Spät *et al.*, 1992; Bezprozvanny and Ehrlich, 1993) and $InsP_3$ binding (Maeda *et al.*, 1991). This effect can also be mimicked by GTP, ADP, and pyrophosphate but not by AMP or phosphate (Maeda *et al.*, 1991). The inhibition by ATP is competitive with $InsP_3$ (Guillemette *et al.*, 1987; Nunn and Taylor, 1990), suggesting that the site of interaction is the $InsP_3$-binding site (Iino, 1991; Bezprozvanny and Ehrlich, 1993). However, Ferris *et al.* (1990) observed that millimolar ATP was still able to inhibit the $InsP_3R$ in the presence of 100 μM $InsP_3$, raising the possibility of another interaction.

2.5.2. Effects of Caffeine

Caffeine has been widely used as a tool for demonstrating the presence of a ryanodine receptor in nonexcitable cells. In these studies, it became apparent that caffeine concentrations in the range of 10–25 mM inhibit Ca^{2+} responses induced by $InsP_3$-producing agonists (e.g., Osipchuk *et al.*, 1990; Wakui *et al.*, 1990; Berridge, 1991; Harootunian *et al.*, 1991; Parker and Ivorra, 1991; Byron and Taylor, 1993; Lynn *et al.*, 1993; Sanchez-Bueno *et al.*, 1994; Diarra *et al.*, 1994; Combettes *et al.*, 1994b). It is now established that caffeine inhibits the $InsP_3R$ (Parker and Ivorra, 1991; Missiaen *et al.*, 1992a, 1994b; Brown *et al.*, 1992; Hirose *et al.*, 1993; Bezprozvanny *et al.*, 1994; Combettes *et al.*, 1994b). The inhibitory effects of caffeine on Ca^{2+} signaling in intact cells may therefore be caused, at least partly, by an inhibition of the $InsP_3R$. This is, however, not the only mechanism involved: caffeine has, for instance, also been reported to inhibit agonist-induced $InsP_3$ production (Toescu *et al.*, 1992; Combettes *et al.*, 1994b).

Caffeine does not affect the binding of $InsP_3$ to its receptor (Toescu *et al.*, 1992; Brown *et al.*, 1992; Missiaen *et al.*, 1994b). Isocaffeine, which differs from caffeine in the position of one methyl group, is ineffective in blocking the release, indicating that the inhibitory effect requires a specific binding site (Missiaen *et al.*, 1994b). ATP prevents the inhibitory effect of caffeine (Missiaen *et al.*, 1994b), suggesting that caffeine interacts with the ATP-binding site of the $InsP_3R$. It is conceivable that such interaction could induce a conformational change of the receptor that then affects the gating properties, resulting in an inhibition of the Ca^{2+} release.

2.6. Interaction with the Cytoskeleton

The interaction between the $InsP_3R$ molecules and the cytoskeleton is still poorly understood but may prove to be of overwhelming importance, e.g., in restricting particular $InsP_3R$ isoforms to determined intracellular localizations. Originally, different lines of evidence were pointing to the possibility of interactions between the $InsP_3R$ and the cytoskeleton. First, depending on the tissue, part of the $InsP_3$ binding activity is resistant to solubilization by nonionic detergents (Maeda *et al.*, 1990; Guillemette *et al.*, 1990; Bond and Taylor, 1991; White *et al.*, 1993; Feng and Kraus-Friedman, 1993). Second, in some tissues $InsP_3$-sensitive Ca^{2+} stores are linked to the cytoskeleton (Tsunada, 1986; Rossier *et al.*, 1991; Hajnóczky *et al.*, 1994). Third, microfilament- and microtubule-disrupting agents inhibit $InsP_3$ binding and/or $InsP_3$-induced Ca^{2+} release in mast cells (Tasaka *et al.*, 1991) and platelets (Bourguignon *et al.*, 1993a).

A breakthrough in understanding came when evidence was presented that the $InsP_3R$-I present in cerebellum (Joseph and Samanta, 1993) as well as in T-lymphoma cells (Bourguignon *et al.*, 1993b) could interact with the cytoskeletal protein ankyrin. Ankyrin interacts with high affinity with $InsP_3R$-I (K_d of 0.2 nM, Bourguignon *et al.*, 1993b), and antibodies raised against ankyrin can immunoprecipitate the $InsP_3R$ (Joseph and Samanta, 1993). In T-lymphoma cells, treatment of $InsP_3$-sensitive Ca^{2+} stores with ankyrin leads to diminished $InsP_3$ binding and diminished $InsP_3$-induced Ca^{2+} release (Bourguignon *et al.*, 1993b). These data indicate that the cytoskeleton can regulate the $InsP_3R$. This regulation can be dependent on the tissue or on the $InsP_3R$ isoform involved. With the exception of T-lymphoma cells (peripheral $InsP_3R$-I), it is not known which $InsP_3R$ isoform(s) are interacting with the cytoskeleton. Although one has to be careful in the interpretation of K_d values determined in different conditions, there is some evidence that in membranes expressing binding sites with very high affinity for $InsP_3$, a larger part of the sites are detergent-insoluble (Guillemette *et al.*, 1990; Bond and Taylor, 1991; White *et al.*, 1993). Moreover, in cerebellum the relatively small part of $InsP_3R$-I that is Triton X-100-insoluble is characterized by an eightfold higher affinity for $InsP_3$ (Joseph and Samanta, 1993). It is not clear if these results indicate that only those isoforms with a high affinity for $InsP_3$ (e.g., peripheral $InsP_3R$-I or $InsP_3R$-II) could interact with the cytoskeleton or instead indicate a regulatory feature of the cytoskeleton on $InsP_3$ binding. Future work is needed to determine which $InsP_3R$ isoform(s) and which cytoskeletal protein(s) are involved and how they interact.

3. HETEROGENEITY OF $InsP_3R$

3.1. Molecular Diversity

There is evidence for several different genes encoding $InsP_3Rs$ as well as for the occurrence of different splice isoforms. Full-length cDNA clones were obtained for the mouse $InsP_3R$-I (Furuichi *et al.*, 1989), for three different rat genes encoding $InsP_3R$-I (Mignery *et al.*, 1990), $InsP_3R$-II (Südhof *et al.*, 1991), and $InsP_3R$-III (Blondel *et al.*, 1993), for $InsP_3R$-I in *Xenopus laevis* (Kume *et al.*, 1993), and for an $InsP_3R$ in *Drosophila melanogaster* (Yoshikawa *et al.*, 1992). There is also partial sequence information for the mouse $InsP_3R$-II and $InsP_3R$-III and for a putative $InsP_3R$-IV (Ross *et al.*, 1992) and $InsP_3R$-V (De Smedt *et al.*, 1994). The human $InsP_3R$-I (Yamada *et al.*, 1994), $InsP_3R$-II (Yamamoto-Hino *et al.*, 1994), and $InsP_3R$-III (Maranto, 1994; Yamamoto-Hino *et al.*, 1994) were only recently cloned.

All these different $InsP_3Rs$ share extensive structural homology. They have the same overall topology with a large hydrophilic region at the N terminal comprising 80% of the protein, followed by a cluster of hydrophobic segments and a short hydrophilic C terminal. There is a significant sequence homology with the ryanodine receptor in the C-terminal region (Furuichi *et al.*, 1989; Mignery *et al.*, 1989; Gill, 1989). Both proteins are tetramers (Wagenknecht *et al.*, 1989; Maeda *et al.*, 1991) and contain an intrinsic Ca^{2+} channel that is gated by a cytosolic signal (Imagawa *et al.*, 1987; Lai *et al.*, 1988; Ferris *et al.*, 1989). A Gly-Asp pair (positions 2549–2550 in the mouse $InsP_3R$-I) is located in a hydrophobic region between the M5 and M6 segments and is highly conserved among all types of $InsP_3Rs$ and ryanodine receptors (Yamamoto-Hino *et al.*, 1994). In addition, all of the $InsP_3R$ subtypes as well as the ryanodine receptors have a cluster of negatively charged amino acid residues flanking this region (Mignery *et al.*, 1990; Yamamoto-Hino *et al.*, 1994). It was therefore proposed that the hydrophobic region around the Gly-Asp pair may act as a pore-forming region and that in this respect the $InsP_3R$ channel shares the basic design of the voltage- and second-messenger-gated ion channels of the plasma membrane. However, the ligand binding site for the latter type of channels is located in the C-terminal region, whereas the $InsP_3$ binding site is located within the N-terminal fourth of the $InsP_3R$ (Mignery and Südhof, 1990; Miyawaki *et al.*, 1991; Mourey *et al.*, 1993). The C-terminal cytosolic domain of the $InsP_3R$ is relatively short, but the C-terminal 144 amino acids were shown by mutation analysis to be involved in the intermolecular association of the $InsP_3R$ to form tetramers (Miyawaki *et al.*, 1991). The importance of the C-terminal cytoplasmic tail of the $InsP_3R$ in regulating the channel activity is also illustrated by the observation that a monoclonal antibody that mapped between amino acids 2736 and 2747 inhibited Ca^{2+} release (Nakade *et al.*, 1991).

Based on dot matrix analysis and sequence alignments, a pattern of 13 conserved regions and 13 short divergent regions was found (Maranto, 1994). There is a considerable degree of similarity between the homologous conserved regions of all known $InsP_3Rs$. It is likely that the conserved regions represent domains that mediate activities common to all $InsP_3Rs$. This is particularly the case with respect to the N-terminal $InsP_3$-binding domain, which overlaps the first and second conserved domains, and with respect to the channel domain, which corresponds to the 12th and to parts of the 11th and 13th conserved domains (Maranto, 1994). The variable regions may represent regions with less functional significance or, on the other hand, regions that confer functional differences among $InsP_3R$ isoforms. In this respect it should be mentioned that the alternatively spliced domains SI and SII, which were described for $InsP_3R$-I (Mignery *et al.*, 1990; Danoff *et al.*, 1991; Nakagawa *et al.*, 1991a,b), correspond to variable regions in the other $InsP_3Rs$. The SI domain consists of 45 nucleotides and is located near the $InsP_3$-binding region. This SI domain is represented to a variable extent among the $InsP_3R$-I messengers in different cell types (Nakagawa *et al.*, 1991a; De Smedt *et al.*, 1994), but all other $InsP_3Rs$ lack this splice domain. The SII domain consists of 120 nucleotides and is localized in the large cytosolic domain. It is inserted only in the $InsP_3R$-I messenger expressed in neuronal cell types (Danoff *et al.*, 1991; Nakagawa *et al.*, 1991a). The SII splice domain is not present in the sequence of the other $InsP_3R$ isoforms, and it remains to be determined whether a similar splice mechanism generates tissue-specific forms of $InsP_3R$-II or -III. By RNase protection analysis of the mouse $InsP_3R$-I, two additional forms were detected that resulted from an exclusion of fragments of the SII domain, suggesting that SII consists of at least two exons (Nakagawa *et al.*, 1991a). These alternative splice isoforms were shown by *in situ* hybridization to be differentially localized in mouse cerebellum and hippocampus (Nakagawa *et al.*, 1991b). An RNase protection and PCR analysis of the rat receptor, however, did not provide evidence for these intermediately spliced forms (Schell *et al.*, 1993).

3.2. Functional Comparison

The $InsP_3Rs$ consist of three functional domains including an N-terminal ligand-binding domain of about 650 amino acids, a central coupling (also called regulatory) domain of more than 1500 amino acids, and a C-terminal domain of about 600 amino acids containing the transmembrane regions that form the channel domain (Furuichi *et al.*, 1989; Mignery and Südhof, 1990; Mignery *et al.*, 1990; Südhof *et al.*, 1991; Ferris and Snyder, 1992; Blondel *et al.*, 1993; Maranto, 1994; Yamamoto-Hino *et al.*, 1994). The ligand binding domain is the most conserved region, whereas the coupling domain is the least similar domain (Südhof *et al.*, 1991; Mikoshiba, 1993; Yamamoto-Hino *et al.*, 1994). Up to now there have been only a limited number of studies in which different $InsP_3R$ isoforms have been compared in identical experimental conditions, and it is therefore not clear whether these isoforms have distinct functional properties.

Differences in affinity for $InsP_3$ have been observed in preparations from different tissues when measured in identical conditions (Varney *et al.*, 1990; White *et al.*, 1993). No correlation could be made, however, with the type of

$InsP_3R$ expressed in these preparations. A direct comparison of ligand affinity and specificity for particular $InsP_3R$ isoforms has been made by expression of the N-terminal ligand-binding domains as a soluble protein in COS cells. The recombinant ligand-binding domain of rat $InsP_3R$-II had a significantly higher affinity for $InsP_3$ than the corresponding domain of $InsP_3R$-I (apparent K_d values of 27 nM and 89.5 nM, respectively), but both isoforms had a similar substrate specificity (Südhof *et al.*, 1991). A similar specificity was also obtained for $InsP_3R$-III (Maranto, 1994; Yamamoto-Hino *et al.*, 1994). The SI splice domain, which is located in the $InsP_3$-binding domain, is not essential for $InsP_3$ binding because it is present only in $InsP_3R$-I, but it could play a modulator role. Deletion experiments indicated that the SI domain and its surrounding region are of critical importance for $InsP_3$ binding, and it probably fulfills conformational requirements (Miyawaki *et al.*, 1991). The insertion of the SI domain is dependent on the cell type and is particularly low in adult cerebellum (Nakagawa *et al.*, 1991b; De Smedt *et al.*, 1994). This is compatible with the observation that the K_d for $InsP_3$ binding was 1.7–10.4 nM in peripheral tissues (Guillemette *et al.*, 1987; Chadwick *et al.*, 1990; Mourey *et al.*, 1990), whereas it was 80–100 nM in cerebellum (Supattapone *et al.*, 1988; Maeda *et al.*, 1990). A low affinity of the cerebellar as compared to the peripheral $InsP_3R$-I isoform, and a much higher affinity for the $InsP_3R$-IV isoform, were recently found by comparing $InsP_3$ binding in cell types where one particular isoform was mainly expressed (Parys *et al.*, 1995).

Several consensus sites located in the long cytosolic regulatory domain may be important for modulating $InsP_3$-induced Ca^{2+} release by various physiological mediators as described in Section 2 of this review. In $InsP_3R$-I, there are two serine residues flanking both ends of the SII splice segment that were shown to be phosphorylated by PKA (Furuichi *et al.*, 1989; Mignery *et al.*, 1989; Danoff *et al.*, 1991). The presence of the SII insert, which is an exclusive property of neuronal cell types (Nakagawa *et al.*, 1991b; De Smedt *et al.*, 1994), changes the preferred site of phosphorylation and could therefore represent a neuronal type of $InsP_3R$ regulation (Danoff *et al.*, 1991). The PKA phosphorylation consensus sites of $InsP_3R$-I are located within variable regions and are not conserved in the other isoforms. The human $InsP_3R$-II (Yamamoto-Hino *et al.*, 1994) and the human (Maranto, 1994; Yamamoto-Hino *et al.*, 1994) and rat (Blondel *et al.*, 1993) $InsP_3R$-III have putative phosphorylation sites in different positions. It remains to be established if the differential presence of phosphorylation consensus sites for PKA, protein kinase C, and Ca^{2+}/calmodulin-dependent protein kinase II (Maranto, 1994) confers physiologically relevant regulation specificity to these different proteins.

Some of the conflicting observations with respect to the effect of PKA on $InsP_3$-mediated Ca^{2+} release could be related to the presence of different isoforms in those preparations (Nakade *et al.*, 1994). ATP-binding sites are located near the phosphorylation sites (Maeda *et al.*, 1991; Ferris *et al.*, 1991). The most C-terminal of the two $InsP_3R$-I consensus sites for ATP binding is completely conserved in all reported $InsP_3Rs$, suggesting that all receptors may bind ATP in the same fashion. In contrast, the second potential nucleotide binding site is present only in $InsP_3R$-I (Yamamoto-Hino *et al.*, 1994), and the third one is unmasked by SII splicing (Marshall and Taylor, 1993b). Rat $InsP_3R$-I and $InsP_3R$-II were also shown to bind Ca^{2+} (Mignery *et al.*, 1992). This potential Ca^{2+}-binding domain is located in the region just preceding the transmembrane sequences. It is not known if a similar interaction with Ca^{2+} also occurs with the homologous domains of $InsP_3R$-III or if this interaction plays any role in the well-documented regulation of $InsP_3$-mediated Ca^{2+} release by Ca^{2+}. It may be relevant that modulation of $InsP_3$-mediated Ca^{2+} release by luminal as well as by cytosolic Ca^{2+} was very similar in permeabilized $C_3H10T\frac{1}{2}$ cells as compared to A7r5 cells, although the major $InsP_3R$ isoform expressed in these cell types differed (De Smedt *et al.*, 1994).

There were slightly higher sequence similarities in the putative Ca^{2+} channel domain among the different $InsP_3R$ subtypes than the overall similarities of the entire receptors (Yamamoto-Hino *et al.*, 1994). The extremely high level of conservation of M5 and M6 sequences suggests their fundamental contribution to the channel pore formation. There is a characteristic distribution of negative charges in the N-terminal half of the luminal loop between M5 and M6, which is possibly involved in efficient Ca^{2+} permeation by accumulating Ca^{2+} near the channel pore (Yamamoto-Hino *et al.*, 1994). Interestingly, the negatively charged regions have significant sequence diversities among the $InsP_3R$ isoforms. It remains to be investigated whether this sequence diversity confers type-specific gating properties.

3.3. Cell-Specific Expression

An *in situ* hybridization analysis using probes for the three main $InsP_3R$ subtypes demonstrated a widespread expression of $InsP_3R$-I in the mouse central nervous system, with much lower levels of the other subtypes (Furuichi *et al.*, 1993). It was therefore suggested that the $InsP_3R$-I homotetramer plays a major role in neuronal Ca^{2+} signaling, whereas other subtypes may function in supplementary or differential signaling, and possibly also as heteromeric complexes. However, in several other studies in which the expression of the messengers of the different $InsP_3R$ isoforms was compared, it was found that the expression pattern is very different for different cell types (Ross *et al.*, 1992; Yamamoto-Hino *et al.*, 1994; De Smedt *et al.*, 1994): $InsP_3R$-I seems to be expressed in most cell types, whereas the other isoforms can be the major form in particular cell types but not be expressed in others. This is particularly evident in studies in which the relative levels of different isoforms were determined by competitive

PCR using primer pairs common for multiple isoforms (Enyedi *et al.*, 1994; De Smedt *et al.*, 1994). $InsP_3R$-III is mainly expressed in epithelial and secretory cell types (Blondel *et al.*, 1993, 1994; Maranto, 1994; De Smedt *et al.*, 1994) and in a number of hematopoietic and lymphoma cell lines (Yamamoto-Hino *et al.*, 1994; Sugiyama *et al.*, 1994), whereas $InsP_3R$-II is much less widespread (Ross *et al.*, 1992; Yamamoto-Hino *et al.*, 1994; De Smedt *et al.*, 1994; Sugiyama *et al.*, 1994). Moreover, recent evidence indicates that the expression of at least $InsP_3R$-III can be regulated by extracellular parameters, e.g., the glucose concentration (Blondel *et al.*, 1994).

Observations at the mRNA level provided evidence for an even larger gene diversity besides the three isoforms that have been fully cloned and sequenced. There seem to be at least two other messengers that, according to the limited sequence information available, deviate slightly from $InsP_3R$-II (Ross *et al.*, 1992; De Smedt *et al.*, 1994). These isoforms very probably correspond to $InsP_3R$-II-like proteins at the amino acid level (De Smedt *et al.*, 1994). Table I gives an overview of the expression patterns of all these isoforms as determined by isoform-specific screening at both the mRNA and the protein level. Isoform-specific detection at the protein level will ultimately allow the characterization, purification, and immunolocalization of the different $InsP_3R$ isoforms. A list of the available antibodies against $InsP_3Rs$ is presented in Table II.

Two or even more isoforms are often coexpressed, even in cloned cultured cell lines (De Smedt *et al.*, 1994). This suggests that these isoforms could be involved in different cellular functions. In this respect, it may be relevant that antibodies against the purified rat brain $InsP_3R$ recognized a slightly smaller protein in a preparation of rat liver nuclei that could be phosphorylated by a nuclear protein kinase C (Matter *et al.*, 1993). Evidence has also been provided for an immunologically related $InsP_3R$-like protein in the plasma membrane of different cell types. In Jurkat T cells, the plasma membrane localization was suggested by surface iodination labeling, immunohistochemical analysis, and capping after stimulation with concanavalin A (Khan *et al.*, 1992b). Moreover, a selective enrichment of sialic acid was found for a presumed plasma membrane $InsP_3R$ in lymphocytes (Khan *et al.*, 1992a). By immunoelectron microscopy it was found that one of a series of monoclonal antibodies raised against $InsP_3R$-I specifically decorated plasmalemmal caveolae in endothelial and smooth muscle cells (Fujimoto *et al.*, 1992).

Subcellular specialization is also suggested by the finding that immunocytochemical localization of $InsP_3R$-III by an antipeptide antibody was confined to the apex of pancreatic acinar cells (Nathanson *et al.*, 1994). This region corresponds to the trigger zone from which Ca^{2+} signals originate (Kasai *et al.*, 1993; Thorn *et al.*, 1993). The $InsP_3R$-III has therefore been suggested to be involved in these polarized signaling patterns (Nathanson *et al.*, 1994). It was not fully excluded, however, that the antibody used in this study could also label a type-II-like $InsP_3R$, which, according to PCR data, is also expressed to a high level in exocrine pancreas (De Smedt *et al.*, 1994). Using two different antibodies against $InsP_3R$-III, Blondel *et al.* (1994) observed that $InsP_3R$-III was restricted in the pancreas to the islets of Langerhans, and in particular to the insulin- and somatostatin-releasing granules of the β and δ cells, respectively. Whatever the case, functional diversity of the different $InsP_3Rs$ could be correlated to a different subcellular localization. It is in this respect remarkable that more than 90% of the $InsP_3$-binding sites in rat cerebellum could be solubilized by 1% Triton X-100 (Supattapone *et al.*, 1988) whereas such treatment resulted in only 10–15% solubilization in other tissues, e.g., bovine adrenal cortex (Guillemette *et al.*, 1990; White *et al.*, 1993). These results may reflect an association with cytoskeletal proteins as discussed in Section 2.6.

4. PARTIAL Ca^{2+} RELEASE IN RESPONSE TO SUBMAXIMAL $InsP_3R$ STIMULATION

Endoplasmic reticulum vesicles from canine cerebellum were incorporated in planar bilayers, and the functional properties of the $InsP_3R$ investigated (Bezprozvanny *et al.*, 1991). Single-channel observations indicate that channel openings increased with increasing [$InsP_3$]. It is therefore expected that both a low and a high dose of $InsP_3$ would eventually empty the whole $InsP_3$-sensitive store; the low dose of $InsP_3$ would do this more slowly than the high dose. Such behavior is, however, never observed for $InsP_3Rs$ in permeabilized cells or microsomes. Even a prolonged application of a submaximal dose of $InsP_3$ is unable to release the same amount of Ca^{2+} as a higher dose of $InsP_3$ (Muallem *et al.*, 1989; Meyer and Stryer, 1990; Taylor and Potter, 1990; Oldershaw *et al.*, 1991; Finch *et al.*, 1991; Menniti *et al.*, 1991; Bootman *et al.*, 1992a; Ferris *et al.*, 1992b; Missiaen *et al.*, 1992b). Low doses of $InsP_3$ will therefore induce only a partial release of the stored Ca^{2+}, indicating that the release process in response to the continuous presence of a submaximal $InsP_3$ stimulus must appreciably slow down or even stop, despite the fact that there is still a large amount of Ca^{2+} left in the $InsP_3$-sensitive store.

Manganese quenching of compartmentalized fluorescent Ca^{2+} indicators has recently been introduced to monitor fluxes through the $InsP_3R$ in permeabilized cells (Glennon *et al.*, 1992). In such experiments, stores are loaded with a fluorescent dye, and Mn^{2+} is added to the cytosolic side. On addition of $InsP_3$, Mn^{2+} will enter the store through the open $InsP_3R$ and quench the dye in the store. Hajnóczky *et al.* (1994) observed that a low dose of $InsP_3$ can induce Mn^{2+} quenching in the whole $InsP_3$-sensitive store in carefully permeabilized hepatocytes, indicating that all stores are luminally connected. In contrast, Short *et al.* (1993) reported that a submaximal [$InsP_3$] had no access to the whole

Table I
Fully or Partially Cloned $InsP_3R$ Isoforms and Their Expression as Detected by Isoform-Specific Screening

Type	Species	Cell and tissue type	Method	Reference
I	Mouse	Cloned from mouse cerebellum	cDNA library	Furuichi *et al.*, 1989
		Central nervous system	*In situ* hybridization	Furuichi *et al.*, 1993
		Nearly all tissues and cell types investigated, particularly cerebellum, brain, spinal cord, thymus, spleen, smooth muscle, etc.	Northern, PCR	Furuichi *et al.*, 1990; Marks *et al.*, 1990; De Smedt *et al.*, 1994
			Western blotting	Maeda *et al.*, 1988, 1989
	Rat	Cloned from rat brain	cDNA library	Mignery *et al.*, 1990
		Nearly all cell types investigated, particularly cerebellum, brain, spinal cord, testis, etc.	PCR	De Smedt *et al.*, 1994; Enyedi *et al.*, 1994
			Western blotting	Mignery *et al.*, 1989, 1990
	Human	Cloned from human uterus and from HL-60 leukemic cells	cDNA library	Yamada *et al.*, 1994
		THP-1 monocytes, HEL, CMK megakaryoblastic cells, ML-1 myeloblastic cells, HL-60 promyelocytic cells	Northern	Yamamoto-Hino *et al.*, 1994; Sugiyama *et al.*, 1994
	Xenopus laevis	Cloned from oocytes	cDNA library	Kume *et al.*, 1993
		Oocytes and eggs	Western blotting	Parys *et al.*, 1992; Kume *et al.*, 1993; Callamaras and Parker, 1994
II	Mouse	Partial clone from placenta	cDNA library	Ross *et al.*, 1992
		Various tissues, particularly brain, lung, placenta	PCR	Ross *et al.*, 1992
	Rat	Cloned from rat brain	cDNA library	Südhof *et al.*, 1991
		AR42J pancreatic tumor cells, exocrine pancreas, adrenal glomerulosa cells	PCR	De Smedt *et al.*, 1994; Enyedi *et al.*, 1994
	Human	Cloned from HUVEC	cDNA library	Yamamoto-Hino *et al.*, 1994
		HEL, CMK megakaryoblastic cells, KU812F basophilic cells, T cells (HPB-ALL, Jurkat), B cells (Namalwa, Jijoye)	Northern	Yamamoto-Hino *et al.*, 1994; Sugiyama *et al.*, 1994

III	Mouse	Partial clone from placenta	cDNA library	Ross *et al.*, 1992
		Various cell types, particularly gastrointestinal tract, brain, thymus	PCR, *in situ* hybridization	Ross *et al.*, 1992
	Rat	Cloned from RINm5F insulinoma cells	cDNA library	Blondel *et al.*, 1993
		Various tissues particularly pancreatic islets, kidney, fat, jejunum	Northern	Blondel *et al.*, 1993
		RINm5F insulinoma cells, PC-12, A7r5, exocrine pancreas	PCR	De Smedt *et al.*, 1994
		RINm5F insulinoma cells, pancreatic islets, exocrine pancreas, enterocytes	Western blotting	Blondel *et al.*, 1993; Maranto, 1994; Nathanson *et al.*, 1994
	Human	Cloned from HT29 adenocarcinoma	cDNA library	Maranto, 1994
		Cloned from HUVEC	cDNA library	Yamamoto-Hino *et al.*, 1994
		Various cell lines, e.g., ML-1 myeloblastic cells, T cells (HPB-ALL, Jurkat, HUT-78), B cells (Namalwa, Raji, RPMI1788, Jijoye), U937 monoblastic cells, K562 erythroblastic cells	Northern	Yamamoto-Hino *et al.*, 1994; Sugiyama *et al.*, 1994
		HT29 adenocarcinoma cells	Western blotting	Maranto, 1994
IV	Mouse	Partial clone from placenta	cDNA library	Ross *et al.*, 1992
		Various tissues particularly brain, spinal cord, testis	PCR	Ross *et al.*, 1992
	Rat	Liver, RBL-2H3 mast cells, AR42J pancreatic tumor cells, exocrine pancreas	PCR	De Smedt *et al.*, 1994
		RBL-2H3 mast cells	Western blotting	Parys *et al.*, 1995
V	Mouse	Partial clone from $C_3H10T^1/_2$ fibroblasts	PCR	De Smedt *et al.*, 1994
		$C_3H10T^1/_2$ fibroblasts, BC_3H1 myoblasts, L cells, Swiss 3T3 cells, pancreatic islets	PCR	De Smedt *et al.*, 1994
		$C_3H10T^1/_2$ fibroblasts	Western blotting	Parys *et al.*, 1995
Drosophila isoform	*Drosophila melanogaster*	Cloned from *Drosophila* head	cDNA library	Yoshikawa *et al.*, 1992

Table II
Antibodies Used for the Detection of the $InsP_3Rs$[a]

Antibody type	Epitope: Amino acid, type	Name of antibody	References
Monoclonal	679–727, mouse cerebellar $InsP_3R$	4C11	Maeda *et al.*, 1989; Furuichi *et al.*, 1989
	943–1237, mouse cerebellar $InsP_3R$	10A6	Maeda *et al.*, 1989; Furuichi *et al.*, 1989
	2736–2747, mouse cerebellar $InsP_3R$	18A10	Maeda *et al.*, 1988; Nakade *et al.*, 1991
	C-terminus, mouse cerebellar $InsP_3R$	IPR.1	Bourguignon *et al.*, 1993b
Polyclonal	Rat cerebellar $InsP_3R$		Ross *et al.*, 1989
	Rat cerebellar $InsP_3R$		Peng *et al.*, 1991
	Rat cerebellar $InsP_3R$		Sharp *et al.* 1992
	Rat cerebellar $InsP_3R$	AP2A	Sharp *et al.*, 1993
	Rabbit brain $InsP_3R$	Goat44	Parys *et al.*, 1992; Parys *et al.*, 1994
	Dog cerebellar $InsP_3R$		Joseph, 1994
	Mouse cerebellar $InsP_3R$	PolyAB-whole	Nakade *et al.*, 1994
Polyclonal anti-peptide	401–414, $InsP_3R$-I	KEEK	Joseph, 1994
	2463–2476, $InsP_3R$-I	V753	Takei *et al.*, 1994
	2504–2523, $InsP_3R$-I	1ML1	Michikawa *et al.*, 1994
	2539–2593, $InsP_3R$-I		Kume *et al.*, 1993
	2690–2700, $InsP_3R$-I	IPR2690	Moschella and Marks, 1993
	30 C-terminal a.a.,[b] $InsP_3R$-I	αH1CT	Nathanson *et al.*, 1994
	19 C-terminal a.a., $InsP_3R$-I	T210	Mignery *et al.*, 1989; Takei *et al.*, 1994
	19 C-terminal a.a., $InsP_3R$-I		Gorza *et al.*, 1993
	19 C-terminal a.a., $InsP_3R$-I		Joseph and Samanta, 1993
	19 C-terminal a.a., $InsP_3R$-I	CT1	Wojcikiewicz *et al.*, 1994
	15 C-terminal a.a., $InsP_3R$-I	Rbt52	Parys *et al.*, 1992; Parys *et al.*, 1994
	15 C-terminal a.a., $InsP_3R$-I	Rbt04	Parys *et al.*, 1995
	2736–2747, $InsP_3R$-I	Anti-pep6	Nakade *et al.*, 1994
	14 C-terminal a.a., $InsP_3R$-I		Lièvremont *et al.*, 1994
	13 C-terminal a.a., $InsP_3R$-II	Rbt02	Parys *et al.*, 1995
	2432–2444, $InsP_3R$-III	AP45	Blondel *et al.*, 1994
	2441–2461, $InsP_3R$-III	$\alpha IIIa_2$	Yamamoto-Hino *et al.*, 1994
	27 C-terminal a.a., $InsP_3R$-III	αH3CT	Maranto, 1994; Nathanson *et al.*, 1994
	15 C-terminal a.a., $InsP_3R$-III	IPR3AB3	Blondel *et al.*, 1993

[a]The $InsP_3R$ isoform indicated is the isoform against which the antibody was raised. Because of sequence similarities among the different isoforms, some antibodies may recognize more than one isoform. Most antibodies will also recognize the $InsP_3R$-isoforms of multiple species. For specific cases, the reader is referred to the original reference(s). Epitopes are listed in the N- to C-terminal direction for each isoform.
[b]Amino acids.

$InsP_3$-sensitive store in DDT_1MF-2 cells measured under comparable conditions. It is therefore premature to abandon the currently held view of partial Ca^{2+} release.

Three main groups of hypotheses have been proposed to explain why the release in response to a low [$InsP_3$] slows down at a time when the stores still contain an appreciable amount of Ca^{2+}. A first hypothesis is that individual store units all have different $InsP_3$ sensitivities and that stimulation with a low [$InsP_3$] discharges only the most sensitive stores while leaving the less sensitive pools more or less untouched (all-or-none mechanism) (Muallem *et al.*, 1989; Oldershaw *et al.*, 1991; Ferris *et al.*, 1992). A second hypothesis is that the $InsP_3$-sensitive Ca^{2+} stores form a more homogeneous population and that a low [$InsP_3$] releases Ca^{2+} until the release is stopped or at least significantly slowed down by some control system, e.g., luminal Ca^{2+} (a steady-state mechanism) (Irvine, 1990, 1991; Tregear *et al.*, 1991; Missiaen *et al.*, 1992b; Loomis-Husselbee and Dawson, 1993). A third proposal is that the $InsP_3R$ itself intrinsically inactivates during the process of Ca^{2+} release. Classical desensitization seems not to be involved because the stores are still able to respond to the subsequent application of a higher [$InsP_3$] (Meyer and Stryer, 1990). It should finally be pointed out that it is very probable that the partial release is in fact obtained by a combination of these different mechanisms.

4.1. All-or-None Ca^{2+} Release

The first hypothesis to explain the partial release behavior is that individual store units present different $InsP_3$ sensitivities so that a low [$InsP_3$] will affect only the more sensitive stores (Muallem *et al.*, 1989; Oldershaw *et al.*, 1991). Heterogeneity in $InsP_3$ sensitivity could result from the expression of different $InsP_3Rs$ in different Ca^{2+} stores or from the above-mentioned modulation of the $InsP_3$ sensitivity by posttranslational mechanisms, e.g., state of phosphorylation. Experimental evidence in favor of this hypothesis came from experiments in which $InsP_3Rs$ purified from cerebellum were reconstituted in lipid vesicles (Ferris *et al.*, 1992b). Each vesicle contained on average less than one $InsP_3R$. The fact that a submaximal [$InsP_3$] affected only some of the vesicles indicates that $InsP_3Rs$ can have a different $InsP_3$ sensitivity.

4.2. Steady-State Ca^{2+} Release

The activation of the release by luminal Ca^{2+} could contribute to the partial Ca^{2+} release in response to submaximal $InsP_3R$ activation if the drop in luminal [Ca^{2+}] slows the release once the pools become partially depleted (Irvine, 1990, 1991; Tregear *et al.*, 1991; Missiaen *et al.*, 1992b; Loomis-Husselbee and Dawson, 1993). The observation by Oldershaw *et al.* (1992) that the partial release behavior in permeabilized rat hepatocytes disappears under conditions in which the luminal free [Ca^{2+}] is kept largely constant by using pyrophosphate as a luminal Ca^{2+} buffer again points to the important modulatory role of luminal Ca^{2+} in the release process. Because the partial release also occurs when the $InsP_3R$ is maximally stimulated by luminal Ca^{2+}, it is possible that in these cases $InsP_3R$ heterogeneity and/or inactivation is also involved (Parys *et al.*, 1993a).

A steady-state mechanism for generating the partial release of course implies that the kinetics of the release process must be slower than the time needed for closure of the $InsP_3$-gated channel. The existence of a slow-release phase in a variety of systems (Champeil *et al.*, 1989; Meyer and Stryer, 1990; Missiaen *et al.*, 1992b; Loomis-Husselbee and Dawson, 1993) suggests that the release process might indeed be quite slow.

The observation of a pacemaker rise in intracellular [Ca^{2+}] culminating in a nearly all-or-none Ca^{2+} release in intact cells (Parker and Ivorra, 1990a; Parker and Yao, 1991) and the very rapid initial Ca^{2+} release at low [$InsP_3$] in permeabilized cells in unbuffered conditions (Meyer and Stryer, 1990; Meyer *et al.*, 1990; Iino and Endo, 1992) point to a more complete emptying of the stores than predicted by the steady-state model. A steady-state model can, however, explain a more complete release if there is a time delay between the decrease of the luminal [Ca^{2+}] below the threshold for $InsP_3R$ activation and the actual closure of the release channel, or if cytosolic Ca^{2+} is allowed to exert its positive feedback on the $InsP_3R$.

4.3. $InsP_3R$ Inactivation

Györke and Fill (1993) described "single channel adaptation" for the ryanodine receptor, which is an intracellular release channel with much resemblance to the $InsP_3R$. Single-channel adaptation means that the channel activity during continuous stimulation progressively decreases, although the channel remains responsive to a more intense stimulation. If the $InsP_3R$ would also intrinsically inactivate, and if such inactivation would be maximal at a time when the $InsP_3$-sensitive store is not completely discharged, then such behavior can also contribute to the partial release of the accumulated Ca^{2+} (Champeil *et al.*, 1989; Finch *et al.*, 1991).

The published single-channel recordings of the $InsP_3R$ do not show any sign of inactivation, although the events immediately following the addition of $InsP_3$ were not recorded. Calcium release experiments in permeabilized cells also show no evidence for inactivation; e.g., Oldershaw *et al.* (1992) observed no inactivation of the release from pyrophosphate-loaded stores. However, in their system, the partial release response was also lost.

Calcium (Ferris *et al.*, 1992b) and Mn^{2+} (Glennon *et al.*, 1992) can permeate through the $InsP_3R$ from the cytoplasmic to the luminal side. The effects of a preincubation with $InsP_3$ on the initial rate of Ca^{2+} and Mn^{2+} entry into stores can be used as a screening for inactivation. The results reported so far seem to be contradictory. The Ca^{2+}-entry experiments by Ferris *et al.* (1992b) indicate that the $InsP_3R$ does not inactivate, because preincubation with $InsP_3$ for up to 5 min at 500 nM Ca^{2+} had no effect on the subsequent phase of Ca^{2+} entry. The Mn^{2+}-entry experiments by Hajnóczky and Thomas (1994), in contrast, suggest that the $InsP_3R$ does inactivate, because preincubation with $InsP_3$ slowed down the subsequent phase of Mn^{2+} entry. Half-maximal inactivation occurred after 15 sec preincubation, and maximal inactivation was reached after 60 sec. The inactivation, which was never complete, was reversible. The $InsP_3$ itself can induce the inactivation, but this process was enhanced by cytosolic Ca^{2+}. Hajnóczky and Thomas (1994) suggested that this behavior may provide a mechanism for adaptation to partially elevated basal levels of $InsP_3$.

5. CONCLUSIONS

It has in the last couple of years become apparent that the $InsP_3R$ is modulated in a complex way by various cellular components. The regulation by Ca^{2+} can appreciably modify the release in the presence of a constant [$InsP_3$] and could therefore be the basis for generating Ca^{2+} oscillations and various types of Ca^{2+} waves. The physiological relevance of other types of regulation still needs to be sorted out. It also became apparent that multiple isoforms of the protein exist. Challenging research topics for the near future will be to find out whether these isoforms would present major functional differences and whether they would be targeted to different compartments in the cell.

ABBREVIATIONS USED IN THIS CHAPTER

$InsP_3$:	inositol-1,4,5-trisphosphate
$InsP_3R$:	$InsP_3$ receptor
PKA:	cyclic-AMP-dependent protein kinase
PKG:	cyclic-GMP-dependent protein kinase
$[Ca^{2+}]$:	Ca^{2+} concentration
$[Ca^{2+}]_i$:	cytosolic Ca^{2+} concentration
PCR:	polymerase chain reaction

6. REFERENCES

Benevolensky, D., Moraru, I. I., and Watras, J., 1994, Micromolar calcium decreases affinity of inositol trisphosphate receptor in vascular smooth muscle, *Biochem J.* **299:**631–636.

Berridge, M. J., 1991, Caffeine inhibits inositol-trisphosphate-induced membrane potential oscillations in *Xenopus* oocytes, *Proc. R. Soc. Lond. [Biol.]* **244:**57–62.

Berridge, M. J., 1993, Inositol trisphosphate and calcium signalling, *Nature* **361:**315–325.

Berridge, M. J., and Irvine, R. F., 1989, Inositol phosphates and cell signalling, *Nature* **341:**197–205.

Bezprozvanny, I., and Ehrlich, B. E., 1993, ATP modulates the function of inositol 1,4,5-trisphosphate-gated channels at two sites, *Neuron* **10:**1175–1184.

Bezprozvanny, I., Watras, J., and Ehrlich, B. E., 1991, Bell-shaped calcium-response curves of $InsP_3$- and calcium-gated channels from endoplasmic reticulum of cerebellum, *Nature* **351:**751–754.

Bezprozvanny, I., Bezprozvannaya, S., and Ehrlich, B. E., 1994, Caffeine-induced inhibition of inositol(1,4,5)-trisphosphate-gated calcium channels from cerebellum, *Mol. Biol. Cell* **5:**97–103.

Bird, G. St. J., Obie, J. F., and Putney, J. W., Jr., 1992, Functional homogeneity of the non-mitochondrial Ca^{2+} pool in intact mouse lacrimal cells, *J. Biol. Chem.* **267:**18382–18386.

Bird, G. St. J., Rossier, M. F., Obie, J. F., and Putney, J. W., Jr., 1993a, Sinusoidal oscillations in intracellular calcium requiring negative feedback by protein kinase C, *J. Biol. Chem.* **268:**8425–8428.

Bird, G. St. J., Burgess, G. M., and Putney, J. W., Jr., 1993b, Sulfhydryl reagents and cAMP-dependent kinase increase the sensitivity of the inositol 1,4,5-trisphosphate receptor in hepatocytes, *J. Biol. Chem.* **268:**17917–17923.

Blondel, O., Takeda, J., Janssen, H., Seino, S., and Bell, G. I., 1993, Sequence and functional characterization of a third inositol trisphosphate receptor subtype, IP3R-3, expressed in pancreatic islets, kidney, gastrointestinal tract, and other tissues, *J. Biol. Chem.* **268:**11356–11363.

Blondel, O., Moody, M., Depaoli, M., Sharp, A. H., Ross, C. A., Swift, H., and Bell, E. I., 1994, Localization of inositol trisphosphate receptor subtype 3 to insulin and somatostatin secretory granules and regulation of expression in islets and insulinoma cells, *Proc. Natl. Acad. Sci. U.S.A.* **91:**7777–7781.

Boitano, S., Dirksen, E. R., and Sanderson, M. J., 1992, Intercellular propagation of calcium waves mediated by inositol trisphosphate, *Science* **258:**292–295.

Bond, J. M., and Taylor, C. W., 1991, Solubilization of rat liver inositol 1,4,5-trisphosphate receptor, *Cell. Signal.* **3:**607–612.

Bootman, M. D., Berridge, M. J., and Taylor, C. W., 1992a, All-or-none Ca^{2+} mobilization from the intracellular stores of single histamine-stimulated HeLa cells, *J. Physiol.* **450:**163–178.

Bootman, M. D., Taylor, C. W., and Berridge, M. J., 1992b, The thiol reagent, thimerosal, evokes Ca^{2+} spikes in Hela cells by sensitizing the inositol 1,4,5-trisphosphate receptor, *J. Biol. Chem.* **267:**25113–25119.

Bootman, M. D., Missiaen L., Parys, J. B., De Smedt, H., and Casteels, R., 1995, Control of inositol 1,4,5-trisphosphate-induced Ca^{2+} release by cytosolic Ca^{2+}, *Biochem J.* **306:**445–451.

Bourguignon, L. Y. W., Iida, N., and Jin, H., 1993a, The involvement of the cytoskeleton in regulating IP_3 receptor-mediated internal Ca^{2+} release in human blood platelets, *Cell Biol.* **17:**751–758.

Bourguignon, L. Y. W., Jin, H., Iida, N., Brandt, N. R., and Zhang, S. H., 1993b, The involvement of ankyrin in the regulation of inositol 1,4,5-trisphosphate receptor-mediated internal Ca^{2+} release from Ca^{2+} storage vesicles in mouse T-lymphoma cells, *J. Biol. Chem.* **268:**7290–7297.

Brass, L. F., and Joseph, S. K., 1985, A role for inositol triphosphate in intracellular Ca^{2+} mobilization and granule secretion in platelets, *J. Biol. Chem.* **260:**15172–15179.

Brown, G. R., Sayers, L. G., Kirk, C. J., Michell, R. H., and Michelangeli, F. , 1992, The opening of the inositol 1,4,5-trisphosphate-sensitive channel in rat cerebellum is inhibited by caffeine, *Biochem. J.* **282:**309–312.

Burgess, G. M., Bird, G. S. J., Obie, J. F., and Putney, J. W., Jr., 1991, The mechanism for synergism between phospholipase C- and adenylylcyclase-linked hormones in liver. Cyclic AMP-dependent kinase augments inositol trisphosphate-mediated Ca^{2+} mobilization without increasing the cellular levels of inositol polyphosphates, *J. Biol. Chem.* **266:**4772–4781.

Byron, K. L., and Taylor, C. W., 1993, Spontaneous Ca^{2+} spiking in a vascular smooth muscle cell line is independent of the release of intracellular Ca^{2+} stores, *J. Biol. Chem.* **268:**6945–6952.

Callamaras, N., and Parker, I., 1994, Inositol 1,4,5-trisphosphate receptors in *Xenopus laevis* oocytes: Localization and modulation by Ca^{2+}, *Cell Calcium* **15:**66–78.

Carroll, J., and Swann, K., 1992, Spontaneous cytosolic calcium oscillations driven by inositol trisphosphate occur during *in vitro* maturation of mouse oocytes, *J. Biol. Chem.* **267:**11196–11201.

Chadwick, C. C., Saito, A., and Fleischer, S., 1990, Isolation and characterization of the inositol trisphosphate receptor from smooth muscle, *Proc. Natl. Acad. Sci. U.S.A.* **87:**2132–2136.

Champeil, P., Combettes, L., Berthon, B., Doucet, E., Orlowski, S., and Claret, M., 1989, Fast kinetics of calcium release induced by myo-inositol trisphosphate in permeabilized rat hepatocytes, *J. Biol. Chem.* **264:**17665–17673.

Cheek, T. R., McGuinness, O. M., Vincent, C., Moreton, R. B., Berridge, M. J., and Johnson, M. H., 1993, Fertilisation and thimerosal stimulate similar calcium spiking patterns in mouse oocytes but by separate mechanisms, *Development* **119:**179–189.

Combettes, L., Claret, M., and Champeil, P., 1992, Do submaximal $InsP_3$ concentrations only induce the partial discharge of permeabilized hepatocyte calcium pools because of the concomitant reduction of intraluminal Ca^{2+} concentration? *FEBS Lett.* **301:**287–290.

Combettes, L., Claret, M., and Champeil, P., 1993, Calcium control on $InsP_3$-induced discharge of calcium from permeabilised hepatocyte pools, *Cell Calcium* **14:**279–292.

Combettes, L., Hannaert-Merah, Z., Coquil, J.-F., Rousseau, C., Claret, M., Swillens, S., and Champeil, P., 1994a, Rapid filtration studies of the effect of cytosolic Ca^{2+} on $InsP_3$-induced $^{45}Ca^{2+}$ release from cerebellar microsomes, *J. Biol. Chem.* **269:**17561–17571.

Combettes, L. Berthon, B., and Claret, M., 1994b, Caffeine inhibits cytosolic calcium oscillations induced by noradrenaline and vasopressin in rat hepatocytes, *Biochem. J.* **301:**737–744.

Danoff, S. K., Ferris, C. D., Donath, C., Fischer, G., Munemitsu, S., Ullrich, A., Snyder, S. H., and Ross, C. A., 1991, Inositol 1,4,5-trisphosphate receptors: distinct neuronal and nonneuronal forms derived by alternative splicing differ in phosphorylation, *Proc. Natl. Acad. Sci. U.S.A.* **88:**2951–2955.

De Lisle, S., Radenberg, T., Wintermantel, M. R., Tietz, C., Parys, J. B., Pittet, D., Welsh, M. J., and Mayr, G. W., 1994, Second messenger specificity of the inositol trisphosphate receptor: Reappraisal based on novel inositol phosphates, *Am. J. Physiol.* **266:**C429–C436.

De Smedt, H., Missiaen, L., Parys, J. B., Bootman, M. D., Mertens, L., Van Den Bosch, L., and Casteels, R., 1994, Determination of the relative amounts of inositol trisphosphate receptor mRNA isoforms by ratio polymerase chain reaction, *J. Biol. Chem.* **269:**21691–21698.

Diarra, A., Wang, R., Garneau, L., Gallo-Payet, N., and Sauvé, R., 1994, Histamine-evoked Ca^{2+} oscillations in HeLa cells are sensitive to methylxanthines but insensitive to ryanodine, *Pflugers Arch.* **426:**129–138.

Ehrlich, B. E., and Watras, J., 1988, Inositol 1,4,5-trisphosphate activates a channel from smooth muscle sarcoplasmic reticulum, *Nature* **336:**583–586.

Enyedi, P., Szabadkai, G., Horváth, A., Szilágyi, L., Gráf, L., and Spät, A., 1994, Inositol 1,4,5-trisphosphate receptor subtypes in adrenal glomerulosa cells, *Endocrinology* **134:**2354–2359.

Feng, L., and Kraus-Friedman, N., 1993, Association of the hepatic IP_3 receptor with the plasma membrane: Relevance to mode of action, *Am. J. Physiol.* **265:**C1588–C1596.

Ferris, C. D., and Snyder, S. H., 1992, Inositol 1,4,5-trisphosphate-activated calcium channels, *Annu. Rev. Physiol.* **54:**469–488.

Ferris, C. D., Huganir, R. L., Supattapone, S., and Snyder, S. H., 1989, Purified inositol 1,4,5-trisphosphate receptor mediates calcium flux in reconstituted lipid vesicles, *Nature* **342:**87–89.

Ferris, C. D., Huganir, R. L., and Snyder, S. H., 1990, Calcium flux mediated by purified inositol 1,4,5-trisphosphate receptor in reconstituted lipid vesicles is allosterically regulated by adenine nucleotides, *Proc. Natl. Acad. Sci. U.S.A.* **87:**2147–2151.

Ferris, C. D., Huganir, R. L., Bredt, D. S., Cameron, A. M., and Snyder, S. H., 1991, Inositol trisphosphate receptor: phosphorylation by protein kinase C and calcium calmodulin-dependent protein kinases in reconstituted lipid vesicles, *Proc. Natl. Acad. Sci. U.S.A.* **88:**2232–2235.

Ferris, C. D., Cameron, A. M., Bredt, D. S., Huganir, R. L., and Snyder, S. H., 1992a, Autophosphorylation of inositol 1,4,5-trisphosphate receptors, *J. Biol. Chem.* **267:**7036–7041.

Ferris, C. D., Cameron, A. M., Huganir, R. L., and Snyder, S. H., 1992b, Quantal calcium release by purified reconstituted inositol 1,4,5-trisphosphate receptors, *Nature* **356:**350–352.

Finch, E. A., and Goldin, S. M., 1994, Response to "Calcium and inositol 1,4,5-trisphosphate-induced Ca^{2+} release," *Science* **265:**813–815.

Finch, E. A., Turner, T. J., and Goldin, S. M., 1991, Calcium as a coagonist of inositol 1,4,5-trisphosphate-induced calcium release, *Science* **252:**443–446.

Fujimoto, T., Nakade, S., Miyawaki, A., Mikoshiba, K., and Ogawa, K., 1992, Localization of inositol 1,4,5-trisphosphate receptor-like protein in plasmalemmal caveolae, *J. Cell Biol.* **119:**1507–1513.

Furuichi, T., Yoshikawa, S., Miyawaki, A., Wada, K., Maeda, N., and Mikoshiba, K., 1989, Primary structure and functional expression of the inositol 1,4,5-trisphosphate-binding protein P_{400}, *Nature* **342:**32–38.

Furuichi, T., Shiota, C., and Mikoshiba, K., 1990, Distribution of inositol 1,4,5-trisphosphate receptor mRNA in mouse tissues, *FEBS Lett.* **267:**85–88.

Furuichi, T., Simon-Chazottes, D., Fujino, I., Yamada, N., Hasegawa, M., Miyawaki, A., Yoshikawa, S., Guénet, J.-L., and Mikoshiba, K., 1993, Widespread expression of inositol 1,4,5-trisphosphate receptor type 1 gene (*Insp3r1*) in the mouse central nervous system, *Receptors Channels* **1:**11–24.

Galione, A., McDougall, A., Busa, W. B., Willmott, N., Gillot, I., and Whitaker, M., 1993, Redundant mechanisms of calcium-induced calcium release underlying calcium waves during fertilization of sea urchin eggs, *Science* **261:**348–352.

Gericke, M., Droogmans, G., and Nilius, B., 1993, Thimerosal induced changes of intracellular calcium in human endothelial cells, *Cell Calcium* **14:**201–207.

Ghosh, T. K., Eis, P. S., Mullaney, J. M., Ebert, C. L., and Gill, D. L., 1988, Competitive, reversible, and potent antagonism of inositol 1,4,5-trisphosphate-activated calcium release by heparin, *J. Biol. Chem.* **263:**11075–11079.

Gill, D. L., 1989, Receptor kinships revealed, *Nature* **342:**16–18.

Glennon, M. C., Bird, G. St. J., Kwan, C.-Y., and Putney, J. W., Jr., 1992, Actions of vasopressin and the Ca^{2+}-ATPase inhibitor, thapsigargin, on Ca^{2+} signaling in hepatocytes, *J. Biol. Chem.* **267:**8230–8233.

Gorza, L., Schiaffino, S., and Volpe, P., 1993, Inositol 1,4-5-trisphosphate receptor in heart: Evidence for its concentration in Purkinje myocytes of the conduction system, *J. Cell Biol.* **121:**345–353.

Guillemette, G., and Segui, J. A., 1988, Effects of pH, reducing and alkylating reagents on the binding and Ca^{2+} release activities of inositol 1,4,5-trisphosphate in the bovine adrenal cortex, *Mol. Endocrinol.* **2:**1249–1255.

Guillemette, G., Balla, T., Baukal, A. J., and Catt, K. J., 1987, Inositol 1,4,5-trisphosphate binds to a specific receptor and releases microsomal calcium in the anterior pituitary gland, *Proc. Natl. Acad. Sci. U.S.A.* **84:**8195–8199.

Guillemette, G., Favreau, I., Boulay, G., and Potier, M., 1990, Solubilization and partial characterization of inositol 1,4,5-trisphosphate receptor of bovine adrenal cortex reveal similarities with the receptor of rat cerebellum, *Mol. Pharmacol.* **38:**841–847.

Györke, S., and Fill, M., 1993, Ryanodine receptor adaptation: control mechanism of Ca^{2+}-induced Ca^{2+} release in heart, *Science* **260:**807–809.

Hajnóczky, G., and Thomas, A. P., 1994, The inositol trisphosphate calcium channel is inactivated by inositol trisphosphate, *Nature* **370:**474–477.

Hajnóczky, G., Gao, E., Nomura, T., Hoek, J. B., and Thomas, A. P., 1993, Multiple mechanisms by which protein kinase A potentiates inositol 1,4,5-trisphosphate-induced Ca^{2+} mobilization in permeabilized hepatocytes, *Biochem. J.* **293:**413–422.

Hajnóczky, G., Lin, C., and Thomas, A . P., 1994, Luminal communication between intracellular calcium stores modulated by GTP and the cytoskeleton, *J. Biol. Chem.* **269:**10280–10287.

Harootunian, A. T., Kao, J. P. Y., Paranjape, S., and Tsien, R. Y., 1991, Generation of calcium oscillations in fibroblasts by positive feedback between calcium and IP_3, *Science* **251:**75–78.

Hilly, M., Piétri-Rouxel, F., Coquil, J. F., Guy, M., and Mauger, J. P., 1993, Thiol reagents increase the affinity of the inositol 1,4,5-trisphosphate receptor, *J. Biol. Chem.* **268:**16488–16494.

Hirata, M., Suematsu, E., Hashimoto, T., Hamachi, T., and Koga, T., 1984, Release of Ca^{2+} from a non-mitochondrial store site in peritoneal macrophages treated with saponin by inositol 1,4,5-trisphosphate, *Biochem. J.* **223:**229–236.

Hirata, M., Narumoto, N., Watanabe, Y., Kanematsu, T., Koga, T., and Ozaki, S., 1994, DL-*myo*-inositol 1,2,4,5-tetrakisphosphate, a potent analog of D-*myo*-inositol 1,4,5-trisphosphate, *Mol. Pharmacol.* **45:**271–276.

Hirose, K., Iino, M., and Endo, M., 1993, Caffeine inhibits Ca^{2+}-mediated potentiation of inositol 1,4,5-trisphosphate-induced Ca^{2+}- release in permeabilized vascular smooth muscle cells, *Biochem. Biophys. Res. Commun.* **194:**726–732.

Iino, M., 1990, Biphasic Ca^{2+} dependence of inositol 1,4,5-trisphosphate-induced Ca release in smooth muscle cells of the guinea pig taenia caeci, *J. Gen. Physiol.* **95:**1103–1122.

Iino, M., 1991, Effect of adenine nucleotides on inositol 1,4,5-trisphosphate-induced calcium release in vascular smooth muscle cells, *J. Gen. Physiol.* **98:**681–698.

Iino, M., and Endo, M., 1992, Calcium-dependent immediate feedback control of inositol 1,4,5-trisphosphate-induced Ca^{2+} release, *Nature* **360:**76–78.

Imagawa, T., Smith, J. S., Coronado, R., and Campbell, K. P., 1987, Purified ryanodine receptor from skeletal muscle sarcoplasmic reticulum is the Ca^{2+}-permeable pore of the calcium release channel. *J. Biol. Chem.* **262:**16636–16643.

Irvine, R. F., 1990, "Quantal" Ca^{2+} release and the control of Ca^{2+} entry by inositol phosphates—a possible mechanism, *FEBS Lett.* **263:**5–9

Irvine, R. F., 1991, Inositol tetrakisphosphate as a second messenger: Confusions, contradictions, and a potential resolution, *BioEssays* **13:**419–427.

Irvine, R. F., Brown, K. D., and Berridge, M. J., 1984, Specificity of inositol trisphosphate-induced calcium release from permeabilized Swiss mouse 3T3 cells, *Biochem. J.* **222:**269–272.

Jacob, R., Merritt, J. E., Hallam, T. J., and Rink, T. J., 1988, Repetitive spikes in cytoplasmic calcium evoked by histamine in human endothelial cells, *Nature* **335:**40–45.

Joseph, S. K., 1994, Biosynthesis of the inositol trisphosphate receptor in WB rat liver epithelial cells, *J. Biol. Chem.* **269:**5673–5679.

Joseph, S. K., and Ryan, S. V., 1993, Phosphorylation of the inositol trisphosphate receptor in isolated rat hepatocytes, *J. Biol. Chem.* **268:**23059–23065.

Joseph, S. K., and Samanta, S., 1993, Detergent solubility of the inositol trisphosphate receptor in rat brain membranes. Evidence for association of the receptor with ankyrin, *J. Biol. Chem.* **268:**6477–6486.

Kasai, H., Li, Y. X., and Miyashita, Y., 1993, Subcellular distribution of Ca^{2+} release channels underlying Ca^{2+} waves and oscillations in exocrine pancreas, *Cell* **74:**669–677.

Khan, A. A., Steiner, J. P., and Snyder, S. H., 1992a, Plasma membrane inositol-trisphosphate receptor of lymphocytes: Selective enrichment in sialic acid and unique binding specificity, *Proc. Natl. Acad. Sci. U.S.A.* **89:**2849–2853.

Khan, A. A., Steiner, J. P., Klein, M. G., Schneider, M. F., and Snyder, S. H., 1992b, IP_3 receptor: Localization to plasma membrane of T cells and cocapping with the T cell receptor, *Science* **257:**815–818.

Kline, J. T., and Kline, D., 1994, Regulation of intracellular calcium in the mouse egg: Evidence for inositol trisphosphate-induced calcium release, but not calcium-induced calcium release, *Biol. Reprod.* **50:**193–203.

Koga, T., Yoshida, Y., Cai, J.-Q., Islam, M. O., and Imai, S., 1994, Purification and characterization of 240-kDA cGMP-dependent protein kinase substrate of vascular smooth muscle. Close resemblance to inositol 1,4,5-trisphosphate receptor, *J. Biol. Chem.* **269:**11640–11647.

Komalavilas, P., and Lincoln, T. M., 1994, Phosphorylation of the inositol 1,4,5-trisphosphate receptor by cyclic GMP-dependent protein kinase, *J. Biol. Chem.* **269:**8701–8707.

Kume, S., Muto, A., Aruga, J., Nakagawa, T., Michikawa, T., Furuichi, T., Nakade, S., Okano, H., and Mikoshiba, K., 1993, The *Xenopus* IP_3 receptor: structure, function, and localization in oocytes and eggs, *Cell* **73:**555–570.

Lai, F. A., Erickson, H. P., Rousseau, E., Liu, Q. Y., and Meissner, G., 1988, Purification and reconstitution of the calcium release channel from skeletal muscle, *Nature* **331:**315–319.

Lechleiter, J., Girard, S., Clapham, D., and Peralta, E., 1991, Subcellular patterns of calcium release determined by G protein-specific residues of muscarinic receptors, *Nature* **350:**505–508.

Lee, H. C., Aarhus, R., and Walseth, T. F., 1993, Calcium mobilization by dual receptors during fertilization of sea urchin eggs, *Science* **261:**352–355.

Lièvremont, J.-P., Hill, A.-M., Hilly, M., and Mauger, J.-P., 1994, The inositol 1,4,5-trisphosphate receptor is localized on specialized sub-regions of the endoplasmic reticulum in rat liver, *Biochem. J.* **300:**419–427.

Lipscombe, D., Madison, D. V., Poenie, M., Reuter, H., Tsien, R. W., and Tsien, R. Y., 1988, Imaging of cytosolic Ca^{2+} transients arising from Ca^{2+} stores and Ca^{2+} channels in sympathetic neurons, *Neuron* **1:**355–365.

Loomis-Husselbee, J. W., and Dawson, A. P., 1993, A steady-state mechanism can account for the properties of inositol 2,4,5-trisphosphate-stimulated Ca^{2+} release from permeabilized L1210 cells, *Biochem. J.* **289:**861–866.

Luttrell, B. M., 1993, The biological relevance of the binding of calcium ions by inositol phosphates, *J. Biol. Chem.* **268:**1521–1524.

Lynn, S., Morgan, J. M., Gillespie, J. I., and Greenwell, J. R., 1993. A novel ryanodine sensitive calcium release mechanism in cultured human myometrial smooth-muscle cells, *FEBS Lett.* **330:**227–230.

Maeda, N., Niinobe, M., Nakahira, K., and Mikoshiba, K., 1988, Purification and characterization of P_{400} protein, a glycoprotein characteristic of Purkinje cell, from mouse cerebellum, *J. Neurochem.* **51:**1724–1730.

Maeda, N., Niinobe, M., Inoue, Y., and Mikoshiba, K., 1989, Developmental expression and intracellular location of P400 protein characteristic of Purkinje cells in mouse cerebellum, *Dev. Biol.* **133:**67–76.

Maeda, N., Niinobe, M., and Mikoshiba, K., 1990, A cerebellar Purkinje cell marker P_{400} protein is an inositol 1,4,5-trisphosphate ($InsP_3$) receptor protein. Purification and characterization of $InsP_3$ receptor complex, *EMBO J.* **9:**61–67.

Maeda, N., Kawasaki, T., Nakade, S., Yokota, N., Taguchi, T., Kasai, M., and Mikoshiba, K., 1991, Structural and functional characterization of inositol 1,4,5-trisphosphate receptor channel from mouse cerebellum, *J. Biol. Chem.* **266:**1109–1116.

Magnussen, A., Haug, L. S., Walaas, S. I., and Østvold, A. C., 1993, Calcium-induced degradation of the inositol (1,4,5)-trisphosphate receptor/Ca^{2+}-channel, *FEBS Lett.* **323:**229–232.

Maranto, A. R., 1994, Primary structure, ligand binding, and localization of the human type 3 inositol 1,4,5-trisphosphate receptor expressed in intestinal epithelium, *J. Biol. Chem.* **269:**1222–1230.

Marks, A. R., Tempst, P., Chadwick, C. C., Riviere, L., Fleischer, S., and Nadal-Ginard, B., 1990, Smooth muscle and brain inositol 1,4,5-trisphosphate receptors are structurally and functionally similar, *J. Biol. Chem.* **265:**20719–20722.

Marshall, I. C. B., and Taylor, C. W., 1993a, Biphasic effects of cytosolic Ca^{2+} on $InsP_3$-stimulated Ca^{2+} mobilization in hepatocytes, *J. Biol. Chem.* **268:**13214–13220.

Marshall, I. C. B., and Taylor, C. W., 1993b, Regulation of inositol 1,4,5-trisphosphate receptors, *J. Exp. Biol.* **184:**161–182.

Marshall, I. C. B., and Taylor, C. W., 1994, Two calcium-binding sites mediate the interconversion of liver inositol 1,4,5-trisphosphate receptors between three conformational states, *Biochem. J.* **301:**591–598.

Matter, N., Ritz, M. F., Freyermuth, S., Rogue, P., and Malviya, A. N., 1993, Stimulation of nuclear protein kinase C leads to phosphorylation of nuclear inositol 1,4,5-trisphosphate receptor and accelerated calcium release by inositol 1,4,5-trisphosphate from isolated rat liver nuclei, *J. Biol. Chem.* **268:**732–736.

Mauger, J.-P., Claret, M., Piétri, F., and Hilly, M., 1989, Hormonal regulation of inositol 1,4,5-trisphosphate receptor in rat liver, *J. Biol. Chem.* **264:**8821–8826.

Mauger, J.-P., Lièvremont, J.-P., Piétri-Rouxel, F., Hilly, M., and Coquil, J.-F., 1994, The inositol 1,4,5-trisphosphate receptor: Kinetic properties and regulation, *Mol. Cell. Endocrinol.* **98:**133–139.

Mayrleitner, M., Chadwick, C. C., Timmerman, A. P., Fleischer, S., and Schindler, H., 1991, Purified IP_3 receptor from smooth muscle forms an IP_3 gated and heparin sensitive Ca^{2+} channel in planar bilayers, *Cell Calcium* **12:**505–514.

Menniti, F. S., Bird, G. St. J., Takemura, H., Thastrup, O., Potter, B. V. L., and Putney, J. W., Jr., 1991, Mobilization of calcium by inositol trisphosphates from permeabilized rat parotid acinar cells. Evidence for translocation of calcium from uptake to release sites within the inositol 1,4,5-trisphosphate- and thapsigargin-sensitive calcium pool, *J. Biol. Chem.* **266:**13646–13653.

Meyer, T., and Stryer, L., 1990, Transient calcium release induced by successive increments of inositol 1,4,5-trisphosphate, *Proc. Natl. Acad. Sci. U.S.A.* **87:**3841–3845.

Meyer, T., Wensel, T., and Stryer, L., 1990, Kinetics of calcium channel opening by inositol 1,4,5-trisphosphate, *Biochemistry* **29:**32–37.

Michikawa, T., Hamanaka, H., Otsu, H., Yamamoto, A., Miyawaki, A., Furuichi, T., Tashiro, Y., and Mikoshiba, K., 1994, Transmembrane topology and sites of N-glycosylation of inositol 1,4,5-trisphosphate receptor, *J. Biol. Chem.* **269:**9184–9189.

Mignery, G. A., and Südhof, T. C., 1990, The ligand binding site and transduction mechanism in the inositol-1,4,5-trisphosphate receptor, *EMBO J.* **9:**3893–3898.

Mignery, G. A., Südhof, T. C., Takei, K., and De Camilli, P., 1989, Putative receptor for inositol 1,4,5-trisphosphate similar to the ryanodine receptor, *Nature* **342:**192–195.

Mignery, G. A., Newton, C. L., Archer, B. T. III, and Südhof, T. C., 1990, Structure and expression of the rat inositol 1,4,5-trisphosphate receptor, *J. Biol. Chem.* **265:**12679–12684.

Mignery, G. A., Johnston, P. A., and Südhof, T. C., 1992, Mechanism of Ca^{2+} inhibition of inositol 1,4,5-trisphosphate ($InsP_3$) binding to the cerebellar $InsP_3$ receptor, *J. Biol. Chem.* **267:**7450–7455.

Mikoshiba, K., 1993, Inositol 1,4,5-trisphosphate receptor, *Trends Pharmacol. Sci.* **14:**86–89.

Missiaen, L., Taylor, C. W., and Berridge, M. J., 1991, Spontaneous calcium release from inositol trisphosphate-sensitive calcium stores, *Nature* **352:**241–244.

Missiaen, L., Taylor, C. W., and Berridge, M. J., 1992a, Luminal Ca^{2+} promoting spontaneous Ca^{2+} release from inositol trisphosphate-sensitive stores in rat hepatocytes, *J. Physiol.* **455:**623–640.

Missiaen, L., De Smedt, H., Droogmans, G., and Casteels, R., 1992b, Ca^{2+} release induced by inositol 1,4,5-trisphosphate is a steady-state phenomenon controlled by luminal Ca^{2+} in permeabilized cells, *Nature* **357:**599–602.

Missiaen, L., De Smedt, H., Droogmans, G., and Casteels, R., 1992c, Luminal Ca^{2+} controls the activation of the $InsP_3$ receptor by cytosolic Ca^{2+}, *J. Biol. Chem.* **267:**22961–22966.

Missiaen, L., De Smedt, H., Parys, J. B., and Casteels, R., 1994a, Co-activation of inositol trisphosphate-induced Ca^{2+} release by cytosolic Ca^{2+} is loading-dependent, *J. Biol. Chem.* **269:**7238–7242.

Missiaen, L., Parys, J. B., De Smedt, H., Himpens, B., and Casteels, R., 1994b, Inhibition of inositol trisphosphate-induced calcium release by caffeine is prevented by ATP, *Biochem. J.* **300:**81–84.

Missiaen, L., Parys, J. B., De Smedt, H., Oike, M., and Casteels, R., 1994c, Partial calcium release in response to submaximal inositol 1,4,5-trisphosphate receptor activation, *Mol. Cell. Endocrinol.* **98:**147–156.

Miyawaki, A., Furuichi, T., Maeda, N., and Mikoshiba, K., 1990, Expressed cerebellar-type inositol 1,4,5-trisphosphate receptor, P_{400}, has calcium release activity in a fibroblast L cell line, *Neuron* **5:**11–18.

Miyawaki, A., Furuichi, T., Ryou, Y., Yoshikawa, S., Nakagawa, T., Saitoh, T., and Mikoshiba, K., 1991, Structure–function relationships of the mouse inositol 1,4,5-trisphosphate receptor, *Proc. Natl. Acad. Sci. U.S.A.* **88:**4911–4915.

Miyazaki, S.-I., Shirakawa, H., Nakada, K., Honda, Y., Yuzaki, M., Nakade, S., and Mikoshiba, K., 1992, Antibody to the inositol trisphosphate receptor blocks thimerosal-enhanced Ca^{2+}-induced Ca^{2+} release and Ca^{2+} oscillations in hamster eggs, *FEBS Lett.* **309:**180–184.

Mohr, F. C., Hershey, P. E. C., Zimányi, I., and Pessah, I. N., 1993, Regulation of inositol 1,4,5-trisphosphate receptors in rat basophilic leukemia cells. I. Multiple conformational states of the receptor in a microsomal preparation, *Biochim. Biophys. Acta* **1147:**105–114.

Moschella, M. C., and Marks, A. R., 1993, Inositol 1,4,5-trisphosphate receptor expression in cardiac myocytes, *J. Cell Biol.* **120:**1137–1146.

Mourey, R. J., Verma, A., Supattapone, S., and Snyder, S. H., 1990, Purification and characterization of the inositol 1,4,5-trisphosphate receptor protein from vas deferens, *Biochem. J.* **272:**383–389.

Mourey, R. J., Estevez, V. A. Marecek, J. F., Barrow, R. K., Prestwich, G. D., and Snyder, S. H., 1993, Inositol 1,4,5-trisphosphate receptors: Labeling the inositol 1,4,5-trisphosphate binding site with photoaffinity ligands, *Biochemistry* **32:**1719–1726.

Muallem, S., Pandol, S. J., and Beeker, T. G., 1989, Hormone-evoked calcium release from intracellular stores is a quantal process, *J. Biol. Chem.* **264:**205–212.

Nakade, S., Maeda, N., and Mikoshiba, K., 1991, Involvement of the C-terminus of the inositol 1,4,5-trisphosphate receptor in Ca^{2+} release analysed using region-specific monoclonal antibodies, *Biochem. J.* **277:**125–131.

Nakade, S., Rhee, S. K., Hamanaka, H., and Mikoshiba, K., 1994, Cyclic AMP-dependent phosphorylation of an immunoaffinity-purified homotetrameric inositol 1,4,5-trisphosphate receptor (type I) increases Ca^{2+} flux in reconstituted lipid vesicles, *J. Biol. Chem.* **269:**6735–6742.

Nakagawa, T., Okano, H., Furuichi, T., Aruga, J., and Mikoshiba, K., 1991a, The subtypes of the mouse inositol 1,4,5-trisphosphate receptor are expressed in a tissue-specific and developmentally specific manner, *Proc. Natl. Acad. Sci. U.S.A.* **88:**6244–6248.

Nakagawa, T., Shiota, C., Okano, H., and Mikoshiba, K., 1991b, Differential localization of alternative spliced transcripts encoding inositol 1,4,5-trisphosphate receptors in mouse cerebellum and hippocampus: *In situ* hybridization study, *J. Neurochem.* **57:**1807–1810.

Nathanson, M. H., Fallon, M. B., Padfield, P. J., and Maranto, A. R., 1994, Localization of the type 3 inositol 1,4,5-trisphosphate receptor in the Ca^{2+} wave trigger zone of pancreatic acinar cells, *J. Biol. Chem.* **269:**4693–4696.

Nunn, D. L., and Taylor, C. W., 1990, Liver inositol 1,4,5-trisphosphate-binding sites are the Ca^{2+}-mobilizing receptors, *Biochem J.* **270:**227–232.

Nunn, D. L., and Taylor, C. W., 1992, Luminal Ca^{2+} increases the sensitivity of Ca^{2+} stores to inositol 1,4,5-trisphosphate, *Mol. Pharmacol.* **41:**115–119.

Oldershaw, K. A., and Taylor, C. W., 1993, Luminal Ca^{2+} increases the affinity of inositol 1,4,5-trisphosphate for its receptor, *Biochem. J.* **292:**631–633.

Oldershaw, K. A., Nunn, D. L., and Taylor, C. W., 1991, Quantal Ca^{2+} mobilization stimulated by inositol 1,4,5-trisphosphate in permeabilized hepatocytes, *Biochem. J.* **278:**705–708.

Oldershaw, K. A., Richardson, A., and Taylor, C. W., 1992, Prolonged exposure to inositol 1,4,5-trisphosphate does not cause intrinsic desensitization of the intracellular Ca^{2+}-mobilizing receptor, *J. Biol. Chem.* **267:**16312–16316.

Ospichuk, Y. V., Wakui, M., Yule, D. I., Gallacher, D. V., and Petersen, O. H., 1990, Cytoplasmic Ca^{2+} oscillations evoked by receptor stimulation, G-protein activation, internal application of inositol trisphosphate or Ca^{2+}: Simultaneous microfluorimetry and Ca^{2+} dependent Cl^- current recording in single pancreatic acinar cells, *EMBO J.* **9:**697–704.

Ozaki, S., Watanabe, Y., Ogasawara, T., Hirata, M., and Kanamatsu, M., 1992, Synthesis and biological properties of 2-substituted *myo*-inositol 1,4,5-trisphosphate analogues directing toward affinity chromatography and photoaffinity labeling, *Carbohydr. Res.* **234:** 189–206.

Parker, I., and Ivorra, I., 1990a, Inhibition by Ca^{2+} of inositol trisphosphate-mediated Ca^{2+} liberation: A possible mechanism for oscillatory release of Ca^{2+}, *Proc. Natl. Acad. Sci. U.S.A.* **87:**260–264.

Parker, I., and Ivorra, I., 1990b, Localized all-or-none calcium liberation by inositol trisphosphate, *Science* **250:**977–979.

Parker, I., and Ivorra, I., 1991, Caffeine inhibits inositol trisphosphate-mediated liberation of intracellular calcium in *Xenopus* oocytes, *J. Physiol.* **433:**229–240.

Parker, I., and Yao, Y., 1991, Regenerative release of calcium from functionally discrete subcellular stores by inositol trisphosphate, *Proc. R. Soc. Lond. [Biol.]* **246:**269–274.

Parys, J. B., Sernett, S. W., DeLisle, S., Snyder, P. M., Welsh, M. J., and Campbell, K. P., 1992, Isolation, characterization, and localization of the inositol 1,4,5-trisphosphate receptor protein in *Xenopus laevis* oocytes, *J. Biol. Chem.* **267:**18776–18782.

Parys, J. B., Missiaen, L., De Smedt, H., and Casteels, R., 1993a, Loading dependence of inositol 1,4,5-trisphosphate-induced Ca^{2+} release in the clonal cell line A7r5. Implications for the mechanism of quantal Ca^{2+} release, *J. Biol. Chem.* **268:**25206–25212.

Parys, J. B., Missiaen, L., De Smedt, H., Droogmans, G., and Casteels, R., 1993b, Bell-shaped activation of inositol 1,4,5-trisphosphate-induced Ca^{2+} release by thimerosal in permeabilized A7r5 smooth-muscle cells, *Pflugers Arch.* **424:**516–522.

Parys, J. B., McPherson, S. M., Mathews, L., Campbell, K. P., and Longo, F., 1994, Presence of inositol 1,4,5-trisphosphate receptor, calreticulin, and calsequestrin in eggs of sea urchins and *Xenopus laevis, Dev. Biol.* **161:**466–476.

Parys, J. B., De Smedt, H., Missiaen, L., Bootman, M. D., Sienaert, I., and Casteels, R., 1995, Rat basophilic leukemia cells as model system for inositol 1,4,5-trisphosphate receptor IV, a receptor of the type II family: functional comparison and immunological detection, *Cell Calcium* **17:**239–249.

Peng, Y.-W., Sharp, A. H., Snyder, S. H., and Yau, K.-W., 1991, Localization of the inositol 1,4,5-trisphosphate receptor in synaptic terminal in the vertebrate retina, *Neuron* **6:**525–531.

Petersen, O. H., 1992, Stimulus–secretion coupling: Cytoplasmic calcium signals and the control of ion channels in exocrine acinar cells, *J. Physiol.* **448:**1–51.

Piétri, F., Hilly, M., and Mauger, J. P., 1990, Calcium mediates the interconversion between two states of the liver inositol 1,4,5-trisphosphate receptor. *J. Biol. Chem.* **265:**17478–17485.

Piétri Rouxel, F., Hilly, M., and Mauger, J. P., 1992 Characterization of a rapidly dissociating inositol 1,4,5-trisphosphate-binding site in liver membranes, *J. Biol. Chem.* **267:**20017–20023.

Poitras, M., Bernier, S., Servant, M., Richard, D. E., Boulay, G., and Guillemette, G., 1993, The high affinity state of inositol 1,4,5-trisphosphate receptor is a functional state, *J. Biol. Chem.* **268:**24078–24082.

Renard, D. C., Seitz, M. B., and Thomas, A. P., 1992, Oxidized glutathione causes sensitization of calcium release to inositol 1,4,5-trisphosphate in permeabilized hepatocytes, *Biochem. J.* **284:**507–512.

Richardson, A., and Taylor, C. W., 1993, Effects of Ca^{2+} chelators on purified inositol 1,4,5-trisphosphate ($InsP_3$) receptors and $InsP_3$-stimulated Ca^{2+} mobilization. *J. Biol. Chem.* **268:**11528–11533.

Rooney, T. A., Renard, D. C., Sass, E. J., and Thomas, A. P., 1991, Oscillatory cytosolic calcium waves independent of stimulated inositol 1,4,5-trisphosphate formation in hepatocytes, *J. Biol. Chem.* **266:**12272–12282.

Ross, C. A., Meldolesi, J., Milner, T. A., Satoh, T., Supattapone, S., and Snyder, S. H., 1989, Inositol 1,4,5-trisphosphate receptor localized to endoplasmic reticulum in cerebellar Purkinje neurons, *Nature* **339:**468–470.

Ross, C. A., Danoff, S. K., Schell, M. J., Snyder, S. H., and Ullrich, A., 1992, Three additional inositol 1,4,5-trisphosphate receptors: Molecular cloning and differential localization in brain and peripheral tissues, *Proc. Natl. Acad. Sci. U.S.A.* **89:**4265–4269.

Rossier, M. F., Bird, G. St. J., and Putney, J. W., Jr., 1991, Subcellular distribution of the calcium-storing inositol 1,4,5-trisphosphate-sensitive organelle in rat liver. Possible linkage to the plasma membrane through the actin microfilaments, *Biochem. J.* **274:**643–650.

Sanchez-Bueno, A., Marrero, I., and Cobbold, P. H., 1994, Caffeine inhibits agonist-induced cytoplasmic Ca^{2+} oscillations in single rat hepatocytes, *Biochem. Biophys. Res. Commun.* **198:**728–733.

Sayers, L. G., Brown, G. R., Michell, R. H., and Michelangeli, F., 1993, The effects of thimerosal on calcium uptake and inositol 1,4,5-triphosphate-induced calcium release in cerebellar microsomes, *Biochem. J.* **289:**883–887.

Schell, M. J., Danoff, S. K., and Ross, C. A., 1993, Inositol (1,4,5)-trisphosphate receptor: Characterization of neuron-specific alternative splicing in rat brain and peripheral tissues, *Mol. Brain Res.* **17:**212–216.

Sharp, A. H., Snyder, S. H., and Nigam, S. K., 1992, Inositol 1,4,5-trisphosphate receptors. Localization in epithelial tissues, *J. Biol. Chem.* **267:**7444–7449.

Sharp, A. H., Dawson, T. M., Ross, C. A., Fotuhi, M., Mourey, R. J., and Snyder, S. H., 1993, Inositol 1,4,5-trisphosphate receptors: Immunohistochemical localization to discrete areas of rat central nervous system, *Neuroscience* **53:**927–942.

Short, A. D., Klein, M. G., Schmeider, M. F., and Gill, D. L., 1993, Inositol 1,4,5-trisphosphate-mediated quantal Ca^{2+} release measured by high resolution imaging of Ca^{2+} within organelles, *J. Biol. Chem.* **268:**25887–25893.

Shuttleworth, T. S., 1992, Ca^{2+} release from inositol trisphosphate-sensitive stores is not modulated by intraluminal [Ca^{2+}]. *J. Biol. Chem.* **267:**3573–3576.

Sitsapesan, R., and Williams, A. J., 1994, Regulation of the gating of the sheep cardiac sarcoplasmic reticulum Ca^{2+}-release channel by luminal Ca^{2+}, *J. Membr. Biol.* **137:**215–226.

Spät, A., Bradford, P. G., McKinney, J. S., Rubin, R. P., and Putney, J. W., Jr., 1986, A saturable receptor for ^{32}P-inositol-1,4,5-trisphosphate in hepatocytes and neutrophils, *Nature* **319:**514–516.

Spät, A., Eberhardt, I., and Kiesel, L., 1992, Low concentrations of adenine nucleotides enhance the receptor binding of inositol 1,4,5-trisphosphate, *Biochem. J.* **287:**335–336.

Südhof, T. C., Newton, C. L., Archer, B. T. III, Ushkaryov, Y. A., and Mignery, G. A., 1991, Structure of a novel $InsP_3$ receptor, *EMBO J.* **10:**3199–3206.
Sugiyama, T., Yamamoto-Hino, M., Miyawaki, A., Furuichi, T., Mikoshiba, K., and Hasegawa, M., 1994, Subtypes of inositol 1,4,5-trisphosphate receptor in human hematopoietic cell lines: Dynamic aspects of their cell-type specific expression, *FEBS Lett.* **349:**191–196.
Supattapone, S., Worley, P. F., Baraban, J. M., and Snyder, S. H., 1988, Solubilization, purification, and characterization of an inositol trisphosphate receptor, *J. Biol. Chem.* **263:**1530–1534.
Swann, K., 1991, Thimerosal causes calcium oscillations and sensitizes calcium-induced calcium release in unfertilized hamster eggs, *FEBS Lett.* **278:**175–178.
Takahashi, M., Tanzawa, K., and Takahashi, S., 1994, Adenophostins, newly discovered metabolites of *Penicillium brevicompactum,* act as potent agonists of the inositol 1,4,5-trisphosphate receptor, *J. Biol. Chem.* **269:**369–372.
Takei, K., Mignery, G. A., Mugnaini, E., Südhof, T. C., and De Camilli, P., 1994, Inositol 1,4,5-trisphosphate receptor causes formation of ER cisternal stacks in transfected fibroblasts and in cerebellar Purkinje cells, *Neuron* **12:**327–342.
Tanaka, Y., and Tashjian, A. H., Jr., 1994, Thimerosal potentiates Ca^{2+} release mediated by both the inositol 1,4,5-trisphosphate and the ryanodine receptors in sea urchin eggs. Implications for mechanistic studies on Ca^{2+} signaling, *J. Biol. Chem.* **269:**11247–11253.
Tasaka, K., Mio, M., Akagi, M., Fujisawa, K., and Aoki, I., 1991, Role of the cytoskeleton in Ca^{2+} release from the intracellular Ca store of rat peritoneal mast cells, *Agents Actions* **33:**44–47.
Taylor, C. W., and Potter, B. V. L., 1990, The size of inositol 1,4,5-trisphosphate-sensitive Ca^{2+} stores depends on inositol 1,4,5-trisphosphate concentration, *Biochem. J.* **266:**189–194.
Taylor, C. W., and Richardson, A., 1991, Structure and function of inositol trisphosphate receptors, *Pharmacol. Ther.* **51:**97–137.
Thorn, P., Brady, P., Llopis, J., Gallacher, D. V., and Petersen, O. H., 1992, Cytosolic Ca^{2+} spikes evoked by the thiol reagent thimerosal in both intact and internally perfused single pancreatic acinar cells, *Pflugers Arch.* **422:**173–178.
Thorn, P., Lawrie, A. M., Smith, P. M., Gallacher, D. V., and Petersen, O. H., 1993, Local and global cytosolic Ca^{2+} oscillations in exocrine cells evoked by agonists and inositol trisphosphate, *Cell* **74:**661–668.
Toescu, E. C., O'Neill, S. C., Petersen, O. H., and Eisner, D. A., 1992, Caffeine inhibits the agonist-evoked cytosolic Ca^{2+} signal in mouse pancreatic acinar cells by blocking inositol trisphosphate production, *J. Biol. Chem.* **267:**23467–23470.
Tregear, R., Dawson, A. P., and Irvine, R. F., 1991, Quantal release of Ca^{2+} from intracellular stores by $InsP_3$: Tests of the concept of control of Ca^{2+} release by intraluminal Ca^{2+}, *Proc. R. Soc. Lond. [Biol.]* **243:**263–268.
Tshipamba, M., De Smedt, H., Missiaen, L., Himpens, B., Van Den Bosch, L., and Borghgraef, R., 1993, Ca^{2+} dependence of inositol 1,4,5-trisphosphate-induced Ca^{2+} release in renal epithelial LLC-PK_1 cells, *J. Cell. Physiol.* **155:**96–103.
Tsukioka, M., Iino, M., and Endo, M., 1994, pH dependence of inositol 1,4,5-trisphosphate-induced Ca^{2+} release in permeabilized smooth muscle cells of the guinea-pig, *J. Physiol.* **475:**369–375.
Tsunada, Y., 1986, An ATP-dependent and inositol trisphosphate-sensitive Ca^{2+} pool linked with microfilaments of the parietal cell, *FEBS Lett.* **207:**47–52.
Van Delden, C., Foti, M., Lew, D. P., and Krause, K. H., 1993, Ca^{2+} and Mg^{2+} regulation of inositol 1,4,5-trisphosphate binding in myeloid cells, *J. Biol. Chem.* **268:**12443–12448.
Varney, M. A., Rivera, J., Lopez Bernal, A., and Watson, S. P., 1990, Are there subtypes of the inositol 1,4,5-trisphosphate receptor? *Biochem. J.* **269:**211–216.
Volpe, P., and Alderson-Lang, B. H., 1990, Regulation of inositol 1,4,5-trisphosphate-induced Ca^{2+} release II. Effect of cAMP-dependent protein kinase, *Am. J. Physiol.* **258:**C1086–C1091.

Wagenknecht, T., Grassucci, R., Frank, J., Saito, A., Inui, M., and Fleischer, S., 1989, Three-dimensional architecture of the calcium channel/foot structure of sarcoplasmic reticulum, *Nature* **338:**167–170.

Wakui, M., Potter, B. V. L., and Petersen, O. H., 1989, Pulsatile intracellular calcium release does not depend on fluctuations in inositol trisphosphate concentration, *Nature* **339:**317–320.

Wakui, M. Osipchuk, Y. V., and Petersen, O. H., 1990, Receptor-activated cytoplasmic Ca^{2+} spiking mediated by inositol trisphosphate is due to Ca^{2+}-induced Ca^{2+} release, *Cell* **63:**1025–1032.

White, A. M., Varney, M. A., Watson, S. P., Rigby, S., Changsheng, L., Ward, J. G., Reese, C. B., Graham, H. C., and Williams, R. J. P., 1991, Influence of Mg^{2+} and pH on n.m.r. spectra and radioligand binding of inositol 1,4,5-trisphosphate, *Biochem. J.* **278:**759–764.

White, A. M., Varney, M. A., Maeda, M., Mikoshiba, K., and Watson, S. P., 1993, Comparison of Ins(1,4,5)P_3 receptors from rat cerebellum and bovine adrenal cortex, *Biochim. Biophys. Acta* **1175:**307–311.

Wilcox, R. A., Whitham, E. M., Liu, C., Potter, B. V. L., and Nahorski, S. R., 1993, *Myo*-inositol 1,3,4,5-tetrakisphosphate can independently mobilise intracellular calcium, via the inositol 1,4,5-trisphosphate receptor: Studies with *myo*-inositol 1,4,5-trisphosphate-3-phosphorothioate and *myo*-inositol hexakisphosphate, *FEBS Lett.* **336:**267–271.

Wilcox, R. A., Safrany, S. T., Lampe, D., Mills, S. J., Nahorski, S. R., and Potter, B. V. L., 1994, Modification at C2 of *myo*-inositol 1,4,5-trisphosphate produces inositol trisphosphates and tetrakisphosphates with potent biological activities, *Eur. J. Biochem.* **223:**115–124.

Wojcikiewicz, R. J. H., Furuichi, T., Nakade, S., Mikoshiba, K., and Nahorski, S. R., 1994, Muscarinic receptor activation down-regulates the type I inositol 1,4,5-trisphosphate receptor by accelerating its degradation, *J. Biol. Chem.* **269:**7963–7969.

Woods, N. M., Cuthbertson, K. S. R., and Cobbold, P. H., 1986, Repetitive transient rises in cytoplasmic free calcium in hormone-stimulated hepatocytes, *Nature* **319:**600–602.

Worley, P. F., Baraban, J. M., Supattapone, S., Wilson, V. S., and Snyder, S. H., 1987, Characterization of inositol trisphosphate receptor binding in brain. Regulation by pH and calcium, *J. Biol. Chem.* **262:**12132–12136.

Yamada, N., Makino, Y., Clark, R. A., Pearson, D. W., Mattel, M.-G., Guénet, J.-L., Ohama, E., Fujino, I., Miyawaki, A., Furuichi, T., and Mikoshiba, K., 1994, Human inositol 1,4,5-trisphosphate type-1 receptor, $InsP_3R1$: Structure, function, regulation of expression and chromosomal localization, *Biochem. J.* **302:**781–790.

Yamamoto-Hino, M., Sugiyama, T., Hikichi, K., Mattei, M. G., Hasegawa, K., Sekine, S., Sakurada, K., Miyawaki, A., Furuichi, T., Hasegawa, M., and Mikoshiba, K., 1994, Cloning and characterization of human type 2 and type 3 inositol 1,4,5-trisphosphate receptors, *Receptors Channels* **2:**9–22.

Yao, Y., and Parker, I., 1992, Potentiation of inositol trisphosphate-induced Ca^{2+} mobilization in *Xenopus* oocytes by cytosolic Ca^{2+}, *J. Physiol.* **458:**319–338.

Yoo, S. H., 1994, pH-dependent interaction of chromogranin A with integral membrane proteins of secretory vesicle including 260-kDa protein reactive to inositol 1,4,5-trisphosphate receptor antibody, *J. Biol. Chem.* **269:**12001–12006.

Yoo, S. H., and Lewis, M. S., 1994, pH-dependent interaction of an intraluminal loop of inositol 1,4,5-trisphosphate receptor with chromogranin A, *FEBS Lett.* **341:**28–32.

Yoshikawa, M., Tanimura, T., Miyawaki, A., Nakamura, M., Yuzaki, M., Furuichi, T., and Mikoshiba, K., 1992, Molecular cloning and characterization of the inositol 1,4,5-trisphosphate receptor in *Drosophila melanogaster, J. Biol. Chem.* **267:**16613–16619.

Zhang, B. X., and Muallem, S., 1992, Feedback inhibition of Ca^{2+} release by Ca^{2+} is the underlying mechanism of agonist-evoked intracellular Ca^{2+} oscillations in pancreatic acinar cells, *J. Biol. Chem.* **267:**24387–24393.

Zhang, B. X., Zhao, H., and Muallem, S., 1993, Ca^{2+}-dependent kinase and phosphatase control inositol 1,4,5-trisphosphate-mediated Ca^{2+} release. Modification by agonist stimulation, *J. Biol. Chem.* **268:**10997–11001.

Zhao, H., Khademazad, M., and Muallem, S., 1990a, Agonist-mediated Ca^{2+} release in permeabilized UMR-106-01 cells. Transport properties and generation of inositol 1,4,5-trisphosphate, *J. Biol. Chem.* **265:**14822–14827.

Zhao, H., Loessberg, P. A., Sachs, G., and Muallem, S., 1990b, Regulation of intracellular Ca^{2+} oscillations in AR42J cells, *J. Biol. Chem.* **265:**20856–20862.

Chapter 4

Regulation of the Actin Cytoskeleton by Inositol Phospholipid Pathways

David E. Kandzari, Jie Chen,
and Pascal J. Goldschmidt-Clermont

1. INTRODUCTION

Changes in the actin cytoskeleton are a fundamental part of the motile response of cells to extracellular ligands. Dynamic alterations in both the structural and biochemical properties of the actin network mediate changes in cell shape and migration (Cooper, 1991; Gips *et al.*, 1994). This meshwork of microfilaments is responsible for many characteristic intracellular structures, including stress fibers, the cortical network, and contractile rings. Actin filament turnover, for example, is associated with the formation of microspikes and lamellipodia at the leading edge of motile cells. In migrating cells, the forward extension of the cellular lamellipodium, a fan-shaped structure filled with a dense, cross-linked meshwork of actin filaments, is driven by the polarized assembly of actin monomers. This assembly is not unique to the lamellipodium, however, because it shares features with other actin-based motile processes occurring beneath the plasma membrane of stimulated platelets, neutrophils, chemotactic amoebas, or

---

David E. Kandzari, Jie Chen, and Pascal J. Goldschmidt-Clermont Bernard Laboratory, Division of Cardiology, Department of Medicine and Department of Cell Biology and Anatomy, The Johns Hopkins University School of Medicine, Baltimore, Maryland 21287.

Subcellular Biochemistry, Volume 26: myo-Inositol Phosphates, Phosphoinositides, and Signal Transduction, edited by B. B. Biswas and Susweta Biswas. Plenum Press, New York, 1996.

even at the rear of the pathogenic bacterium *Listeria monocytogenes* during its propulsive movement through the host cell cytoplasm (Theriot *et al.*, 1994).

Many questions regarding actin-based motility remain unanswered. For example, what are the biochemical steps connecting the signal at the plasma membrane to the assembly of actin filaments into an integrated and well-orchestrated motile response? Specifically, for this review on the role of inositol phospholipids in this process, how metabolically active membrane phospholipids affect the organization of the actin cytoskeleton remains unknown. In spite of much progress in the study of actin biochemistry *in vitro*, it remains a challenge for cell biologists to integrate these findings in the dynamic function of actin in living, motile cells. Surely, several regulatory mechanisms must exist to mediate responses occurring over a wide range of time scales because filamentous actin assembly occurs within seconds of growth factor stimulation, whereas minutes to hours are required for formation of actin-association focal adhesions to the substratum (Cooper, 1991; Gips *et al.*, 1994).

Although the assembly of actin and actin-binding proteins can occur in reconstituted systems in the absence of membranes, membrane lipids are necessary to mediate the cytoarchitectural changes orchestrated by surface receptors such as the receptor tyrosine kinases (RTK). Most likely, membranes surrounding intracellular organelles and vesicles contribute with the plasma membrane in this dynamic signaling process to the cytoskeleton. Filaments of actin assemble by the addition of ATP-bound monomers (G actin) primarily to the barbed end of the filament, at or near the leading edge of the cell, followed by transport of some actin filaments in retrograde fashion toward the nucleus. Although nucleation of filaments takes place primarily at the advancing edge of motile cells, and most filaments are stationary, addition of actin monomers also occurs within the cytoplasm to the barbed end of "released" filaments, thus creating a gradient of polymerization from the cell edge to the nucleus (Theriot and Mitchison, 1992). Such a gradient can be modulated by cell activation.

Growth factors bind to membrane receptors, whose tyrosine kinase activity affects several signaling pathways to culminate in substantial reorganization of the actin cytoskeleton within seconds (Schlessinger and Ullrich, 1992; Fantl *et al.*, 1993). Growth factor stimulation induces dramatic redistribution of the filamentous actin as central stress fibers disappear and filaments accumulate in the cell periphery at sites of membrane ruffling. On binding to their cell surface ligands, the growth factor receptors, which contain a tyrosine kinase located within their cytoplasmic domain, oligomerize, a process leading to the clustering of effector proteins within functional units beneath the plasma membrane. The clustering occurs as a result of binding of modular domains called *src* homology domain 2 (SH2) displayed by effector proteins to tyrosine-phosphorylated residues of the receptor itself and clustered ligands (Schlessinger and Ullrich, 1992; Fantl *et al.*, 1993; Carpenter, 1992). On reaching a critical mass, RTK effector units send out signals through the metabolism of membrane phospholipids and other second messenger molecules that regulate the interaction of actin with its satellite regulatory proteins. Several actin-binding proteins interact transiently with small clusters of membrane inositol phospholipids, a process that can either up-regulate or down-regulate the interaction of these proteins with actin (Gips *et al.*, 1994; Janmey, 1994). The phospholipids involved in the control of the actin cytoskeleton are metabolically very active and abundant in cells and are therefore poised to mediate reactions linking signal transduction molecules to the reorganization of the actin cytoskeleton on cell activation (Goldschmidt-Clermont and Janmey, 1991).

The focus of this review is to integrate recent information obtained on the metabolism of membrane phospholipids as it relates to the localization and activity of actin regulatory proteins.

2. EFFECT OF MEMBRANES ON ACTIN AND SATELLITE PROTEINS

Recently, a key mechanism has been discovered that explains how activated RTK can elicit substantial changes in cell activity in spite of their limited kinase response on binding an extracellular ligand. It turns out that the clustering of effector proteins represents a fundamental process of amplification in signal transduction (Figure 1). In actuality, the tyrosine kinase reaction generated by RTK serves as the anchoring mechanism to promote binding of key effector molecules containing one or several *src* homology 2 (SH2) domain(s) to the activated RTK. Several enzymes or adaptor proteins regulating biochemical pathways contain such SH2 domains, including phospholipase $C_{\gamma1}$ ($PLC_{\gamma1}$) and $C_{\gamma2}$, GTPase-activating protein (GAP), ash, and cytoplasmic tyrosine kinases (Schlessinger and Ullrich, 1992; Fantl *et al.*, 1993; Carpenter, 1992). SH2 domains form a pocket that binds to the phosphate group of tyrosine-phosphorylated residues specifically, as the pocket is too deep to allow binding to other phosphorylated residues (e.g., phosphorylated serine or threonine) (Neel, 1993). The binding site also recognizes the three residues immediately C terminal to this phosphorylated tyrosine. A similar mechanism of specificity exists for the catalytic sites of protein tyrosine phosphatases (PTP) and their phosphotyrosine substrate. Once a critical mass of these effector signaling molecules is achieved, the signaling unit triggers a response whose potency is determined by the molecular size of the effector unit, the density of these units at the plasma membrane, and existing cellular conditions such as the presence of tyrosine phosphatases.

This clustering mechanism of signal transduction is exemplary of multiple cellular processes. For example, within focal adhesions, transmembrane inte-

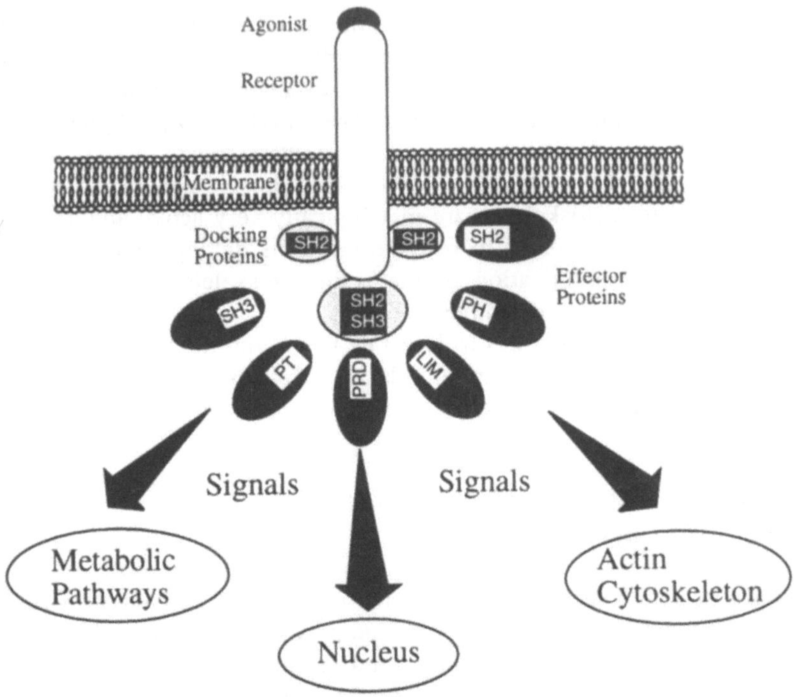

FIGURE 1. Membrane receptor signaling through clustering of effector molecules. On binding to its extracellular activating ligand, a generic membrane receptor undergoes posttranslational modification (such as tyrosine autophosphorylation), resulting in the display of binding sites for modular domains expressed by docking proteins (such as GRB2 or the p85 subunit of PI3-kinase). Depending on the receptor involved (receptor tyrosine kinases, integrins, and others), the modular domains can correspond to SH2, SH3, phosphotyrosine residues (PT), LIM (Schmeichel and Beckerle, 1994; Arber *et al.*, 1994), proline-rich domains (PRD), pleckstrin homology domain (PH), and others, whose role is to target signaling molecules to strategic areas of the cell (such as an activated receptor). For clarity of the diagram, only one or two modular domains are shown per docking molecule, while it is known that docking proteins often display multiple modular domains. Once attached to the receptor molecule, the docking proteins promote further clustering of various effector signaling enzymes and other proteins, such as kinases, phosphatases, GTPase activating proteins, GTP-exchanger proteins, GTPase, and cytoskeletal proteins. Once a critical concentration of these signaling molecules is reached, a wave of signals is generated targeted to specific cellular systems, such as the transcription machinery or the actin cytoskeleton. It is likely that membrane phospholipids contribute to the signaling cascade by binding to signaling molecules and thereby modulating their concentration near the signaling epicenter and/or altering their conformation. Phospholipid by-products also contribute to the generation of second messengers by the signaling unit.

grins, cytoskeletal proteins, and cytoplasmic tyrosine kinases appear to be organized in a similar fashion and involve the binding of specific modular protein domains (Figure 1). Clustering of phospholipids is also necessary to provide the binding sites for actin-binding proteins such as profilin (about five phospholipid molecules per profilin molecule) (Goldschmidt-Clermont *et al.*, 1990) or gel-

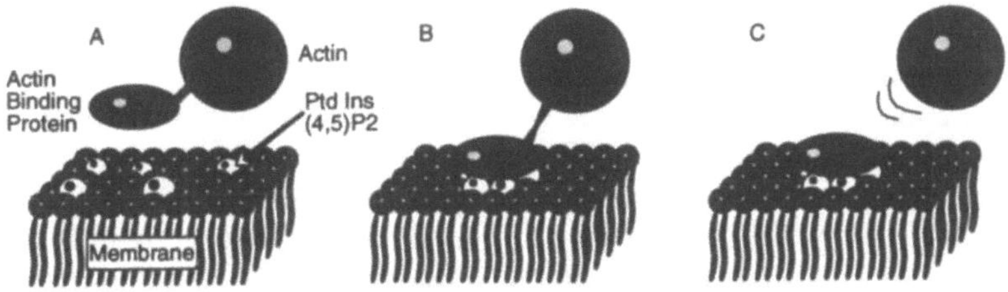

FIGURE 2. Polyphosphoinositide-mediated surface catalysis for proteins of the cytoskeleton. Several actin binding proteins have been shown to interact with membrane polyphosphoinositides (Stossel, 1993) which function towards the actin cytoskeleton as surface catalysts, facilitating reactions by lowering the energy required for reactions to occur. In the absence of actin binding proteins, the polyphosphoinositides (solid circles) are scattered within the membrane bilayer. The head groups of the polyphosphoinositides can be clustered at the interface between membrane and cytoplasm on interacting with actin binding proteins. Binding to these clusters destabilizes (or enhances in the case of α-actinin) the interaction between the regulatory proteins and actin. Interaction between the polyphosphoinositides and actin binding proteins is usually transient, and, therefore, the actin binding proteins cycle back to the cytoplasm and are free to interact with other actin molecules (Finkel *et al.*, 1994). Although the polyphosphoinositides are not metabolized as a result of interaction with the actin binding proteins, they represent substrates for specific kinases, phosphatases, and phospholipases, and their availability for these enzymes can be altered by interaction with actin binding proteins.

solin (about ten phospholipid molecules per gelsolin) (Janmey and Stossel, 1989) or even downstream signaling molecules such as protein kinases C (one diacylglycerol and four phosphatidylserine molecules) (Mosior and Epand, 1994). In turn, the presence of binding proteins promotes the aggregation of phospholipids. When phospholipid clusters interact with an actin satellite protein, the free energy required to dissociate the regulatory protein from actin decreases, allowing the actin-binding protein to participate in additional regulatory reactions of the cytoskeleton (Figure 2). Alternatively, binding to a phospholipid cluster can also promote the interaction between a regulatory protein and actin, as is the case for α-actinin (Fukami *et al.*, 1993). Such catalysis by membrane inositol phospholipids of reactions involving proteins of the actin cytoskeleton has been compared to a process of surface catalysis (Crawford *et al.*, 1993; Friend, 1993).

3. ras AND THE CYTOSKELETON

One of the signaling pathways affected by growth factor stimulation is that of the p21ras (ras) family of small GTP-binding proteins (Schlessinger and Ullrich, 1992; Fantl *e al.*, 1993). That proteins of the ras family are involved in the assembly of the actin cytoskeleton was expected, because many small GTP-binding proteins of the ras family are known to regulate the assembly of other amultisubunit protein and nucleic acid complexes in cells (Bourne *et al.*, 1990, 1991). Like actin (Goldschmidt-Clermont *et al.*, 1992) ras binds its respective nucleotide (adenine nucleotide for actin, guanine nucleotide for ras) with high affinity and a K_d well below the intracellular nucleotide concentration. As ras proteins cycle between the active, growth-promoting GTP-bound form and the inactive GDP-bound state, ras modulation by RTK corresponds to a dynamic state controlled by two antagonistic systems (Schlessinger and Ullrich, 1992; Fantl *et al.*, 1993). The RTK acts as a promoter of ras function by influencing the association of ras with GTP through both proteins controlling the exchange of the nucleotide bound to ras and proteins limiting the GTPase hydrolytic capacity of the active GTP-bound form. In contrast with this promoter effect, RTK acts also as an inhibitor of ras function. For example, proteins such as the GTPase-activating proteins p120GAP and neurofibromin act as molecular switches activated by RTK to down-regulate the function of the small GTP-binding proteins by increasing GTP hydrolysis and thereby promoting GDP binding, producing the inactive form of ras. Interestingly, other GAPs unrelated in sequence to rasGAP, such as rhoGAP, share structurally related domains with several other signaling proteins including the product of the break-point cluster region gene and the p120GAP target candidate, tyrosine-phosphorylated p190 (Ridley *et al.*, 1993). Microinjection of isolated domains of p190GAP inhibit rho-mediated stress fiber formation, whereas bcrGAP microinjection specifically inhibits rac-induced membrane ruffling (Ridley *et al.*, 1993.

Recently, information regarding the mechanism by which RTKs control the actin cytoskeleton provided insight into the role played by a family of small GTP-binding proteins in the biology of the cell. A key to intracellular signaling mechanisms is the identification of molecules that link RTK phosphorylation and small GTP-binding proteins. Like most substrates for these RTK, one highly conserved 25 to 28-kDa protein named ash (also GRB2/sem5) interacts with the epidermal growth factor and platelet-derived growth factor receptors through the binding of an SH2 domain to a tyrosine residue on the cytoplasmic aspect of the receptor, a site phosphorylated by the tyrosine kinase of the receptor itself (Lowenstein *et al.*, 1992). Ash functions as a protein coupler between the RTK and the son of sevenless (sos), a family of exchanger proteins, which markedly increases the off-rate of the nucleotide bound to ras (Simon *et al.*, 1991). The sos subsequently unlocks ras from its GDP-bound inactive state (in the presence of a large excess of GTP over GDP, as is usually in the case in normal cells) (Bourne *et al.*, 1990, 1991). Additionally, ash links receptor-associated phosphorylation events to two other members of the ras family, rac and rho, which control changes in the reorganization of the actin cytoskeleton by stabilization of actin filaments within membrane rufflers or stress fibers, respectively (Goldschmidt-Clermont *et al.*, 1992; Ridley and Hall, 1992; Ridley *et al.*, 1992; Heldman *et al.*, 1996). The essential role played by both ash and ras in these growth-factor-triggered cellular responses is further emphasized when microinjection of antibody against ash not only prevents cell entry into S phase following RTK activation but also inhibits reorganization of the actin cytoskeleton (Matuoka *et al.*, 19930.

Rho induces the formation of actin stress fibers, perhaps by activating a phosphatidylinositol 4-kinase (PI4-kinase), and thereby raising the phosphatidylinositol-4,5-bisphosphate [PtdIns(4,5)P_2] concentration at the site of focal adhesion, where integrins bind to their extracellular ligand (Chong *et al.*, 1994). The mechanism responsible for rac stabilization of actin filaments within ruffles is unclear but may involve the production of superoxide by oxidases (Kreck *et al.*, 1994; Crawford *et al.*, 1996).

4. PHOSPHOLIPASE A_2 AND ARACHIDONATE METABOLISM

Ras functions upstream of a series of kinase kinases, whose up-regulation results in the activation of mitogen-activated protein kinase (MAP kinase) (Hattori *et al.*, 1992). That the activity of MAP kinase kinase is itself dependent on serine threonine phosphorylation suggests that there is at least one further protein serine and threonine kinase between it and the RTK; evidence from use of dominant negative raf proteins has implicated the serine kinase of this protein downstream of ras and upstream of the AP-1/cjun transcription complex (Hower *et al.*, 1992). Moreover, raf can directly phosphorylate and activate MAP kinase kinase *in vitro*, introducing itself as a potential MAP kinase kinase kinase. MAP kinase appears to be downstream to ras in signaling through the RTK pathway because experiments in which dominant negative ras proteins were expressed in cells stimulated by growth factors did not result in activation of MAP kinase (Leevers *et al.*, 1994), which normally occurs on growth factor stimulation. Also, MAP kinase is constitutively activated in some of the ras-transformed cell lines, indicating that the sustained function of these kinases may be necessary for maintaining the ras-transformed state (Williams *et al.*, 1993). Phosphorylated MAP kinase is translocated from the cytoplasm to the nucleus, where it promotes the entry of cells into mitosis (Lenormand *et al.*, 1993a). A cytoplasmic phosphatase down-regulates MAP kinase but does not access the karyoplasm (Lenormand *et al.*, 1993b). Thus, the level of activity of MAP kinase is the product of a dual regulatory cytoplasmic system of competing enzymes: the agonist kinase cascade and the antagonist phosphatase. Activated MAP kinase functions as a serine/threonine-specific kinase, although some investigations suggest that it may also autophosphorylate on tyrosine residues. Several important signal transduction proteins serve as *in vitro* substrates for MAP kinases, including myosin light chain kinase, the EGF receptor, and various DNA-binding proteins. Arachidonyl-specific cytoplasmic phospholipase A_2 (PLA_2) is also a substrate for MAP kinase (Lin *et al.*, 1993), which, on phosphorylation, is able to generate arachidonate by hydrolysis of phospholipids of the plasma membrane and other cellular membranes. The mechanism of this enzyme has been characterized such that when PLA_2 interacts with the cytoplasmic aspect of membranes, it selectively tweezes out individual phospholipid substrates to hydrolyze them within the catalytic domain of the phospholipase, releasing arachidonate and a lysophospholipid (White *et al.*, 1990).

Arachidonate serves as the metabolic precursor for synthesis of both leukotrienes and prostaglandins. Leukotriene synthesis is necessary to allow actin polymerization beneath the cell membrane (cortical actin) in response to EGF stimulation, whereas the breakdown of actin stress fibers in activated cells does not occur in the absence of prostaglandins (Peppelenbosch *et al.*, 1993). Although the specific mechanism(s) by which leukotrienes and prostaglandins induce actin network reorganization are not known, several hypotheses have been offered. One possibility is a calcium-current-mediated pathway, as leukotriene production has been shown to modulate signal transduction in EGF-activated cells by activating a Ca^{2+} channel (Peppelenbosch *et al.*, 1991). Alternatively, induction of c-*fos*, *jun*B, *egr*-a, and c-*myc* cellular oncogene expression by EGF is sensitive to inhibition of arachidonate synthesis (Handler *et al.*, 1990). However, because of the time course of the actin cytoskeletal response to growth factors and the large concentration of cytoskeletal proteins involved in the response, it is unlikely that the regulation of the cytoskeleton operates at a transcriptional level (Chinkers *et al.*, 1979). Finally, it may also be that the products of leukotriene and prostaglandin synthesis disrupt the regulation of the actin cytoskeleton by actin satellite proteins through a process that would involve alteration by arachidonate products of the binding of actin satellite proteins to membrane lipids (Goldschmidt-Clermont and Janmey, 1991). Whatever the mechanism implicated, both autocrine and paracrine regulation of the actin cytoskeleton by arachidonate metabolism appear to play a critical role in the cytoskeleton response to growth factor stimulation (Peppelenbosch *et al.*, 1993).

5. PHOSPHOLIPASE C AND THE CYTOSKELETAL RESPONSE

On binding to their cell surface receptors, many extracellular signaling molecules, including hormones, peptide growth factors, neurotransmitters, and immunoglobulins, elicit several intracellular responses by activating the phosphoinositide pathway. Activated PLC catalyzes the hydrolysis of PtdIns(4,5)P_2 to generate diacyglycerol (DAG) and inositol-1,4,5-trisphosphate (IP_3) (Berridge, 1993). Diacylglyceol then activates protein kinase C (PKC), and IP_3 triggers the release of calcium from intracellular stores, a process ubiquitously used by transmembrane signal transduction pathways that regulate a myriad of cellular processes including proliferation, metabolism, secretion, and migration.

Following receptor tyrosine autophosphorylation in response to ligand binding, increased ruffling and peripheral redistribution of actin filaments occur within cells. Yet despite concurrent phosphoinositide turnover during actin cytoskeletal reorganization, our understanding of the interaction of actin regulatory proteins and inositol phospholipids remains limited *in vivo*. One substrate for the RTK is $PLC_{\gamma1}$ (Meisenhelder *et al.*, 1989; Rhee, 1991), which itself is activated by phosphorylation of tyrosine residues (Goldschmidt-Clermont *et al.*, 1991; Wahl *et al.*, 1992).

Although $PLC_{\gamma1}$ is both an abundant and ubiquitous molecule within cells, it remains unclear whether its role is simply the hydrolysis of PtdIns(4,5)P_2 or if the enzyme has additional roles (Smith *et al.*, 1994) involving its increased affinity for PtdIns(4,5)P_2 on tyrosine phosphorylation (Wahl *et al.*, 1992b), which might then allow competition with other PtdIns(4,5)P_2-binding proteins for binding to polyphosphoinositides (Goldschmidt-Clermont and Janmey, 1991).

Although many studies have suggested that cellular PLC activity increases after growth factor treatment of cells, only recently have workers more clearly demonstrated the effects of tyrosine phosphorylation on $PLC_{\gamma1}$ catalytic activity. Several groups have shown that the addition of epidermal growth factor or platelet-derived growth factor to cells produces rapid increases in the amount of both phosphotyrosine and phosphoserine on $PLC_{\gamma1}$ (Rhee, 1991). Four sites of ligand-induced tyrosine phosphorylation on $PLC_{\gamma1}$ have been identified both *in vitro* and *in vivo* (Kim *et al.*, 1991). Overexpression of $PLC_{\gamma1}$ cDNA in NIH 3T3 fibroblasts produces elevated growth-factor-induced IP_3 levels *in vivo*, although the calcium response is unaffected by $PLC_{\gamma1}$ overexpression in these cells, in spite of the increased production of IP_3 (Margolis *et al.*, 1990). Interestingly, the V_{max} of $PLC_{\gamma1}$ is not affected by growth-factor-induced tyrosine phosphorylation of this enzyme (Wahl *et al.*, 1992a,b). Instead, the K_m of $PLC_{\gamma1}$ for PtdIns(4,5)P_2 within membranes is lowered by tyrosine phosphorylation, and this effect may be responsible for the increased PtdIns(4,5)P_2 breakdown observed after RTK activation by growth factors (Wahl *et al.*, 1992a).

The hydrolytic activity of this phospholipase is controlled by the actin regulatory protein profilin (Goldschmidt-Clermont *et al.*, 1990), which binds PtdIns(4,5)P_2 (Lassing and Lindberg, 1985). Interestingly, plant profilin can also affect the activity of plant membrane PLC in a similar fashion to mammalian cells (Drobak *et al.*, 1994), indicating that such an interaction between this actin-binding regulatory protein and PtdIns(4,5)P_2 has been conserved throughout evolution. The inhibitory effect of profilin can be overcome, however, by tyrosine phosphorylation of $PLC_{\gamma1}$ (Goldschmidt-Clermont *et al.*, 1991), suggesting that by affecting the affinity of $PLC_{\gamma1}$ for polyphosphoinositides (Wahl *et al.*, 1992b), RTK may, in turn, through the hydrolytic activity of $PLC_{\gamma1}$ on membrane inositol phospholipids, modulate the interaction between profilin and actin and subsequently the equilibrium of polymerized actin (Goldschmidt-Clermont and Janmey, 1991).

Interestingly, in EJ-ras-transformed cells, the assembly of actin filaments and formation of membrane ruffles can be mediated by RTK-triggered signaling events despite a damped-down tyrosine kinase activity and absent $PLC_{\gamma 1}$ activity and Ca^{2+} transients (Marshall, 1993; Heldman *et al.*, 1996). The actin response of EJ-ras-transformed cells to PDGF-BB is dependent, however, on some residual tyrosine kinase activity and also on the activity of phosphatidylinositol 3-kinase (P13-kinase).

It was suggested that $PLC_{\gamma 1}$ induces DNA synthesis in cells independently of its lipase activity (Smith *et al.*, 1994). It is possible that other properties of $PLC_{\gamma 1}$ are based on activity related to its src homology domain 3 (SH3). Although phosphorylation is not directly involved in SH3-mediated reactions, recent genetic and biochemical evidence suggests that SH3 domains target regulatory proteins, for example, proteins that control small GTP-binding proteins, to strategic areas of the cell (Bar-Sagi *et al.*, 1993).

Unlike $PLC_{\gamma 1}$, PLC_{β} is not sensitive to profilin interaction with PtdIns(4,5)P_2 (Goldschmidt-Clermont *et al.*, 1991), and the effect of profilin on PLC_{δ}, whose regulatory mechanism remains incompletely understood (Rhee, 1991), has not yet been investigated. Receptor-mediated activation of PLC_{β} isozymes appears to be elicited by two distinct mechanisms, one mediated by the α subunit of G_q of heterotrimeric conventional G proteins (an α isoform insensitive to pertussin toxin), and the other through the $\beta\gamma$ subunits (Taylor *et al.*, 1991; Park *et al.*, 1993). A constitutively activated G_q mutant that couples to PLC_{β} is able to transform NIH 3T3 cells (Kalinec *et al.*, 1992). Unlike other PLC isozymes, PLC_{δ} isozymes lack src homology domains 2 and 3 and are, therefore, unlikely to serve as substrates for tyrosine kinases (Rhee, 1991). Alternatively, their calcium-binding EF-hand motif may allow intracellular calcium levels to regulate PLC_{δ}-mediated hydrolysis of phosphoinositides *in vivo*. Mutagenesis of PLC_{δ} indicates that the amino-terminal domain of the enzyme is important for binding both Ins(1,4,5)P_3 and PtdIns(4,5)P_2 (Yagisawa *et al.*, 1994). Thrombin-induced activation of CHO cells overexpressing the mammalian isoform of PLC_{δ} demonstrated an enhanced InsP production compared to control cells, and the increase was modulated by receptor-coupled G proteins, intracellular calcium concentrations, and the activation of PKC (Banno *et al.*, 1994).

Yeast cells null for the inositol-phospholipid-specific PLC_1, which is most similar to the mammalian PLC isoform δ, display an abnormal phenotype that includes morphological changes and an abnormal actin network reminiscent of the changes observed in profilin-deficient *S. cerevisiae* (Flick and Thorner, 1993). Similar changes have also been observed in yeast cells null for the C terminus of the ras pathway protein cyclase-associated protein (cap) (Vojtek *et al.*, 1991). Profilin isoforms able to bind to PtdIns(4,5)P_2 are able to rescue the abnormal phenotype induced by the C-terminus cap deletion (Vojtek *et al.*, 1991; Machesky *et al.*, 1990). Thus, it is possible that in yeast, PLC_{δ}, ras, and cap participate in a regulatory pathway for the actin cytoskeleton, using the actin-binding protein profilin as a mediator (Goldschmidt-Clermont and Janmey, 1991). Another actin-binding protein, gelsolin, which has also been implicated in a complex containing both actin and $PLC_{\gamma 1}$ (Banno *et al.*, 1992), may participate as well in a pathway linking ras and $PLC_{\gamma 1}$; a flat revertant clonal line from EJ-ras-transformed cells contains a gelsolin mutant that inhibits $PLC_{\gamma 1}$ more efficiently than wild-type gelsolin *in vitro* (Fujita *et al.*, 1993).

Except for genetic experiments performed in yeast with cap and in mammalian cells with gelsolin, evidence of interaction of actin-binding proteins (like profilin and gelsolin) with PtdIns(4,5)P_2 and its precursor, phosphatidylinositol-4-phosphate, is based solely on *in vitro* experiments. In fact, unsuccessful attempts to detect a correlation between actin–gelsolin interaction and bulk phospholipid turnover are consistent with the hypothesis that actin-binding protein interaction with membrane phospholipids is not relevant *in vivo* (Dabaday *et al.*, 1991). It remains possible, however, that observed changes in bulk phospholipid concentration do not reflect alterations in the turnover of relatively small pools of inositol phospholipids that are responsible for interaction with actin-binding proteins (King *et al.*, 1989). Alternatively, it may also be that the functional alteration of one pathway can be overcome by either the increased activity of a synergistic pathway or the suppression of an antagonistic pathway. Nevertheless, evidence for a role for tyrosine phosphorylation of $PLC_{\gamma 1}$ in the growth-factor-induced motile response is provided by genetic experiments on the PDGF and EGF receptors, where mutations of the autophosphorylated residue necessary for $PLC_{\gamma 1}$ binding leads to an abnormal cytoskeletal response (Fantl *et al.*, 1993).

6. PHOSPHOLIPID KINASES AND CYTOSKELETAL REORGANIZATION

The turnover rate of membrane phospholipids is remarkably rapid, reflecting the highly dynamic character of the hydrolytic and synthetic enzymes involved in the metabolism of these membrane constituents.

The PI3-kinase is a heterodimeric enzyme consisting of a catalytic subunit (p110) bound to a p85 docking subunit, which itself possesses receptor-binding SH2 domains but has no PI3-kinase activity (Graziani *et al.*, 1991; Skolnick *et al.*, 1991). Ligand-activated RTK communicate through binding SH2 domains of the p85 subunit to a phosphotyrosine residue of the activated EGF and PDGF receptors, and of the polymer middle T/pp60c-src complex (Schlessinger and Ullrich, 1992; Fantl *et al.*, 1993).

Despite the relatively small fraction of inositol phospholipids that are phosphorylated on the third carbon of the inositol ring, several lines of evidence implicate PI3-kinase as a pivotal regulatory enzyme in growth-factor-induced cell activation. Recent studies suggest that binding of tyrosine-phosphorylated insulin receptor substrate (IRS-1) to the two SH2 domains of PI3-kinase is, by itself, sufficient to stimulate catalytic activity (Skolnick *et al.*, 1993; Myers *et al.*, 1992; Hadari *et al.*, 1992). To date, the pathway downstream of PI3-kinase has not yet been determined, although it appears that the kinase may serve as an important regulatory protein in the actin response to growth factors (Eberle *et al.*, 1990). Phosphatidylinositol-3,4, 5-trisphosphate [PtdIns(3,4,5)P_3] is not a good substrate for inositol-specific PLC, and it has not been established whether PtdIns(4,5)P_2 protein ligands, particularly ligands acting as regulatory proteins for actin, also bind to phospholipid clusters containing PtdIns(3,4,5)P_3. Interestingly, wortmannin, a selective inhibitor of PI3-kinase activity, effectively blocks the formation of actin-rich membrane ruffles, suggesting that PtdIns(3,4,5)P_3 accumulation may be necessary for this cell response to RTK activation, a finding supported by genetic studies on RTK (Heldman *et al.*, 1996; Wennstrom *et al.*, 1994; Wymann and Arcaro, 1994; Kotani *et al.*, 1994).

Phosphatidylinositol 4 phosphate, 5 kinase (Pi4P, 5 kinase) is also responsive to extracellular signaling (Chong *et al.*, 1994). Contact between the actin cytoskeleton and adhesion receptors (integrins), and between these adhesion receptors and the extracellular matrix proteins, must be tightly regulated to provide an anchorage to the extracellular substratum for forward movement of the leading edge of the cell (Cooper, 1991). Focal adhesion integrins function as receptors for specific extracellular matrix proteins or adhesive molecules on the surface of other cells, thereby anchoring cells to the extracellular matrix or to adjacent cells (Burridge *et al.*, 1992). Adhesion to fibronectin stimulates the local accumulation of membrane PtdIns(4,5)P_2 (McNamee *et al.*, 1993) mediated by the activation of a Pi4P, 5 kinase (Kreck *et al.*, 1994), suggesting that focal adhesion proteins may play an integral role in providing the cell with information concerning the interaction with adjacent cells, the extracellular matrix, or even the intensity of shear stress applied to the cellular surface (Davies, 1993). In turn, production of PtdIns(3,4)P_2 may signal to the actin cytoskeleton in a fashion that modulates the formation and disruption of specific actin structures (Goldschmidt-Clermont and Janmey, 1991; Goldschmidt-Clermont *et al.*, 1992).

In addition to phospholipid kinase activity, focal adhesions also contain the tyrosine kinase $pp60^{src}$ and pp125 focal adhesion kinase ($pp125^{FAK}$), which likely contribute to the signal transduction potential of focal adhesions (Kornberg *et al.*, 1992). The adhesive-protein-induced, filamentous-actin-dependent clustering of integrins such as GPIIb-IIIa induces the activation of $pp125^{FAK}$ and also of a tyrosine kinase that functions upstream from $pp125^{FAK}$ (Shattil, 1993). Although the pathways downstream of these kinases are not entirely characterized, the potential role of $pp125^{FAK}$ in preventing apoptosis of attached cells is interesting, as it may link integrins, the actin cytoskeleton, and cell cycle regulation with programmed cell death (Ruoslahti and Reed, 1994).

7. EFFECTS OF OTHER PHOSPHOLIPIDS ON THE ACTIN STRUCTURE

Membrane lipids that do not possess an inositol ring are also capable of effecting changes in the actin cytoskeleton. Lysophosphatidic acid bound to albumin, and to a lesser extent phosphatidic acid, have been shown to rapidly induce stress fiber formation (Ridley and Hall, 1992). Through activation of PKC and serine and threonine phosphorylation, DAG can modify specific interactions of actin with actin-binding proteins such as MARCKS proteins that shuttle between the membrane and cytosolic actin (Hartwig *et al.*, 1992). DAG is also able to increase the activity of a nucleating protein for actin filament formation independent of PKC involvement (Pollard *et al.*, 1991). Phosphatidylserine, together with other negatively charged membrane lipids, stabilizes the interaction of actin-binding proteins with the membrane, an effect that may play an important role for the association of myosin 1 with the surface of membrane organelles (Pollard *et al.*, 1991).

8. CONCLUSIONS

The actin cytoskeleton is a highly dynamic superstructure, constantly remodeled in response to extracellular stimuli. The actin reorganization is controlled by the concurrent metabolism of membrane inositol phospholipids. In turn, proteins of the actin cytoskeleton can modulate the turnover of these phospholipids. The orchestration of these complex reactions is dependent on the highly specific clustering of molecules at strategic sites of the cell, clustering that is provided by the binding of modular protein domains such as SH2 and SH3, pleckstrin homology domains (PH), and LIM domains to their respective ligands. The stability of these protein and phospholipid clusters is controlled by phosphorylation reactions and by the activity of small GTP-binding proteins of the $p21^{ras}$ family.

9. REFERENCES

Arber, S., Halder, G., and Caroni, P., 1994, Muscle LIM protein, a novel essential regulator of myogenesis, promotes myogenic differentiation, *Cell* **79**:221–231.

Banno, Y., Nakashima, T., Kumada, T., Ebisawa, K., Nonomura, Y., and Nozawa, Y., 1992, Effects of gelsolin on human platelet cytosolic phosphoinositide-phospholipase C isozymes, *J. Biol. Chem.* **267**:6488–6394.

Banno, Y., Okano, Y., and Nozawa, Y., 1994, Thrombin-mediated phosphoinositide hydrolysis in Chinese hamster ovary cells overexpressing phospholipase C-δ1, *J. Biol. Chem.* **269:**15846–15852.

Bar-Sagi, D., Rotin, D., Batzer, A., Mandiyan, V., and Schlessinger, J., 1993, SH3 domains direct cellular localization of signalling molecules, *Cell* **74:**83–90.

Berridge, M. J., 1993, Inositol trisphosphate and calcium signalling, *Nature* **361:**315–325.

Bourne, H. R., Sanders, D. A., and McCormick, F., 1990, The GTPase superfamily: A conserved switch for diverse cell functions, *Nature* **348:**125–132.

Bourne, H. R., Sanders, D. A., and McCormick, F., 1991, The GTPase superfamily: Conserved structure and molecular mechanism, *Nature* **349:**117–127.

Burridge, K., Petch, L. A., and Romer, L. H., 1992, Signals from focal adhesions, *Curr. Biol.* **2:**537–539.

Carpenter, G., 1992, Receptor tyrosine kinase substrates: src homology domains and signal transduction, *FASEB J.* **6:**3283–3289.

Chinkers, M., McKanna, J. A., and Cohen, S., 1979, Rapid induction of morphological changes in human carcinoma cells A-431 by epidermal growth factor, *J. Cell Biol.* **83:**260–265.

Chong, L. D., Traynor-Kaplan, A., Bokoch, G. M., and Schwartz, M. A., 1994, The small GTP-binding protein rho regulates a phosphatidylinositol 4-phosphate 5-kinase in mammalian cells, *Cell* **79:**507–513.

Cooper, J. A., 1991, The role of actin polymerization in cell motility, *Annu. Rev. Physiol.* **53:**585–605.

Crawford, L. E., Tucker, R. W., Heldman, A., and Goldschmidt-Clermont, P. J., 1994, Actin regulation and surface catalysis, in *Actin: Biophysics, Biochemistry, and Cell Biology* (J. E. Estes and P. J. Higgins, eds.), pp. 105–112, Plenum Press, New York.

Dadabay, C. Y., Patton, E., Cooper, J. A., and Pike, L., 1991, Lack of correlation between changes in polyphosphoinositide levels and actin/gelsolin complexes in A431 cells treated with epidermal growth factor, *J. Cell Biol.* **112:**1151–1156.

Davies, P. F., 1993, Endothelium as a signal transduction interface for flow forces: Cell surface dynamics, *Thromb. Haemstas.* **70:**124–128.

Drobak, B. K., Watkins, P.A.C., Valenta, R., Dove, S. K., Lloyd, C. W., and Staiger, C. J., 1994, Inhibition of plant plasma membrane phosphoinositide phospholipase C by the actin-binding protein profilin, *Plant J.* **6:**389–400.

Eberle, M., Traynor-Kaplan, A. E., Sklar, L. A., and Norgauer, J., 1990, Is there a relationship between phosphatidylinositol trisphosphate and F-actin polymerization in human neutrophils? *J. Biol. Chem.* **265:**16725–16728.

Fantl, W. J., Johnson, D. E., and Williams, L. T., 1993, Signalling by receptor tyrosine kinases, *Annu. Rev. Biochem.* **62:**453–481.

Finkel, T., Theriot, J. A., Dise, K., Tomaselli, G. F., and Goldschmidt-Clermont, P. J., 1994, Dynamic actin structures stabilized by profilin, *Proc. Natl. Acad. Sci. USA.* (in press).

Flick, J. S., and Thorner, J. 1993, Genetic and biochemical characterization of a phosphatidylinositol-specific phospholipase C in *Saccharomyces cerevisiae, Mol. Cell Biol.* **13:**5861–5876.

Friend, C. M. 1993, Catalysis on surfaces, *Sci. Am.* **268:**74–80.

Fujita, H., Banno, Y., Mullauer, L. Ishizaki, A. Nozawa, Y., and Kuzumaki, N., 1993, Enhanced inhibition of phospholipase $C_{\gamma 1}$ by mutated gelsolin derived from a flat revertant of EJ-ras oncogene-transformed NIH/3T3 cells, in *Cold Spring Harbor Laboratory Meeting on the Cytoskeleton and Cell Function,* p. 201, Cold Spring Harbor Laboratory, New York.

Fukami, K., Furuhashi, K., Inagaki, M., Endo, T., Hatano, S., and Takenawa, T., 1993, Requirement of phosphatidylinositol 4,5 bisphosphate for alpha-actinin function, *Nature* **359:**150–152.

Gips, S. J., Kandzari, D. E., and Goldschmidt-Clermont, P. J., 1994, Growth factor receptors, phospholipases, phospholipid kinases, and actin reorganization, *Semin. Cell Biol.* **5:**201–208.

Goldschmidt-Clermont, P. J., and Janmey, P. A., 1991, Profilin: A weak CAP for actin and ras, *Cell* **66:**419–421.

Goldschmidt-Clermont, P. J., Machesky, L. M., Baldassare, J. J., and Pollard, T. D., 1990, The actin-binding protein profilin binds to PIP_2 and inhibits its hydrolysis by phospholipase-C, *Science* **247:**1575–1578.

Goldschmidt-Clermont, P. J., Kim, J. W., Machesky, L. M., Rhee, S. G., and Pollard, T. D., 1991, Regulation of phospholipase $C_{\gamma 1}$ by profilin and tyrosine phosphorylation, *Science* **251:**1231–1233.

Goldschmidt-Clermont, P. J., Mendelsohn, M. E., and Gibbs, J. B. 1992, Rac and rho in control, *Curr. Biol.* **2:**668–671.

Graziani, A., Kapeller, R., and Soltoff, S., 1991, Oncogenes and signal transduction, *Cell* **64:**281–302.

Hadari, Y. R., Tzahar, E., Nadiv, O., Rothenberg, P., Roberts, C. T., LeRoith, D., Yarden, Y., and Zick, Y., 1992, Insulin and insulinomimetic agents induce activation of phosphatidylinositol 3′-kinase upon its association with pp185 (IRS-1) in intact rat livers, *J. Biol. Chem.* **267:**17483–17486.

Handler, J. A., Danilowicz, R. M., and Eling, T. E., 1990, Mitogenic signalling by epidermal growth factor (EGF), but not platelet derived growth factor, requires arachidonic acid metabolism in BALB/c 3T3 cells, *J. Biol. Chem.* **265:**3669–3673.

Hartwig, J. H., Thelen, M., Rosen, A., Janmey, P. A., Nairn, A. C., and Aderem, A., 1992, MARCKS is an actin filament crosslinking protein regulated by protein kinase C and calcium-calmodulin, *Nature* **356:**618–622.

Hattori, S., Fukuda, M., Yamashita, T., Nakamura, S., Gotoh, Y., and Nishida, E., 1992, Activation of mitogen-activated protein kinase and its activator by ras in intact cells and in a cell-free system, *J. Biol. Chem.* **267:**20346–20351.

Heldman, A. W., and Goldschmidt-Clermont, P. J., 1993, Cell signalling and motile response, in *Cell Adhesion and Motility* (R. M. Warn, ed.), pp. 317–324, The Society of Experimental Biology, London.

Heldman, A. W., Kandzari, D. E., Tucker, R. W., Crawford, L. E., Fearon, E. R., Koblan, K. S., and Goldschmidt-Clermont, P. J., 1996, EJ-ras inhibits phospholipase $C_{\gamma 1}$, but not actin polymerization induced by PDGF-BB via phosphatidylinositol-3 kinase, *Circ. Res.*, in press.

Howe, L. R., Leevers, S. J., Gomez, N., Nakielny, S., Cohen, P., and Marshall, C. J., 1992, Activation of the MAP kinase pathway by the protein kinase raf, *Cell* **71:**335–342.

Janmey, P. A., 1994, Phosphoinositides and calcium as regulators of cellular actin assembly and disassembly, *Annu. Rev. Physiol.* **56:**169–191.

Janmey, P. A., and Stossel, T. P., 1989, Gelsolin–polyphosphoinositide interaction, *J. Biol. Chem.* **264:**4825–4831.

Kalinec, G., Nazarali, A. J., Hermouet, S., Xu, N., and Gutkind, J. S., 1992, Mutated alpha subunit of the G_q protein induces malignant transformation in NIH 3T3 cells, *Mol. Cell. Biol.* **12:**4687–4693.

Kim, H. K., Kim, J. W., Zilberstein, A. Margolis, B., Kim, J. G., Schlessinger, J., and Rhee, S. G., 1991, PDGF stimulation of inositol phospholipid hydrolysis requires PLC-gamma 1 phosphorylation on tyrosine residues 783 and 1254, *Cell* **65:**435–441.

King, C. E., Hawkins, P. T., Stephens, L. A., and Michell, R. H ., 1989, Determination of the steady-state turnover rates of the metabolically active pools of phosphatidylinositol 4-phosphate and phosphatidylinositol 4,5 bisphosphate in human erythrocytes, *Biochem. J.* **259:**893–896.

Kornberg, L., Earp, H. S., Parson, J. T., Schaller, M., Juliano, R. L., 1992, Cell adhesion or integrin clustering increases phosphorylation of a focal adhesion-associated tyrosine kinase, *J. Biol. Chem.* **267:**23439–23442.

Kotani, K., Yonezawa, K., Hara, K., Ueda, H., Kitamura, Y., Sakaue, H., Ando, A., Chavanieu, A., Calas, B., and Grigorescu, F., 1994, Involvement of phosphoinositide 3-kinase in insulin- or IGF-1-induced membrane ruffling, *EMBO J.* **13:**2313–2321.

Kreck, M. L., Uhlinger, D. J. Tyagi, W. R., Inge, K. L., and Lambeth, J. D., 1994, Participation of the small molecular weight GTP-binding protein rac1 in cell-free activation and assembly of the respiratory burst oxidase, *J. Biol. Chem.* **269:**4161–4168.

Lassing, I., and Lindberg, U., 1985, Specific interaction between phosphatidylinositol 4,5 bisphosphate and profilactin, *Nature* **318:**472–474.

Leevers, S. J. Paterson, H. F., and Marshall, C. J., 1994, Requirement for ras in raf activation is overcome by targeting raf to the plasma membrane, *Nature* **369:**411–414.

Lenormand, P., Pages, G., Sardet, C., L'Allemain, G., Meloche, S., and Pouyssegur, J., 1993a, MAP kinases: Activation, subcellular localization, and role in the control of cell proliferation, *Adv. Second Messenger Phosphoprotein Res.* **28:**237–244.

Lenormand, P., Sardet, C., Pages, G., L'Allemain, G., Brunet, A., and Pouyssegur, J., 1993b, Growth factors induce nuclear translocation of MAP kinases (p42mapk and p44mapk) but not of their activator MAP kinase kinase (p45mapkk) in fibroblasts, *J. Cell. Biol.* **122:**1079–1088.

Lin, L. L., Wartmann, M., Lin, A. Y., Knopf, J. L., Seth, A., and Davis, R. J., 1993, $cPLA_2$ is phosphorylated and activated by MAP kinase, *Cell* **72:**269–278.

Lowenstein, E. J., Daly, R. J., Batzer, A. G., Li, W., Margolis, B., Lammers, R. Ullrich, A., Skolnik Bar-Sagi, D., and Schlessinger, J., 1992, The SH2 and SH3 domain-containing protein GRB2 links receptor tyrosine kinases to ras signalling, *Cell* **70:**431–442.

Machesky, L. M., Goldschmidt-Clermont, P. J., and Pollard, T. D. 1990, The affinities of human platelet and *Acanthamoeba* profilin isoforms for polyphospho-inositides account for their relative abilities to inhibit phospholipase C, *Cell Regul.* **1:**937–950.

Margolis, B., Zilberstein, A., Franks, C., Felder, S., Kremer, S., Ullrich, A., Rhee, S. G., Skorecki, K., and Schlessinger, J., 1990, Effect of phospholipase C-gamma overexpression on PDGF-induced second messengers and mitogenesis, *Science* **248:**607–610.

Marshall, M. S., 1993, The effector interactions of p21ras, *Trends Biochem. Sci.* **18:**250–254.

Matuoka, K., Shibasaki, F., Shibata, M., and Takenawa, T., 1993, Ash-Grb-2, a SH2/SH3-containing protein, couples to signalling for mitogenesis and cytoskeletal reorganization by EGF and PDGF, *EMBO J.* **12:**3467–3473.

McNamee, H. P., Ingber, D. E., and Schwartz, M. A., 1993, Adhesion of fibronectin stimulates inositol lipid synthesis and enhances PDGF-induced inositol lipid breakdown, *J. Cell Biol.* **121:**673–678.

Meisenhelder, J., Pann-Ghill, S., Rhee, S. G., and Hunter, T., 1989, Phospholipase $C_{\gamma 1}$ is a substrate for the PDGF and EGF receptor protein-tyrosine kinases *in vivo* and *in vitro*, *Cell* **57:**1109–1116.

Mosior, M., and Epand, R. M., 1994, Characterization of the calcium-binding site that regulates the association of protein kinase C with phospholipid bilayers, *J. Biol. Chem.* **269:**13798–13805.

Myers, M. G., Backer, J. M., Sun, X. J., Shoelson, S., Hu, P., Schlesinger J., Yoakim, M., Schaffhausen, B., and White, M. 1992. IRS-1 activates phosphatidylinositol 3′-kinase by associating with src homology 2 domains of p85, *Proc. Natl. Acad. Sci. USA.* **89:**10350–10354.

Neel, B. G., 1993, Structure and function of SH2-domain containing tyrosine phosphatases, *Semin. Cell Biol.* **4:**419–432.

Park, D., Jhon, D. Y., Lee, C. W., Lee, K. H., and Rhee, S. G., 1993, Activation of phospholipase C isozymes by G protein beta gamma subunits, *J. Biol. Chem.* **268:**4573–4576.

Peppelenbosch, M. P., Tertoolen, L.G.J., and de Laat, S. W., 1991, Epidermal growth factor-activated calcium and potassium channels, *J. Biol. Chem.* **266:**19938–19944.

Peppelenbosch, M. P., Tertoolen, L.G.J., Hage, W. J., and de Laat, S. W., 1993, Epidermal growth factor-induced actin remodeling is regulated by 5-lipoxygenase and cyclooxygenase products, *Cell* **74**:565–575.

Pollard, T. P., Doberstein, S. K., and Zot, H. G., 1991, Myosin I, *Annu. Rev. Physiol.* **53**:653–681.

Rhee, S. G., 1991, Inositol phospholipid-specific phospholipase C: Interaction of the g1 isoform with tyrosine kinase, *Trends Biochem. Sci.* **16**:297–301.

Ridley, A. J., and Hall, A., 1992, The small GTP-binding protein rho regulates the assembly of focal adhesions and actin stress fibers in response to growth factors, *Cell* **70**:389–399.

Ridley, A. J., Paterson, H. F., Johnston, C. L., Diekmann, D., and Hall, A., 1992, The small GTP-binding protein rac regulates growth factor-induced membrane ruffling, *Cell* **70**:401–410.

Ridley, A. J., Self, A. J., Kasmi, F., Paterson, H. F., Hall, A., Marshall, C. J., and Ellis, C., 1993, Rho family GTPase activating proteins p190, bcr and rhoGAP show distinct specificities *in vitro* and *in vivo*, *EMBO J.* **12**:5151–5160.

Ruoslahti, E., and Reed, J. C., 1994, Anchorage dependence, integrins, and apoptosis, *Cell* **77**:477–478.

Schlessinger, J., and Ullrich, A., 1992, Growth factor signalling by receptor tyrosine kinases, *Neuron* **9**:383–391.

Schmeichel, K. L., and Beckerle, M. C., 1994, The LIM domain is a modular protein-binding interface, *Cell* **79**:211–219.

Shattil, S. J., 1993, Regulation of platelet anchorage and signalling by integrin aIIbb3, *Thromb. Haemostas.* **70**:224–228.

Simon, M. A., Bowtell, D.D.L., Dodson, G. S., Laverty, T. R., and Rubin, G. M., 1991, Ras1 and putative guanine nucleotide exchange factor perform crucial steps in signalling by the sevenless protein tyrosine kinases, *Cell* **67**:701–716.

Skolnik, E. Y., Margolis, B., Mohammadi, M., Lowenstein, E., Fischer, R., Drepps, A., Ullrich, A., and Schlessinger, J., 1991, Cloning of PI3-kinase-associated p85 utilizing a novel method for expression/cloning of target proteins for receptor tyrosine kinases, *Cell* **65**:83–90.

Skolnik, E. Y., Lee, C. H., Batzer, A., Vicentini, L. M., Zhou, M., Daly, R., Myers, M. J., Backer, J. M., Ullrich, A., and White, M. F., 1993, The SH2/SH3 domain-containing protein GRB2 interacts with tyrosine-phosphorylated IRS-1 and Shc: Implications for insulin control of ras signalling, *EMBO J.* **12**:1929–1936.

Smith, M. R., Liu, Y., Matthews, N. T., Rhee, S. G., Sung, W. K., and Kung, H., 1994, Phospholipase C-g1 can induce DNA synthesis by a mechanism independent of its lipase activity, *Proc. Natl. Acad. Sci. U.S.A.* **91**:6554–6558.

Stossel, T. P., 1993, On the crawling of animal cells, *Science* **260**:1086–1094.

Taylor, S. J., Chae, H. Z., Rhee, S. G., and Exton, J. H., 1991, Activation of the beta 1 isozyme of phospholipase C by alpha subunits of the G_q class of G proteins, *Nature* **350**:516–518.

Theriot, J. A., and Mitchison, T. J ., 1992, The nucleation–release model of actin filament dynamics in cell motility, *Trends Cell Biol.* **2**:219–222.

Theriot, J. A., Rosenblatt, J., Portnoy, D. A., Goldschmidt-Clermont, P. J., and Mitchison, T. J., 1994, Involvement of profilin in the actin-based motility of *L. monocytogenes in* cells and cell-free extracts, *Cell* **76**:505–517.

Vojtek, A., Haarer, B., Field, J., Gerst, J., Pollard, T. D., Brown, S., and Wigler, M., 1991, Evidence for a functional link between profilin and CAP in yeast *S. cerevisiae*, *Cell* **66**:497–505.

Wahl, M. I., Jones, G. A., Nishibe, S., Rhee, S. G., and Carpenter, G., 1992a, Growth factor stimulation of phospholipase C-gammal activity. Comparative properties of control and activated enzymes. *J. Biol. Chem.* **267**:10447–10456.

Wahl, M. I., Jones, G. A., Nishibe, S., Rhee, S. G., and Carpenter, G., 1992b, Growth factor stimulation of phospholipase C-g1 activity. *J. Biol. Chem.* **267**:10447–10456.

Wennstrom, S., Siegbahn, K., Yokote, A., Arvidson, K., Heldin, C. H., and Mori, S., 1994,

Membrane ruffling and chemotaxis transduced by the PDGF beta-receptor require the binding site for phosphatidylinositol 3′ kinase, *Oncogene* **9:**651–660.

White, S. P., Scott, D. L., Otwinowski, Z., Gelb, M. H., and Sigler, P. B., 1990, Crystal structure of cobra-venom phospholipase A_2 in a complex with a transition-state analogue, *Science* **250:**1560–1563.

Williams, N. G., Paradis, H., Agarwal, S., Charest, D. L., Pelech, S. L., and Roberts, T. M., 1993, Raf-1 and p21v-ras cooperate in the activation of mitogen-activated protein kinase, *Proc. Natl. Acad. Sci. U.S.A.* **90:**5772–5776.

Wymann, M., and Arcaro, A., 1994, Platelet derived growth factor-induced phosphatidylinositol 3-kinase activation mediates actin rearrangements in fibroblasts, *Biochem. J.* **298:**517–520.

Yagisawa, H., Hirata, M., Kanematsu, T., Watanabe, Y., Ozaki, S., Sakuma, K., Tanaka, H., Yabuta, N., Kamata, H., Hirata, H., and Nojima, H., 1994, Expression and characterization of an inositol 1,4,5-trisphosphate binding domain of phosphatidylinositol-specific phospholipase C-δ1, *J. Biol. Chem.* **31:**20179–20188.

Chapter 5

Protein Phosphorylation and Signal Transduction

Sailen Barik

1. INTRODUCTION

The phenotype and behavior of a cell, just as those of the organism in which it resides, are constantly influenced by a variety of external signals that eventually determine such apparently diverse phenomena as cellular taxis, proliferation, differentiation, and even death (apoptosis). The corresponding extracellular signals are equally varied, e.g., nutrient flux, hormones, peptide growth factors, interferons and other cytokines, and antigens. The term "signal transduction" refers to the acquisition and subsequent transmission of a given signal.

In analogy to electronic data-processing operations, signal transduction in living cells can be conceptually dissected into the following basic steps: *receiving* of the signal; *conversion* of the signal to the macromolecular language of the cell; *amplification* and *conduction* of the primary signal through a series of interactive transducers, sometimes followed by *dampening* of the signal to restore the pre-excited state; and finally *execution*. The pathway, in general, begins with cell surface receptors and ends in the cell nucleus with proteins that regulate gene transcription. In between, each protein component of the pathway integrates

Sailen Barik Department of Biochemistry and Molecular Biology, College of Medicine, University of South Alabama, Mobile, Alabama 36688-0002.

Subcellular Biochemistry, Volume 26: myo-Inositol Phosphates, Phosphoinositides, and Signal Transduction, edited by B. B. Biswas and Susweta Biswas. Plenum Press, New York, 1996.

signals from the upstream ones and transduces them on to the downstream targets or effector proteins.

Some of the original studies of signal transduction quickly led to the realization that primary gene activation (or repression) in eukaryotes does not require *de novo* protein synthesis and must, therefore, involve posttranslational modification of existing ones. Subsequent studies have shown that an overwhelming majority of proteins involved in signal transduction are regulated by reversible phosphorylation (and dephosphorylation) catalyzed by specific protein kinases (and phosphatases), many of which are themselves phosphoproteins.

This chapter is not intended to accomplish the impossible task of cataloguing all known instances of signal transduction by phosphoproteins, the list of which is continually expanding. Rather, emphasis is placed on bringing harmony to the various systems on mechanistic and molecular grounds. In the recent past, excellent reviews have appeared on some of the individual signal transduction phenomena and related phosphorylations, including oncogenesis (Cantley *et al.*, 1991; Hunter, 1991; Bishop, 1991), growth factors (Cross and Dexter, 1991), receptor tyrosine kinases (Fantl *et al.*, 1993; Perrimon, 1993), protein tyrosine phosphatases (Walton and Dixon, 1993), inositol phosphate metabolism (Rhee and Choi, 1992; Majerus, 1992), MAP kinase cascade (Crews and Erikson, 1993), phosphorylation of transcription factors (Bohman, 1990; Hunter and Karin, 1992), and bacterial signal transduction (Bourret *et al.*, 1991; Parkinson, 1993).

2. PRIMARY SIGNAL TRANSDUCTION MECHANISMS IN EUKARYOTES

The dominant mechanism by which cells respond to extracellular signals can be summed up, somewhat simplistically, as follows. Signaling commences on binding of the effector molecule (i.e., growth factor, hormone, etc.) to its cognate receptor on the cell surface. The binding activates a protein tyrosine kinase. The tyrosine kinase activity can either be an intrinsic property of the receptor itself (e.g., growth factor receptors) or may reside in a distinct and separate kinase molecule associated with the receptor (e.g., cytokine receptors); the former has come to be known as receptor tyrosine kinase (RTK). This leads to autophosphorylation of the receptor or other substrates, and downstream activation of a number of common signaling molecules, either directly or through a cascade of biochemical reactions. Frequently activated proteins (enzymes), many of which are cytoplasmic, include cyclic-AMP-dependent protein kinase, protein kinase C, GTPase activating protein, p34cdc2/cyclin, a variety of oncogenes such as p60src, p21ras, Raf, Erk kinase (MAP kinase kinase or Mek kinase), phospholipase C_{γ}, phosphatidylinositol 3-kinase, and ribosomal protein S6 kinase (RSK). Many of these latter signaling molecules are themselves protein kinases that are activated on phosphorylation. A number of other pathways, such as those originating from G proteins, merge into this scheme at various points. Although phosphorylation is found on Ser and Tyr as well as on Asp and His residues, the vast majority occur on Ser and Tyr. Thus, protein kinases are mainly of two kinds, tyrosine kinases and serine/threonine kinases. Eventually, this chain of events results in the activation of transcription factors through phosphorylation, in turn effecting transcription of specific genes. Finally, once the signaling is complete, a given phosphoprotein must revert back to its pre-excited resting state, a reaction often catalyzed by the simple removal of the phosphate groups by protein phosphatases.

There are a number of remarkable features about this overall scheme. First, the functional domains of the various signaling molecules (receptors, kinases, and phosphatases) are highly conserved through evolution. Second, and perhaps more surprisingly, all effectors appear to activate the same set of signaling molecules listed above. Hence, instead of classifying the subtopics in terms of the primary effectors (growth factors, hormones, etc.), which is the established practice, I have experimented with an organization based on the transducers that are shared among them. Prokaryotes are presented in a separate section.

2.1. Receptor Tyrosine Kinases

Based on their subunit structure, the nature of the various domains, and their relative placement on the receptors, the transmembrane RTKs are classified into nine (I through IX) families (Ullrich and Schlessinger, 1990). Table I lists these families and their best-studied members. The extracellular regions of the receptors, which bind the appropriate agonist (such as a growth factor or a hormone), consist of various combinations of the following modular domains: cysteine-rich domain, immunoglobulin-like domain, acid box domain, and fibronectin type III domain (Table I). For example, the extracellular region of epidermal growth factor receptor (EGFR) consists of two cysteine-rich domains, whereas those of the receptors of platelet-derived growth factor (PDGFR), fibroblast growth factor (FGFR), and vascular endothelial cell growth factor (VEGFR) are made almost entirely of immunoglobulin-like domains. In addition, all RTKs contain a transmembrane region and, by definition, a tyrosine kinase (TK) domain (Hanks *et al.*, 1988) in the cytoplasmic tail. In type III, IV, and V RTKs, the TK domain is divided by a spacer region called "kinase insert" (Section 2.1.1a).

Our knowledge of the molecular mechanism of RTK-mediated signaling derives primarily from studies of PDGF and EGF receptors. On binding the ligand, RTKs undergo dimerization followed by self-phosphorylation. This phenomenon is variously called autophosphorylation or transphosphorylation, as it is

Table I
Receptor Tyrosine Kinase Families and Members[a]

RTK family: Members	Structure	Ligand	Downstream proteins	Effect
ax1 (IX)	140 kDa; FNIII, IgG; related to IR/Eph?	?	?	Cell adhesion?
eph family (VIII)				
eph, eck, elk, eek, erk, cek5, cek4 (mek4, hek)	130–135 kDa; FNIII, Cys-rich	?	?	Cell adhesion
EGFR family (I)	Cys-rich			
EGFR	170 kDa	EGF, TGF-α, etc.	PLC_γ, grb2, shc, GAP	Ca^{2+}, pH, cell division
neu	185 kDa	NDF, NEL-GF		?
der	?	?	?	Cell fate, eye
lin-3, let-23	?	?	?	Vulva growth
FGFR (1–4) (IV)	Ig	FGF family	PLC_γ raf, erk, rsk	pH, Ca^{2+}, PI turnover
HGFR (met) (VI)	α(50 kDa), β(145 kDa)	HGF, SF	PLC_γ, GAP, PI3-K, src-related kinase	Mitogenesis, cell adhesion, motility

IR family (II)				
IR	$\alpha_2\beta_2$	Insulin	IRS-1, PI-3 kinase, FABP, ATPase, erk, ISPK	Glucose, glycogen, lipid, protein, division, DNA, differentiation
IGFR	$\alpha_2\beta_2$	IGF-1, -2	?	?
IRR	$\alpha_2\beta_2$	?	?	?
sevenless	$\alpha_2\beta_2$	?	Sos; GAP, rap, ras	Vision
NTR family (trk) (VII)	140–145 kDa; Ig? FNIII?	NGF, BDNF, NT 3–5	PLC_γ, erk, raf, mek	Ca^{2+}, PI, membrane
PDGFR family (III)				
PDGFR α and β	Ig	PDGF	GAP, raf, src, PI3-K, PLC_γ	Mitogen
MCSF-1R	Ig	MCSF-1	?	?
steel R (c-kit)	?	?	?	?
VEGFR family (V)	Ig	VEGF		Mitogen

[a]The different families and their numbering (roman numerals) are based on the classification proposed by Ullrich and Schlessinger (1990). Only the major functions and some unique domains are indicated (Fantl *et al.*, 1993; Perrimon, 1993). Abbreviations commonly referred in the text are: R, receptor; Ig, Immunoglobulin-like domains; FNIII, fibronectin III domain; erk, extracellular (signal-)regulated kinase, also known as MAP kinase (mitogen-activated protein kinase or mapk); mek (mapk/erk kinase), MAP kinase kinase; EGF, epidermal growth factor; PLC, phospholipase C; TGF, transforming growth factor; NDF, "neu" differentiation factor; NEL-GF, neu/erbB ligand growth factor; GAP, GTPase-activating protein [*Drosophila* EGFR (der) is a GAP)]; FGF, fibroblast growth factor; rsk, ribosomal protein S6 kinase; PI, phosphoinositol; HGF, hepatocyte growth factor; SF, scatter factor; IR, insulin receptor; IRS, IR substrate; FABP, fatty acid binding protein; IGF, insulin-like growth factor; IRR, insulin-related receptor; sos, son of sevenless; NT, neurotrophin; NGF, nerve growth factor; BDNF, brain-derived growth factor; PDGF, platelet-derived growth factor; MCSF, macrophage colony-stimulating factor; VEGF, vascular endothelial growth factor. A question mark indicates unknown or weak evidence.

not clear whether each subunit phosphorylates itself or is phosphorylated by the kinase domain of the other subunit of the dimer. Following autophosphorylation, individual phosphotyrosine (Tyr-P) residues located on the cytoplasmic domain of the receptor serve as highly selective binding sites for specific cytoplasmic molecules, which then participate in the signaling process.

2.1.1. Determinants of Specificity

2.1.1a. Binding Sequences on RTK. The first hint that signaling molecules directly associate with specific domains of RTKs came from the observation that autophosphorylated PDGF β-receptor binds phosphotidylinositol 3′-kinase (PI3-kinase). A mutant PDGF β-receptor with a deletion in the cytoplasmic "kinase insert" region did not bind PI3-kinase even though it still possessed TK activity and was capable of activating another signaling molecule, phospholipase C_γ (PLC_γ). Detailed mutational analysis later showed that PI3-kinase can bind to either or both phosphotyrosines at positions 708 and 719 of the kinase insert region. Another signaling molecule, GTPase-activating protein (GAP), is bound to Tyr^{739}; still another, PLC_γ, is bound to Tyr^{977} and Tyr^{989}. Thus, it was apparent that sequences surrounding these Tyr residues contribute to the specificity of interaction with the various signaling molecules. Studies using synthetic peptides have supported this assumption (Fantl *et al.*, 1993; Perrimon, 1993). Tyrosine-phosphorylated peptides representing the relevant sites on RTK block the interaction of the corresponding signaling molecules with RTKs by competitively binding to the sites on the signaling molecules. The binding is highly specific and occurs at low concentrations (1–10 mM) of the peptide. For example, the peptide Y^pVPML or Y^pMDMS (where Y_p is phosphotyrosine), representing the two PI3-kinase binding sites on PDGFR, blocks binding of PI3-kinase to PDGFR but does not block the other signaling molecules, GAP and PLC_γ. Shorter versions of the peptides do not block binding, suggesting that five amino acids may be the minimum length required for high-affinity binding of downstream signaling molecules to the RTK.

2.1.1b. Binding Sequences on Signaling Molecules: SH2 and SH3 Domains. Specific sequences of the signaling molecules that interact with the short phosphotyrosine-containing motifs described above are commonly known as SH2 domains. Originally identified as regions with homology to a 100-amino acid stretch of src oncogene region 2 (src homology 2 or SH2), these sequences are widely distributed among phosphotyrosine-binding proteins (Koch *et al.*, 1991; Pawson and Gish, 1992). Proteins with SH2 domains include many enzymes such as phospholipase C (PLC), Ras, GTPase activating protein (GAP), nonreceptor tyrosine kinases (src, fyn, fgr, lck, lyn, fps, abl, syk, and ZAP-70) (Section 2.2), and a number of protein tyrosine phosphatases (PTPs; Section

2.7). In addition, SH2 and SH3 domains also constitute integral parts of non-receptor tyrosine kinases (Section 2.2). A second group of SH2-containing proteins lack obvious enzymatic activity and may, therefore, serve as linkers between RTKs and specific enzymatic subunits. A prototype of this group is the p85 subunit of phosphatidylinositol 3-kinase which has an SH2 domain but no catalytic activity. The p110 catalytic subunit of PI3-kinase communicates with the RTKs through p85. Other SH2-domain proteins lacking enzymatic activity include transforming proteins SHC; the crk family proteins, viz., c-crk, nck, vav, and sem-5/grb2/drk; and cytoskeletal actin-binding protein, tensin. SHC proteins have been found to associate with activated EGFRs *in vivo* and are themselves phosphorylated on Tyr residues, although the kinase responsible for their phosphorylation has not been identified. Proteins in the crk family, on the other hand, consist entirely of SH2 (and SH3, see below) domains. Thus, it appears that some signaling proteins have evolved to encode both Tyr-P binding and enzymatic activities, but in other pathways, the two functions are assigned to two different polypeptides.

Signaling by insulin receptor (IR) involves the SH2 domain in a unique way (Shoelson *et al.*, 1992). On binding of insulin to the extracellular α subunit of the IR, the intracellular β subunits undergo autophosphorylation primarily within the catalytic domain. However, this phosphorylation is mainly involved in stimulating its own tyrosine kinase activity rather than binding signaling proteins. Indeed, the main substrate for IR is a 160- to 185-kDa protein called insulin receptor substrate 1 (IRS-1), tyrosine phosphorylation of which (by IR) creates binding sites for downstream SH2 proteins, such as p85 of PI3-kinase.

Sequence alignment of SH2 regions of various signaling molecules resulted in the identification of four highly conserved basic residues; their locations in the src protein itself are Arg^{155}, Arg^{175}, His^{201}, and Lys^{203}. X-ray and NMR analysis and determination of solution structure of the SH2 domains of src and c-abl, either free or complexed with small phosphopeptides (see Section 2.1.1a), provided a mechanism for the binding specificity (Pawson and Gish, 1992; Birge and Hanafusa, 1993). Among the four basic residues, Arg^{155} (of src), which is strictly conserved in all SH2 domains (except that of PTP IC), was seen to form tight hydrogen bonds with two oxygen atoms of the phosphate group while two of the other basic residues interact with the aromatic ring of tyrosine, thus accounting for the high specificity for phosphotyrosine. Interestingly, the overall three-dimensional structures of all three SH2 domains were very similar. Thus, the specificity of interaction in any given pair of SH2 and RTK must be determined through the recognition of variable amino acids in the interacting proteins. Indeed, amino acid residues $+1$ to $+3$ relative to phosphotyrosine appeared to be in direct contact with the species-specific unique residues of the cognate SH2 region.

Many cytosolic signaling proteins that contain an SH2 domain also contain an adjacent SH3 domain (Ciccetti *et al.*, 1992). Although not studied in detail, SH3 domains, like SH2, are modular in design and appear to be responsible for protein–protein interaction. Examples include the following SH3 domain proteins (the corresponding binding protein being shown in parentheses): grb2, drk (*Drosophila* son-of-sevenless); abl (GAP-like proteins 3BP-1 and 3BP-2); crk (C3G and c-abl); nck (Ser/Thr protein kinase); src (actin cytoskeletal proteins p110 and paxillin); and PLC_γ (a variety of cytoskeletal proteins). Recent structural analyses indicate that SH3, like SH2, recognizes short specific peptide sequences, although phosphorylation is not directly involved in SH3-mediated interactions (Ciccetti *et al.*, 1992). Genetic and biochemical evidence suggests that SH3 domains are involved in regulating enzymes that control small guanine-nucleotide-binding proteins (Schlessinger and Ullrich, 1992). The occurrence of SH3 domains in many actin-binding proteins implied a role of this domain in interaction with cytoskeleton and cell structure (Drubin *et al.*, 1990). The exact mechanism of action of SH3 domains and their possible synergy with SH2 domains in the same polypeptide remain largely unexplored.

2.2. Nonreceptor Tyrosine Kinases

Essentially all members of this class are products of protooncogenes, the best known member being the src kinase (pp60src); other notable members are: yes, fgr, JAK, lck, and lyn. In addition to binding to the cytoplasmic domain of the RTKs (e.g., src, fyn, and yes binding to PGFR), these kinases bind to supplement specific receptors that lack a tyrosine kinase activity of their own, including the hematopoietic receptors, that is, CD receptors (e.g., CD3 CD4, CD48), and receptors for interleukins (IL), G-CSF. Thus, despite the lack of an extracellular domain, many of them are activated by growth factors and other cellular activators (Gutkind *et al.*, 1990; Kypta *et al.*, 1990). In addition to a tyrosine kinase domain, they contain SH2 and SH3 domains as well as a short amino terminal sequence where myristoylation occurs. The myristic acid is responsible for membrane attachment, probably via a specific 32-kDa receptor in the membrane (Resh and Ling, 1990).

Important clues to the mechanism of src activation (Cooper and Howell, 1993) came from the discovery that c-src is phosphorylated *in vivo* at two tyrosine residues: at Tyr^{416} by autophosphorylation and at Tyr^{527}, which is near the C terminus and outside of the kinase domain, apparently by a distinct tyrosine kinase and not by autophosphorylation. The Tyr^{527} is missing in the oncogenic v-src, which is constitutively active; moreover, mutation of this Tyr to Phe increased the tyrosine kinase activity of c-src, suggesting that phosphorylated Tyr^{527} may play a negative regulatory role. The existing model for the activation of c-src (Matsuda *et al.*, 1990) is included in Figure 1, which proposes that when Tyr^{527} of c-src is phosphorylated, it binds to the SH2 domain of the same

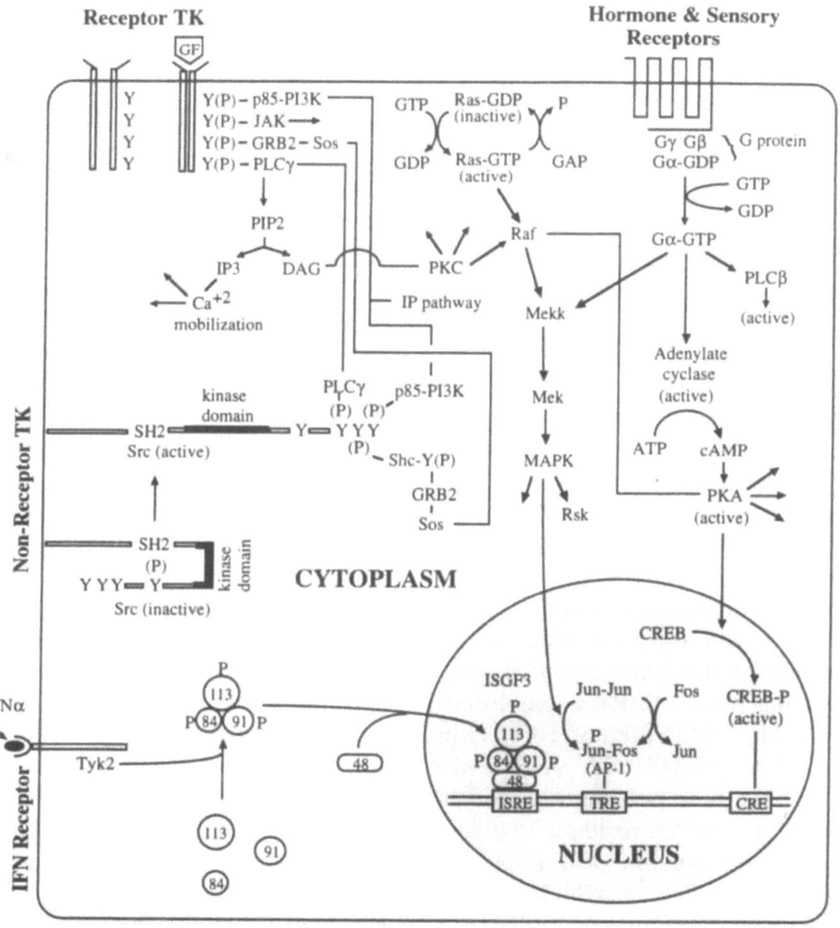

FIGURE 1. Representative intracellular signal transduction pathways. Three classes of receptors are depicted (see text for details): RTKs of growth factors (Section 2.1); nonreceptor tyrosine kinases (src) (Section 2.2); heptahelical hormone and sensory receptors signaling through heterotrimeric G proteins (Section 2.3.5c); and ISGF3α receptor (Section 2.6.1c). Y(P) indicates phosphotyrosine residues, drawn in arbitrary order; c-jun undergoes positive (as shown) as well as negative regulation (by CKII and GSK-3; Section 2.6.2a). The unphosphorylated form of CREB can bind DNA; phosphorylation only stimulates its transactivation property.

molecule in a head-to-tail fashion, resulting in the folding and hence the inhibition of the intervening tyrosine kinase. Thus, in principle, prevention of this self-association by inhibition of Tyr^{527} phosphorylation or by blocking either the Tyr^{527} or the SH2 domain should activate the src tyrosine kinase. This model is analogous to the autoinhibitory domain model that has been proposed for protein serine kinases (Soderling, 1990) and can explain a variety of experimental observations (Cooper and Howell, 1993; Fantl *et al.*, 1993).

The targets of nonreceptor tyrosine kinases have generally been elusive. Their high efficiency of transformation despite a low level of self-phosphorylation argues that the immediate downstream targets are likely to be amplifiers, such as enzymes (Figure 1). A number of such amplifiers that interact with various members of the src family have been identified in the recent past (Cantley *et al.*, 1991); these include PI3-kinase, PLC_{γ}, GAP proteins, raf (see Section 2.3.5b), and PP2A (protein Ser/Thr phosphatase; see Section 2.7.2b). The lesser known proteins that are found to be phosphorylated in src-transformed cells and, therefore, likely to be targets of src kinase include ezrin (Gould *et al.*, 1989), a 42-kDa ser/thr kinase (Ely *et al.*, 1990), the lipocortin/calpactin family of proteins, and others (Hunter *et al.*, 1988).

The importance of the membrane association of the src-family proteins and hence the role of myristoylation have been emphasized. Mutations that prevent myristoylation also abolish the transformation activity of src, suggesting that src may function to localize downstream signaling molecules to critical sites on the membrane (Cross *et al.*, 1984; Kamps *et al.*, 1985).

A different class of nonreceptor tyrosine kinases includes the ones that are associated with the membrane only through specialized receptors. The prototype members of this class are p56lck and p59fyn, which associate with activated T-cell receptors (TCR), exemplified by CD3 (reviewed by Perlmutter *et al.*, 1993). The CD3 receptor consists of an $\alpha\beta$ heterodimer (involved in antigen binding) associated with γ, δ, and ϵ chains and a disulfide-linked pair of ζ chains. All these polypeptides contain similar "activation motifs" consisting of paired Tyr and Leu residues in a precise orientation. Binding of antigen stimulates phosphorylation of Tyr residues in these motifs, most likely by p59fyn acting directly and by p56lck recruited through CD4 and CD8 receptors. This is followed by the recruitment of protein tyrosine kinase ZAP-70, the SH2 domains of which bind to paired phosphate groups of the ζ chains (Arpaia *et al.*, 1994). The binding activates the ZAP-70 kinase that in turn acts on a myriad of targets to provoke the lymphocyte activation process. A recently discovered family of Janus kinase (also known as just another kinase) subfamily that includes closely related JAK1 and JAK2 (Firmbach-Kraft *et al.*, 1990). JAK2 is recruited by the erythropoietin receptor in response to stimulation by erythropoietin (Witthuhn *et al.*, 1993), which is essential for the differentiation and proliferation of the erythroid cells.

Perhaps the truly soluble tyrosine kinases are those exemplified by Met and Trk oncogene products, ligands of which remain unidentified. Some oncogenic tyrosine kinases (e.g., fps and abl) are found associated with membranes as well as elsewhere. A population of abl protein, for example, is found in the nucleus, although its cytoplasmic translocation is required for transformation (Van Etten *et al.*, 1989).

2.3. Protein Serine/Threonine Kinases

The majority of protein phosphates in an adult tissue are located on Ser and Thr, with only a small fraction being on Tyr. A variety of serine/threonine kinases, generally cytoplasmic in nature, act as intermediary transducers of signals between the membrane and the final site of execution, often the nucleus. Some members important in signal transduction are cyclic-AMP- and cyclic-GMP-dependent protein kinases (PKA and PKG), casein kinases I and II, glycogen synthase kinase (GSK-3), CA^{2+}/calmodulin-dependent protein kinases, protein kinase C (PKC), raf, and those belonging to the MAP kinase cascade. A number of agonists activate adenylate cyclase through seven transmembrane G-protein-linked receptors (Lefkowitz, 1993); these include the neurotransmitters norepinephrine, dopamine, and enkephalin; thyroid- and follicle-stimulating hormones (TSH, FSH). Guanylate cyclase is activated by receptors such as the transmembrane atrial natriuretic peptide (ANP) receptor. These enzymes catalyze the synthesis of cyclic AMP and cyclic GMP, which in turn activate PKA and PKG, respectively. The involvement of PKA in two classic biochemical pathways (glucose metabolism and muscle contraction) is described below.

2.3.1. Cyclic-AMP-Dependent Protein Kinase and Glycogen Metabolism

The PKA is a tetramer (R_2C_2) composed of two catalytic subunits (C) and two regulatory subunits (R) that inhibit the catalytic subunits (Taylor *et al.*, 1990). Binding of cyclic CAMP to the R subunits alters their conformation such that they cannot bind to the catalytic subunits any more, which results in the release and activation of the catalytic subunits. The R subunits are of two kinds, RI and RII. RI is expressed in most tissues, but the two RII subunits, RII-H and RII-B, although similar in sequence, are expressed in a tissue-specific manner. Biochemically, the most important difference between RI and RII is that RII is (auto)phosphorylated by the C subunits, whereas RI is not. The phosphorylation slows the rate of reassociation of RII with the C subunits and, therefore, may have a role in regulation. The consensus phosphorylation motif of PKA is $RRXS^PX$, and that of PKG is $RXXS^PRX$, where the S^P is a phosphorylated Ser (Kennelly and Krebs, 1991).

Historically, glycogenolysis (breakdown of stored glycogen into glucose) in mammalian skeletal muscle was the first system in which protein phosphorylation was identified as a regulatory mechanism. The many facets of glycogenolysis that continue to provide insights into the mechanism of action of protein kinases and phosphatases are briefly summarized here. Glycogenolysis, catalyzed by the enzyme glycogen phosphorylase, is activated by epinephrine (via cyclic AMP) in resting muscle and by Ca^{2+} during contraction (Cohen, 1989). The effects of Ca^{2+} and cyclic AMP are integrated by phosphorylase kinase, an enzyme that phosphorylates glycogen phosphorylase, resulting in the activation of the latter. The phosphorylase kinase (PK) is composed of α, β (regulatory), γ (catalytic), and δ (calmodulin) subunits. Association of dephosphorylated α and β subunits inhibits the catalytic activity of the δ subunit. Inhibition is relieved (1) by the interaction of Ca^{2+} with the δ subunit, (2) by the interaction of troponin C with the β subunit, or (3) by the phosphorylation of the α and β subunits, the phosphorylated forms being free of their inhibitory effect on δ. Increased steady-state phosphorylation of α and β can in turn be achieved by either of two general mechanisms—stimulation of phosphorylation or inhibition of dephosphorylation—as described below.

The first mechanism involves phosphorylation of α and β by PKA in response to epinephrine. The phosphorylated form of PK can be activated by a much lower concentration of Ca^{2+} compared to the dephosphorylated enzyme. Although the exact mechanism of this enhancement of Ca^{2+} sensitivity remains unknown, detailed studies of Ca^{2+} binding have offered a possible explanation. There are four Ca^{2+}-binding sites in calmodulin (δ subunit), two high-affinity and two low-affinity ones, in the N-terminal and C-terminal domains, respectively (Section 2.3.4). In its dephosphorylated state, activation of PK requires occupancy of all four sites by Ca^{2+}. Thus, it is possible that phosphorylation by PKA may allow activation when only the high-affinity sites are occupied by Ca^{2+}. This may explain why phosphorylation sensitizes the PK holoenzyme to Ca^{2+} without altering the Ca^{2+}-binding properties of its calmodulin subunit.

The second mechanism involves inactivation of PP1G, a member of the protein Ser/Thr phosphatase 1 (PP1) family, which inactivates both glycogen phosphorylase and phosphorylase kinase through dephosphorylation. Association of PP1G with glycogen is mediated by a glycogen-binding protein called the G subunit (also see Section 2.7.2a). The PKA phosphorylates the G subunit at a specific Ser residue near the N-terminus (site 2), which then triggers the dissociation of PP1 from the G subunit, thus preventing dephosphorylation and inactivation of the phosphorylase. The PP1 released into the cytoplasm may be further inactivated by inhibitor-1, a protein that is transformed into a potent inhibitor of PP1 on phosphorylation by PKA (Huang and Glinsmann, 1976). Glycogen metabolism thus exemplifies an intricate network of phosphorylation and dephosphorylation controlling a biochemical pathway of fundamental importance. One other aspect of this pathway, the stimulation of glycogen synthesis by insulin, is described below.

2.3.2. Insulin-Stimulated Protein Kinase

Insulin stimulates uptake of nutrients (glucose, fatty acids) by cells and their conversion into complex storage molecules (glycogen, lipid). Epinephrine has an opposing effect: acting via cyclic AMP, it promotes the breakdown of glycogen (and lipid). As described in the previous section, epinephrine activates PKA via the cyclic AMP pathway, which phosphorylates site 2 of the G subunit (the protein that mediates association of phosphatase PP1G with glycogen). On the other hand, binding of insulin to its receptor (IR) results in receptor activation through autophosphorylation. This in turn activates a Ser kinase cascade, resulting in the activation of "insulin-stimulated protein Ser/Thr kinase" (ISPK, a mammalian homologue of ribosomal protein S6 kinase), which phosphorylates on site 1 of the G subunit. Unlike the site 2 phosphate, the site 1 phosphate stabilizes the association of the G subunit with PP1G and thus dephosphorylates both glycogen phosphorylase and glycogen synthase. This results in the inhibition of glycogen phosphorylase and activation of glycogen synthase. Thus, phosphorylation of the same protein (G subunit) at two different sites produced opposing effects on its activity, which formed part of the basis of antagonistic effects of the two hormones.

2.3.3. Protein Kinase A and Cardiac Muscle Contraction

Troponin I and phospholamban are the two major PKA substrates in the heart. Troponin phosphorylation increases the rate of dissociation of Ca^{2+} from troponin C, presumably the rate-limiting step in cardiac relaxation. Phospholamban inhibits the sarcoplasmic reticulum (SR) ATPase responsible for Ca^{2+} transport and hence the rate of Ca^{2+} uptake in the SR. Phosphorylation of Ser^{16} of phospholamban by PKA relieves the inhibition and thus increases the sacroplasmic uptake of the Ca^{2+} released by troponin C. Stimulation of heart by epinephrine results in increased cyclic AMP synthesis and hence activation of PKA. The dual effect of phosphorylation of troponin and phospholamban thus enables epinephrine to rapidly lower the cytosolic concentration of Ca^{2+}, effecting a quick relaxation of the cardiac muscle. This mechanism is unique to the cardiac muscle because phospholamban is absent in the fast-twitch skeletal muscle. Moreover, the skeletal muscle isoform of troponin I lacks the N-terminal region of the cardiac troponin that contains Ser residues phosphorylated by PKA, which further vindicates the model suggested above.

Increased cytoplasmic Ca^{2+} level in the SR also appears to activate a calmodulin-dependent protein kinase in the SR membrane, which then phosphorylates Thr^{17} of phospholamban, activating the Ca^{2+}-ATPase activity and increasing the Ca^{2+} uptake even more. The level of phosphorylation of phospholamban may be amplified further through the inhibition of PP1 by a mechanism essentialy identical to the one adopted by epinephrine (described in Section 2.3.1).

2.3.4. Calcium/Calmodulin-Dependent Protein Kinases

The 15-kDa protein calmodulin (CaM), discovered in 1970, is the most important Ca^{2+}-binding protein found in all eukaryotic cells (Kennedy, 1989).

Each polypeptide of calmodulin binds four Ca^{2+} ions in pockets called "EF hand" (because the perpendicularly oriented E and F helices distantly resemble the open thumb and index finger of an otherwise closed fist) when the concentration of the latter rises above 1 μM, thus converting calmodulin into a multifunctional activator. The multitude of proteins that are activated by the Ca^{2+}-calmodulin complex include adenylate cyclase, Ca^{2+}-ATPase, calcineurin (PP2B), Ca^{2+}/calmodulin-dependent protein kinase (CaMK), myosin light chain kinase (MLCK), and nitric oxide synthase. These proteins can vary widely in their binding affinity constants for Ca^{2+}-calmodulin, which may allow regulated, stepwise recruitment of a different set of proteins as the intracellular Ca^{2+} concentration changes. The CaMKs are perhaps more varied in structure and function than any other class of kinases and play important roles in nerve signaling. In brain, several CaMKs, including MLCK and CaMK I and II, exhibit highly specific substrate requirements. The only known substrate of CaMK I is the vesicle-associated protein synapsin I, which is phosphorylated at the same Ser residue (Ser^9) by CaMK I and by PKA. Phosphorylation by either enzyme modulates the affinity of synapsin for actin filaments and may serve to regulate the availability of synaptic vesicles for transmitter release.

The CaMK II is highly concentrated in the neurons of the forebrain and constitutes nearly 2% of the total protein of the hippocampus. In fractions enriched for postsynaptic membranes, the kinase consists ~25% of the total protein, making it a likely target for Ca^{2+} entering through postsynaptic ligand-gated ion channels. A large variety of neural proteins act as substrates for CaMK II, e.g., Tyr and Trp hydroxylases, tubulin, and synapsin I. The CaMKII-mediated phosphorylation of synapsin I occurs at sites (Ser^{566} and Ser^{603}) different from the CaMK I site and may also regulate the affinity of synapsin for actin and synaptic vesicles. The CaMK II is a family of large oligomers of ten homologous subunits. An intriguing aspect of CaMK II regulation is its autophosphorylation on activation by the Ca^{2+}-calmodulin complex, allowing it to function as a molecular switch in a complex series of allosteric conformational changes. On autophosphorylation for a few seconds in the presence of Ca^{2+}-calmodulin, the enzyme becomes Ca^{2+} independent. The location of the site of autophosphorylation, Thr^{286}, suggests that the phosphorylation may mimic the effect of Ca^{2+} binding and prevent the refolding of the protein to the inactive conformation even after Ca^{2+} levels fall. Activation by autophosphorylation is a highly cooperative process; autophosphorylation of one or two subunits produces a rapid activation of the other subunits of the holoenzyme, likely by allosteric interactions. In addition, autophosphorylation within the holoenzyme becomes Ca^{2+} independent. When Ca^{2+} is removed after autophosphorylation of Thr^{286}, another Thr (Thr^{305}) located within the calmodulin-binding domain is exposed and becomes autophosphorylated. As a result, the kinase becomes insensitive to stimulation by Ca^{2+}-calmodulin. This mechanism is corroborated by the finding that following the dephosphorylation of the Thr-phosphates, the Ca^{2+}-dependent and -independent activities operate *in vivo*.

2.3.5. The MAP Kinase Cascade and raf: The ras Pathway

GTP-binding proteins, such as heterotrimeric G proteins and ras, encode GTPase but no protein kinase activity. However, as exemplified by ras, they are often at the crossroads of signaling pathways, connecting growth stimulants and their cognate RTKs with other protein kinases, including raf, which eventually culminates in the phosphorylation and activation of transcription factors in the nucleus (Figure 1). Thus, we provide a brief overview of this exciting area of research using ras as the prototype model (Boguski and McCormick, 1993; Crews and Erikson, 1993; Lowy and Willumsen, 1993; Satoh *et al.*, 1992). Signaling of heterotrimeric G proteins via mekk is also presented here (Section 2.3.5c).

2.3.5a. ***ras*.** In 1980, only two *ras* genes were known, both of them in rat sarcoma viruses: v-H-*ras*, the transforming gene (oncogene) of Harvey, BALB, and Rasheed sarcoma viruses; and v-K-*ras*, the transforming gene of Kirsten sarcoma virus. Within the next decade, more than 50 distinct ras-like proteins controlling many different aspects of cellular behavior have been identified in mammals, frogs, flies, fungi, and nematodes. It is now known that mammalian cells express four "true" ras proteins—H-ras, N-ras, K-rasA, and K-rasB—each of which function as an oncogene following mutational activation.

The normal product of a *ras* gene binds guanine nucleotides, GDP and GTP, and functions as a binary switch (Figure 1). In its *off* state, ras binds GDP and rests on the inner surface of the cell membrane. After receiving an activating signal, it releases the GDP and binds GTP, which represents the *on* state. This is mediated by proteins known as guanine nucleotide exchange factors (GEFs), less commonly called guanine nucleotide releasing proteins (GNRPs) or guanine nucleotide dissociation stimulators (GDSs), and inhibited by guanine nucleotide dissociation inhibitors (GDIs). The GTP-bound form is, therefore, converted back to the GDP-bound form; however, because the intrinsic GTPase activity of ras is much lower than that of heterotrimeric G proteins, this process needs to be accelerated by GTPase-activating proteins (GAPs). Oncogenic ras proteins, containing mutations at sites 12, 13, or 61, are resistant to GAP-mediated GTPase stimulation and are, therefore, constitutively locked in their GTP-bound, active state (Hall, 1990). The GTP-binding (and -hydrolyzing) property of ras proteins and their association with cytoplasmic membrane through modification of lipids suggest that, like G proteins, ras serve as transducers in transmembrane signaling systems.

Despite intensive search, the immediate downstream effector of mammalian ras in the signaling pathway remains elusive. Nevertheless, the complex mediators described above must play important roles in ras-induced phosphorylation of signaling molecules. A ras-GEF has recently been shown to be responsible for activating ras in response to EGF. As described in Section 2.1, binding of EGF induces dimerization and autophosphorylation of its receptor, creating binding sites for SH2 domain proteins (Table I). One such protein in humans is grb2 (homologous to sem-5 in *C. elegans* and drk in flies), which, in turn, binds to the ras-GEF known as sos1 via its SH3 domain (Figure 1). These protein–protein interactions thus serve to recruit cytoplasmic sos to the membrane, inducing ras activation. Nonreceptor tyrosine kinases may participate in ras activation in a similar way (e.g., src in Figure 1); for example, they may directly or indirectly phosphorylate shc, which then binds the grb2–sos1 complex, presumably bringing sos1 to ras at the membrane. It is not known how the other GEF proteins, which do not contain recognizable SH2 domains, communicate with ras; in such proteins, other motifs may be involved in this role, e.g., SH3 domains in cdc25 and scd25, or the pleckstrin homology (PH) regions in ras-GEF. Although p120-GAP, the first GAP protein to be characterized at molecular level, contains SH2, SH3, PH, and phospholipase A_2 calB-like domains, many other GAPs identified afterwards lack these motifs.

In addition to GAP and raf, PI3-kinase has recently been shown to transduce signals directly from activated ras (Rodriguez-Viciana *et al.*, 1994) in a GTP-dependent reaction. A dominant negative ras mutant (N17) inhibited growth-factor-induced production of $PI(3,4)P_2$ and $PI(3,4,5)P_3$, transfection of ras, but not raf, corrected the defect.

2.3.5b. raf. Genetic evidence supporting a ras-downstream location of raf is as follows: (1) activated raf transformed cells that had reverted to a flat phenotype in spite of having two activated *ras* alleles (Noda *et al.*, 1983); (2) in the reciprocal experiment, activated ras failed to overcome the block caused by a dominant negative raf mutant (Kolch *et al.*, 1991). Thus, raf remains the most likely effector of ras activity and the closest known protein Ser/Thr kinase downstream of ras (Figure 1). Use of the yeast two-hybrid system has shown that the noncatalytic N-terminal domain is sufficient for ras–raf interaction. *In vitro*, ras protein immobilized on an affinity matrix binds raf as well as MAP kinase and mek in a GTP-dependent manner (Moodie *et al.*, 1993). However, it is still unclear how ras activates raf; two attractive possibilities are that ras does it by stimulating the kinase activity of raf or by recruiting raf to the membrane.

Among the few downstream signaling molecules known to interact with raf, mek appears to be the most important, as it is at the crossroad of signaling originating from heterotrimeric G proteins and ras (Figure 1). Although most enzyme–substrate interactions are transient, raf and mek form an extremely tight complex that cannot be dissociated with 70% ethylene glycol, 3 M NaCl, or 3%

n-octyl glucoside (Huang *et al.*, 1993). Catalytic functions of raf or mek are not required for this binding.

Originally detected as a product of the transforming genes of murine and avian sarcoma viruses, raf represents a family of related protein kinases that include the p74raf-1, p68rafA, and the p95rafB (Heidecker *et al.*, 1992). Homologues of raf have since been found in mouse, chicken (the *mil* gene), *Xenopus, Drosophila* (the *polehole* gene), and *C. elegans;* raf-1 is ubiquitously expressed in adult tissues, whereas rafA expression is highest in cells of the urogenital system, and rafB expression in brain and testes. A large variety of mitogens (e.g., growth factors) rapidly stimulate raf activity, which is accompanied by increased raf phosphorylation, leading to characteristic shift in mobility. Among the principal upstream effectors that transmit the activated receptor signals to raf are PKC and ras; raf is, therefore, strongly implicated in cellular growth control. Curiously, the only mitogen that has failed to activate raf is interleukin 4 (IL-4), indicating that IL-4 may be an incomplete mitogen.

Phylogenetic studies suggest that raf, together with mos, are the only known Ser/Thr kinases that belong to the src superfamily, which otherwise encodes protein tyrosine kinases (Section 2.2). All raf proteins share three highly conserved regions (CR): CR1 with Cys residues in a Zn-finger configuration; CR2 with a high content of Ser, Thr, and Pro; and CR3 corresponding to the catalytic domain. The C-terminal 20 amino acids following the CR3 region are dispensable for transformation and kinase activity. Phosphorylation of raf occurs at Ser, Thr, and Tyr residues, the exact sites and nature of the kinase(s) remaining largely unknown. Recently, three sites of phosphorylation have been mapped: Ser^{43}, belonging to the sequence RRASDDGK with homology to PKC consensus site (RX_2S: Kennelly and Krebs, 1991); Ser^{259} in the amino-terminal half of CR2 having the sequence (^{253}QRQRSTS^{259}TPN); and Ser^{621}, near the C-terminal end of CR3. The viral versions of raf (v-raf) are constitutively active because of the deletion of the N-terminal regulatory half of c-raf. Mutation of Ser^{621} to Ala abrogates the ability of v-src and v-ras to activate raf, suggesting a role of phosphorylation in raf function (Morrison *et al.*, 1993).

2.3.5c. MAP Kinase Cascade. Protein kinases belonging to the MAP kinase cascade are MAP kinase kinase kinase (mek kinase), MAP kinase kinase (mek), and MAP kinase (mapk or erk) (Nishida and Gotoh, 1993). Each member of this cascade is a protein Ser/Thr kinase that phosphorylates the one downstream of it. In addition, MAP kinases can autophosphorylate on Tyr residues. Activation of mek by raf-1 or v-mos occurs through phosphorylation of Ser^{218} and Ser^{222} (Zheng and Guan, 1994). The mek (MAP kinase kinase), in turn, phosphorylates mapk on neighboring Thr^{183} and Tyr^{185} residues, which are both required for mapk activation. Thus, mek is a "dual specificity" kinase that phosphorylates both Thr and Tyr; mek seems to occupy a strategic position in raf downstream signaling because it integrates two converging pathways, one involving mekk (MAP kinase kinase kinase), the other involving ras and raf (Figure 1). Very recently, ras has also been shown to activate Mekk directly on growth factor stimulation (Lange-Carter and Johnson, 1994).

The existence of a G-protein stimulatory pathway of mek activation has been unraveled (Figure 1). For example, heterotrimeric G-protein-linked thrombin receptor and the GTPase-inhibited G_{i2} mutant, gip2, both activate mek in a raf-independent manner, likely via mek kinase (mekk) (Gardner *et al.*, 1993).

The 40- to 46-kDa MAP kinases that have been reported include erk1 (mapk2), erk2 (mapk1), erk3, erk4, and hera in mammals; fus3, kss1, and spk1 in yeast; and mpk in sea stars. The optimum recognition sequence of MAP kinases seem to be Pro-X-(Ser/Thr)-Pro, where X is a basic or neutral amino acid. *In vitro,* several important signaling proteins serve as substrates for MAP kinases, including ribosomal protein S6 kinases p90rsk and p70rsk, myosin light chain kinase (MLCK), the EGF receptor, and DNA-binding proteins c-fos, c-jun, c-myb, c-myc, and TCF. Activation of rsk stimulates protein synthesis.

At least three systems have been dissected genetically in detail and therefore serve as models for a reasonably complete signal transduction pathway extending from the membrane to the nucleus (reviewed by Perrimon, 1993; Brown and Hartley, 1994); the *torso* and *sevenless* pathways of *Drosophila* and the let-23 pathway of *C. elegans*. Because their overall designs are very similar, only the two *Drosophila* systems are described here. In the sevenless–sos pathway, the R7 photoreceptor cell is induced through the sevenless receptor (tyrosine) kinase by its ligand *boss*. Activated sevenless then recruits the SH2–SH3 domain protein drk (*d*ownstream of *r*eceptor *k*inases, previously called enhancer of sevenless) and nucleotide exchange factor sos to the membrane. This leads to the activation of ras and in turn Draf (*Drosophila* raf). The cascade continues through mek (encoded by *Dsor1*), and mapk (encoded by *rolled*), which translocates into the nucleus to phosphorylate two ETS-like transcriptional regulators, inactivating a negative one (*yan*) and activating a positive one (*pointed*). This leads to simultaneous activation and derepression of target gene *Rh4*. *Pointed* was also found to be a target for EGF receptor signaling during midline glia development, consistent with the general requirement of the ras pathway in RTK signaling.

Torso is a receptor tyrosine kinase whose structure is reminiscent of the PDGF receptor. Analysis of gain-of-function *torso* mutations and expression of torso protein by injection of its mRNA into syncytial embryos have demonstrated that torso is necessary and sufficient for terminal cell fates (Sprenger and Nusslein-Volhard, 1992). On activation by the torso ligand, the receptor undergoes dimerization and autophosphorylation and recruits the sos protein (son of sevenless) via an unidentified adaptor, possibly drk. The rest of the pathway then follows the cascade phosphorylation scheme as described for the *sevenless* pathway above, resulting in the activation of mapk. Activated mapk in turn phosphorylates the transcription factor encoded by gene Y, which controls spatial expression of the terminal genes *tailless* and *huckebein*. Interestingly, in yeast, the three kinases mekk, mek, and mapk associate together to form a multikinase complex, implicated in mating pheromone signaling (Choi *et al.*, 1994).

In a recent study, constitutively active mek mutants were made in which Ser^{218} and Ser^{222} (sites of phosphorylation) were substituted with acidic residues (Glu/Asp), and a putative negative regulatory domain (residues 32–51) was deleted (Mansour *et al.*, 1994). Such mutants exhibited basal kinase activities up to 400 times higher than that of the unphosphorylated wild-type mek. Expression of these mutants in mammalian cells resulted in the transformation of the cells and activation of transcription mediated by AP-1, further strengthening the model in Figure 1.

2.3.6. Transmembrane Ser/Thr Kinase Receptors: TGF-β Signaling

The transforming growth factor β (TGF-β) signals by contacting two distantly related transmembrane protein Ser/Thr kinases, named receptors I and II. The TGF-β first binds directly to receptor II, which is constitutively active. Bound TGF-β is then recognized by receptor I, which is recruited into the complex and becomes phosphorylated by receptor II. Phosphorylated receptor I in turn propagates the signal to downstream substrates. This is a unique mechanism whereby a cytokine directly activates the first step of a signaling cascade (Wrana *et al.*, 1994). Decapentaplegic, the TGF-β homologue in *Drosophila*, signals through two members of the type I receptor family, saxophone (sax) and thickvein (tkv), to generate body patterns (Nellen *et al.*, 1994).

Because of the extensive literature on protein kinase C activation by the inositol signaling pathway, it is treated separately in the following section. Other Ser/Thr kinases are described in appropriate places (such as cdc2/cdc28 in Section 2.6.1a).

2.4. Protein Kinase C and Inositol Phosphates

Our knowledge of the inositol phosphates and their role in cell signaling has expanded tremendously in the last decade. These reactions form the cornerstone of a ubiquitous signal transduction pathway that underlies a large array of processes such as cell growth, metabolism, secretion, and neural and contractile responses. It was recognized early on that stimulation of a number of cell surface receptors initiates hydrolysis of the membrane-bound inositol lipid, phosphotidylinositol-4,5-bisphosphate (PIP_2) to produce two second messengers, inositol-1,4,5-trisphosphate (IP_3) and 1,2-diacylglycerol (DAG) (for recent reviews, see Berridge and Irvine, 1989; Rana and Hokin, 1990; Majerus, 1992;

Nishizuka, 1992; and other chapters of this volume). These messengers are generated by a membrane transduction process involving three major components: the agonist receptor, a coupling G protein, and phosphoinositidase C (phospholipase C, PLC). Subsequently, IP_3 mobilizes calcium from internal cellular stores, and DAG, together with Ca^{2+}, stimulates protein kinase C (PKC). This textbook diagram of the classic inositol phosphate pathway (Figure 1) is now intertwined with a variety of regulatory circuits. The different arms of the pathway can be activated by Ca^{2+} ionophores and exogenous activators of PKC such as phorbol esters. Recent findings have demonstrated the existence of a nuclear inositol cycle, entirely separate from the more familiar one in the plasma membrane and operating essentially independently, raising the possibility that phosphoinositides may play a more direct role in nuclear transcription (Divecha *et al.*, 1993).

The myriad of inositol phosphates and inositol-containing lipids that are found in the cell undergo interconversions in every imaginable way, as detailed in other chapters of this volume. Nevertheless, IP_3 and DAG continue to remain the major entities of the inositol pathway, and hence, in keeping with the primary theme of this review, this section will focus on PKC and its activation through these messengers.

2.4.1. Structure and Activation of PLC

Although agonist-mediated activation of PLC occurs in a variety of tissues, two major classes of receptors are involved in the process: (1) the "serpentine" receptors with seven transmembrane domains, which function through heterotrimeric G proteins (Figure 1), e.g., receptors for angiotensin II, α-adrenergic agonists, and chemotactic peptides; (2) the RTKs that are activated by growth factors, e.g., EGF, FGF, and PDGF receptors (Section 2.1). The three major isoforms of PLC—β, γ, and δ—have molecular masses of 150, 145, and 85 kDa, respectively, and show an overall low degree of similarity. Greater homology is seen in two regions, referred to as X and Y, which consist of 170 and 260 residues, respectively, and are likely to be involved in phosphoinositide binding and/or catalytic activity. The γ isoform additionally contained recognizable SH2 and SH3 domains (see Section 2.1.1b). Thus, whereas PLC_β is activated by G-protein-linked receptors, PLC_γ is activated by autophosphorylated growth factor receptors (Rhee and Choi, 1992). A novel class of heterotrimeric G proteins, named the G_q/G_{11} superfamily, has recently been identified (Blank *et al.*, 1991; Smrcka *et al.*, 1991). The α subunits of these G proteins, e.g., $G\alpha_q$, have been shown to activate PLC_β, in particular $PLC_{\beta 1}$. The crystal structure of a Gα subunit has been solved (Coleman *et al.*, 1994).

Growth factor-induced phosphorylation of $PLC_{\gamma 1}$ takes place at $Tyr^{771,783,1254}$ (Kim *et al.*, 1991). *In vitro,* phosphorylated and unphosphorylated

PLCs show similar activity in standard assays. However, the unphosphorylated enzyme was selectively inhibited in the presence of Triton X-100 or profilin, a small protein that binds PIP_2 and actin. Because only the phosphorylated enzyme hydrolyzes profilin-bound PIP_2, thereby releasing PIP_2 and altering the polymerization state of actin, profilin has been implicated as a link between transmembrane signaling and cytoskeletal responses (Goldschmidt-Clermont *et al.*, 1991).

2.4.2. Molecular Structure and Activation of the PKC Family

Protein kinase C interacts directly and indirectly with many other pathways that make use of second messengers such as cyclic AMP, cyclic GMP, and Ca^{2+} and play important roles in nerve signaling, smooth muscle contraction, cardiac function, multidrug resistance, neoplastic transformation, differentiation, and possibly apoptosis (Mahoney and Huang, 1994).

2.4.2a. PKC Isoforms. The mammalian PKC family consists of ten known members, α, βI, βII, γ, δ, ϵ, ζ, η, θ, and λ (Nishizuka, 1992), some of which are products of alternate splicing of the same mRNA. Each isoform consists of a single polypeptide with highly conserved domains; the C-terminal half contains the catalytic domain and the ATP-binding site (that includes a $GXGXXGX_{16}K$ motif), and the N-terminal half encodes the regulatory domain containing binding sites for Ca^{2+}, Zn^{2+}, phospholipid, and DAG/phorbol esters. The hinge region that joins these two domains is susceptible to proteolysis, which may provide a natural means to down-regulate PKC activity. The Zn-finger sequences found in all PKC proteins are responsible for binding Zn^{2+} and consist of the motif C_6H_2 ($HX_{12}CX_2CX_{10-14}CX_2CX_4HX_2CX_7C$), which is unique and different from other Zn-finger motifs such as those found in TFIIIA (C_2H_2) and estrogen receptors ($C_4 + C_4$). The PKC isozymes show distinct subcellular, cellular, and tissue distributions. For instance, PKC_γ is expressed only in brain and spinal cord, whereas η, θ, and λ are most abundant in lung, skin, and skeletal muscle. The fact that no tissue is totally devoid of PKCs is indicative of their essential role.

2.4.2b. Protein Kinase Substrates. Numerous cellular proteins have been identified as substrates for PKC, such as cell surface receptors, ion channels, cytoskeletal components, and cytosolic and nuclear proteins, including raf (Section 2.3.5b) and transcription factors (Section 2.6). The choice of substrates often correlates with the localization of specific PKC isoforms in the corresponding regions of the cell. Depending on the particular substrate, phosphorylation may effect either a positive or a negative regulation. The consensus phosphorylation site of PKC is a Ser or Thr with basic residues on either side (e.g., K/RXS/TXK/R). In the last few years, a group of CaM-binding proteins such as MARCKS, neuromodulin, and neurogranin have been identified as the major substrates of PKC in neural tissues.

Phosphorylation of MARCKS (myristoylated alanine-rich C-kinase substrate) has been variously used as an index of PKC activation. MARCKS is considered to have critical roles in cellular morphology and cytoskeletal function (Hartwig *et al.*, 1992; Aderem, 1992). Three Ser residues within a highly basic 25-amino-acid stretch of MARCKS appear to be potential sites of phosphorylation by PKC. Phosphorylation of MARCKS by PKC leads to its translocation from the membrane to the cytoplasm, abrogates its ability to cross-link actin, and reduces its affinity for CaM. Thus, PKC would induce a local destabilization of actin skeleton through the phosphorylation of MARCKS. Recent cloning and sequence analysis of the MARCKS (see Harlan *et al.*, 1991, and references therein) have revealed that the myristoylation, CaM-binding, and PKC-phosphorylation domains are all located within the N-terminal half of the protein.

Neuromodulin (also called GAP-43, B-50, and F-1) is a 24-kDa CNS-specific phosphoprotein that appears to play a role in neuronal growth, neurotransmitter release, and synaptic plasticity (Coggins and Zwiers, 1991). Phosphorylation of neuromodulin by PKC at Ser^{41}, which is adjacent to the CaM-binding site, reduces its affinity for CaM. It has been suggested that a function of neuromodulin is to concentrate CaM on the growth cone membrane and that activation of PKC releases high local concentrations of CaM in the growth cone. Interactions between the released CaM and cytoskeletal proteins may effect polymerization and cross-linking of the cytoskeleton, thus triggering the initial events in filopedia formation and neurite extension. Neurogranin (RC3) is a small acidic protein that is phosphorylated *in vitro* by PKC and not by any other kinase (Baudier *et al.*, 1991). *In vivo,* it is phosphorylated in response to phorbol esters, suggesting phosphorylation by PKC. Because its sequence around the site of phosphorylation (KIQA*S*FRGH) is similar to that of neuromodulin, the two proteins may be regulated by PKC in a similar manner.

Protein kinase C phosphorylates a number of transcription factors including ADR1, C/EBP α/β, CREB, fos, HSF, jun, oct2, gal4, myb, max, myogenin, SP1, SRF, and the vitamin D receptor, some of which have been presented in Section 2.6.

2.5. Soluble Guanylate Cyclase and Nitric Oxide

The discovery in 1987 of the formation of nitric oxide (NO) in vascular epithelial cells opened up what can be considered a new era in signal transduction (reviewed by Moncada *et al.*, 1991; Lowenstein and Snyder, 1992; Snyder, 1992). Nitric oxide, a highly reactive gas, is identical to endothelium-derived relaxing factor (EDRF) and is produced from arginine by the enzyme nitric oxide synthase (NOS). A flurry of new research has implicated NO in a variety of inter- and intracellular signaling processes, e.g., vasodilation, immune reaction, and neurotransmission. The predominant pathway of NO function so far involves stimulation of soluble guanylate cyclase (a heme-containing protein), leading to the formation of cyclic GMP, activation of PKG, and protein phosphorylation. Although a direct effect of NO on protein kinases or phosphatases is yet to be demonstrated, the fact that NOS is phosphorylated by PKA, PKC, and CaMK, suggests possible interactions of the NO pathway with a variety of others. Recently, NO-dependent phosphorylation of the 46- to 56-kDa thrombocyte protein designated "vasodilator-stimulated protein" (VASP) has been shown to be important in the inhibition of platelet function during coronary passage (Pohl *et al.*, 1994). The immunosuppressants FK506 and cyclosporin A prevent the dephosphorylation of NOS and its subsequent activation, presumably by inhibiting calcineurin (PP2B), which may underlie the mechanism behind the protective effect of these drugs against the neurotoxic effects of glutamate (Dawson *et al.*, 1993).

2.6. Phosphorylation of Transcription Factors

The majority of eukaryotic transcription factors are found to be phosphorylated, although the exact role of phosphorylation and the identity of the relevant protein kinase(s) are often unknown. In principle, phosphorylation can affect and regulate any or all of the following properties of a transcription factor, either positively or negatively: nuclear translocation, DNA binding, and transactivation (Bohman, 1990; Hunter and Karin, 1992). In most transcription factors, these domains are distinct and separate, and function as relatively independent modules. A few representative transcription factors, and their mechanisms of regulation by phosphorylation are listed in Table II. Because gene expression constitutes the final phase of most kinds of signal transduction, some of the well-documented examples are described here in detail.

2.6.1. Regulation of Nuclear Import

Nuclear import of transcription factors on demand provides a quick means of transcriptional activation not available to prokaryotes. Transcription factors SW15 and NF-κB are the best examples of this type of regulation.

2.6.1a. SW15. The yeast transcription factor SW15 is required for the expression of the HO site-specific endonuclease, which initiates switching of the mating type (a/α). SW15 is phosphorylated by cdc28, the yeast homologue of cdc2, in conjunction with an activating cyclin subunit, on specific Ser residues near or within the nuclear localization signal (NLS). The resultant phosphorylated form is inactive and resides in the cytoplasm. However, in the G_1 phase of the cell cycle, cdc28 is inactivated, resulting in the dephosphorylation and nuclear entry of SW15 to activate HO transcription. In contrast to SW15, nuclear import of SV40 large T antigen is stimulated by phosphorylation of Ser^{111} and Ser^{112} by casein kinase II (CKII).

Table II
Regulation of Transcription Factors by Phosphorylation

Mode of regulation	Transcription factor	Nature of phosphorylation[a]
Nuclear translocation	NF-κB	Phosphorylation of inhibitory subunit, IκB, by PKC
	ISGF3	Tyr phosphorylation of ISGF3α
	SW15	Phosphorylation of $Ser^{552,646,664}$ by cdc28
DNA binding		
Stimulation	SRF	CKII phosphorylation upstream of binding domain
	E2F, E4F	Phosphorylation increases DNA-binding
Inhibition	c-jun	Phosphorylation within 231–249 by GSK3 or CKII
	c-myb	Phosphorylation of $Ser^{11/12}$ by CKII
	ets1	Ca^{2+}-dependent phosphorylation
	GHF1/Pit1	Phosphorylation of Thr^{220} by PKA or PKC
	Max	CKII phosphorylation inhibits dimer formation
	Myogenin	Phosphorylation by PKC in DNA-binding domain
	Oct1	Phosphorylation of DNA-binding domain (Ser^{385})
	C/EBPα	Phosphorylation by PKC on Ser^{299}
Transactivation		
Stimulation	C/EBPβ	Phosphorylation of Ser^{276} by CaMK II
	c-jun	Phosphorylation of $Ser^{63,73}$
	c-myc	Phosphorylation of Ser^{62}
	CREB	Phosphorylation of Ser^{133} by PKA or CaMK I/II
Inhibition	ADR1	Phosphorylation of Ser^{230} by PKA
	v-erbA	Phosphorylation of $Ser^{16,17}$ by PKA or PKC
Transrepression		
	c-fos	Phosphorylation near C terminus

[a]Only the major phosphorylation sites are listed, as detailed in the text (Bohman, 1990; Hunter and Karin, 1992).

2.6.1b. NK-κB. NF-κB is a member of the rel family of transcription factors that include p50NF-κB, p65NF-κB, c-rel, relB, p49, and dorsal (Schmitz *et al.*, 1991). These transcription factors possess conserved domains that determine DNA binding, dimerization, and transactivation. They bind DNA as homo- and heterodimers of various combinations through response elements (binding sites) related to the NF-κB recognition site. These proteins also share a similar mode of regulation in that they are normally sequestered in the cytoplasm in an inactive form and, on activation by appropriate stimulus, translocate into the nucleus. For example, the prototype member of this group, NF-κB, is held in the cytoplasm in an inactive complex with the inhibitory protein, IκB. In response to stimuli such as phorbol ester TPA, the NF-κB–IκB complex dissociates, releasing active NF-κB, which then enters the nucleus and activates transcription. *In vitro,* the dissociation can be mimicked by a number of means, e.g., treatment with deoxycholate (Baeuerle and Baltimore, 1988). The precise mechanism of inhibition by IκB remains to be determined, although it is possible that IκB simply interferes with the NLS of NF-κB. Curiously, unlike NF-κB, c-rel can bind DNA even when complexed with its inhibitor, p40, the avian homologue of IκB.

If sequestration by IκB is the principal mode of regulation of NF-κB, growth stimuli must somehow inactivate IκB. *In vitro,* phosphorylation by PKC, PKA, or heme-regulated eIF2α kinase activates NF-κB, likely as a result of phosphorylation of IκB (Kerr *et al.*, 1991; Ghosh and Baltimore, 1990). However, it remains undecided whether the same also happens *in vivo.* Although both IκB and p40 are phosphorylated *in vivo,* there is no evidence that they are dephosphorylated by stimuli that activate NF-κB. Moreover, IκB-β, but not IKB-α, is inactivated by dephosphorylation *in vitro* (Link *et al.*, 1992). An additional layer of regulation may be provided by the phosphorylation of the rel factors themselves; c-rel, for example, is phosphorylated at multiple sites, and mutation of Ser^{266} (a putative PKA site) to acidic residues (Asp/Glu) results in constitutive nuclear localization, suggesting that the role of the phosphate group was to increase the net negative charge (Mosialos *et al.*, 1991).

2.6.1c. ISGF3. Recent studies have provided a detailed picture of the pathway by which interferon (IFN) signals are transmitted from the membrane to the nuclear genes (David and Larner, 1992; Fu, 1992; Schindler *et al.*, 1992; Velazquez *et al.*, 1992). Interferons are small polypeptides that exhibit diverse biological activities such as inhibition of virus multiplication and inhibition of cell growth. Type I interferons, viz., IFN-α and IFN-β, activate transcription of a variety of cellular genes. Regulatory sequences near the promoter of these genes, termed "interferon-simulated response elements" (ISRE), are necessary and sufficient to confer inducibility by IFN. A multisubunit protein complex, ISGF3, appears to be the major effector of IFN action. The DNA-binding subunit of the complex, $ISGF3_{\gamma}$ (48 kDa), is made constitutively in the nucleus and has been shown to interact directly with ISRE. Another component of the ISGF3 complex, $ISGF3_{\alpha}$, is composed of three proteins (84, 91, and 113 kDa) that are present in latent form in the cytoplasm. Interferon α treatment of cells resulted in rapid phosphorylation of these three proteins in tyrosine residues. This is followed by their association with each other to form the $ISGF3_{\alpha}$ complex, which enters the nucleus, complexes with $ISGF3_{\gamma}$, and binds to ISRE to activate transcription (Figure 1). The process could also be reproduced in cell-free extracts, where incubation of a plasma-membrane-enriched fraction with IFN-α was sufficient to cause the activation of ISGF3 within minutes (David and Larner, 1992).

Unlike the RTKs, the IFN-α receptor lacks a tyrosine kinase activity. Thus, the exact identity of the kinases that carry out the phosphorylation of ISGF3α is under active investigation. Screening of gene libraries resulted in the identification of a tyrosine kinase gene, *tyk2,* a member of the JAK family of kinases (Section 2.2), that could complement a class of cellular mutants unresponsive to

IFN-α, suggesting that tyk2 may be the relevant kinase. In analogy to the p56lck scenario (Section 2.2), it is possible that the IFN receptor, on stimulation by IFN, recruits tyk2 to the membrane. It is not known whether tyk2 itself is regulated by reversible autophosphorylation–dephosphorylation; however, it is curious to note that all three subunits of ISGF3$_{\alpha}$ contain SH2 domains, raising the possibility that they may recognize specific phosphotyrosine-containing signaling molecules.

One such example is already known for the p91 subunit of ISGF3$_{\alpha}$ that can also function as a transcription factor for the c-*fos* promoter and can be phosphorylated through a second pathway, i.e., by the RTK activity of the activated EGF receptor. This occurs through a direct binding of p91 with the stimulated (autophosphorylated) EGFR via the SH2 domain of the former, resulting in tyrosine phosphorylation and nuclear translocation of p91 (Fu and Zhang, 1993). Protooncogene c-*fos* is rapidly induced by a number of growth factors, and its promoter contains several regulatory elements, namely, SRE (see Section 2.5.2b) and SIE. While the transcription factor SRF recognizes the SRE sequence (Section 2.5.2b), phosphorylated p91 recognizes the SIE sequence, providing dual control of c-*fos* transcription. Thus, p91 provides an example in which a growth factor activates nuclear gene expression in a direct manner. Interestingly, a portion of the SIE sequence (TTCCCGTCAA) shows some similarity with the intereferon-activated site (TTC/ACNNNAA).

2.6.1d. Nuclear Protein Kinases. Finally, there are protein kinases that are themselves residents of the nucleus or translocate to the nucleus in response to stimulation; examples (phosphorylation sites in parentheses) include c-abl IV/1b (Y), CKI and II (S/T), cyclin/cdk (S/T), fer (Y), rsk (S/T), MAP kinases (S/T), sty (S/T/Y), and the catalytic subunit of PKA (S). Thus, they are not transcription factors on their own merit, but they can potentially activate *bona fide* transcription factors through phosphorylation. The subcellular localization of the c-abl tyrosine kinase, for instance, is specified by its unique C terminus. In the cytoplasm, c-abl binds to actin filaments via an actin-binding domain in the C terminus; in the nucleus, it binds to specific DNA sequences via a DNA-binding domain, also located in the C-terminal region. A role of c-abl in cell cycle progression is suggested by its interaction with the retinoblastoma (RB) protein and its phosphorylation by the cdk/cyclin complex, the activity of which is regulated with the cell cycle (Welch and Wang, 1993, and references therein). A domain in the C terminus of RB binds to the ATP-binding lobe of c-abl, resulting in the inhibition of the tyrosine kinase activity. Hyperphosphorylation of RB near the c-abl-binding domain by cdk/cyclin results in the release and activation of c-abl in the S phase. The activated c-abl may regulate the expression of specific genes necessary for the progression through the S phase. It is postulated that at the G_2/M transition, phosphorylation of c-abl causes its dissociation from DNA, which contributes to the cessation of gene expression, thus allowing a new round of regulation in the next cell cycle.

The carboxy-terminal domain (CTD) of RNA polymerase II appears to be a c-abl substrate. The CTD of mammalian RNAPII contains 52 tandem repeats of a heptapeptide with the consensus sequence YSPTSPS, at least 30 of which are phosphorylated specifically by c-abl *in vitro*. Phosphorylation of CTD has been shown to promote the transition of RNAPII from the initiation to the elongation mode, which may explain in part the enhanced gene transcription induced by activated c-abl. Other protein kinases have also been reported to phosphorylate the CTD; these include the DNA-dependent protein kinase (Peterson *et al.*, 1992), a kinase activity associated with yeast initiation factor b (Feaver *et al.*, 1991) and mammalian TFIIH (Lu *et al.*, 1992), and a trimeric CTD kinase in yeast, which has a 58-kDa kinase subunit (Lee and Greenleaf, 1991). It is tempting to speculate that the various CTD kinases differ in the exact sites and extent of phosphorylation and hence that specific signals may result in the recruitment of specific CTD kinases, thus fine-tuning the expression of each gene.

2.6.2. Regulation of DNA Binding

Inhibition of DNA binding by phosphorylation at or near the DNA-binding domain is a common mechanism affecting a variety of transcription factors. The mechanism may involve either a conformational change or a decrease in the net positive charge of the DNA-binding domain, resulting in a reduction of affinity toward the negatively charged DNA.

2.6.2a. Inhibition of DNA Binding. The majority of effects of phosphorylation on DNA binding fall into this group. Phosphorylation of c-myb by CKII at $Ser^{11,12}$ results in a large decrease in its DNA-binding affinity that is reversed by phosphatase treatment. In the constitutively active v-myb protein, these Ser residues are deleted (Luscher *et al.*, 1990). Phosphorylation by multiple kinases (e.g., GSK-3, CKII, cyclin/cdc2) has been postulated for c-jun, a component of the AP-1 transcription complex. Dephosphorylation of the CKII and GSK-3 (glycogen synthase kinase 3) sites of c-jun occurs following PKC activation and expression of several oncogenes such as v-Ha-*ras*, v-*sis*, and v-*src*, which correlates with restoration of AP-1 activity. The DNA-binding activity of the basic helix–loop–helix (bHLH) proteins max and myogenin is inhibited on phosphorylation by CKII and PCK, respectively. The specificity of the effect is suggested by the finding that phosphorylation inhibits DNA binding by the max–max homodimer; however, the myc–max heterodimer remains unaffected.

2.6.2b. Stimulation of DNA Binding. Examples of stimulation of DNA-binding activity by phosphorylation are rarer. A well-studied example is the serum response factor (SRF) that binds and activates the c-*fos* promoter through the SRE (serum response element) sequence (Treisman, 1992). The SRF is phosphorylated on several Ser residues *in vivo* ($Ser^{77,79,83,85,103}$), the majority of which are potential CKII sites. *In vitro*, phosphorylation of SRF by recombinant CKII stimulates DNA binding only when tested in the absence of Mg^{2+} and spermidine, whereas in the presence of these two effectors both the *on* and the *off* rates are stimulated. Mutational analysis suggests that phosphates at Ser^{83} and Ser^{85} play the most critical role in the kinetic effect on DNA binding, while phosphorylation of Ser^{103}, presumably by PKA, is essential for DNA binding.

The binding of SRF to the c-*fos* promoter is additionally modulated by the SRF accessory protein, elk-1 (and related protein SAP-1), which contains an N-terminal DNA-binding domain and a C-terminal activation domain, named "C box." The MAP-kinase-catalyzed phosphorylation at multiple sites in the C box has no effect on the DNA-binding property of elk-1 but apparently stabilizes its interaction with SRF, thus indirectly contributing to increased c-*fos* transcription on growth factor stimulation (Marais *et al.*, 1993).

2.6.3. Regulation of Transactivation

The effect of phosphorylation on the activation domain of transcription factors is generally positive, i.e., stimulatory in nature, ADR1 being one of the few known examples of inhibition.

2.6.3a. CREB. Transcription factor CREB (cyclic-AMP-responsive element-binding protein) is a member of the basic leucine zipper (bZIP) family of dimeric transcription factors, which also includes jun (see below), fos, and NFIL-6 (Gonzalez *et al.*, 1991, and references therein). Transcriptional activation of cyclic-AMP-responsive genes by CREB requires phosphorylation of CREB by PKA, predominantly on Ser^{133}, which lies downstream of the N-terminal Gln-rich transactivation domain. This site contains the consensus RRPSTR and is also phosphorylated by purified PKA *in vitro*. In addition, maximal transactivation may require phosphorylation of Ser residues within region 106–122, possibly catalyzed by CKII. The effect of phosphorylation is primarily on transactivation and not on DNA binding or dimerization. The DNA-binding domain of CREB can be replaced with heterologous DNA-binding domains to produce chimeric transactivators that are still activated by PKA (Lee *et al.*, 1990). Replacement of Ser^{133} with an acidic residue results in an inactive protein, suggesting that the phosphate group plays a specific role in changing the conformation of the protein. Phosphorylation of Ser^{133} can be attenuated through dephosphorylation by protein phosphatase 1 (PP1), which is itself a substrate for PKA.

2.6.3b. c-jun and c-myc. The activation domain of c-jun, a relative of CREB and a component of the AP-1 family, is regulated in a similar manner.

Phosphorylation of c-jun occurs at $Ser^{63,73}$ and is stimulated by a number of oncogenes, such as H-ras, v-src, raf-1, and by TPA (Smeal *et al.*, 1991). The exact nature of the protein kinase responsible remains debated, although MAPK was implicated in one study (Pulverer *et al.*, 1991). As in the case of CREB, the DNA-binding domain of c-jun can be replaced with heterologous DNA-binding domains without loss of activation by phosphorylation; however, substitution of $Ser^{63,73}$ with Leu or Ala results in loss of transactivation. An essentially similar regulation has been reported for c-myc, the site of phosphorylation being Ser^{62} (Alvarez *et al.*, 1991).

The regulation of signal transduction by dephosphorylation of transcription factors is included within the protein phosphatase section that follows.

2.7. Protein Phosphatases

Although protein kinases and protein phosphorylation have received paramount emphasis in studies of signal transduction, it is gradually being recognized that protein phosphatases must be equally important, based on the following premise: in order that phosphorylation may play a regulatory role, it ought to be a reversible process, because otherwise a signaling phosphoprotein would not revert back to its nonphosphorylated state. This is particularly true of signaling proteins (e.g., c-jun) that are phosphorylated by constitutively expressed, unregulated kinases such as CKII (Lin *et al.*, 1992). A direct role of protein phosphatases has been unveiled in a number of signal transduction pathways in recent years (Walton and Dixon, 1993; Tan, 1993; Charbonneau and Tonks, 1992; Mumby and Walter, 1991). Like their kinase counterparts, protein phosphatases are generally specific for either P-Tyr (protein tyrosine phosphatase or PTP) or P-Ser/P-Thr (protein Ser/Thr phosphatase or PP). A few phosphatases with dual specificity have been reported, e.g., MAP kinase phosphatase-1 (Sun *et al.*, 1993), p80cdc25 (Millar and Russell, 1992), and the unique bacteriophage lambda orf221 phosphatase, PP-λ, that can additionally dephosphorylate P-His residues in appropriate sequence context (Barik, 1993; Zhuo *et al.*, 1993). More often than not, however, they exhibit pronouncedly higher specific activity toward one residue over another; PP-λ, for example, exhibits K_{cat}/K_m values that are 10–20 times higher for P-Ser than for P-Tyr substrates (Barik, 1993; Zhuo *et al.*, 1993). Unlike the kinases, the majority of the phosphatases are not regulated by second messengers. Furthermore, they appear to be more promiscuous in their choice of substrates, giving rise to speculation that they may not be as numerous as the kinases. This plurality of action, at the same time, adds uncertainty in identifying the "true" physiological substrate of a given phosphatase by the use of inhibitors.

Nevertheless, recent studies have hinted that selection of specific substrates may be regulated, at least in part, by the cellular localization of the phosphatase, spatially as well as temporally. The two transcripts of the cytosolic phosphatase STEP, for instance, are transcribed only in the brain; the larger transcript is found in many regions of the brain, but the smaller one is detected only in the striatum (Lombroso *et al.*, 1991). Similarly, PTP-1 expression is more pronounced in the hippocampus than in other brain regions (Guan *et al.*, 1990). The HePTP, a cytoplasmic PTP, is expressed abundantly in thymus but poorly in spleen; on stimulation of spleen lymphocytes by concanavalin A, LPS, or anti-CD3 antibodies, HePTP expression can be significantly enhanced (Zanke *et al.*, 1992). A truncated form of PTP-1B lacking the regulatory C-terminal domain and therefore no longer restricted to its normal intracellular locale has been implicated in platelet aggregation response (Frangioni *et al.*, 1993); PTP-1C and SHP/HCP, two SH2-containing tyrosine phosphatases, are expressed almost exclusively in cells of the hematopoietic system.

Within a given cell, a phosphatase may move between the various compartments in response to external stimuli. For instance, on treatment of the myeloid leukemia cell line with phorbol esters, its PTP activity redistributes between particulate and soluble fractions (Butler *et al.*, 1990). When T cells are activated, the CD45 phosphatase leaves the Golgi complex (Minami *et al.*, 1991).

2.7.1. Protein Tyrosine Phosphatases

Historically, CD45 was probably the first phosphatase gene to be cloned and sequenced; however, it was not recognized as a phosphatase until later, when its sequence was compared with that of a recognized phosphatase, PTP-1B (Tonks *et al.*, 1988). Since then, nearly 40 different PTPs have been isolated, and CD45 has served as the prototype for the general structure of the receptor-type phosphatases (Walton and Dixon, 1993). Similarly, PTP-1B has served as the prototype for the soluble class of phosphatases. All PTPs have a common catalytic domain that is characterized by an 11-residue "signature" motif: I/V HCXAGX-XR(S/T)G. In a pioneering study, the first crystal structure of a PTP has been determined through the inclusion of tungstane ion (WO_4) in the phosphate-binding loop (Barford *et al.*, 1994). The structure confirmed the roles of the invariant His, Cys, and Arg residues of the signature motif in forming the activity pocket. Biochemical studies had earlier shown that PTPs function through a covalent Cys-P intermediate (Guan and Dixon, 1991). Additional sequence features found in PTPs include one or more of the following: transmembrane (membrane anchor) domain (in receptor types, e.g., CD45, HLAR, HPTPα, DPTP10D), immunoglobulin repeat (in HLAR), fibronectin type III repeat (in HLAR and DPTP10D), cytoskeletal domain (mediating attachment to cytoskeleton, e.g., in PTPMEG), and SH2 domains (in PTP-1C and SHP/HCP).

Recent work has illustrated the importance of PTPs in T-cell signaling. Stimulated T cells lacking CD45 are unable to generate the PI-derived second messengers IP_3 and Ca^{2+} (Koretzky *et al.*, 1990). Lymphokines that regulate the differentiation of hematopoietic system also activate a PTP activity (Mire-Sluis and Thorpe, 1991). For example, addition of IL-4 to leukemic cell lines causes a rapid and dose-dependent dephosphorylation of the Tyr-phosphorylated protein p80, which is reversed by the addition of vanadate, a specific inhibitor of PTPs. Mice carrying the *motheaten* mutation show marked abnormalities in hematopoiesis and immune function; this mutation has recently been traced to the *hcph* gene, coding for hematopoietic cell protein tyrosine phosphatase (HCP) (Schultz *et al.*, 1993). Because wild-type HCP is rapidly phosphorylated on stimulation by CSF-1 (colony-stimulating factor), HCP dysfunction may be responsible for loss of CSF signaling (Yeung *et al.*, 1992).

A group of soluble PTPs in yeast play direct roles in cell cycle progression. The maturation-promoting factor (MPF) of yeast is a complex of the Ser/Thr kinase p34cdc2 and the regulatory subunit cyclin B. Initiation of mitosis requires the activation of this complex via the dephosphorylation of P-Tyr^{15} (and possibly P-Thr^{14}) of cdc2. Genetic and biochemical evidence has shown that cdc25 is the phosphatase responsible for this conversion (Gould *et al.*, 1990). Interestingly, the primary sequence of cdc25 resembles that of VH1, a 20-kDa phosphatase from vaccinia virus, which is composed entirely of the catalytic domain and dephosphorylates both Tyr- and Ser-phosphorylated proteins. In human, a similar mitosis-specific component of MPF, p65, also exhibits both Ser and Tyr phosphatase activities (Meikrantz *et al.*, 1991). The *corkscrew* (*csw*) gene of *Drosophila*, which encodes a putative nonreceptor PTP with two SH2 domains, operates positively in torso signaling (Perrimon, 1993; Section 2.3.5c). A mammalian PTP (named SH-PTP2, PTP-1D, or syp), with sequence homology to csw, binds activated RTKs and becomes phosphorylated on Tyr residues, providing a potential for downstream signaling.

2.7.2. Protein Ser/Thr Phosphatases

Although many Tyr phosphatases are localized in the membrane, a large majority of the Ser/Thr phosphatases are soluble in nature. Primarily on the basis of their divalent cation requirement, sensitivity to heat-stable inhibitors 1 and 2, and ability to dephosphorylate the α or β subunit of phosphorylase kinase, the Ser/Thr phosphatases have been classified as either PP1, PP2A, PP2B, or PP2C (Ingebritsen and Cohen, 1983). In addition, they exhibit differential sensitivity to the marine sponge toxin okadaic acid (OA). At a concentration of 1 nM OA, for example, PP2A is completely inhibited, whereas much higher concentrations are required to inhibit PP1 and PP2B; PP2C is resistant. However, it is important to remember that as new species of phosphatases are being discovered (e.g., PPV, PPX, PPY, PPZ1, PPZ2, PPλ, RdgC, SIT4), examples with intermediate properties are coming to light, generating novel branch points in the classification scheme (Barik, 1993; Honkanen *et al.*, 1991; Honkanen and Boynton, 1994). The active site of Ser/Thr phosphatases remain unmapped, although sequence alignments among PP1, PP2A, PP2B, and PP-λ show extensive similarity in the central region containing blocks of conserved amino acid residues such as GD(L/Y)VDRG and RGNHE (Cohen and Cohen, 1989) that appear to be essential for function (S. Barik and L. Dupuy, unpublished data).

2.7.2a. PP1. Roles of PP1 in various signaling pathways have been established. Genetic experiments in fungi have indicated that PP1 is required for the separation of daughter chromosomes at late M phase (reviewed by Cyert and Thorner, 1989). Similarly, a deficiency in the PP1 gene of *Drosophila* causes a block in mitosis at the larval stage (Axton *et al.*, 1990). A function of PP1 during mitosis is also suggested by the observation that microinjection of PP1, but not PP2A, results in dephosphorylation of nuclear lamins, the major *in vivo* substrate for the cell cycle control kinase p34cdc (Brautigan *et al.*, 1989).

A regulatory mechanism of PP1 is provided through the association of its catalytic subunits with one of several high-molecular-weight "targeting" subunits that bind it to specific cellular structures and inhibit its activity. The best-studied targeting subunits are the G subunit that binds skeletal PP1 to glycogen particles (Sections 2.3.1 and 2.3.2) and the M subunit that binds it to myofibrils. Moreover, PP1 in the cytosol usually exists as an inactive complex with heat-stable inhibitor-2 (I-2). Phosphorylation of the targeting subunits or I-2 by various second-messenger-regulated protein kinases releases active PP1 catalytic subunits. In contrast, inhibitor-1 (I-1), only when phosphorylated by PKA, binds and inhibits the catalytic subunit of PP1 in a highly selective manner (Nimmo and Cohen, 1978). Conversely, on dephosphorylation by PP2B, I-1 loses its inhibitory activity and releases the catalytic subunit of PP1. The level of I-2 has been shown to oscillate during the cell cycle (Brautigan *et al.*, 1990), suggesting a possible role of PP1 in cell cycle progression. A neuronal isoform of PP1 inhibitor-1, DARPP-32 (*d*opamine and cyclic *A*MP-*r*egulated *p*hospho*p*rotein of 32 kDa), is found concentrated in dopaminoceptive neurons and, like I-1, is a substrate for PKA (at Thr^{34}) and is dephosphorylated by PP2B (King *et al.*, 1984). Thus, it is likely that the activity of PP1 can be modulated by agents (e.g., dopamine) that elevate cyclic AMP. Indeed, treatment of brain tissue with forskolin leads to the phosphorylation of DARPP-32 on Thr^{34}, and this is antagonized by stimulation of NMDA-specific glutamate receptor, which is linked to Ca^{2+} channels (Halpain *et al.*, 1990). Although direct evidence is lacking, it is possible that this antagonism results from the activation of PP2B, leading in turn to the activation of PP1. Thus, signals that generate two major intracellular second messengers (cyclic AMP and Ca^{2+}) that do not directly affect PP1 may modulate its activity through the reversible phosphorylation–dephosphorylation of I-1 and I-2 (Hemmings *et al.*, 1984).

A novel mechanism of cellular regulation of protein phosphatases has been suggested by the recent finding that PP1, but not PP2A, is inhibited through phosphorylation by cdc2/cydlinA and cdc2/cyclinB (Dohadwala *et al.*, 1994). Phosphopeptide mapping and site-directed mutagenesis documented Thr^{320} as the site of phosphorylation. Furthermore, PP1, but not PP2A, activity was found to oscillate during the cell cycle of a human osteosarcoma cell line. Thus, it was speculated that this may be related to the known oscillation of cyclin-dependent kinases during the cell cycle.

2.7.2b. PP2A. In contrast to yeast MPF (Section 2.7.1), in *Xenopus* oocytes, dephosphorylation of cdc2 by INH, a form of PP2A Ser/Thr phosphatase, is responsible for the inactivation of the cdc2 kinase activity, resulting in inhibition of the MPF. The exact site of dephosphorylation has not been identified but probably corresponds to Thr^{167} in *S. pombe*. Thus, MPF represents an example in which dephosphorylation at different sites of the same polypeptide produces apparantly opposing effects on activity. Incidentally, the two genes in *S. pombe* (*ppal* and *ppa2*) that code for phosphatases with ~78% homology to mammalian PP2A are essential but redundant, because disruption of either gene results in viable cells, whereas disruption of both is lethal (Kinoshita *et al.*, 1990).

Other protein kinases involved in cell growth may also be regulated by PP2A (reviewed by Mumby and Walter, 1991). The catalytic subunit of PP2A (C subunit) is found in complex with a variety of regulatory subunits, including A, B, and B′, generating holoenzymes such as AC, BC. Although the physiological significance of so many holoenzymes remains unknown, they may represent a mechanism of regulation of PP2A. Substrates include rsk, p53, and members of the MAP kinase family. PP2A dephosphorylates only the P-Thr residue in MAPK (Section 2.3.5c) and leaves the P-Tyr residue intact. Recently, a novel interaction between the tumor antigens of small DNA tumor viruses and PP2A has been suggested. For instance, nearly all of the PP2A in SV40-infected cells is found associated with the SV40 small t antigen, an interaction that was found to inhibit the ability of PP2A to inactivate ERK and MEK, resulting in the stimulation of the MAP kinase signaling pathway and induction of cell proliferation (Mumby and Walter, 1991; Sontag *et al.*, 1993). The growth suppressor protein p53 is readily dephosphorylated *in vitro* by PP2A at specific sites, likely near the amino terminus; this is also inhibited by SV40 small t antigen (Scheidtmann *et al.*, 1991). It is to be noted that small t inhibits only the dephosphorylation of p53 by the AC form of PP2A, while dephosphorylation by the free C subunit or the ABC form is unaffected, suggesting a role of the regulatory subunits in the interaction. Recently, mutation of the *twins* gene encoding a regulatory subunit of PP2A has been shown to produce pattern duplication in the imaginal disk of *Drosophila* (Uemura *et al.*, 1993).

2.7.2c. PP2B. Specific functions of PP2B (calcineurin) remain unknown; however, it shows a much narrower substrate specificity *in vitro*, the majority of its substrates being regulators of other kinases and phosphatases (Klee and Cohen, 1988). Unlike members of the other three classes, PP2B shows an absolute requirement for Ca^{2+}-calmodulin. Thus, PP2B may serve to regulate intermediates in the signal transduction pathway that are yet unknown.

Analysis of the retinal degeneration gene C (rdgC), required to prevent light-induced retinal degeneration in *Drosophila* (Steele *et al.*, 1992), indicates that it codes for a phosphatase with ~30% identity with the PP1/PP2A/PP2B family. Although the mechanism of action of the rdgC phosphatase is not known, it seems to be the only phosphatase with an "EF hand" motif (cf. calmodulin, Section 2.3.4), suggesting that it may be regulated by intracellular Ca^{2+}.

The *SIT4* gene in *S. cerevisiae* was originally defined by mutations that restore transcription of the *HIS4* gene in the absence of the normal transactivators of the gene (Arndt *et al.*, 1989). Later, the gene was implicated in various phases of the cell cycle and their normal progression (Sutton *et al.*, 1991; Fernandez-Sarabia *et al.*, 1992). The SIT4 protein shows significant similarity with PP1 and PP2A, although its exact role as a phosphatase remains unexplored.

An important issue is whether phosphatases directly control transcription factors. Transcription attenuation following cyclic AMP induction requires PP1-mediated dephosphorylation of CREB (Section 2.6.3a) at Ser^{133} (Hagiwara *et al.*, 1992). In addition, the phosphorylation and activation of ISGF3 as described earlier (Section 2.6.1c) are inhibited by vanadate, phenylarsine oxide, and zinc chloride, which are diagnostic inhibitors of PTPs (Tan, 1993). Perhaps the kinase (Tyk2?) responsible for phosphorylation of ISGF3 is itself activated by dephosphorylation.

3. SIGNAL TRANSDUCTION SYSTEMS IN PROKARYOTES

Bacteria are capable of sensing and responding to a wide variety of environmental signals, such as changes in chemical concentration and variation of physical parameters (temperature, light, osmolarity). This is essential for their ability to locate the most favorable (or the least unfavorable) condition in any given situation. In this section, we briefly review the three most extensively studied bacterial signal transduction systems that serve as paradigms for other, related ones (Bourret *et al.*, 1991; Parkinson, 1993). In spite of their technical and operational variety, however, all these systems embody some common molecular strategies that are reminiscent of the growth factor signaling pathways. The major effectors consist of the following: a *detector* or *transducer* protein activating an autophosphorylating *kinase* that is phosphorylated at a His residue and the subsequent transfer of the phosphoryl group to an Asp residue of a *regulator* protein, which controls the final *target* protein(s). Because signal transduction in such systems involves transient phosphorylation of a pair of components, the kinase and the regulator, these systems are often referred as "two-component regulatory systems." However, this description tends to be confusing and is avoided here because the systems are almost always multicomponent in nature.

3.1. Bacterial Chemotaxis

E. coli swims toward or away from a variety of compounds, which are empirically classified as "attractants" or "repellants," respectively. Swimming is effected by flagellar rotation (Bourret *et al.*, 1991; Parkinson, 1993). Normal behavior in a homogeneous environment consists of smooth swimming interrupted by brief periods of tumbling, which serve to reorient the cell to search for a better environment in a new direction. In general, clockwise rotation of the flagella produces tumbling, and counterclockwise rotation leads to smooth swimming. Although a number of individual steps of the chemotactic pathway have been elucidated in molecular detail, exactly how they transmit the signal to the flagellar motor remains unknown. It is also unexplained how the bacteria respond to changes in chemoeffector concentrations that may range over a factor of 10^5. Furthermore, they can integrate signals from a complex mixture of attractants and repellents and do so with extraordinary rapidity, usually within a fraction of a second. Nonetheless, more than a dozen proteins, originally defined by mutations, are now recognized to be involved in the process; these include membrane-bound transducers (tar, tsr, trg, and tap), cytoplasmic proteins encoded by the *che* genes (cheA, B, R, W, Y, and Z), and flagellar components encoded by *fli* genes (fliG, M, and N). Although methylation is the major signal-induced modification of the transducer proteins, the che proteins downstream of them are controlled primarily through phosphorylation (reviewed by Bourret *et al.*, 1991; Parkinson, 1993), some of which is summarized here.

The proteins cheA and cheY are the two central components of the chemotactic phosphorylation pathway. On excitation, cheA undergoes autophosphorylation, which has been located at the N-3 position of His^{48} by proteolytic and mutational analyses. Genetic and mutational probing of the cheA protein have indicated that it consists of two functional domains, corresponding to the N- and C-terminal regions. Although the middle and C-terminal regions seem to determine autophosphorylation activity, the N-terminal part is involved in the transfer of phosphate groups to cheY and cheB. The ability of a purified N-terminal fragment of cheA-phosphate to transfer phosphate but not to autophosphorylate is supportive of the modular structure. Sedimentation equilibrium studies indicate that cheA may be a dimer (Gegner and Dahlquist, 1991), suggesting the possibility that autophosphorylation may be a result of the two subunits cross-phosphorylating each other, a scenario also proposed for RTK dimerization and autophosphorylation (Section 2.1).

Phosphorylated cheA rapidly transfers the phosphate group to conserved

Asp residues (see below) in cheY and cheB. Both phospho-cheY and phospho-cheB autodephosphorylate with a half-life of seconds and are thus much more unstable than would be predicted for acyl phosphates, suggesting that the proteins themselves catalyze the dephosphorylation reactions. Denatured cheB and cheY lose this autophosphatase activity, exhibiting half-times of hydrolysis of over 1 hr; cheZ stimulates dephosphorylation of cheY but has no effect on phospho-cheB. The findings that cheA and cheY mutants are smooth-swimming whereas cheZ mutants tumble supported the hypothesis that cheY phosphate may be one of the tumble regulators. The interplay between the methylation and phosphorylation of the che proteins plays a crucial link between adaptation and excitation. Phosphorylation of cheB within the N-terminal domain, for example, has been shown to greatly stimulate its methyltransferase activity, which resides at the C-terminal region. Thus, the instability of cheB phosphates may account for the transient nature of the increased demethylation rate. The autophosphorylation and phosphotransferase activities of cheA as well as the autodephosphorylation activity of cheY requires divalent cations, such as Mg^{2+}.

Significant advances have been made in the structural analysis of cheY. The N-terminal domain of cheY has strong sequence homology to cheB as well as to the N-terminal domains of a large number of bacterial regulatory proteins, including ompA, and spo. X-ray diffraction analysis (Stock *et al.*, 1989) showed that cheY consists of a compact single domain formed from a core of five parallel β-strands alternately surrounded by five α-helices. Three Asp residues ($Asp^{12,13,57}$), which are conserved among regulator proteins of the "two-component" families, cluster on the same face of the three-dimensional structure to form an "acid pocket." The site of cheY phosphorylation has been mapped to Asp^{57}, although substitutions of any of the three Asp residues resulted in loss of chemotaxis (Parkinson, 1993; Sanders *et al.*, 1989). Allele-specific suppression of mutations in the flagellar gene *fliG* by *cheY* mutants suggests an interaction between the two proteins, although it is not known whether phosphorylation affects the interaction.

3.2. Nitrogen Assimilation

The key enzyme in the assimilation of elemental nitrogen is glutamine synthetase (encoded by the *glnA* gene), whose activity and quantity in a variety of bacterial species are regulated in response to nitrogen via the *ntr* gene system (Bourret *et al.*, 1991; Parkinson, 1993). The availability of nitrogen influences the ratio of glutamine (Gln) and 2-ketoglutarate (2-KG) in the cell, which in turn regulates the activity of the uridylyltransferase/uridylyl-removing (UT/UR) enzyme system. Under nitrogen-limiting conditions, 2-KG levels increase and stimulate uridylylation of P_{II} protein; conversely, when nitrogen is abundant, P_{II} remains unmodified. Unmodified P_{II}, but not the modified species, accelerates the dephosphorylation of ntrC in the presence of ATP and ntrB protein. The P_{II}

protein thus acts as the communicating signal between the nitrogen-sensing machinery and the downstream phosphorylation pathway.

In the ntr phosphorylation cascade, the major kinase appears to be ntrB (NR_{II}), and the regulator is ntrC (NR_I): ntrB autophosphorylates on His^{139} (Ninfa and Bennett, 1991), followed by transfer of the phosphate to Asp^{54} of ntrC (Sanders *et al.*, 1992); ntrC phosphates are spontaneously lost with a half-life of about 4 min. As mentioned above, this process is accelerated by the combined presence of ATP, ntrB, and P_{II}. Finally, the phosphorylation status of ntrC regulates the transcriptional activity of the major promoter (*glnAp2*) of the *glnA* gene via a mechanism that resembles enhancer function in eukaryotic transcription. Specific ntrC-binding sites upstream of the transcriptional start site of the *glnAp2* promoter recruits the phosphorylated form of ntrC, which in turn serves to increase the concentration of the RNA polymerase at the promoter, thus stimulating transcription. In the absence of the ntrC-binding sites, activation of *glnAp2* occurs but requires a much higher amount of ntrC protein. Like enhancer elements, the ntrC-binding sites remain fully functional when moved various distances to either side of the promoter. Furthermore, transcription activation takes place even when the ntrC-binding sites and *glnAp2* are on two different but interwound DNA molecules. It appears that ntrC promotes the conversion of the promoter-bound polymerase from the closed to the open form and then leaves the transcription complex (Popham *et al.*, 1989). Electron microscopic studies have indeed revealed the physical proximity of ntrC and RNA polymerase in the open complex (Su *et al.*, 1990). Sequence comparison with other regulators and mutational data suggested the presence of the following domains in the ntrC protein (Stock *et al.*, 1990, and references therein): (1) the phosphorylated N-terminal domain with homology to other regulators such as cheB and cheY, as mentioned earlier; (2) the central domain that interacts with RNA polymerase and exhibits similarity to transcriptional regulators; and (3) the C-terminal helix–loop–helix (HLH) domain that is responsible for DNA binding.

Although ntrB is necessary to rapid activation and deactivation of glnAp2 transcription, steady-state levels of glutamine synthetase under nitrogen starvation remain normal even in an ntrB mutant. However, a complete reduction of glutamine synthetase level under nitrogen excess conditions requires ntrB. Thus, it has been proposed that the major function of ntrB is to hydrolyze ntrC phosphates, whereas the phosphorylation of ntrC is carried out primarily by another bacterial protein kinase.

3.3. Outer Membrane Protein Synthesis

In response to changes in osmolarity, bacteria synthesize specialized membrane proteins, called "porins," which are then incorporated into membranes. In *E. coli*, two major outer membrane porins, ompC and ompF, are regulated in a reciprocal manner, being induced at high and low osmolarity, respectively (Bourret *et al.*, 1991; Parkinson, 1993). In this system, the kinase is envZ, the regulator is ompR, and the overall design is very similar to that of the nitrogen assimilation system described above. The external osmolarity is sensed by envZ, a membrane protein whose topology is similar to that of other transducers; i.e., it contains a cytoplasmic domain, a periplasmic domain, and two transmembrane domains. The cytoplasmic domain appears to interact directly with the downstream signaling molecule ompR, because this fragment as well as the intact envZ protein can autophosphorylate and transfer the phosphates to the N-terminal domain of ompR. At lower levels of ompR phosphorylation, *ompF* is preferentially transcribed; as phosphorylation of ompR increases, *ompF* transcription is reduced, and at the same time *ompC* transcription is activated (Aiba and Mizuno, 1990). It remains unknown whether this differential transactivation of the two *omp* genes results from a quantitative or qualitative change in the phosphorylation status of the ompR protein.

Spontaneous dephosphorylation of ompR occurs with a half-life of about 1.5 hr and is, therefore, a much slower process than autodephosphorylation of other regulators such as cheY and cheB proteins. *In vivo* as well as *in vitro,* the natural dephosphorylation rate of ompR is accelerated by envZ in the presence of ATP. The observation that bacteria expressing autophosphorylation-defective envZ mutant proteins are apparently normal in the phosphorylation of ompR in an osmolarity-dependent manner has led to the suggestion that the primary role of envZ may be to dephosphorylate ompR, while phosphorylation of ompR is catalyzed by some other kinase (Forst *et al.*, 1990).

In addition to the three model systems described above, a number of bacterial regulatory circuits have been characterized that function in various signal transduction pathways (Bourret *et al.*, 1991; Parkinson, 1993). Some of the major regulatory systems (along with the *kinase* and *regulator,* in that order within parentheses) of each system are as follows (studied in *E. coli* unless otherwise mentioned): motility (frzE, frzG in *Myxococcus*); sporulation (kinA/B, spoOA/OF in *Bacillus*); dicarboxylic acid transport (dctB, dctD in *Rhizobium*); phosphoglycerate transport (pgtB, pgtA in *Salmonella*); phosphate assimilation (phoR, phoB); virulence (virA, virG in *Agrobacterium;* phoQ, phoP in *Salmonella;* bvgS, bvgA in *Bordetella*); carbohydrate transport (uhpB, uhpA); capsule synthesis (rcsC; rcsB); nitrate reduction (narX, narL); competence (comP, comA in *Bacillus*); and degradation (degS, degU in *Bacillus*). Detailed studies of a number of these systems have led to the notion that the rate of dephosphorylation of the regulator protein generally reflects the rapidity with which the system must respond to environmental changes. It has been mentioned earlier that ompR dephosphorylates more slowly than the che regulators. This fits well with the fact that although bacterial response to chemotactic signals is extremely quick, it takes nearly two cell-division cycles for the porin composition of the outer membrane to alter appreciably, in part because of the slow turnover rate of the existing porins. Extreme cases are responses such as symbiosis and sporulation, which persist over long time spans; this may explain the complete lack of autophosphatase activity in their corresponding regulator phosphoproteins, exemplified by virG and spoOF. There are many regulatory systems for which the environmental signal remains undefined (Mekalanos, 1992; Taha *et al.*, 1991).

Expression of a large variety of genes in prokaryotes is induced or repressed in response to specific chemicals in the growth medium, the classical example of which is the induction of *lac* operon transcription by lactose. Traditionally these are not categorized under signal transduction; however, many of them embody the elements of a typical chemical sensory pathway, e.g., a receptor, a second messenger, and a transcriptional repressor or activator. Although the majority of them do not involve protein phosphorylation, a unique example involving both phosphorylation and dephosphorylation of a regulator by the same enzyme is briefly described here.

Growth of *E. coli* on acetate requires activation of the so-called "glycolytic bypass" operon, and the flow of isocitrate through this pathway is regulated via the phosphorylation–dephosphorylation of isocitrate dehydrogenase (IDH). When bacteria are grown on glucose, IDH is unphosphorylated and fully active. Conversely, when cells are cultured on acetate, IDH becomes phosphorylated with a drastic loss of activity. This serves to divert isocitrate from the Krebs cycle and prevent its loss as CO_2. Interestingly, both the phosphorylation and dephosphorylation of IDH are catalyzed by the same enzyme, IDH kinase/phosphatase. Thus, the intriguing question is how the two opposing activities peacefully coexist in the 578-residue polypeptide and whether their active sites are physically separated or not. Although the final answers are yet to come, recent mutational analyses have shed some light on the mechanism (Ikeda *et al.*, 1992, and references therein). It was known that ATP was required not only for the kinase but for the phosphatase activity as well. However, the amino acid sequence revealed the presence of only one putative ATP-binding site. Mutation of the invariant Lys (Lys^{236}) of this motif to Met abrogated both the kinase and phosphatase activities, leading to the hypothesis that both reactions occur at the same active site and that the phosphatase reaction results from the "back reaction" of the kinase coupled to ATP hydrolysis. On the other hand, mutations were selected that specifically destroyed the phosphatase activity only; such mutations mapped within a 113-residue (365–477) region near the center of the polypeptide. Some of these mutants had an enhanced kinase activity (up to three to four times that of the wild type), which suggested that these mutations may define a regulatory domain for both activities.

4. MALFUNCTION AND DISEASE

Although lesions in signaling molecules have been incriminated in a colossal variety of human and nonhuman diseases, conclusive evidence that they are direct results of defects in protein phosphorylation *per se* is often lacking. As alluded to earlier, the uncertainty stems mainly from the multifunctional and interactive nature of each pathway.

By far the largest class of defective signal transduction molecules is represented by oncogenes. Originally discovered as retroviral genes that produced tumors in birds and rodents, oncogenes described to date number in the hundreds and represent all the major groups of signal transduction molecules: growth factors (v-*sis*); growth factor receptors (v-*erbB*, v-*fms*, v-*kit*); transducers (v-*src*, v-*ras*, v-*raf*) (Cantley *et al.*, 1991); transcription factors (v-*jun*, v-*fos*); and G proteins ($G\alpha_s$ in pituitary tumor and $G\alpha_{i2}$ in sex cord stromal and adrenal cortical tumor) (Lyons *et al.*, 1990). A given class can be subdivided further; e.g., some transducers are protein kinases or phosphatases.

Mutations in *ras* are prevalent in a variety of cancers; furthermore, such mutations occur in "hot spots" on the ras protein, e.g., codon 12 (Section 2.3.5a), and generally result in constitutively active ras proteins. Consistent with a central role of ras in signaling, defects in the various effectors of ras have also been implicated in many abnormalities (Boguski and McCormick, 1993). In the fission yeast, loss of function of lte1, a putative GEF, results in cold sensitivity. A defective allele of the ubiquitously expressed protein neurofibromin (NF1), a rasGAP, is responsible for von Recklinghausen neurofibromatosis in humans.

In the recent past, a number of disorders have been directly traced to defects in protein kinases and phosphatases. A type of selective T-cell deficiency (STD) characterized by persistent infections reminiscent of AIDS results from a mutation in the zap-70, a tyrosine kinase indispensable for T-cell signaling (Arpaia *et al.*, 1994). BCR-abl, a chimeric oncoprotein generated by the translocation of *c-abl* sequences on chromosome 9 onto BCR sequences on chromosome 22, has been implicated in Philadelphia chromosome (Ph^1)-positive human leukemias (Pendergast *et al.*, 1993); oncogenesis by BCR-abl requires its association with the SH2/SH3-domain protein grb2 (Section 2.1.1b), suggesting the involvement of the signal transduction pathway. Essentially all nonreceptor PTKs encode protooncogenes (Section 2.2), src being a classic example. The unique viral protooncogene product crk apparently stimulates PTKs (Bishop, 1991). Both mos and raf are prominent examples of protein Ser/Thr kinases involved in oncogenesis. Malfunctions of protein phosphatases are exemplified by the *rdgC* mutation in the fruit fly, *SIT4* mutants of yeast, and *motheaten* mutation in mice (Section 2.7.1). Finally, many human mammary carcinomas display increased levels of EGFR and PLC_γ, providing a potential to amplify signals through the inositol pathway (Rhee and Choi, 1992).

5. FUTURE DIRECTIONS

The single biggest question in the studies of phosphorylation and signal transduction is one of specificity. This complexity arises as a result of cross-talk among different pathways, i.e., more than one pathway signaling through the same molecule (e.g., mek in Figure 1), and conversely, one molecule (e.g., second messengers) feeding into multiple pathways. Different signals are integrated to produce additive, synergistic, or antagonistic effects, making it difficult to pinpoint a causal relationship. Thus, although there is convincing evidence for the role of specific protein kinases and phosphatases in regulating their signaling substrates *in vitro,* proof of their *in vivo* role is generally lacking. Inhibitors are rarely specific for one enzyme, and their use very likely perturbs more than one pathway. Unfortunately, there is no simple solution to this problem; one can only hope that a larger data base derived from a combination of quantitative biochemistry, genetics, and mutagenesis in different systems will make the big picture clearer. Such studies would ideally require conditional lethal enzyme and substrate mutants that are specifically defective in their interactions; however, there is no easy selection for such mutants. Microinjection of antikinase antibody or excess of substrate peptides have sometimes provided useful but limited information.

At the molecular level, an important question is how phosphorylation affects the structure of a protein. In principle, phosphate groups may simply increase the net negative charge of the protein and/or alter its structure, and there is precedence for both. In mek and c-rel, acidic amino acids could mimic or outperform phosphoserines (Mansour *et al.*, 1994; Mosialos *et al.*, 1991), suggesting that the PO_4 group provided a negative charge. On the other hand, there is ample evidence that phosphorylation often produces allosteric changes in protein conformation, judged by physical as well as biochemical parameters. Experiments with model RNA viral phosphoproteins, which are transactivators of viral transcription, have suggested an allosteric change in conformation following phosphorylation by cellular casein kinase II (Barik and Banerjee, 1992b; Banerjee and Barik, 1992). The biological relevance of this is underscored by the discovery that phosphorylation is essential for the RNA transcriptional activity of the phosphoprotein (Barik and Banerjee, 1992a,b; Mazumder and Barik, 1994; Mazumder *et al.*, 1994). Site-directed mutagenesis and comparison of the crystal structure of a given phosphoprotein with that of its unphosphorylated counterpart should shed light on the exact role phosphorylation plays in protein function. Such studies may eventually lead to a rational design of inhibitory drugs targeted against the site of phosphorylation.

The discovery of membrane-bound PTPs has raised the question whether they, like the RTKs, may also serve as receptors. Clearly, the answer must await the identification of their ligands and downstream substrates.

Deciphering the sensory signals (Ngai *et al.*, 1993) appears to be another major frontier of the future. It remains a mystery how the olfactory (smell) signals, generated by nearly a thousand receptors in the mammalian nose, are integrated within fractions of a second, a situation reminiscent of the bacterial chemotactic response. Perhaps the most grandiose of all challenges is to solve the processes of signal transduction in the human brain, in particular the mechanisms of cognition and long-term memory, which may be ever too complex for the human brain to solve (Tully *et al.*, 1994 and the two accompanying papers).

ACKNOWLEDGMENTS. Thanks are due to Dr. Richard Honkanen for constructive suggestions and to Dr. Barsanjit Mazumder, Tammy McLean, and Ameshia Bush for assisting in the literature search. Apology is offered to the hundreds of researchers whose original work could not be cited because of space constraints. Research in the author's laboratory was supported in part by the National Institutes of Health (NIH) grant AI37938.

6. REFERENCES

Aderem, A., 1992, Signal transduction and the actin cytoskeleton: The roles of MARCKS and profilin, *Trends Biochem. Sci.* **17:**438–443.

Aiba, H., and Mizuno, T., 1990, Phosphorylation of a bacterial activator protein, OmpR, by a protein kinase, EnvZ, stimulates the transcription of the *ompF* and *ompC* genes in *Escherichia coli*, *FEBS Lett.* **261:**19–22.

Alvarez, E., Northwood, I. C., Gonzalez, F. A., Latour, D. A., Seth, A., Abate, C., Curran, T., and Davis, R. J., 1991, Pro-Leu-Ser/Thr-Pro is a consensus primary sequence for substrate protein phosphorylation. Characterization of the phosphorylation of c-myc and c-jun proteins by an epidermal growth factor receptor threonine 669 protein kinase, *J. Biol. Chem.* **266:**15277–15285.

Arndt, K. T., Styles, C. A., and Fink, G. R., 1989, A suppressor of a *HIS4* transcriptional defect encodes a protein with homology to the catalytic subunit of protein phosphatases, *Cell* **56:**527–537.

Arpaia, E., Shahar, M., Dadi, H., Cohen, A., and Roifman, C. M., 1994, Defective T cell receptor signaling and $CD8^+$ thymic selection in humans lacking zap-70 kinase, *Cell* **76:**947–958.

Axton, J. M., Dombradi, V., Cohen, P. T., and Glover, D. M., 1990, One of the protein phosphatase 1 isoenzymes in *Drosophila* is essential for mitosis, *Cell* **63:**33–46.

Baeuerle, P. A., and Baltimore, D., 1988, IκB: a specific inhibitor of the NF-κB transcription factor, *Science* **242:**540–546.

Banerjee, A. K., and Barik, S., 1992, Minireview: Gene expression of vesicular stomatitis virus genome RNA, *Virology* **188:**417–428.

Barford, D., Flint, A. J., and Tonks, N. K., 1994, Crystal structure of human protein tyrosine phosphatase 1B, *Science* **263:**1397–1403.

Barik, S., 1993, Expression and biochemical properties of a protein serine/threonine phosphatase encoded by bacteriophage λ, *Proc. Natl. Acad. Sci. U.S.A.* **90:**10633–10637.

Barik, S., and Banerjee, A. K., 1992a, Phosphorylation by cellular casein kinase II is essential for transcriptional activity of vesicular stomatitis virus phosphoprotein P, *Proc. Natl. Acad. Sci. U.S.A.* **89:**6570–6574.

Barik, S., and Banerjee, A. K., 1992b, Sequential phosphorylation of the phosphoprotein of vesicular stomatitis virus by cellular and viral protein kinases is essential for transcription activation, *J. Virol.* **66:**1109–1118.

Baudier, J., Deloulme, J. C., Van Dorsselaer, A., Black, D., and Matthes, H. W., 1991, Purification and characterization of a brain-specific protein kinase C substrate, neurogranin (p17). Identification of a consensus amino acid sequence between neurogranin and neuromodulin (GAP43) that corresponds to the protein kinase C phosphorylation site and the calmodulin-binding domain, *J. Biol. Chem.* **266:**229–237.

Berridge, M. J., and Irvine, R. F., 1989, Inositol phosphates and cell signaling, *Nature* **341:**197–205.

Birge, R. B., and Hanafusa, H., 1993, Closing in on SH2 specificity, *Science* **262:**1522–1524.

Bishop, J. M., 1991, Molecular themes in oncogenesis, *Cell* **64:**235–248.

Blank, J. L., Ross, A. H., and Exton, J. H., 1991, Purification and characterization of two G proteins that activate the β1 isozyme of phosphoinositide-specific phospholipase C. Identification as members of the G_q class, *J. Biol. Chem.* **266:**18206–18216.

Boguski, M. S., and McCormick, F., 1993, Proteins regulating Ras and its relatives, *Nature* **366:**643–654.

Bohman, D., 1990, Transcription factor phosphorylation: A link between signal transduction and the regulation of gene expression, *Cancer Cells* **2:**337–344.

Bourret, R. B., Borkovich, K. A., and Simon, M. I., 1991, Signal transduction pathways involving protein phosphorylation in prokaryotes, *Annu. Rev. Biochem.* **60:**401–441.

Brautigan, D. L., Fernandez, A., and Lamb, N.J.C., 1989, Microinjection of protein phosphatase type-1 selectively alters fibroblast functions and *in vivo* phosphorylation, *Adv. Prot. Phosphatases* **5:**547–566.

Brautigan, D. L., Sunwoo, J., Labbe, J. C., Fernandez, A., and Lamb, N. J., 1990, Cell cycle oscillation of phosphatase inhibitor-2 in rat fibroblasts coincident with $p34^{cdc2}$ restriction, *Nature* **344:**74–78.

Brown, N. H., and Hartley, D. A., 1994, Exploring signaling pathways, *Nature* **370:**414–415.

Butler, T. M., Ziemiecki, A., and Friis, R. R., 1990, Megakaryocytic differentiation of K562 cells is associated with changes in the cytoskeletal organization and the pattern of chromatographically distinct forms of phosphotyrosyl-specific protein phosphatases, *Cancer Res.* **50:**6323–6329.

Cantley, L. C., Auger, K. R., Carpenter, C., Duckworth, B., Graziani, A., Kapeller, R., and Soltoff, S., 1991, Oncogenes and signal transduction, *Cell* **64:**281–302.

Charbonneau, H., and Tonks, N. K., 1992, 1002 protein phosphatases? *Annu. Rev. Cell Biol.* **8:**463–493.

Choi, K.-Y., Satterberg, B., Lyons, D. M., and Elion, E. A., 1994, Ste5 tethers multiple protein kinases in the MAP kinase cascade required for mating in *S. cerevisiae*, *Cell* **78:**499–512.

Ciccetti, P., Mayer, B. J., Thiel, G., and Baltimore, D., 1992, Identification of a protein that binds to the SH3 region of Abl and is similar to Bcr and GAP-rho, *Science* **257:**803–806.

Coggins, P. J., and Zwiers, 1991, B-50 (GAP-43): Biochemistry and functional neurochemistry of a neuron-specific phosphoprotein, *J. Neurochem.* **56:**1095–1106.

Cohen, P., 1989, The structure and regulation of protein phosphatases, *Annu. Rev. Biochem.* **58:**453–508.

Cohen, P. T., and Cohen, P., 1989, Discovery of a protein phosphatase activity encoded in the genome of bacteriophage lambda. Probable identity with open reading frame 221, *Biochem. J.* **260:**931–934.

Coleman, D. E., Berghuis, A. M., Lee, E., Linder, M. E., Gilman, A. G., and Sprang, S. R., 1994, Structures of active conformations of $G_{i\alpha 1}$ and the mechanism of GTP hydrolysis, *Science* **265:**1405–1412.

Cooper, J. A., and Howell, B., 1993, The when and how of Src regulation, *Cell* **73:**1051–1054.

Crews, C. M., and Erikson, R. L., 1993, Extracellular signals and reversible protein phosphorylation: What to Mek of it all, *Cell* **74:**215–217.

Cross, F. R., Garber, E. A., Pellman, D., and Hanafusa, H., 1984, A short sequence in the $p60^{src}$ N terminus is required for $p60^{src}$ myristoylation and membrane association and for cell transformation, *Mol. Cell. Biol.* **4:**1834–1842.

Cross, M., and Dexter, T. M., 1991, Growth factors in development, transformation, and tumorigenesis, *Cell* **64:**271–280.

Cyert, M. S., and Thorner, J., 1989, Putting it on and taking it off: Phosphoprotein phosphatase involvement in cell cycle regulation, *Cell* **57**:891–893.

David, M., and Larner, A. C., 1992, Activation of transcription factors by interferon-alpha in a cell-free system, *Science* **257**:813–815.

Dawson, T. M., Steiner, J. P., Dawson, V. L., Dinerman, J. L., Uhl, G. R., and Snyder, S. H., 1993, Immunosuppressant FK506 enhances phosphorylation of nitric oxide synthase and protects against glutamate neurotoxicity, *Proc. Natl. Acad. Sci. U.S.A.* **90**:9808–9812.

Divecha, N., Banfic, H., and Irvine, R. F., 1993, Inositides and the nucleus and inositides in the nucleus, *Cell* **74**:405–407.

Dohadwala, M., da Cruz e Silva, E. F., Hall, F. L., Williams, R. T., Carbonaro-Hall, D. A., Nairn, A. C., Greengard, P., and Berndt, N., 1994, Phosphorylation and inactivation of protein phosphatase 1 by cyclin-dependent kinases, *Proc. Natl. Acad. Sci. U.S.A.* **91**:6408–6412.

Drubin, D. G., Mulholland, J., Zhu, Z., and Botstein, D., 1990, Homology of a yeast actin-binding protein to signal transduction proteins and myosin-I, *Nature* **343**:288–290.

Ely, C. M., Oddie, K. M., Litz, J. S., Rossomando, A. J., Kanner, S. B., Sturgill, T. W., and Parsons, S. J., 1990, A 42-kD tyrosine kinase substrate linked to chromaffin cell secretion exhibits an associated MAP kinase activity and is highly related to a 42-kD mitogen-stimulated protein in fibroblasts, *J. Cell Biol.* **110**:731–742.

Fantl, W. J., Johnson, D. E., and Williams, L. T., 1993, Signaling by receptor tyrosine kinases, *Annu. Rev. Biochem.* **62**:453–481.

Feaver, W. J., Gileadi, O., Li, Y., and Kornberg, R. D., 1991, CTD kinase associated with yeast RNA polymerase II initiation factor b, *Cell* **67**:1223–1230.

Fernandez-Sarabia, M. J., Sutton, A., Zhong, T., and Arndt, K. T., 1992, SIT4 protein phosphatase is required for the normal accumulation of SW14, CLN1, CLN2 and HCS26 RNAs during late G_1, *Genes Dev.* **6**:2417–2428.

Firmbach-Kraft, I., Byers, M., Shows, T., Dalla-Favera, R., and Krolewski, J. J., 1990, *tyk2*, prototype of a novel class of non-receptor tyrosine kinase genes, *Oncogene* **5**:1329–1336.

Forst, S., Delgado, J., Rampersaud, A., and Inouye, M., 1990, *In vivo* phosphorylation of OmpR, the transcription activator of the *ompF* and *ompC* genes in *Escherichia coli, J. Bacteriol.* **172**:3473–3477.

Frangioni, J. V., Oda, A., Smith, M., Salzman, E. W., and Neel, B. G., 1993, Calpain-catalyzed cleavage and subcellular relocation of protein phosphotyrosine phosphatase 1B (PTP-1B) in human platelets, *EMBO J.* **12**:4843–4856.

Fu, X. Y., 1992, A transcription factor with SH2 and SH3 domains is directly activated by an interferon α-induced cytoplasmic protein tyrosine kinase(s), *Cell* **70**:323–335.

Fu, X. Y., and Zhang, J. J., 1993, Transcription factor p91 interacts with the epidermal growth factor receptor and mediates activation of the c-*fos* gene promoter, *Cell* **74**:1135–1145.

Gardner, A. M., Vaillancourt, R. R., and Johnson, G. L., 1993, Activation of mitogen-activated protein kinase/extracellular signal-regulated kinase kinase by G protein and tyrosine kinase oncoproteins, *J. Biol. Chem.* **268**:17896–17901.

Gegner, J. A., and Dahlquist, F. W., 1991, Signal transduction in bacteria: CheW forms a reversible complex with the protein kinase CheA, *Proc. Natl. Acad. Sci. U.S.A.* **88**:750–754.

Ghosh, S., and Baltimore, D., 1990, Activation *in vitro* of NF-κB by phosphorylation of its inhibitor IκB, *Nature* **344**:678–682.

Goldschmidt-Clermont, P. J., Kim, J. W., Machesky, L. M., Rhee, S. G., and Pollard, T. D., 1991, Regulation of phospholipase C-gamma 1 by profilin and tyrosine phosphorylation, *Science* **251**:1231–1233.

Gonzalez, G. A., Menzel, P., Leonard, J., Fischer, W. H., and Montminy, M. R., 1991, Characterization of motifs which are critical for activity of the cyclic AMP-responsive transcription factor CREB, *Mol. Cell. Biol.* **11**:1306–1312.

Gould, K. L., Bretscher, A., Esch, F. S., and Hunter, T., 1989, cDNA cloning and sequencing of the protein-tyrosine kinase substrate, ezrin, reveals homology to band 4.1, *EMBO J.* **8:**4133–4142.

Gould, K. L., Moreno, S., Tonks, N. K., and Nurse, P., 1990, Complementation of the mitotic activator, $p80^{cdc25}$, by a human protein-tyrosine phosphatase, *Science* **250:**1573–1576.

Guan, K. L., and Dixon, J. E., 1991, Evidence for protein-tyrosine-phosphatase catalysis proceeding via a cysteine-phosphate intermediate, *J. Biol. Chem.* **266:**17026–17030.

Guan, K. L., Huan, R. S., Watson, S. J., Geahlen, R. L., and Dixon, J. E., 1990, Cloning and expression of a protein-tyrosine-phosphatase, *Proc. Natl. Acad. Sci. U.S.A.* **87:**1501–1505.

Gutkind, J. S., Lacal, P. M., and Robbins, K. C., 1990, Thrombin-dependent association of phosphatidylinositol-3 kinase and $p60^{c\text{-}src}$ and $p59^{fyn}$ in human platelets, *Mol. Cell. Biol.* **10:**3806–3809.

Hagiwara, M., Alberts, A., Brindle, P., Meinkoth, J., Feramisco, J., Deng, T., Karin, M., Shenolikar, S., and Montminy, M., 1992, Transcriptional attenuation following cAMP induction requires PP-1-mediated dephosphorylation of CREB, *Cell* **70:**105–113.

Hall, A., 1990, ras and GAP—who's controlling whom? *Cell* **61:**921–923.

Halpain, S., Girault, J. A., and Greengard, P., 1990, Activation of NMDA receptors induces dephosphorylation of DARPP-32 in rat striatal slices, *Nature* **343:**369–372.

Hanks, S. K., Quinn, A. M., and Hunter, T., 1988, The protein kinase family: Conserved features and deduced phylogeny of the catalytic domains, *Science* **241:**42–52.

Harlan, D. M., Graff, J. M., Stumpo, D. J., Eddy, R. L., Jr., Shows, T. B., Boyle, J. M., and Blackshear, P. J., 1991, The human myristoylated alanine-rich C kinase substrate (MARCKS) gene (MACS), *J. Biol. Chem.* **266:**14399–14405.

Hartwig, J. H., Thelen, M., Rosen, A., Jammey, P. A., Nairn, A. C., and Aderem, A., 1992, MARKS is an actin filament crosslinking protein regulated by protein kinase C and calcium-calmodulin, *Nature* **356:**618–622.

Heidecker, G., Kolch, W., Morrison, D. K., and Rapp, U. R., 1992, The role of Raf-1 phosphorylation in signal transduction, *Adv. Cancer Res.* **58:**53–73.

Hemmings, H. C., Jr., Greengard, P., Tung, H.Y.L., and Cohen, P., 1984, DARPP-32, a dopamine regulated neuronal phosphoprotein is a potent inhibitor of protein phosphatase 1, *Nature* **310:**503–505.

Honkanen, R. E., and Boynton, A. L., 1994, Serine/threonine protein phosphatases and their inhibitors, in *Protein Kinase C* (J. F. Kuo, ed.), pp. 305–324, Oxford University Press, New York.

Honkanen, R. E., Zwiller, J., Daily, S. L., Khatra, B. S., Dukelow, M., and Boynton, A. L., 1991, Identification, purification, and characterization of a novel serine/threonine protein phosphatase from bovine brain, *J. Biol. Chem.* **266:**6614–6619.

Huang, F. L., and Glinsmann, W. H., 1976, Separation and characterization of two phosphorylase phosphatase inhibitors from rabbit skeletal muscle, *Eur. J. Biochem.* **70:**419–426.

Huang, W., Alessandrini, A., Crews, C. M., and Erikson, R. L., 1993, Raf-1 forms a stable complex with Mek1 and activates Mek1 by serine phosphorylation, *Proc. Natl. Acad. Sci. U.S.A.* **90:**10947–10951.

Hunter, T., 1991, Cooperation between oncogenes, *Cell* **64:**249–270.

Hunter, T., and Karin, M., 1992, The regulation of transcription by phosphorylation, *Cell* **70:**375–387.

Hunter, T., Angel, P., Boyle, W. J., Chiu, R., Freed, E., Gould, K. L., Isacke, C. M., Karin, M., Lindberg, R. A., and van der Geer, P., 1988, Targets for signal-transducing protein kinases, *Cold Spring Harbor Symp. Quant. Biol.* **53:**131–142.

Ikeda, T. P., Houtz, E., and LaPorte, D. C., 1992, Isocitrate dehydrogenase kinase/phosphatase: Identification of mutations which selectively inhibit phosphatase activity, *J. Bacteriol.* **174:**1414–1416.

Ingebritsen, T. S., and Cohen, P., 1983, Protein phosphatases: properties and role in cellular regulation, *Science* **221**:331–338.

Kamps, M. P., Buss, J. E., and Sefton, B. M., 1985, Mutation of NH_2-terminal glycine of $p60^{src}$ prevents both myristoylation and morphological transformation, *Proc. Natl. Acad. Sci. U.S.A.* **82**:4625–4628.

Kennedy, M., 1989, Regulation of neuronal function by calcium, *Trends Neurosci.* **12**:417–420.

Kennelly, P. J., and Krebs, E. G., 1991, Consensus sequences as substrate specificity determinants for protein kinases and protein phosphatases, *J. Biol. Chem.* **266**:15555–15558.

Kerr, L. D., Inoue, J., Davis, N., Link, E., Baeuerle, E. A., Bose, H. R., Jr., and Verma, I. M., 1991, The Rel-associated pp40 protein prevents DNA-binding of Rel and NF-κB: Relationship with IκBβ and regulation by phosphorylation, *Genes Dev.* **5**:1464–1476.

Kim, H. K., Kim, J. W., Zilberstein, A., Margolis, B., Kim, J. G., Schlessinger, J., and Rhee, S. G., 1991, PDGF stimulation of inositol phospholipid hydrolysis requires PLC-γ1 phosphorylation on tyrosine residues 783 and 1254, *Cell* **65**:435–441.

King, M. M., Huang, C. Y., Chock, P. B., Nairn, A. C., Hemmings, H. C., Jr., Jesse Chan, K. F., and Greengard, P., 1984, Mammalian brain phosphoproteins as substrates for calcineurin, *J. Biol. Chem.* **259**:8080–8083.

Kinoshita, N., Ohkura, H., and Yanagida, M., 1990, Distinct, essential roles of type 1 and 2A protein phosphatases in the control of the fission yeast cell division cycle, *Cell* **63**:405–415.

Klee, C. B., and Cohen, P., 1988, The calmodulin-regulated protein phosphatase, *Mol. Aspects Cell. Regul.* **5**:225–248.

Koch, C. A., Anderson, D., Moran, M. F., Ellis, C., and Pawson, T., 1991, SH2 and SH3 domains: Elements that control interactions of cytoplasmic signaling proteins, *Science* **252**:668–674.

Kolch, W., Heidecker, G., Lloyd, P., and Rapp, U. R., 1991, Raf-1 protein kinase is required for growth of induced NIH/3T3 cells, *Nature* **349**:426–428.

Koretzky, G. A., Picus, J., Thomas, M. L., and Weiss, A., 1990, Tyrosine phosphatase CD45 is essential for coupling T-cell antigen receptor to the phosphatidyl inositol pathway, *Nature* **346**:66–68.

Kypta, R. M., Goldberg, Y., Ulug, E. T., and Courtneidge, S. A., 1990, Association between the PDGF receptor and members of the *src* family of tyrosine kinases, *Cell* **62**:481–492.

Lange-Carter, C. A., and Johnson, G. L., 1994, Ras-dependent growth factor regulation of MEK kinase in PC12 cells, *Science* **265**:1458–1461.

Lee, C. Q., Yun, Y. D., Hoeffler, J. P., and Habener, J. F., 1990, Cyclic-AMP-responsive transcriptional activation of CREB-327 involves interdependent phosphorylated subdomains, *EMBO J.* **9**:4455–4465.

Lee, J. M., and Greenleaf, A. L., 1991, CTD kinase large subunit is encoded by CTK1, a gene required for normal growth of *Saccharomyces cerevisiae, Gene Expr.* **1**:149–167.

Lefkowitz, R. J., 1993, G protein-coupled receptor kinases, *Cell* **74**:409–412.

Lin, A., Frost, J., Deng, T., Smeal, T., Al-Alawi, N., Kikkawa, U., Hunter, T., Brenner, D., and Karin, M., 1992, Casein kinase II is a negative regulator of c-Jun DNA binding and AP-1 activity, *Cell* **70**:777–789.

Link, E., Kerr, L. D., Schreck, R., Zabel, U., Verma, I. M., and Baeuerle, P. A., 1992, Purified IκB-β is inactivated upon dephosphorylation, *J. Biol. Chem.* **267**:239–246.

Lombroso, P. J., Murdoch, G., and Lerner, M., 1991, Molecular characterization of a protein-tyrosine-phosphatase enriched in striatum, *Proc. Natl. Acad. Sci. U.S.A.* **88**:7242–7246.

Lowenstein, C. J., and Snyder, S. H., 1992, Nitric oxide, a novel biologic messenger, *Cell* **70**:705–707.

Lowy, D. R., and Willumsen, B. M., 1993, Function and regulation of RAS, *Annu. Rev. Biochem.* **62**:851–891.

Lu, H., Zawel, L., Fisher, L., Egly, J.-M., and Reinberg, D., 1992, Human general transcription factor IIH phosphorylate the C-terminal domain of RNA polymerase II, *Nature* **358**:641–645.

Luscher, B., Christenson, E., Litchfield, D. W., Krebs, E. G., and Eisenman, R. N., 1990, Myb DNA binding inhibited by phosphorylation at a site deleted during oncogenic activation, *Nature* **344:**517–522.

Lyons, J., Landis, C. A., Harsh, G., Vallar, L., Grunewald, K., Feichtinger, H., Duh, Q.-Y., Clark, O. H., Kawasaki, E., Bourne, H. R., and McCormick, F., 1990, Two G protein oncogenes in human endocrine tumors, *Science* **249:**655–659.

Mahoney, C. H., and Huang, K.-P., 1994, Molecular and catalytic properties of protein kinase C, in *Protein Kinase C* (J. F. Kuo, ed.), pp. 16–63, Oxford University Press, New York.

Majerus, P. W., 1992, Inositol phosphate biochemistry, *Annu. Rev. Biochem.* **61:**225–250.

Mansour, S. J., Matten, W. T., Hermann, A. S., Candia, J. M., Rong, S., Fukasawa, K., Vande Wonde, R. F., and Ahn, N. G., 1994, Transformation of mammalian cells by constitutively active MAP kinase kinase, *Science* **265:**966–969.

Marais, R., Wynne, J., and Treisman, R., 1993, The SRF accessory protein Elk-1 contains a growth factor-regulated transcriptional activation domain, *Cell* **73:**381–393.

Matsuda, M., Mayer, B. J., Fukui, Y., and Hanafusa, H., 1990, Binding of transforming protein, $p47^{gag\text{-}crk}$, to a broad range of phosphotyrosine-containing proteins, *Science* **248:**1537–1539.

Mazumder, B., and Barik, S., 1994, Requirement of casein kinase II-mediated phosphorylation for the transcriptional activity of human respiratory syncytial virus phosphoprotein P: Transdominant negative phenotype of phosphorylation-defective P mutants, *Virology* **205:**104–111.

Mazumder, B., Adhikary, G., and Barik, S., 1994, Bacterial expression of human respiratory syncytial virus phosphoprotein P and identification of Ser^{237} as the site of phosphorylation by cellular casein kinase II, *Virology* **205:**93–103.

Meikrantz, W., Smith, D. M., Sladicka, M. M., and Schlegel, R. A., 1991, Nuclear localization of an O-glycosylated protein phosphotyrosine phosphatase from human cells, *J. Cell Sci.* **98:**303–307.

Mekalanos, J. J., 1992, Minireview: Environmental signals controlling expression of virulence determinants in bacteria, *J. Bacteriol.* **174:**1–7.

Millar, J.B.A., and Russell, P., 1992, The cdc25 M-phase inducer: An unconventional protein phosphatase, *Cell* **68:**407–410.

Minami, Y., Stafford, F. J., Lippincott-Schwartz, J., Yuan, L. C., and Klausner, R. D., 1991, Novel redistribution of an intracellular pool of CD45 accompanies T cell activation, *J. Biol. Chem.* **266:**9222–9230.

Mire-Sluis, A. R., and Thorpe, R., 1991, Interleukin-4 proliferative signal transduction involves the activation of a tyrosine-specific phosphatase and the dephosphorylation of an 80-kDa protein, *J. Biol. Chem.* **266:** 18113–18118.

Moncada, S., Palmer, R.M.J., and Higgs, E. A., 1991, Nitric oxide: Physiology, pathophysiology, and pharmacology, *Pharmacol. Rev.* **43:**109–142.

Moodie, S. A., Willumsen, B. M., Weber, M. J., and Wolfman, A., 1993, Complexes of Ras.GTP with Raf-1 and mitogen-activated protein kinase kinase, *Science* **260:**1658–1661.

Morrison, D. K., Heidecker, G., Rapp, U. R., and Copeland, T. D., 1993, Identification of the major phosphorylation sites of the Raf-1 kinase, *J. Biol. Chem.***268:**17309–17316.

Mosialos, G., Hamer, P., Capobianco, A. J., Laursen, R. A., and Gilmore, T. D., 1991, A protein kinase-A recognition sequence is structurally linked to transformation by $p59^{v\text{-}rel}$ and cytoplasmic retention of $p68^{c\text{-}rel}$, *Mol. Cell. Biol.* **11:**5867–5877.

Mumby, M. C., and Walter, G., 1991, Protein phosphatases and DNA tumor viruses: Transformation through the back door? *Cell Regul.* **2:**589–598.

Nellen, D., Affolter, M., and Basler, K., 1994, Receptor serine/threonine kinases implicated in the control of *Drosophila* body pattern by *decapentaplegic*, *Cell* **78:**225–237.

Ngai, J., Dowling, M. M., Buck, L., Axel, R., and Chess, A., 1993, The family of genes encoding odorant receptors in the channel fish, *Cell* **72:**657–666.

Nimmo, G. A., and Cohen, P., 1978, The regulation of glycogen metabolism: Phosphorylation of inhibitor-1 from rabbit skeletal muscle, and its interaction with protein phosphatases-III and II, *Eur. J. Biochem.* **87**:353–365.

Ninfa, A. J., and Bennett, R. L., 1991, Identification of the site of autophosphorylation of the bacterial protein kinase/phosphatase NRII, *J. Biol. Chem.* **266**:6888–68893.

Nishida, E., and Gotoh, Y., 1993, The MAP kinase cascade is essential for diverse signal transduction pathways, *Trends Biochem. Sci.* **18**:128–131.

Nishizuka, Y., 1992, Intracellular signaling by hydrolysis of phospholipids and activation of protein kinase C, *Science* **258**:607–614.

Noda, M., Selinger, Z., Scolnick, E. M., and Bassin, R. H., 1983, Flat revertants isolated from Kirsten sarcoma virus-transformed cells are resistant to the action of specific oncogenes, *Proc. Natl. Acad. Sci. U.S.A.* **80**:5602–5606.

Parkinson, J. S., 1993, Signal transduction schemes of bacteria, *Cell* **73**:857–871.

Pawson, T., and Gish, G. D., 1992, SH2 and SH3 domains: From structure to function, *Cell* **71**:359–362.

Pendergast, A. M., Quilliam, L. A., Cripe, L. D., Bassing, C. H., Dai, Z., Li, N., Batzer, A., Rabun, K. M., Der, C. J., Schlessinger, J., and Gishizky, M. L., 1993, BCR-ABL-induced oncogenesis is mediated by direct interaction with the SH2 domain of the GRB-2 adaptor protein, *Cell* **75**:175–185.

Perlmutter, R. M., Levin, S. D., Appleby, M. W., Anderson, S. J., and Alberola-Ila, J., 1993, Regulation of lymphocyte function by protein phosphorylation, *Annu. Rev. Immunol.* **11**:451–499.

Perrimon, N., 1993, The torso receptor protein-tyrosine kinase signaling pathway: An endless story, *Cell* **74**:219–222.

Peterson, S. R., Dvir, A., Anderson, C. W., and Dynan, W. S., 1992, DNA-binding provides a signal for phosphorylation of the RNA polymerase II heptapeptide repeats, *Genes Dev.* **6**:426–438.

Pohl, U., Nolte, C., Bunse, A., Eigenthaler, M., and Walter, U., 1994, Endothelium-dependent phosphorylation of vasodilator-stimulated protein in platelets during coronary passage, *Am. J. Physiol.* **266**:606–612.

Popham, D., Szeto, D., Keener, J., and Kustu, S., 1989, Function of a bacterial protein that binds to transcriptional enhancers, *Science* **243**:629–635.

Pulverer, B. J., Kyriakis, J. M., Avruch, J., Nikolakaki, E., and Woodgett, J. R., 1991, Phosphorylation of c-Jun mediated by MAP kinases, *Nature* **353**:670–674.

Rana, R. S., and Hokin, L. E., 1990, Role of phosphoinositides in transmembrane signaling, *Physiol. Rev.* **70**:115–164.

Resh, M. D., and Ling, H. P., 1990, Identification of a 32K plasma membrane protein that binds to the myristylated amino terminal sequence of $p60^{v\text{-}src}$, *Nature* **346**:84–86.

Rhee, S. G., and Choi, K. D., 1992, Regulation of inositol phospholipid-specific phospholipase C isozymes, *J. Biol. Chem.* **267**:12393–12396.

Rodriguez-Viciana, P., Warne, P. H., Dhand, R., Vanhaesebroeck, B., Gout, I., Fry, M. J., Waterfield, M. D., and Downward, J., 1994, Phosphotidylinositol-3-OH kinase as a direct target of Ras, *Nature* **370**:527–532.

Sanders, D. A., Gillece-Castro, B. L., Stock, A. M., Burlingame, A. L., and Koshland, D. E., Jr., 1989, Identification of the site of phosphorylation of the chemotaxis response regulator protein, *J. Biol. Chem.* **264**:21770–21778.

Sanders, D. A., Gillece-Castro, B. L., Burlingame, A. L., and Koshland, D. E., Jr., 1992, Phosphorylation site of NtrC, a protein phosphatase whose covalent intermediate activates transcription, *J. Bacteriol.* **174**:5117–5122.

Satoh, T., Nakafuku, M., and Kaziro, Y., 1992, Function of Ras as a molecular switch in signal transduction, *J. Biol. Chem.* **267**:24149–24152.

Scheidtmann, K. H., Mumby, M. C., Rundell, K., and Walter, G., 1991, Dephosphorylation of simian virus 40 large-T antigen and p53 protein by protein phosphatase 2A: Inhibition by small-t antigen, *Mol. Cell. Biol.* **11:**1996–2003.

Schindler, C., Shuai, K., Prezioso, V. R., and Darnell, J. E., 1992, Interferon-dependent tyrosine phosphorylation of a latent cytoplasmic transcription factor, *Science* **257:**809–813.

Schlessinger, J., and Ullrich, A., 1992, Growth factor signaling by receptor tyrosine kinases, *Neuron* **9:**383–391.

Schmitz, J. L., Henkle, T., and Baeuerle, P. A., 1991, Proteins controlling the nuclear uptake of NF-κB, Rel and Dorsal, *Trends Cell Biol.* **1:**130–137.

Shoelson, S. E., Chatterjee, S., Chaudhuri, M., and White, M. F., 1992, YMXM motifs of IRS-1 define substrate specificity of the insulin receptor kinase, *Proc. Natl. Acad. Sci. U.S.A.* **89:**2027–2031.

Shultz, L. D., Schweitzer, P. A., Rajan, T. V., Yi, T., Ihle, J. N., Matthews, R. J., Thomas, M. L., and Beier, D. R., 1993, Mutations at the murine motheaten locus are within the hematopoietic cell protein-tyrosine phosphatase (Hcph) gene, *Cell* **73:**1445–1454.

Smeal, T., Binetruy, B., Mercola, D. A., Birrer, M., and Karin, M., 1991, Oncogenic and transcriptional cooperation with Ha-Ras requires phosphorylation of c-Jun on serines 63 and 73, *Nature* **354:**494–496.

Smrcka, A. V., Hepler, J. R., Brown, K. O., and Sternweis, P. C., 1991, Regulation of polyphosphoinositide-specific phospholipase C activity by purified G_q, *Science* **251:**804–807.

Snyder, S. H., 1992, Nitric oxide: First in a new class of neurotransmitters? *Science* **257:**494–496.

Soderling, T. R., 1990, Protein kinases: Regulation by autoinhibitory domains, *J. Biol. Chem.* **265:**1823–1826.

Sontag, E., Fedorov, S., Kamibayashi, C., Robbins, D., Cobb, M., and Mumby, M., 1993, The interaction of SV40 small tumor antigen with protein phosphatase 2A stimulates the map kinase pathway and induces cell proliferation, *Cell* **75:**887–897.

Sprenger, F., and Nusslein-Volhard, C., 1992, Torso receptor activity is regulated by a diffusible ligand produced at the extracellular terminal regions of the *Drosophila* egg, *Cell* **71:**987–1001.

Steele, F. R., Washburn, T., Rieger, R., and O'Tousa, J. E., 1992, *Drosophila* retinal degeneration C (rdgC) encodes a novel serine/threonine protein phosphatase, *Cell* **69:**669–676.

Stock, A. M., Mottonen, J. M., Stock, J. B., and Schutt, C. E., 1989, Three-dimensional structure of CheY, the response regulator of bacterial chemotaxis, *Nature* **337:**745–749.

Stock, J. B., Stock, A. M., and Mottonen, J. M., 1990, Signal transduction in bacteria, *Nature* **334:**395–400.

Su, W., Porter, S., Kustu, S., and Echols, H., 1990, DNA-looping and enhancer activity: Association between DNA-bound NtrC activator and RNA polymerase at the bacterial glnA promoter, *Proc. Natl. Acad. Sci. U.S.A.* **87:**5504–5508.

Sun, H., Charles, C. H., Lau, L. F., and Tonks, N. K., 1993, MKP-1 (3CH134), an immediate early gene product, is a dual specificity phosphatase that dephosphorylates MAP kinase *in vivo, Cell* **75:**487–493.

Sutton, A., Immanuel, D., and Arndt, K. T., 1991, The SIT4 protein phosphatase functions in late G_1 for progression into S phase, *Mol. Cell Biol.* **11:**2133–2148.

Taha, M. K., Dupuy, B., Saurin, W., So, M., and Marchal, C., 1991, Control of pilus expression in *Neisseria gonorrhoeae* as original system in the family of two-component regulators, *Mol. Microbiol.* **5:**137–148.

Tan, Y. H., 1993, Yin and yang of phosphorylation in cytokine signaling, *Science* **262:**376–377.

Taylor, S. S., Buechler, J. A., and Yonemoto, W., 1990, cAMP-dependent protein kinase: Framework for a diverse family of regulatory enzymes, *Annu. Rev. Biochem.* **59:**971–1005.

Tonks, N. K., Charbonneau, H., Diltz, C. D., Fischer, E. H., and Walsh, K. A., 1988, Demonstration that the leucocyte common antigen CD45 is a protein tyrosine phosphatase, *Biochem.* **27:**8695–8701.

Treisman, R., 1992, The serum response element, *Trends Biochem. Sci.* **17:**423–426.

Tully, T., Preat, T., Boynton, S. C., and Del Vecchio, M., 1994, Genetic dissection of consolidated memory in *Drosophila melanogaster*, *Cell* **79:**35–48.

Uemura, T., Shiomi, K., Togashi, S., and Takeichi, M., 1993, Mutation of *twins* encoding a regulator of protein phosphatase 2A leads to pattern duplication in *Drosophila* imaginal discs, *Genes Dev.* **7:**429–440.

Ullrich, A., and Schlessinger, J., 1990, Signal transduction by receptors with tyrosine kinase activity, *Cell* **61:**203–212.

Van Etten, R. A., Jackson, P., and Baltimore, D., 1989, The mouse type IV c-abl gene product is a nuclear protein, and activation of transforming ability is associated with cytoplasmic localization, *Cell* **58:**669–678.

Velazquez, L., Fellous, M., Stark, G. R., and Pellagrini, S., 1992, A protein tyrosine kinase in the interferon α/β signaling pathway, *Cell* **70:**313–322.

Walton, K. M., and Dixon, J. E., 1993, Protein tyrosine phosphatases, *Annu. Rev. Biochem.* **62:**101–120.

Welch, P. J., and Wang, J. Y., 1993, A C-terminal protein-binding domain in the retinoblastoma protein regulates nuclear c-Abl tyrosine kinase in the cell cycle, *Cell* **75:**779–790.

Witthuhn, B. A., Quelle, F. W., Silvennoinen, O., Yi, T., Tang, B., Miura, O., and Ihle, J. N., 1993, JAK2 associates with the erythropoietin receptor and is tyrosine phosphorylated and activated following stimulation with erythropoietin, *Cell* **74:**227–236.

Wrana, J. L., Attisano, L., Wieser, R., Ventura, F., and Massague, J., 1994, Mechanism of activation of the TGF-β receptor, *Nature* **370:**341–347.

Yeung, Y. G., Berg, K. L., Pixley, F. J., Angeletti, R. H., and Stanley, E. R., 1992, Protein tyrosine phosphatase-1C is rapidly phosphorylated in tyrosine in macrophages in response to colony stimulating factor-1, *J. Biol. Chem.* **267:**23447–23450.

Zanke, B., Suzuki, H., Kishihara, K., Mizzen, L., Minden, M., Pawson, A., and Mak, T. W., 1992, Cloning and expression of an inducible lymphoid-specific protein tyrosine phosphatase (HePTPase), *Eur. J. Immunol.* **22:**235–239.

Zheng, C. F., and Guan, K. L., 1994, Activation of MEK family kinases requires phosphorylation of two conserved Ser/Thr residues, *EMBO J.* **13:**1123–1131.

Zhuo, S., Clemens, J. C., Hakes, D. J., Barford, D., and Dixon, J. E., 1993, Expression, purification, crystallization, and biochemical characterization of a recombinant protein phosphatase, *J. Biol. Chem.* **268:**17754–17761.

Chapter 6

Structural and Functional Roles of Glycosylphosphoinositides

Alan R. Saltiel

1. INTRODUCTION

Many proteins are anchored to the cell surface by hydrophobic interactions with the membrane bilayer. Several structural motifs are utilized for membrane anchorage in mammalian cells, including the covalent attachment of proteins to lipids (Saltiel *et al.*, 1991). One such mechanism involves the linkage of proteins to a glycosylated form of phosphatidylinositol (PI), termed glycosylphosphatidylinositol (GPI). This mechanism has been detected in a variety of cell types and is utilized by over 100 membrane proteins (Ferguson and Williams, 1988; Low and Saltiel, 1988). Additionally, a glycophospholipid with several structural similarities to the GPI anchor has been described in a number of cell types (Saltiel and Cuatrecasas, 1986, 1988; Saltiel *et al.*, 1986). This lipid, which shares the core structure of the membrane anchor and may represent a biosynthetic intermediate, undergoes a phospholipase C (PLC)-catalyzed hydrolysis in response to insulin and functionally related growth factors, generating an aqueous glycan in cells that mediates some of the actions of the hormone (Saltiel and Cuatrecasas, 1986; Saltiel *et al.*, 1986). Although the precise relationship be-

---

Alan R. Saltiel Department of Signal Transduction, Parke-Davis Pharmaceutical Research, Division of Warner-Lambert Co., Ann Arbor, Michigan 48105.

Subcellular Biochemistry, Volume 26: myo-Inositol Phosphates, Phosphoinositides, and Signal Transduction, edited by B. B. Biswas and Susweta Biswas. Plenum Press, New York, 1996.

tween the protein anchored and the hormone-sensitive free form of the lipid remains unknown, numerous similarities suggest that these glycolipids may participate in related functions in cellular regulation. This chapter focuses on the unusual properties and important roles of this unique class of lipids in cellular regulation.

2. STRUCTURE AND DISTRIBUTION OF GPI MOLECULES

The anchorage of proteins to membranes via GPI has been detected in a variety of eukaryotic organisms in many cell types. The distribution and identification of these proteins have been extensively reviewed (Ferguson and Williams, 1988; Low and Saltiel, 1988). Over 100 proteins of considerable functional and evolutionary diversity are attached to biological membranes by this mechanism. This list includes a number of hydrolytic enzymes, adhesion molecules, coat proteins, mammalian antigens, and additional proteins, many of unknown function. The only feature common to these proteins is their location on the extracellular face of the plasma membrane or (in some cases) on the luminal surface of intracellular organelles.

Compositional and structural analyses of the GPI anchors of several proteins have revealed certain conserved and variant features. Most of the structural information on GPI has emerged from studies on the variant surface glycoprotein (VSG) of *Trypanosoma brucei* (Ferguson *et al.*, 1988), acetylcholinesterase (Roberts *et al.*, 1988b), and the Thy-1 antigen (Homans *et al.*, 1988). These studies have revealed a common basic structure in which the C-terminal amino acid of the protein is linked to an ethanolamine through an amide bond. The ethanolamine, in turn, is attached via a phosphodiester linkage to an oligosaccharide consisting of a trimannose backbone linked to glucosamine. The presence of this nonacetylated glucosamine is one of the unique features of this molecule, possibly serving as an important recognition site in the biosynthesis and metabolism of GPI. The glucosamine is glycosidically linked to the 6-OH position of the inositol ring of PI. This linkage allows for the solubilization of the anchored protein by nitrous acid deamination (causing hydrolysis of the glucosamine–inositol bond) and by bacterial PI-specific phospholipases C (PLCs). The conserved core structure of GPI is shown in Figure 1.

A number of variations in this basic core structure have been observed, perhaps contributing to alterations in functional roles. The presence of varied oligosaccharide side chains for most of the anchors of known structure has been demonstrated. In addition, the Thy-1 anchor appears to possess an additional phosphoethanolamine. Another variation in structure concerns the fatty acid composition of the glycerolipid. In general, the inositol phospholipid moiety of most GPI anchors does not resemble the predominant, mature form of PI (1-

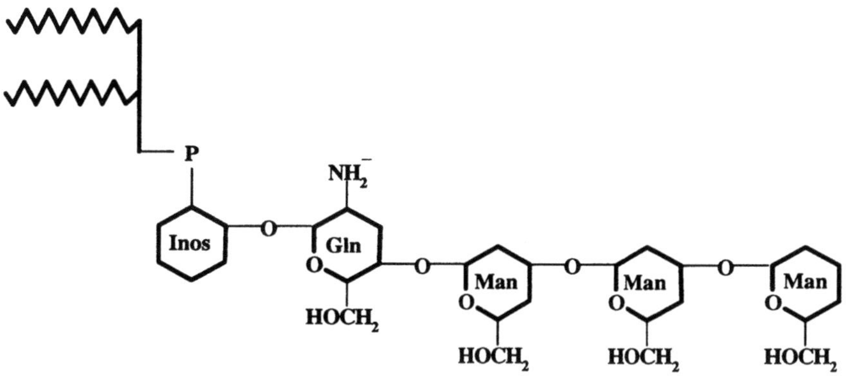

FIGURE 1. Structure of glycosylphosphatidylinositol. The core conserved structure of GPI is illustrated. Inos, inositol; Gln, glucosamine; Man, mannose.

stearyl, 2-arachidonyl) present in most cells but is more likely to contain a saturated fatty acid or a fatty alkyl group in the 2-position (Ferguson and Williams, 1988; Homans *et al.*, 1988; Low and Saltiel, 1988). Another interesting modification is the addition of an extra fatty acid ester linked to one of the hydroxyls of the inositol moiety. The resulting acylated GPI is resistant to cleavage by the PI- and GPI-specific PLCs (Homans *et al.*, 1988).

Considerable interest has been focused on the precise structure and biosynthesis of the free, insulin-sensitive GPI precursor and on its relationship to the protein anchor. Comparison of compositional analyses from several laboratories has indicated certain conserved and variant features. The precursor glycolipid appears to contain a basic core structure of PI-hexosamine. Species identified in T-lymphocyte and BC_3H1 cells appear to contain diacylglycerol as the major glycerolipid moiety (Farese *et al.*, 1988a; Gaulton *et al.*, 1988; Saltiel *et al.*, 1986, 1987), although in hepatoma cells a 1,2-alkylacylglycerol structure has been suggested (Mato *et al.*, 1987b). More recent studies have indicated the presence of significant but variable amounts of *chiro*-inositol, which perhaps accounts for the apparent lack of [^{3}H]*myo*-inositol labeling in some cell types (Larner *et al.*, 1988; Mato *et al.*, 1987a). In fact, deficiencies in cellular and circulating *chiro*-inositol levels have recently been proposed to contribute to insulin resistance associated with NIDDM (Huang *et al.*, 1993; Ortmeyer *et al.*, 1993a,b). Distal to the hexosamine, there appears to be considerable heterogeneity in glycan composition, although the structural basis of this heterogeneity is unclear (Gaulton, 1991).

Most of the analytic techniques developed for quantitation of GPI metabolism have relied on metabolic labeling in tissue culture cells. Hormone-sensitive cells have been labeled with inositol (Gaulton, 1991; Saltiel *et al.*, 1986), glucosamine (Gaulton, 1991; Mato *et al.*, 1987a; Saltiel *et al.*, 1986), mannose

(Larner *et al.*, 1988), galactose (Mato *et al.*, 1987b), phosphate (Mato *et al.*, 1987b), and fatty acids (Saltiel *et al.*, 1986, 1987). In addition, immunostaining techniques have been developed for quantitation of the lipid from thin-layer chromatographs using anti-GPI specific antisera. Interestingly, with either immunostaining or metabolic labeling detection methods, chromatography of these lipids on TLC or HPLC has revealed substantial heterogeneity, even within a single cell type.

Although direct structural analysis of the non-protein-bound, hormone-sensitive pool of GPI molecules awaits completion, numerous studies indicate that these lipids share the major structural elements of the GPI anchor and may represent biosynthetic intermediates for the mature anchor (Tartakoff and Singh, 1992). An antibody raised against a degalactosylated glycan fragment from the *T. brucei* VSG anchor selectively inhibits certain actions of insulin (Romero *et al.*, 1990). Conversely, an antibody raised against a rat-liver-derived insulin-sensitive GPI recognizes the Thy-1 GPI anchor (Gaulton, 1991). This affinity-purified antibody requires a $(Man)_3$ epitope for binding (Gaulton, 1991). Finally, a glycan fragment derived from the *T. brucei* VSG anchor mimics some of the metabolic actions of insulin in rat adipocytes and hepatocytes (Misek and Saltiel, 1992) via a mechanism involving protein phosphatase 1 (Misek and Saltiel, 1994).

3. BIOSYNTHESIS AND ATTACHMENT OF GPI

3.1. Biosynthesis of GPI

Biosynthetic pathways for GPIs have been largely elucidated in *Trypanosoma brucei* (Doering *et al.*, 1990). Metabolic labeling studies have revealed a basic scheme in which N-acetylglucosamine (GlcNac) is first transferred from UDP-GlcNac to PI to form GlcNac-PI. This lipid is short lived and undergoes a rapid deacetylation to glucosamine (Gln)-PI. In many cells, this Gln-PI species may undergo fatty acylation of the inositol ring (Urakaze *et al.*, 1992), which may be removed in later steps. The Gln-PI is subsequently glycosylated in stepwise fashion with three mannoses, using dolichol-mannose as donors. This glycosylated species is often subject to fatty acid remodeling. It is likely that these early glycosylation steps all take place at the cytoplasmic face of the endoplasmic reticulum (ER) (Vidugiriene and Menon, 1994). Prior to its addition to protein in the lumen of the ER, one (or in some cases up to three) ethanolamine phosphate is added to the glycolipid. Although the precise location for ethanolamine phosphate addition is not known, it is likely that, minimally, the PI-Gln-$(Man)_3$ species is translocated across the ER membrane.

Although very little is known about the reactions involved in biosynthesis of GPI molecules, the initial step in PI glycosylation has been genetically characterized by complementation analyses in a series of T-cell mutants. These variants were discovered by the lack of expression of Thy-1 on the cell surface. Six complementation classes have been identified in GPI biosynthesis, three of which (classes A, C, and H) cannot transfer UDP-GlcNac to PI, suggesting that at least three genes are involved in this early step in GPI synthesis (Steven and Raetz, 1991; Urakaze *et al.*, 1992). None of these genes showed significant homology to other glycosyltransferases, and the proteins encoded by these DNAs have not yet been studied. Similar defects, possibly corresponding to the class A mutation, have been found in patients with the rare disorder paroxysmal nocturnal hemoglobinuria, which arises from the defective anchoring of GPI-anchored proteins such as decay-accelerating factor and CD-59 in stem cells (Selveraj *et al.*, 1988; Stafford *et al.*, 1988).

3.2. Attachment of GPI Anchors

The anchoring of proteins to GPI appears to occur by the attachment of the proteins to the preformed lipid precursor. This attachment requires the removal of a relatively short C-terminal hydrophobic domain from the protein. The formation of this linkage appears to occur cotranslationally or posttranslationally in the ER, as judged by sensitivity to cleavage by PI-PLC or reactivity with antibodies that recognize the glycan moiety of the anchor (Bailey *et al.*, 1989; Bangs *et al.*, 1985; Conzelmann *et al.*, 1987; Ferguson *et al.*, 1986). The rapid kinetics of anchor addition suggest that cleavage of the hydrophobic C-terminal peptide and GPI attachment may be catalyzed by the same enzyme (transpeptidase or transamidase) and that a presynthesized GPI moiety is added *en bloc*.

The protein signals that direct attachment to GPI appear to be localized to C-terminal hydrophobic regions. Initial studies comparing the amino acid sequence predicted from cDNAs of Thy-1 or trypanosomal VSG with the actual peptide sequence demonstrated the loss of the predicted C-terminal peptide prior to replacement with the GPI moiety (Boothroyd *et al.*, 1980; Tse *et al.*, 1985). Several GPI-anchored proteins are alternatively expressed with transmembrane hydrophobic peptide regions [e.g., N-CAM (Hemperly *et al.*, 1986), FcR III (Selveraj *et al.*, 1988; Simmons and Seed, 1988), and LFA-3 (Dustin *et al.*, 1987)] or with sequences that produce secreted proteins [e.g., DAF (Caras *et al.*, 1987a), Qa-2 (Stroynowski *et al.*, 1987), and N-CAM (Gower *et al.*, 1981)]. These variations are derived from differences in the C-terminal regions generated either by alternate splicing of a single gene or transcription of alternative genes.

Recent studies have revealed several structural features of this carboxy-terminal sequence that are involved in GPI attachment. Sequence analysis of a number of proteins has suggested that a relatively short hydrophobic domain of 20–30 amino acids is contained within this COOH-terminal region. The GPI attachment occurs 8–12 residues to the NH_2 terminus of this hydrophobic domain, in a region containing short side-chain amino acids. A number of studies employing progressive deletion mutants of the GPI-anchored enzyme placental alkaline phosphatase have indicated that the length of this hydrophobic domain may play an important role. Reduction of the hydrophobic domain to 17 amino acids resulted in a mutant that was still GPI-anchored, but a deletion mutant that contained only a 13-amino-acid hydrophobic domain was secreted rather than GPI-anchored (Berger *et al.*, 1988, 1989). Although the precise sequence of the hydrophobic domain that is replaced by the GPI anchor may not be highly constrained, in one case (the MHC class I protein Qa2), a single charged amino acid (Asp) within the hydrophobic domain was demonstrated to be essential for GPI attachment (Waneck *et al.*, 1988).

Further studies have defined two different regions within the C-terminal GPI-attachment "signal" of DAF: a 17-amino-acid C-terminal hydrophobic region and the 20 amino acids located just upstream. Removal of the 17-amino-acid C-terminal hydrophobic segment of decay-accelerating factor prevented GPI anchoring, although the addition of this hydrophobic domain to a secreted protein (human growth hormone) did not confer GPI anchoring, indicating that this hydrophobic segment is necessary but not sufficient to direct lipid attachment (Caras and Weddell, 1989). Thus, additional information (cleavage and attachment sites) must reside within the adjacent 20 upstream amino acids. Moreover, it is likely that the signal nature of the 17-amino-acid C-terminal hydrophobic region depends on overall hydrophobicity, because replacement with a random hydrophobic sequence or by the N-terminal ER signal sequence of human growth hormone still allowed GPI attachment (Caras *et al.*, 1987b). Similar deletion or chimeric gene experiments employing GPI-anchored placental alkaline phosphatase, Qa-2, and Thy-1 have also localized the signal to C-terminal hydrophobic regions (Berger *et al.*, 1988; Crise *et al.*, 1989; Waneck *et al.*, 1988).

The amino acid that is amide-linked to the GPI anchor is highly constrained. A survey of a number of GPI-anchored proteins reveals that Ser, Cys, Asp, Asn, Gly, and Ala are found at this position and that hydrophobic or bulky-side-chain amino acids are excluded. Recently, studies have indicated that a single amino acid at the GPI attachment site may determine which membrane anchor is to be utilized. A human IgG Fc receptor expressed on granulocytes (FcRIII-1) is GPI linked, but a protein with >95% identity, expressed in macrophages and NK cells, utilizes a transmembrane amino acid domain (FcRIII-2). Substitution of a Ser in FcγRIII-1 for a Phe in III-2 at the GPI attachment site reverses the membrane anchors for these proteins (Hibbs *et al.*, 1989; Kurosaki and Ravetch, 1989; Lanier *et al.*, 1989). Mutational analyses of the hydrophobic transmembrane domain and cytoplasmic domain have revealed that the amino acid at the GPI attachment site is dominant over other structural features. These studies suggest that the hydrophobicity of the transmembrane domain as well as the length of the cytoplasmic sequence may influence the efficiency of GPI anchoring.

In several proteins, alternative forms of membrane anchoring have been found. Two genes encode FcRIII: one of these is expressed exclusively in granulocytes, encoding the GPI-linked form; the second is expressed in NK cells and macrophages, encoding a transmembrane protein (Kurosaki and Ravetch, 1989). Regulation of GPI attachment for FcRIII is transcriptional, determined by the selective expression of these two genes. Transcriptional regulation for these genes must be tightly controlled, as coexpression of these transcripts has not been detected. In contrast, the regulation of GPI attachment for N-CAM is posttranscriptional. The N-CAM can exist as a GPI-anchored or transmembrane attached protein, with each form exhibiting identical extracellular domains. These differences result from alternative mRNA splicing through the substitution of a hydrophobic domain for one that is preceded by a GPI attachment domain that observes the consensus sequence (Gower *et al.*, 1988; He *et al.*, 1987; Hemperly *et al.*, 1986). These transcripts are differentially expressed, suggesting that regulation of GPI attachment for N-CAM occurs posttranscriptionally. In contrast to the cases cited above, in which distinct cells express either a GPI-anchored or transmembrane protein, LFA-3 has been shown to be expressed on the surface of the same cell in both forms (Dustin *et al.*, 1987). A single gene encodes the LFA-3 protein, and alternative cDNAs have been described that differ in the length of a putative cytoplasmic domain. However, because both molecules predicted by these cDNAs have identical GPI attachment and transmembrane domains, the significance of these two cDNAs in determining the membrane topology of LFA-3 has yet to be determined.

4. GLYCOSYLPHOSPHATIDYLINOSITOL IN INSULIN ACTION

The search for an insulin-dependent “second messenger” has been under way since the early 1970s. An insulin-sensitive substance was first detected in skeletal muscle that could acutely modulate glycogen synthetase activity *in vitro* by stimulating the dephosphorylation of the enzyme (Larner *et al.*, 1979). Similar kinds of extracts or substances of elusive chemical identity were subsequently identified in a variety of cell types and reported capable of regulating the activities of several insulin-sensitive enzymes (Kiechle *et al.*, 1980; Parker *et al.*, 1982; Saltiel *et al.*, 1981, 1982, 1983; Seals and Czech, 1980; Suzuki *et al.*, 1984). Although these enzyme-modulating activities were detected in several laboratories, there were inconsistencies in their reported properties, and little progress was made in the elucidation of their chemical composition or structure. The search for a defined chemical substance that could mediate some of the metabolic actions of insulin resulted initially in the purification of two structurally similar substances that were released from hepatic plasma membranes in response to insulin (Saltiel and Cuatrecasas, 1986). Their relative hydrophilicity, negative charge, apparent molecular weight of 800–1000 and sensitivity to periodate and nitrous acid suggested an oligosaccharide phosphate structure. Preliminary compositional analysis revealed inositol as a component, although the known inositol phosphates did not share the chemical properties, chromatographic or electrophoretic behavior, or insulin sensitivity of this factor, indicating that it was a glycosylated derivative containing inositol phosphate.

The first indication that the insulin-dependent enzyme modulators might arise from the hydrolysis of a glycolipid structurally similar to GPI emerged from the demonstration that the PI-specific bacterial PLC generated a substance from rat liver plasma membranes that was chemically identical to that produced in response to insulin (Saltiel and Cuatrecasas, 1986; Saltiel *et al.*, 1986, 1987). Moreover, a potential precursor of the PI-PLC-generated glycan was identified in a number of cell types, characterized by enzymatic and chemical means (Gaulton, 1991; Mato *et al.*, 1987a; Saltiel, 1987; Saltiel and Cuatrecasas, 1986) as well as by sensitivity to hormones (Saltiel and Cuatrecasas, 1986; Saltiel *et al.*, 1986, 1987). In addition to insulin, other hormones and growth factors, including IGF-1, EGF, NGF, ACTH, and TSH, have been shown to stimulate the turnover of the lipid in a variety of cell types (Chan *et al.*, 1989; Eardley and Koshland, 1991; Kojima *et al.*, 1990; Martiny *et al.*, 1990; Merida *et al.*, 1990; Represa *et al.*, 1991).

Although insulin is known to cause increased labeling of several phospholipids (Detorrontegui and Berthet, 1966; Farese *et al.*, 1988b; Manchester, 1963; Pennington and Martin, 1985; Stein and Hale, 1974), it does not acutely stimulate the hydrolysis of the polyphosphoinositides, and it does not induce calcium mobilization through the generation of inositol trisphosphate (Farese *et al.*, 1985; Pennington and Martin, 1985). These observations suggested that the insulin-sensitive hydrolysis of GPI results from the activation of a GPI-specific PLC. Such an enzyme was isolated from a plasma membrane fraction of liver, using as an assay the liberation of diacylglycerol from the GPI-anchored variant surface glycoprotein (VSG) from *Trypanosoma brucei* (Fox *et al.*, 1987). The catalytic activity of this enzyme appears to reside in a single polypeptide with an apparent molecular mass of about 52 kDa. The enzyme is calcium-independent, and it is specific for GPI; no hydrolysis of PI, PIP_2, or other phospholipids is observed under a variety of conditions.

Although the topological distribution of the insulin-sensitive glycolipid in the plasma membrane remains uncertain, Alvarez *et al.* (1988) have suggested an extracellular location for the lipid, based on surface labeling experiments. In addition, Romero *et al.* (1990) reported that exposure of BC_3H1 myocytes to an antibody raised against a glycan fragment from a *Trypanosoma brucei* VSG membrane anchor selectively inhibited some actions of insulin, suggesting that the glycan is released from cells on insulin addition. In this regard, PI-PLC can exert insulin-like effects on intact cells (Saltiel and Sorbara-Cazan, 1987), exogenous addition of purified inositol glycan to intact cells mimics many of the effects of insulin (Alemany *et al.*, 1987; Alvarez *et al.*, 1987; Gottschalk and Jarett, 1988; Kelly *et al.*, 1987a,b; Saltiel and Sorbara-Cazan, 1987; Witters and Watts, 1988; Witters *et al.*, 1987), and insulin can stimulate the release of certain GPI-anchored proteins from the cell surface (Chan *et al.*, 1988; Lisanti *et al.*, 1989b). However, the notion that GPI hydrolysis and release of the inositol glycan occur extracellularly remains to be conclusively demonstrated. Moreover, recent studies (Vidugiriene and Menon, 1994) have shown that glycosylation of PI to form the PI-Gln-$(Man)_3$ intermediate occurs on the cytoplasmic face of the ER, making it likely that the insulin-sensitive lipid is primarily cytoplasmically oriented.

Initial studies on the biological activities of the inositol glycan focused on the insulin-sensitive, low-K_m cyclic AMP phosphodiesterase in fat cell membranes (Saltiel and Steigerwalt, 1986). The activity of this cyclic AMP phosphodiesterase was stimulated acutely by the glycan, as reflected by an increase in the V_{max} of the enzyme with no appreciable effect on its K_m (Saltiel, 1987). The purified glycan could also modify *in vitro* other insulin-sensitive enzymes assayed in subcellular fractions (Kelly *et al.*, 1986; Saltiel, 1987; Saltiel and Cuatrecasas, 1986). Although the precise biochemical mechanism(s) involved in the regulation of these enzymes were not determined, some data suggested that phosphorylation of the enzyme or closely related regulatory factors might be involved.

To evaluate the insulin-mimetic properties of the inositol glycans in more detail, these compounds have been added to intact fat, liver, or muscle cells. Inositol glycans isolated from insulin-treated cells or membranes mimic the lipogenic (Machicao *et al.*, 1990; Saltiel and Sorbara-Cazan, 1987) and antilipolytic (Kelly *et al.*, 1987a; Misek and Saltiel, 1992) actions of insulin, stimulate pyruvate kinase (Alvarez *et al.*, 1987) and pyruvate dehydrogenase (Gottschalk and Jarett, 1988), inhibit PEPCK transcription (Alvarez *et al.*, 1991), hepatic glucose output (Misek and Saltiel, 1992), phosphorylase a (Alvarez *et al.*, 1987), and cyclic AMP levels (Alvarez *et al.*, 1987), and produce alterations in protein phosphorylation patterns similar to that observed with insulin (Alemany *et al.*, 1987; Misek and Saltiel, 1994).

In order to evaluate the structure–function relationship of these compounds, a structurally similar glycan derived from the VSG GPI anchor has been utilized. This glycan mimics the antilipolytic and antigluconeogenic actions of insulin in fat and liver cells (Misek and Saltiel, 1992). Like insulin, this inositol glycan specifically attenuated the phosphorylation of the major isoproterenol-stimulated 70-kDa phosphoprotein in adipocytes (Misek and Saltiel, 1994). The differential sensitivity of this dephosphorylation reaction to various inhibitors in intact cells and cellular homogenates indicated that dephosphorylation of the 70-kDa protein results from the specific activation of a type 1 serine/threonine phosphatase in adipocytes. Interestingly, this 70-kDa protein was the only phosphoprotein dephosphorylated in response to the glycan (Misek and Saltiel, 1994).

Despite the progress made in identifying the structure and biogenesis of the inositol glycans, it is still premature to regard these compounds as second messengers for any of the actions of insulin. The apparent insulin dependence, rapidity of action, and extent of generation of the inositol glycans are consistent with the properties expected for second messengers. However, many questions remain concerning their precise structures as well as the nature of their insulin-mimetic properties. Although the insulin-mimetic actions of the inositol glycan on intact cells suggest that this substance might exert its metabolic effects after entering the cell, its chemical structure suggests that passive transport is unlikely. Interestingly, at millimolar concentrations, mannose, glucosamine, and inositol monophosphate block the lipogenic effect of insulin and the inositol glycan in intact adipocytes (Machicao *et al.*, 1990; Saltiel and Sorbara-Cazan, 1987) but are ineffective in attenuating the stimulation of cyclic AMP phosphodiesterase activity in adipocyte membranes (Saltiel and Sorbara-Cazan, 1987). This observation suggests the possible existence of an active transport system that specifically recognizes the inositol glycan.

Although the inositol glycan appears to be a promising candidate for a second messenger of insulin action, some issues remain to be resolved. The ultimate proof of a role for these compounds as second messengers will depend critically on, among other things, the determination of their precise structures. This may be complicated by the chromatographic resolution of multiple species of these molecules, perhaps reflecting the existence of distinct forms with different enzyme-modulating functions. It will also be important to produce these compounds in large quantity and homogeneous form, perhaps by organic synthesis (Plourde *et al.*, 1992). The biosynthetic route, mode of production and degradation, relationship to the insulin receptor, and the precise biochemical actions of these molecules need further exploration. Moreover, the evaluation of the metabolism of GPI in pathological states of insulin resistance may lead to the identification of possible critical postreceptor defects in diabetes. Diabetic GK rates were shown to respond poorly to insulin, perhaps because of a defect in GPI synthesis (Farese *et al.*, 1994). Additionally, in recent studies (Lazar *et al.*, 1994) employing mutant K562 cells, GPI synthesis was shown to be absolutely required for one of the best-established metabolic effects of insulin, glycogen synthesis, without modifying early tyrosine phosphorylation events (Lazar *et al.*, 1994). These studies suggest that GPI may provide an important clue toward understanding the biochemical basis of insulin resistance.

5. FUNCTIONAL ROLES OF GPI ANCHORS

Although the basic structural features of the GPI anchor are known, the broad diversity of the anchored proteins makes assignment of a function for the anchor quite difficult. Nevertheless, the unusual properties of this anchor, and the evolution of a complex mechanism to remove a hydrophobic peptide domain for lipid attachment, suggest that the utilization of this mechanism of anchoring merely as an inert structural anchor is unlikely. Three possible functional roles for this GPI anchor are discussed.

5.1. Protein Distribution and Targeting

Most of the GPI-anchored proteins identified thus far are located at the cell surface and face the extracellular space. A few GPI-anchored proteins exhibit intracellular localizations, e.g., GP-2 in the zymogen granules of the exocrine pancreas (Le Bel and Beattie, 1988), elongation factor EF-1a in the endoplasmic reticulum of V79-UF Chinese hamster fibroblasts (Hayashi *et al.*, 1989), and a fraction of PH-20 in the acrosome of guinea pig sperm (Phelps *et al.*, 1988); however, each of these is distributed in the lumen of intracellular vesicles and not on the cytoplasmic face. Interestingly, some of these proteins (i.e., GP-2 and PH-20) undergo regulated secretion, suggesting a potential mechanism for release involving the GPI anchor. The observed asymmetric orientation of all GPI-anchored proteins is compatible with their incorporation into the lumen of the ER via N-terminal signal sequences with subsequent processing and transfer to GPI.

In experiments with COS and MDCK cells (Caras *et al.*, 1987a; Caras and Weddell, 1989), attachment of the C-terminal peptide of a GPI-anchored protein to a secreted protein resulted in GPI addition and targeting to the plasma membrane. These and similar experiments suggested that GPI-anchored proteins must require additional signals for intracellular retention and targeting to a pathway for regulated secretion, as has been shown for regulated secretory proteins such as insulin, growth hormone, and preprosomatostatin. Although this observation may seem trivial, because in most cells transport to the cell surface is by default, and retention at specific compartments along the secretory pathway is signal-mediated, it is especially relevant to protein transport in polarized cells that contain multiple plasma membrane domains of distinct lipid and protein composition. In a variety of polarized epithelial cell lines (MDCK I and II, LLC-PK_1, Caco-2, and SK-CO15), GPI-anchored proteins are selectively enriched in the apical domain and depleted or absent from the basolateral cell surface (Lisanti *et al.*, 1988, 1989a, 1990). This correlation between GPI anchoring and apical localization is highly conserved across species (pig, dog, and human) and tissue type (renal and intestinal), suggesting that the anchor might act as an apical transport signal to target the attached protein to the apical cell surface. Recombinant transfer of GPI attachment signals to proteins known to exhibit basolateral distribution on polarized cells, such as the viral envelope glycoproteins $HSV_{\gamma}D$-1 or VSV G protein, or to a regulated secretory protein such as human growth hormone, resulted in GPI anchoring and targeting to the apical membrane. Other polarized cells also exhibit an asymmetric distribution of GPI-anchored proteins. For example, NCAM120, F3, 5′-nucleotidase and acetylcholinesterase are localized to neuronal outgrowths and neuronal ramifications. The distribution of PH-20 varies with the degree of sperm maturation: this protein is randomly distributed in testicular sperm, localized to the posterior head region in epididymal sperm, and undergoes redistribution to the anterior head region after the acrosome reaction. Paradoxically, PH-20 is "immobile" when randomly distributed but freely diffusible when localized to the head region (posteriorly or anteriorly).

Recent studies have revealed that many GPI-anchored proteins have been found to be clustered in caveolae, small plasma-membrane-associated vesicles that are rich in cholesterol and sphingomyelin (Anderson, 1993). These structures appear to exclude clathrin, do not merge with other endocytotic vesicles, and can open or close to process or deliver material to the cell (Anderson, 1993). Interestingly, numerous signaling molecules are also found in caveolae, including the tandemly acylated *src*-family kinases that are frequently found associated with GPI-anchored proteins (Brown and Rosy, 1992; Rodgers *et al.*, 1994). It has been suggested that these structures may function as signaling organelles (Anderson, 1993), perhaps because of the unique concentration of these signaling proteins directed by the GPI-rich lipid environment. The recent finding (C. C. Mastick and A. R. Saltiel, unpublished data) that insulin-dependent tyrosine phosphorylation is detected in caveolae from 3T3-L1 adipocytes adds further support to this contention. Thus, linkage to GPI may convey specific patterns of cell surface localization in different polarized cell types or membrane microdomains, with important consequences for cellular regulation.

5.2. Regulation of Lateral Mobility

An obvious consequence of lipid anchoring is an inherent increase in mobility in the plane of the membrane. Diffusion coefficients on the order of $1–4 \times 10^{-9}$ cm^2/sec are observed for Thy-1, alkaline phosphatase, DAF, and PH-20 (Dragsten *et al.*, 1979; Noda *et al.*, 1987; Phelps *et al.*, 1988; Thomas *et al.*, 1987). These are lower than values determined for freely diffusing lipid probes ($0.5–1 \times 10^{-9}$) but much higher than those for transmembrane glycoproteins ($0.5–6 \times 10^{-10}$). Paradoxically, some GPI-anchored proteins possess significant immobile fractions, up to 50% in the case of Thy-1. Certain GPI-anchored proteins may require increased lateral mobility for function, as in the case of DAF (for rapid inhibition of the complement cascade) or AChE (for deactivation of acetylcholine at the synapse). Other proteins require a high degree of immobility, such as the trypanosomal VSG (diffusion constant 1×10^{-10}) and PH-20. Interestingly, the low lateral mobility of VSG is not a result of some other factor in the trypanosomal cell surface, because implantation of the protein in the BHK cell membrane did not increase its mobility (Bulow *et al.*, 1989). The diffusion of PH-20 is dependent on the state of differentiation of guinea pig sperm (see above).

5.3. Regulated Release at the Cell Surface

One of the unique features of the structure of the GPI anchor is the presence of sites for enzymatic modification. Within the glycolipid attachment domain, there are a number of sites that represent potential substrates for GPI-specific anchor-degrading enzymes (proteases, glycosidases, and/or phospholipases), the actions of which may result in release of the attached protein from the cell surface. Such an enzyme-mediated release mechanism could be regulated or constitutive. Regulated degradation of the GPI anchor, perhaps under hormonal control, could potentially provide a unique mechanism for down-regulation of the concentration of the anchored protein at the cell surface or for up-regulation of the protein in the circulation, for its actions at a local or downstream target tissue. Similarly, a constitutive degradation of the GPI anchor could provide a mechanism for secretion of proteins such as GP-2 after translocation to the cell surface.

A number of GPI-anchored proteins such as alkaline phosphatase, 5′-nucleotidase, DAF, lipoprotein lipase, GP-2, CEA, Qa-2 antigen, Thy-1, FcγRIII-1 from neutrophils, and 34-kDa placental growth factor have been detected in soluble forms (Ishihara *et al.*, 1987a). These soluble forms may result from anchor degradation or differential splicing of mRNAs to produce non-GPI-anchored proteins, as documented for N-CAM, Qa-2, and DAF. The detection of free GPI molecules that are hydrolyzed in response to hormones suggested the possibility that the GPI protein anchor might undergo a similar hormone-sensitive hydrolysis reaction (Saltiel and Cuatrecasas, 1988). The acute release of certain GPI-anchored proteins from tissue culture cells in response to hormones has been observed, including lipoprotein lipase (Chan *et al.*, 1988), heparan sulfate proteoglycan (Ishihara *et al.*, 1987b), 5′-nucleotidase (Klip *et al.*, 1988), and alkaline phosphatase (Romero *et al.*, 1988). Interestingly, the cell surface or circulating concentrations of each of these proteins is known to be altered in diabetes (Karnieli *et al.*, 1987; Skillen *et al.*, 1987). The FcγRIII receptor was released on neutrophil activation in response to chemotactic peptide or phorbol esters (Huizinga *et al.*, 1988). A survey of the total cell-surface GPI-anchored proteins in insulin-sensitive BC_3H1 myocytes indicated that a number of the PLC-releasable proteins were depleted by prior treatment of cells with insulin or serum (Lisanti *et al.*, 1989b). Interestingly, exposure to insulin caused the loss of only some of the GPI-anchored proteins, while others remained unchanged. These observations suggest the existence of hormone-sensitive and insensitive "structural" pools of the GPI anchor. One intriguing possibility is that structural modifications of the GPI moiety play a role in dictating susceptibility to enzymatic degradation. As discussed above, the presence of an ester-linked fatty acid on the inositol ring renders the anchor insensitive to cleavage by PI- or GPI-specific PLCs (Roberts *et al.*, 1988a).

Although the experiments described above are indirect, they suggest that hormonal treatment results in the activation of an anchor-degrading enzyme(s). However, the nature of this enzyme remains unknown, and it could be a specific phospholipase, protease, or glycosidase. GPI-specific phospholipase Cs with similar properties have been isolated from *Trypanosoma brucei* (Fox *et al.*, 1986) and mammalian liver (Fox *et al.*, 1987). Both the *T. brucei* and liver enzymes exhibit similar peptide maps, are membrane-associated and calcium-independent, and specifically catalyze the hydrolysis of GPI but not other phospholipids (Fox *et al.*, 1986, 1987). The cDNA for the trypanosomal enzyme has been cloned and sequenced (Hereld *et al.*, 1988). This predicted sequence revealed no homology to other phospholipases and gave no indication of a signal sequence or glycosylation sites that would be indicative of an extracellular or transmembrane protein. In addition, immunohistochemical studies indicated an intracellular localization (Bulow *et al.*, 1989). Thus, it seems unlikely that the mammalian GPI-PLC is involved in anchor degradation under normal conditions. However, the possibility remains that other GPI-specific phospholipases exist. A GPI-specific phospholipase D was found in plasma derived from a number of sources (Davitz *et al.*, 1987, 1989; Low and Prasad, 1988). This enzyme specifically hydrolyzes the GPI anchor for purified membrane-bound proteins, although it cannot remove cell surface GPI-anchored proteins from intact cells.

6. CONCLUSIONS

Recent investigations into the structural details of membrane proteins have yielded surprising insights into the dynamics of protein–membrane interaction. Although the covalent modification of proteins with lipids has now gained wide acceptance, the functional consequences of these novel interactions are only now emerging. It is interesting to note that the proteins that undergo these modifications exhibit wide diversity of tissue and species distribution, with little or no evolutionary or functional similarity. Similarly, although the regulation of GPI metabolism has been studied primarily in response to insulin, it is likely that other stimuli may utilize this unique signaling system. It is likely that continued investigation in this area will lead to new advances in our understanding of the dynamics of biological membranes.

7. REFERENCES

Alemany, S., Mato, J. M., and Stralfors, P., 1987, Phosphodephospho-control by insulin is mimicked by a phosphooligosaccharide in adipocytes, *Nature* **330**:77–79.

Alvarez, J. F., Cabello, M. A., Feliu, J. E., and Mato, J. M., 1987, A phosphooligosaccharide mimics insulin action on glycogen phosphorylase and pyruvate kinase activities in isolated rat hepatocytes, *Biochem. Biophys. Res. Commun.* **147**:765–771.

Alvarez, J. F., Varela, I., Ruiz-Albusac, J. M., and Mato, J. M., 1988, Localisation of the insulin-sensitive phosphatidylinositol glycan at the outer surface of the cell membrane, *Biochem. Biophys. Res. Commun.* **152:**1455–1462.

Alvarez, L., Avila, M. A., Mato, J. M., Castano, J. G., and Varela-Nieto, I., 1991, Insulin-like effects of inositol phosphate-glycan on messenger RNA expression in rat hepatocytes, *Mol. Endocrinol.* **5:**1062–1068.

Anderson, R.G.W., 1993, Caveolae: Where incoming and outgoing messengers meet, *Proc. Natl. Acad. Sci. U.S.A.* **90:**10909–10913.

Bailey, C. A., Gerber, L., Howard, A. D., and Udenfriend, S., 1989, Processing at the carboxyl terminus of nascent placental alkaline phosphatase in a cell-free system: Evidence for specific cleavage of a signal peptide, *Proc. Natl. Acad. Sci. U.S.A.* **86:**22–26.

Bangs, J. D., Hereld, D., Krakow, J. L., Hart, G. W., and Englund, P. T., 1985, Rapid processing of the carboxyl terminus of a trypanosome variant surface glycoprotein, *Proc. Natl. Acad. Sci. U.S.A.* **82:**3207–3211.

Berger, J., Howard, A. D., Brink, L., Gerber, L., Hauber, J., Cullen, B. R., and Udenfriend, S., 1988, COOH-terminal requirements for the correct processing of a phosphatidylinositol-glycan anchored membrane protein, *J. Biol. Chem.* **263:**10016–10021.

Berger, J., Micanovic, R., Greenspan, R. J., and Udenfriend, S., 1989, Conversion of placental alkaline phosphatase from a phosphatidylinositol-glycan-anchored protein to an integral transmembrane protein, *Proc. Natl. Acad. Sci. U.S.A.* **86:**1457–1460.

Boothroyd, J. C., Cross, G.A.M., Hoeijmakers, J.H.J., and Borst, P., 1980, Variant surface glycoprotein of *Trypanosoma brucei* synthesized with a C-terminal hydrophobic "tail" absent from purified glycoprotein, *Nature* **288:**624–626.

Brown, D. A., and Rosy, J. K., 1992, Sorting of GPI-anchored proteins to glycolipid-enriched membrane subdomains during transport to the apical cell surface, *Cell* **68:**533–544.

Bulow, R., Griffiths, G., Webster, P., Stierhof, Y. D., Opperdoes, F. R., and Overath, P., 1989, Intracellular localization of the GPI-specific phospholipase C of *Trypanosoma brucei*, *J. Cell Sci.* **93:**233–240.

Caras, I. W., and Weddell, G. N., 1989, Signal peptide for protein secretion directing glycophospholipid membrane anchor attachment, *Science* **243:**1196–1198.

Caras, I. W., Davitz, M. A., Rhee, L., Weddell, G., Martin, D. W., and Nussenzweig, V., 1987a, Cloning of decay-accelerating suggests novel use of splicing to generate two proteins, *Nature* **325:**545–548.

Caras, I. W., Weddell, G. N., Davitz, M. A., Nussenzweig, V., and Martin, D. W., 1987b, Signal for attachment of a phospholipid membrane anchor in decay accelerating factor, *Science* **238:**1280–1283.

Chan, B. L., Lisanti, M. P., Rodriguez-Boulan, E., and Saltiel, A. R., 1988, Insulin-stimulated release of lipoprotein lipase by metabolism of its phosphatidylinositol anchor, *Science* **241:**1670–1672.

Chan, B. L., Chao, M. V., and Saltiel, A. R., 1989, Nerve growth factor stimulates the hydrolysis of glycosyl-phosphatidylinositol in PC-12 cells: A mechanism of protein kinase C regulation, *Proc. Natl. Acad. Sci. U.S.A.* **86:**1756–1760.

Conzelmann, A., Spiazzi, A., and Bron, C., 1987, Glycolipid anchors are attached to Thy-1 glycoprotein rapidly after translocation, *Biochem. J.* **246:**605–610.

Crise, B., Ruusala, A., Zagouras, P., Shaw, A., and Rose, J. K., 1989, Oligomerization of glycolipid-anchored and soluble forms of the vesicular somatitis virus glycoprotein, *J. Virol.* **63:**5328–5333.

Davitz, M. A., Hereld, D., Shak, S., Krakow, J., Englund, P. T., and Nussenzweig, V., 1987, A glycan-phosphatidylinositol-specific phospholipse D in human serum, *Science* **238:**81–84.

Davitz, M. A., Hom, J., and Schenkman, S., 1989, Purification of a glycosyl-PI-specific phospholipase D from human plasma, *J. Biol. Chem.* **264:**13760–13764.

Deterrontegui, G., and Berthet, J., 1966, The action of insulin on the incorporation of [^{32}P]phosphate in the phospholipids of rat adipose tissue, *Biochim. Biophys. Acta* **116:**477–481.

Doering, T. L., Masterson, W. J., Hert, G. W., and Englund, P. T., 1990, Biosynthesis of glycosyl phosphatidylinositol membrane anchors, *J. Biol. Chem.* **265:**611–614.

Dragsten, P., Henkart, P., Blumenthal, R., Weinstein, J., and Schlessinger, J., 1979, Lateral diffusion of surface immunoglobulin, Thy-1 antigen, and a lipid probe in lymphocyte plasma membranes, *Proc. Natl. Acad. Sci. U.S.A.* **76:**5163–5167.

Dustin, M. L., Selveraj, P., Mattaliano, R. J., and Springer, T. A., 1987, Anchoring mechanisms for LFA-3 cell adhesion glycoprotein at membrane surface, *Nature* **329:**846–848.

Eardley, D. D., and Koshland, M. E., 1991, Glycosylphosphatidylinositol: A candidate system for interleukin-2 signal transduction, *Science* **251:**78–81.

Farese, R. V., Davis, J. T., Barnes, D. E., Standaert, M. L., Bebishkin, J., Hock, R., Rosie, N. K., and Pollet, R. J., 1985, The *de novo* phospholipid effect of insulin is associated with increases in diacylglycerol, but not inositol phosphates or cytosolic Ca, *Biochem. J.* **231:**269–278.

Farese, R. V., Cooper, D. R., Konda, T. S., Nair, G., Standart, M. L., Davis, J. S., and Pollet, R. J., 1988a, Mechanisms whereby insulin increases diacylglycerol in BC3H1 myocytes, *Biochem. J.* **256:**175–184.

Farese, R. V., Cooper, D. R., Konda, T. S., Nair, G., Standart, M. L., and Pollet, R. J., 1988b, Insulin provokes coordinated increases in the synthesis of phosphatidylinositol, phosphatidylinositolphosphates and the phosphatidylinositol-glycan in BC3H1 myocytes, *Biochem. J.* **256:**185–188.

Farese, R. V., Standaert, M. L., Yamada, K., Huang, L. C., Zhang, C., Cooper, D. R., Wang, Z., Yang, Y., Suzuki, S., Toyota, T., and Larner, J., 1994, Insulin-induced activation of glycerol-3-phosphate acyltransferase by a chiro-inositol-containing insulin mediator is defective in adipocytes of insulin resistant, type II diabetic Goto–Kakizaki rats, *Proc. Natl. Acad. Sci. U.S.A.* **91:**11040–11044.

Ferguson, M.A.J., and Williams, A. F., 1988, Cell-surface anchoring of proteins via glycosylphosphatidylinositol structures, *Annul Rev. Biochem.* **57:**285–320.

Ferguson, M.A.J., Duszenko, M., Lamont, G. S., Overath, P., and Cross, G.A.M., 1986, Biosynthesis of *Trypanosoma brucei* variant surface glycoproteins, *J. Biol. Chem.* **261:**356–362.

Ferguson, M.A.J., Homans, S. W., Dwek, R. A., and Rademacher, T. W., 1988, Glycosylphosphatidylinositol moiety that anchors *Trypanosoma brucei* variant surface glycoprotein to the membrane, *Science* **239:**753–759.

Fox, J. A., Duszenko, M., Ferguson, M.A.J., Low, M. G., and Cross, G.A.M., 1986, Purification and characterization of a novel glycan-PI specific phospholipase C from *T. brucei, J. Biol. Chem.* **261:**15767–15771.

Fox, J. A., Soliz, N. M., and Saltiel, A. R., 1987, Purification of a phosphatidylinositol-glycan specific phospholipase C from liver plasma membranes: A possible target of insulin action, *Proc. Natl. Acad. Sci. U.S.A.* **84:**2663–2667.

Gaulton, G. N., 1991, Differential regulation of glycosylated phosphatidylinositol subtypes by insulin, *Diabetes* **40:**1297–1304.

Gaulton, G. N., Kelly, K. L., Pawlowski, J., Mato, J. M., and Jarett, L., 1988, Regulation and function of an insulin-sensitive glycosylphosphatidylinositol during T lymphocyte activation, *Cell* **538:**963–970.

Gottschalk, W. K., and Jarett, L., 1988, The insulinomimetic effects of the polar head group of an insulin-sensitive glycophospholipid on pyruvate dehydrogenase in both subcellular and whole cell assays, *Arch. Biochem. Biophys.* **261:**175–185.

Gower, H. J., Barton, C. H., Elsom, V. L., Thompson, J., Moore, S. E., Dickson, G., and Walsh,

F. S., 1988, Alternative splicing generates a secreted form of N-CAM in muscle and brain, *Cell* **55**:955–964.

Hayashi, Y., Urade, R., Utsumi, S., and Kito, M., 1989, Anchoring of peptide elongation factor EF-1a by phosphatidylinositol at the endoplasmic reticulum membrane, *Biochem. J.* **106**:560–563.

He, H. T., Finne, J., and Goridis, C., 1987, Biosynthesis, membrane association, and release of N-CAM-120, a phosphatidylinositol-linked form of the neural cell adhesion molecule, *J. Cell Biol.* **105**:2489–2500.

Hemperly, J. J., Edelman, G. N., and Cunningham, B. A., 1986, cDNA clones of N-CAM lacking a membrane-spanning region consistent with evidence for membrane attachment via a phosphatidylinositol intermediate, *Proc. Natl. Acad. Sci. U.S.A.* **83**:9822–9826.

Hereld, D., Hart, G. W., and Englund, P. T., 1988, cDNA encoding the glycosyl-PI-specific phospholipase C of *T. brucei*, *Proc. Natl. Acad. Sci. U.S.A.* **85**:8914–8918.

Hibbs, M. L., Selvaraj, P., Carpen, O., Springer, T. A., Kuster, H., Jouvin, M. H. E., and Kinet, J. P., 1989, Mechanisms for regulating expression of membrane isoforms of Fc gamma R III (CD 16), *Science* **246**:1608–1611.

Homans, S. W., Ferguson, M.A.J., Dwek, R. A., Rademacher, T. W., Anand, R., and Williams, A. F., 1988, Complete structure of the glycosyl phosphatidylinositol membrane anchor of rat brain Thy-1 glycoprotein, *Nature* **333**:269–272.

Huang, L. C., Fonteles, M. C., Houston, D. B., Zhang, C., and Larner, J., 1993, Chiroinositol deficiency and insulin resistance. III. Acute glycogenic and hypoglycemic effects of two inositol phosphoglycan insulin mediators in normal and streptozotocin-diabetic rats *in vivo*, *Endocrinology* **132**:652–657.

Huizinga, T.W.J., van der Schoot, C. E., Jost, C., Klaassen, R., Kleijer, M., von dem Borne, A.E.G.K., Roos, D., and Tetteroo, P.A.T., 1988, The PI-linked receptor FcRIII is released on stimulation of neutrophils, *Nature* **333**:667–669.

Ishihara, A., Hou, Y., and Jacobson, K., 1987a, The Thy-1 antigen exhibits rapid lateral diffusion in the plasma membrane of rodent lymphoid cells and fibroblasts, *Proc. Natl. Acad. Sci. U.S.A.* **84**:1290–1293.

Ishihara, M., Fedearko, N. S., and Conrad, H. E., 1987b, Involvement of phosphatidylinositol and insulin in the coordinate regulation of proteoheparin sulfate metabolism and hepatocyte growth, *J. Biol. Chem.* **262**:4708–4716.

Karnieli, E., Armoni, M., Cohen, P., Kanter, Y., and Rafaeloff, R., 1987, Reversal of insulin resistance in diabetic rat adipocytes by insulin therapy: Restoration of pool of glucose transporters and enhancement of glucose transport activity, *Diabetes* **36**:925–931.

Kelly, K. L., Mato, J. M., and Jarett, L., 1986, The polar head group of a novel insulin-sensitive glycophospholipid mimics insulin action on phospholipid methyltransferase, *FEBS Lett.* **209**:238–242.

Kelly, K. L., Mato, J. M., Merida, I., and Jarett, L., 1987a, Glucose transport and antilipolysis are differentially regulated by the polar head group of an insulin-sensitive glycophospholipid, *Proc. Natl. Acad. Sci. U.S.A.* **84**:6404–6407.

Kelly, K. L., Merida, I., Wong, E.H.A., DiCenzo, D., and Mato, J. M., 1987b, A phospho-oligosaccharide mimics the effect of insulin to inhibit isoproterenol-dependent phosphorylation of phospholipid methyltransferase in isolated adipocytes, *J. Biol. Chem.* **262**:15285–15290.

Kiechle, F. L., Jarett, L., Kotagal, N., and Popp, D. A., 1980, Isolation from rat adipocytes of a chemical mediator for insulin activation of pyruvate dehydrogenase, *Diabetes* **29**:852–855.

Klip, A., Ramlal, T., Douen, A. G., Burdett, E., Young, D., Cartee, G. D., and Holloszy, J. O., 1988, Insulin-induced decrease in 5′-nucleotidase activity in skeletal muscle membranes, *FEBS Lett.* **238**:419–423.

Kojima, I., Kitaoka, M., and Ogata, E., 1990, Insulin-like growth factor-I stimulates diacylglycerol production via multiple pathways in Balb/c 3T3 cells, *J. Biol. Chem.* **265:**16846–16850.

Kurosaki, T., and Ravetch, J., 1989, A single amino acid in the glycosyl phosphatidylinositol attachment domain determines the membrane topology of Fc gamma R III, *Nature* **342:**805–807.

Lanier, L. L., Cwirla, S., Yu, G., Testi, R., and Phillips, J. H., 1989, Membrane anchoring of a human IgG Fc receptor (CD 16) determined by a single amino acid, *Science* **246:**1611–1613.

Larner, J., Galasko, G., Cheng, K., DePaoli-Roach, A. A., Huang, L., Daggy, P., and Kellog, J., 1979, Generation by insulin of a chemical mediator that controls protein phosphorylation and dephosphorylation, *Science* **206:**1408–1410.

Larner, J., Huang, L. C., Schwartz, C.F.W., Oswald, A. S., Shen, T.-Y., Kinter, M., Tang, G., and Zeller, K., 1988, Rat liver insulin mediator which stimulates pyruvate dehydrogenase phosphatase contains galactosamine and D-*chiro*-inositol, *Biochem. Biophys. Res. Commun.* **151:**1416–1426.

Lazar, D. F., Knez, J. J., Medof, M. E., Cuatrecasas, P., and Saltiel, A. R., 1994, Stimulation of glycogen synthesis by insulin in human erythroleukemia cells requires the synthesis of glycosylphosphatidylinositol, *Proc. Natl. Acad. Sci. U.S.A.* **91:**9665–9669.

Le Bel, D., and Beattie, M., 1988, The major protein of pancreatic zymogen granule membranes (GP-2) is anchored via covalent bonds to phosphatidylinositol, *Biochem. Biophys. Res. Commun.* **154:**818–823.

Lisanti, M. P., Sargiacomo, M., Graeve, L., Saltiel, A. R., and Rodriguez-Boulan, E., 1988, Polarized apical distribution of glycosylphosphatidylinositol anchored proteins in a renal epithelial cell line, *Proc. Natl. Acad. Sci. U.S.A.* **85:**9557–9561.

Lisanti, M. P., Caras, I. W., Davitz, M. A., and Rodriguez-Boulan, E., 1989a, A glycophospholipid membrane anchor acts as an apical targeting signal in polarized epithelial cells, *J. Cell Biol.* **109:**2145–2156.

Lisanti, M. P., Darnell, J. C., Chan, B. L., Rodriguez-Boulan, E., and Saltiel, A. R., 1989b, The distribution of glycosyl-phosphoinositol anchored proteins is differentially regulated by serum and insulin, *Biochem. Biophys. Res. Commun.* **164:**824–832.

Lisanti, M. P., Le Bivic, A., Saltiel, A. R., and Rodriguez-Boulan, E., 1990, Preferred apical distribution of glycosyl-phosphatidylinositol (GPI) anchored proteins: A highly conserved feature of the polarized epithelial cell phenotype, *J. Membr. Biol.* **113:**155–167.

Low, M. G., and Prasad, A.R.S., 1988, A phospholipase D for the PI-anchor of cell-surface proteins is abundant in plasma, *Proc. Natl. Acad. Sci. U.S.A.* **85:**980–984.

Low, M. G., and Saltiel, A. R., 1988, Structural and functional roles of glycosyl-phosphatidylinositol in membranes, *Science* **239:**268–275.

Machicao, F., Mushack, J., Seffer, E., Ermel, B., and Haring, H. U., 1990, Mannose, glucosamine and inositol monophosphate inhibit the effects of insulin on lipogenesis. Further evidence for a role for inositol phosphate-oligosaccharides in insulin action, *Biochem. J.* **266:**909–916.

Manchester, K. L., 1963, Stimulation by insulin of incorporation of [^{32}P]phosphate and [^{14}C]acetate into lipid and protein of isolated rat diaphragm, *Biochim. Biophys. Acta* **70:**208–210.

Martiny, L., Antonicelli, F., Thuilliez, B., Lambert, B., Jacquemin, C., and Haye, B., 1990, Control by thyrotropin of the production by thyroid cells of an inositol phosphate-glycan, *Cell Signalling* **2:**21–27.

Mato, J. M., Kelly, K. L., Abler, A., and Jarett, L., 1987a, Identification of a novel insulin-sensitive glycophospholipid from H35 hepatoma cells, *J. Biol. Chem.* **262:**2131–2137.

Mato, J. M., Kelly, K. L., Abler, A., Jarett, L., Corkey, B. E., Cashel, J. A., and Zopf, D., 1987b, Partial structure of an insulin-sensitive glycophospholipid, *Biochem. Biophys. Res. Commun.* **146:**764–770.

Merida, I., Pratt, J. C., and Gaulton, G. N., 1990, Regulation of interleukin 2-dependent growth responses by glycosylphosphatidylinositol molecules, *Proc. Natl. Acad. Sci. U.S.A.* **87:**9421–9425.

Misek, D. E., and Saltiel, A. R., 1992, An inositol phosphate glycan derived from a *Trypanosoma brucei* glycosyl-phosphatidylinositol mimics some of the metabolic actions of insulin, *J. Biol. Chem.* **267:**16266–16273.

Misek, D. E., and Saltiel, A. R., 1994, An inositol phosphate glycan derived from a *Trypanosoma brucei* glycosyl phosphatidylinositol promotes protein dephosphorylation in rat epididymal adipocytes, *Endocrinology* **135:**1869–1876.

Noda, M., Yoon, K., Rodan, G. A., and Koppel, D. E., 1987, High lateral mobility of endogenous and transfected alkaline phosphatase: A phosphatidylinositol-anchored membrane protein, *J. Cell Biol.* **105:**1671–1677.

Ortmeyer, H. K., Bodkin, N. L., Lilley, K., Larner, J., and Hansen, B. C., 1993a, Chiroinositol deficiency and insulin resistance. I. Urinary excretion rate of chiroinositol is directly associated with insulin resistance in spontaneously diabetic rhesus monkeys, *Endocrinology* **132:**640–645.

Ortmeyer, H. K., Huang, L. C., Zhang, L., Hansen, B. C., and Larner, J., 1993b, Chiroinositol deficiency and insulin resistance. II. Acute effects of *d*-chiroinositol administration in streptozotocin-diabetic rats, normal rats given a glucose load, and spontaneously insulin-resistance rhesus monkeys, *Endocrinology* **132:**646–651.

Parker, J. C., Kiechle, F. L., and Jarett, L., 1982, Partial purification from hepatoma cells of an intracellular substance which mediates the effects of insulin on pyruvate dehydrogenase and low K_m cAMP phosphodiesterase, *Arch. Biochem. Biophys.* **215:**339–344.

Pennington, S. R., and Martin, B. R., 1985, Insulin-stimulated phosphoinositide metabolism in isolated fat cells, *J. Biol. Chem.* **260:**11039–11045.

Phelps, B. M., Primakoff, P., Koppel, D. E., Low, M. G., and Myles, D. G., 1988, Restricted lateral diffusion of PH-20, a PI-anchored sperm membrane protein, *Science* **240:**1780–1782.

Plourde, R., d'Alarcao, M., and Saltiel, A. R., 1992, Synthesis and characterization of insulin-mimetic disaccaride, *J. Organ. Chem.* **57:**2606–2610.

Represa, J., Avila, M. A., Miner, C., Giraldez, F., Romero, G., Clemente, R., Mato, J. M., and Varela-Nieto, I., 1991, Glycosylphosphatidylinositol/inositol phosphoglycan: A signaling system for the low-affinity nerve growth factor receptor, *Proc. Natl. Acad. Sci. U.S.A.* **88:**8016–8019.

Roberts, W. L., Myher, T. J., Kuksis, A., Low, M. G., and Rosenberry, T. L., 1988a, Lipid analysis of the glycoinositol phospholipid membrane anchor of human erythrocyte acetylcholinesterase. Palmioylation of inositol results in resistance to phosphatidyl-specific phospholipase C, *J. Biol. Chem.* **263:**18766–18775.

Roberts, W. L., Santikarn,, S., Reinhold, V. N., and Rosenberry, T. L., 1988b, Structural characterization of the glycoinositol phospholipid membrane anchor of human erythrocyte acetylcholinesterase by fast atom bombardment mass spectrometry, *J. Biol. Chem.* **263:**18776–18784.

Rodgers, W., Crise, B., and Rose, J. K., 1994, Signals determining protein tyrosine kinase and GPI-anchored protein targeting to a glycolipid-enriched membrane fraction, *Mol. Cell. Biol.* **14:**-5384–5391.

Romero, G., Luttrell, L., Rogol, A., Zeller, K., Hewlett, E., and Larner, J., 1988, Phosphatidylinositol-glycan anchors of membrane proteins: Potential precursors of insulin mediators, *Science* **240:**509–511.

Romero, G., Gamez, G., Huang, L. C., Lilley, K., and Luttrell, L.,, 1990, Antiinositol glycan antibodies selectively block some of the actions of insulin in intact BC3H1 cells, *Proc. Natl. Acad. Sci. U.S.A.* **87:**1476–1480.

Saltiel, A. R., 1987, Insulin generates an enzyme modulator from hepatic plasma membranes: Regulation of adenosine 3′,5′-monophosphate phosphodiesterase pyruvate dehydrogenase, and adenylate cyclase, *Endocrinology* **120:**967–972.

Saltiel, A. R., and Cuatrecasas, P., 1986, Insulin stimulates the generation from hepatic plasma membranes of modulators derived from an inositol glycolipid, *Proc. Natl. Acad. Sci. U.S.A.* **83:**5793–5797.

Saltiel, A. R., and Cuatrecasas, P., 1988, In search of a second messenger for insulin, *Am. J. Physiol.* **255:**C1–C11.

Saltiel, A. R., and Sorbara-Cazan, L. R., 1987, Inositol glycan mimics the action of insulin on glucose utilization in rat adipocytes, *Biochem. Biophys. Res. Commun.* **149:**1084–1092.

Saltiel, A. R., and Steigerwalt, R. W., 1986, Purification of a putative insulin-sensitive cAMP phosphodiesterase or its catalytic domain from rat adipocytes, *Diabetes* **35:**698–704.

Saltiel, A. R., Jacobs, S., Siegel, M., and Cuatrecasas, P., 1981, Insulin stimulates the release from liver plasma membranes of a chemical modulator of pyruvate dehydrogenase, *Biochem. Biophys. Res. Commun.* **102:**1041–1047.

Saltiel, A. R., Siegel, M. I., Jacobs, S., and Cuatrecasas, P., 1982, Putative mediators of insulin action: Regulation of pyruvate dehydrogenase and adenylate cyclase activities, *Proc. Natl. Acad. Sci. U.S.A.* **79:**3513–3517.

Saltiel, A. R., Doble, A., Jacobs, S., and Cuatrecasas, P., 1983, Putative mediators of insulin regulate hepatic acetyl CoA carboxylase activity, *Biochem. Biophys. Res. Commun.* **110:**789–795.

Saltiel, A. R., Fox, J. A., Sherline, P., and Cuatrecasas, P., 1986, Insulin-stimulated hydrolysis of a novel glycolipid generates modulators of cAMP phosphodiesterase, *Science* **233:**967–972.

Saltiel, A. R., Sherline, P., and Fox, J. A., 1987, Insulin-stimulated diacylglycerol production results from the hydrolysis of a novel phosphatidylinositol glycan, *J. Biol. Chem.* **262:**1116–1121.

Saltiel, A. R., Ravetch, J., and Aderem, A. A., 1991, Functional consequences of lipid-mediated protein-membrane interactions, *Biochem. Pharmacol.* **42:**1–11.

Seals, J. R., and Czech, M. P., 1980, Evidence that insulin activates an intrinsic plasma membrane protease in generating a secondary chemical mediator, *J. Biol. Chem.* **255:**6529–6531.

Selveraj, P., Roose, W. F., Siber, R., and Springer, T., 1988, The major Fc receptor in blood has a phosphatidylinositol anchor and is deficient in paroxysmal nocturnal hemoglobinuria, *Nature* **333:**565–568.

Simmons, D., and Seed, B., 1988, The Fc gamma receptor of natural killer cells is a phospholipid-linked membrane protein, *Nature* **333:**568–570.

Skillen, A. W., Hawthorne, G. C., and Turner, G. A., 1987, Serum alkaline phosphatase in rats with streptozotocin-induced diabetes, *Horm. Metab. Res.* **19:**505–506.

Stafford, H. A., Tykocinski, M. L., Wolin, D. M., Holeus, V. M., Roose, W. F., Atkinson, J. P., and Medof, E., 1988, Normal polymorphic variations and transcription of the decay accelerating factor gene in paroxysmal noctural hemoglobinuria cells, *Proc. Natl. Acad. Sci. U.S.A.* **85:**880–884.

Stein, J. M., and Hale, C. N., 1974, The effect of insulin on ^{32}P incorporation into rat-fat cell phospholipids, *Biochim. Biophys. Acta* **337:**41–49.

Steven, V. L., and Raetz, C.K.H., 1991, Detection GPI biosynthesis in extracts of three Thy-1 negative lymphoma cell mutants, *J. Biol. Chem.* **266:**10039–10042.

Stroynowski, I., Soloski, M., Low, M. G., and Hood, L., 1987, A single gene encodes soluble and membrane-bound forms of the major histocompatibility Qa-2 antigen: Anchoring of the product by a phospholipid tail, *Cell* **50:**759–768.

Suzuki, S., Toyoto, T., Suzuki, H., and Goto, Y., 1984, A putative second messenger of insulin action regulates hepatic microsomal glucose-6-phosphatase, *Biochem. Biophys. Res. Commun.* **118:**40–46.

Tartakoff, A. M., and Singh, N., 1992, How to make a glycoinositol phospholipid anchor, *Trends Biochem. Sci.* **17:**470–473.

Thomas, J., Webb, W., Davitz, M. A., and Nussenzweig, V., 1987, Decay-accelerating factor diffuses rapidly on Hela (AE) cell surfaces, *Biophys. J.* **51**(issue 2, part 2):522a.

Tse, A.G.D., Barclay, A. N., Watts, A., and Williams, A. F., 1985, A glycophospholipid tail at the carboxyl terminus of the Thy-1 glycoprotein of neurons and thymocytes, *Science* **230**:1003–1008.

Urakaze, M., Kamitoni, T., DeGasperi, R., Sugiyama, E., Chang, H., and Yeh, E.T.H., 1992, Identification of a missing link in GPI anchor biosynthesis in mammalian cells, *J. Biol. Chem.* **267**:6459–6462.

Vidugiriene, J., and Menon, A. K., 1994, The GPI anchor of cell-surface proteins is synthesized on the cytoplasmic face of the endoplasmic reticulum, *J. Cell Biol.* **127**:333–341.

Waneck, G. L., Stein, M. E., and Flavell, R. A., 1988, Conversion of a PI-anchored protein to an integral membrane protein by a single amino acid mutation, *Science* **241**:697–699.

Witters, L. A., and Watts, T. D., 1988, An autocrine factor from Reuber hepatoma cells that stimulates DNA synthesis and acetyl-CoA carboxylase: Characterization of biologic activity and evidence for a glycan structure, *J. Biol. Chem.* **263**:8027–8036.

Witters, L. A., Watts, T. D., Gould, G. W., Lienhard, G. E., and Gibbs, E. M., 1987, Regulation of protein phosphorylation by insulin and an insulin-mimetic oligosaccharide in 3T3-L1 adipocytes and FaO hepatoma cells, *Biochem. Biophys. Res. Commun.* **153**:992–998.

Chapter 7

Inositol Pentakis- and Hexakisphosphate Metabolism Adds Versatility to the Actions of Inositol Polyphosphates

Novel Effects on Ion Channels and Protein Traffic

Stephen B. Shears

1. INTRODUCTION

The discovery of receptor-dependent regulation of inositol-1,4,5-trisphosphate [Ins(1,4,5)P_3*]-mediated cellular Ca^{2+} mobilization was a pivotal breakthrough in the field of signal transduction (Berridge and Irvine, 1989). One consequence of this development has been the unraveling of a complex metabolic network of inositol phosphates emanating from Ins(1,4,5)P_3 (Figure 1). As these studies progressed, it emerged that Ins(1,3,4,5,6)P_5 and InsP_6 usually comprise the bulk

*All inositol phosphates are designated according to IUPAC nomenclature, where Ins represents D-*myo*-inositol, with substituents numbered accordingly, and "P" is italic.

Stephen B. Shears Inositol Lipid Section, Laboratory of Cellular and Molecular Pharmacology, National Institute of Environmental Health Sciences, National Institutes of Health, Research Triangle Park, North Carolina 27709.

Subcellular Biochemistry, Volume 26: myo-Inositol Phosphates, Phosphoinositides, and Signal Transduction, edited by B. B. Biswas and Susweta Biswas. Plenum Press, New York, 1996.

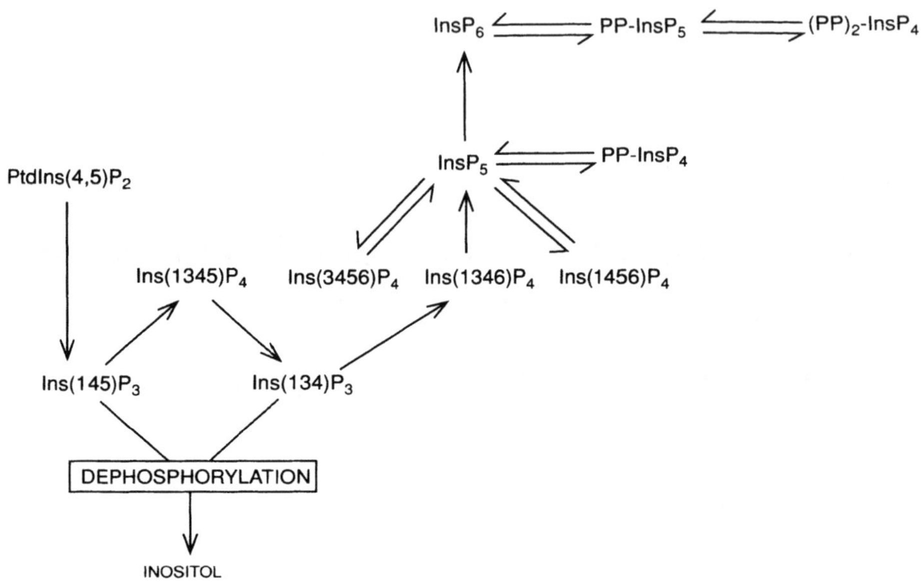

FIGURE 1. Metabolic interrelationships of inositol polyphosphates. This figure depicts those pathways of inositol phosphate turnover that are known to be widespread, at least in animal cells. There are additional reactions that are not shown here that may not be as ubiquitous; these are described in the text.

of the mammalian cell's inositol polyphosphate content (for reviews, see Berridge and Irvine, 1989; Shears, 1992). At first, it was generally believed that Ins-(1,3,4,5,6)P_5 and InsP_6 were somewhat metabolically inert compounds (e.g., Michell *et al.*, 1988), and their existence in animal cells caused little excitement. The renaissance of interest in the functions of these metabolites (Hughes and Michell, 1993; Menniti *et al.*, 1993b) owes much to the discovery (Stephens *et al.*, 1993; Menniti *et al.*, 1993a) that the cellular pools of these polyphosphates turn over rapidly in animal cells. The purpose of this chapter is to update these recent metabolic developments and also to evaluate the growing body of evidence suggesting that there are important roles in cell biology for Ins(1,3,4,5,6)P_5, InsP_6, and their metabolites. An underlying theme is the considerable versatility of these inositol phosphates in the control of diverse cellular activities.

2. SYNTHESIS OF Ins(1,3,4,5,6)P_5

The pathways to and from Ins(1,3,4,5,6)P_5 constitute a significant metabolic hub (Figure 1). As evidence has accumulated that the products of Ins(1,3,4,5,6)P_5

turnover play important roles in cell biology (see below), so studies into the control of metabolic flux through Ins(1,3,4,5,6)P_5 have assumed a greater importance. There is a well-characterized pathway of Ins(1,3,4,5,6)P_5 synthesis in eukaryotes that begins with the phospholipase-C-mediated hydrolysis of PtdIns(4,5)P_2 (Figure 1). A certain amount of this hydrolytic activity is ongoing even in unstimulated cells, and is generally sufficient to maintain levels of Ins(1,4,5)P_3 in the 0.1–0.2 μM range (Palmer *et al.*, 1988) even though it has a metabolic half-live *in vivo* of approximately 10 sec (Irvine *et al.*, 1985; Hughes and Putney, 1989). The Ins(1,4,5)P_3 can either be dephosphorylated by a 5-phosphatase or phosphorylated by a 3-kinase to Ins(1,3,4,5)P_4 (see Shears, 1992, for a review). The competition between these two routes of Ins(1,4,5)P_3 metabolism positions the 3-kinase at an important early stage in the commitment of precursor material to Ins(1,3,4,5,6)P_5 synthesis (Figure 1). While this may be considered a crucial function of the Ins(1,4,5)P_3 3-kinase, this enzyme does have additional roles: It also terminates the signaling activities of Ins(1,4,5)P_3, and there is substantial (albeit sometimes controversial) evidence that the reaction product, Ins(1,3,4,5)P_4, also has an important role in Ca^{2+} mobilization (see Irvine, 1992). Indeed, Irvine (1991) has stated that effects of Ins(1,3,4,5)P_4 upon cellular Ca^{2+} fluxes provide ". . . the reason for the selective pressure that caused InsP_4 to evolve." Whatever side one takes in the ongoing controversy regarding the physiological activity of Ins(1,3,4,5)P_4 (see Irvine, 1991, 1992; Putney, 1992), perhaps it would be more circumspect to leave open the question of which of the physiological functions of the Ins(1,4,5)P_3 3-kinase evolved first.

The Ins(1,3,4,5)P_4 is dephosphorylated to Ins(1,3,4)P_3 (Figure 1). The latter sits at yet another critical branch point of inositol phosphate metabolism: Ins(1,3,4)P_3 is attacked by specific phosphatases so that the cell can salvage free inositol for resynthesis of inositol lipids (see Shears, 1992 for a review). In addition, experiments with a variety of cell-free systems have shown that Ins-(1,3,4)P_3 is phosphorylated first to Ins(1,3,4,6)P_4 and then to Ins(1,3,4,5,6)P_5 (Balla *et al.*, 1987, 1989; Shears *et al.*, 1987a; Shears, 1989; Hansen *et al.*, 1988; Stephens *et al.*, 1988a; Mattingly *et al.*, 1991). The two-stage phosphorylation of Ins(1,3,4)P_3 to Ins(1,3,4,5,6)P_5 is a relatively slow process. This is evident from experiments in which cultured cells are labeled with [^{3}H]inositol; Ins(1,3,4)P_3 becomes radiolabeled to equilibrium within 24 hr, but it is 2–3 days before Ins(1,3,4,5,6)P_5 attains equilibrium radiolabeling (Stewart *et al.*, 1987; Menniti *et al.*, 1990; Oliver *et al.*, 1992; Sasakawa *et al.*, 1992). This sluggish metabolic flux from Ins(1,3,4)P_3 to Ins(1,3,4,5,6)P_5 *in vivo* is consistent with the relatively meager rate at which phosphorylation of Ins(1,3,4)P_3 is observed *in vitro*. At the levels of Ins(1,3,4)P_3 that typically prevail in cells (resting, 1–4 μM; stimulated 10–60 μM, Tarver *et al.*, 1987; Michell *et al.*, 1988: Hughes *et al.*, 1989), the 6-kinase usually has a much lower capacity to metabolize Ins(1,3,4)P_3 compared to the Ins(1,3,4)P_3 phosphatases (Hughes *et al.*, 1989,

1994; King *et al.*, 1990; Mattingly *et al.*, 1991), with the one known exception of avian salt glands (Hilderbrandt and Shuttleworth, 1992). The activity of the 6-kinase is additionally constrained by product inhibition (Abdullah *et al.*, 1992; Hughes *et al.*, 1994). Arguably the most striking restraint on 6-kinase activity, *in vitro* at least, is its near complete inhibition by physiologically relevant levels of Ins(3,4,5,6)P_4 (Hughes *et al.*, 1989, 1994; Hilderbrandt and Shuttleworth, 1992; Vajanaphanich *et al.*, 1994). Thus, upon receptor-activation, it is no surprise to observe that the increase in the level of Ins(1,3,4)P_3 tends to be severalfold higher than the elevation in concentration of Ins(1,3,4,6)P_4 (Balla *et al.*, 1987; 1989; Stephens *et al.*, 1988a; Hansen *et al.*, 1988; Hughes *et al.*, 1989; King *et al.*, 1990). What is puzzling about these studies is that so much Ins(1,3,4,6)P_4 is formed so quickly, in view of the severity of the constraints that appear to be placed upon the Ins(1,3,4)P_3 6-kinase activity. It therefore seems likely that there is something important about the short-term regulation of 6-kinase activity *in vivo* that has not been revealed in experiments with cell-free systems. Perhaps the process of metabolic compartmentation (Srere, 1987) maintains a pool of Ins(1,3,4)P_3 substrate that does not readily mix with the inhibitory pool of Ins(3,4,5,6)P_4. Unfortunately, this is an area of research that does not readily lend itself to the empirical testing, and even the theorists do not agree that compartmentation can significantly reduce a local concentration of a metabolic intermediate (see Cornish-Bowden, 1991). Another possible explanation is that the 6-kinase activity may be increased following receptor-activation by some rapidly-acting mechanism. Several second messengers associated with mediating rapid cellular responses—cAMP, Ca^{2+}/calmodulin, protein kinase C—all have no effect upon the purified 6-kinase (Abdullah *et al.*, 1992; Hughes *et al.*, 1994). However, what makes this whole problem even more intriguing is the doubling of the rate of angiotensin-stimulated phosphorylation of Ins(1,3,4)P_3 to Ins(1,3,4,6)P_4 in adrenal glomerulosa cells if they are first pretreated for 2–16 hr with adrenocorticotrophin or 8-bromo cyclic AMP (Balla *et al.*, 1989). In other words, whatever the factors are that enable the 6-kinase activity to proceed *in vivo*, they are even more effective after long-term cAMP treatment. Moreover, this cAMP effect was sensitive to cycloheximide (Balla *et al.*, 1989), so there may even be control over the *levels* of the Ins(1,3,4)P_3 6-kinase. This result not only consolidates the basic role of the 6-kinase in the pathway of synthesis of Ins(1,3,4,6)P_4 and Ins(1,3,4,5,6)P_5, but it also indicates that such a function is sufficiently important for it to be a target for cellular regulation. This demonstration by Balla *et al.*, (1989) deserves more recognition and further study.

Another suggestion has been put forward in an effort to account for the size of the Ins(1,3,4,6)P_4 signal during short-term receptor activation. This is the possibility that there may be a supplementary mechanism by which cells synthesize Ins(1,3,4,6)P_4 (Hughes *et al.*, 1989). Indeed, under certain circumstances the Ins(1,3,4)P_3 and Ins(1,3,4,6)P_4 in cultured cell populations incorporate exogenous radiolabeled inositol at different rates (Barker *et al.*, 1992; Balla *et al.*, 1994); this observation could be interpreted as evidence that Ins(1,3,4)P_3 is not the sole precursor of Ins(1,3,4,6)P_4 and Ins(1,3,4,5,6)P_5. Of course, it is also likely that differential rate of labeling of Ins(1,3,4)P_3 and Ins(1,3,4,6)P_4 merely reflects the kinetic constraints imposed upon the 6-kinase activity. Nevertheless, the possibility that there may be alternate routes of Ins(1,3,4,6)P_4 synthesis affects our evaluation of the importance of the Ins(1,3,4)P_3 6-kinase.

The avian erythrocyte is an example of a cell in which the contribution of the Ins(1,3,4)P_3 6-kinase to Ins(1,3,4,6)P_4 synthesis has been addressed (Stephens and Downes, 1990). The latter workers investigated product–precursor relationships of inositol phosphates by assessing the specific radioactivity of individual phosphate groups of the inositol phosphates after short-term labeling of cells with [^{32}P]P_i. The underlying strategy for these experiments relies upon the γ-phosphate of ATP being the sole donor for the various inositol phosphate kinase activities. The specific radioactivity of the donor phosphate of ATP increased for the duration of the experiment, such that the later in a synthetic sequence that a phosphate was added to the inositol ring, the greater was its specific radioactivity in the pool of end-product. One result of this study was the determination that the 3-phosphate of Ins(1,3,4,6)P_4 had the highest specific radioactivity, which is consistent with the major precursor of Ins(1,3,4,6)P_4 being Ins(1,4,6)P_3 (Stephens and Downes, 1990). If Ins(1,4,6)P_3 were to be formed by sequential phosphorylation of inositol, then this would be independent of inositol lipid turnover (see Figure 1). However, as Stephens and Downes (1990) pointed out, there is an alternative explanation for these data: Ins(1,4,6)P_3 might instead be produced by dephosphorylation of Ins(1,3,4,6)P_4 (Stephens *et al.*, 1989). In other words, there may be a substrate cycle between Ins(1,3,4,6)P_4 and Ins(1,4,6)P_5 which could be a closed branch-point to the more general pathway of Ins(1,4,6)P_3 synthesis described in Figure 1. Thus, these results are consistent with the Ins(1,3,4)P_3 6-kinase being entirely responsible for *de novo* Ins(1,3,4,6)P_4 synthesis in avian erythrocytes.

The possibility has been raised that at least some Ins(1,3,4,6)P_4 might be produced by dephosphorylation of Ins(1,3,4,5,6)P_5 (Balla *et al.*, 1994). Indeed, *in vitro* at least, Ins(1,3,4,6)P_4 is a minor side-product of Ins(1,3,4,5,6)P_5 dephosphorylation by the multiple inositol polyphosphate phosphatase (S. B. Shears, unpublished data), although Ins(1,4,5,6)P_4 is the primary product (see Section 3.2). Even so, much of the available evidence argues against there being any significant dephosphorylation of Ins(1,3,4,5,6)P_5 to Ins(1,3,4,6)P_4 *in vivo*. This was, for example, one of the outcomes of the study of avian erythrocytes by Stephens and Downes (1990). Menniti *et al.* (1990) and Barker *et al.* (1992) reached the same conclusion by applying different methods to receptor-activated mammalian cells: various radiolabeling protocols were employed to differentially label the Ins(1,3,4,5,6)P_5 pool relative to the pool of Ins(1,3,4)P_3. During receptor activation, the specific radioactivity of Ins(1,3,4,6)P_4 was independent of that of Ins(1,3,4,5,6)P_5, and instead mirrored the specific radioactivity of Ins(1,3,4)P_3. Thus, Ins(1,3,4,6)P_4 seemed to be more closely metabolically related to Ins(1,3,4)P_3 than to Ins(1,3,4,5,6)P_5. The results of Balla *et al.* (1994) also seem to rule out a role of Ins(1,3,4,5,6)P_5 in generating Ins(1,3,4,6)P_4 upon receptor activation: These workers were studying the effects of overexpression of Ins(1,4,5)P_3 3-kinase upon inositol phosphate profiles in [^{3}H]inositol-labeled 3T3 fibroblasts. Mass levels of Ins(1,3,4,5,6)P_5 were unaffected by the increase in Ins(1,4,5)P_3 3-kinase activity, but, surprisingly, the Ins(1,3,4,5,6)P_5 pool then incorporated [^{3}H]inositol more slowly, so its specific activity was two-fold below that of control cells. Balla *et al.* (1994) further found that 10-min stimulation with platelet derived growth factor evoked similar absolute increases in levels of [^{3}H]Ins(1,3,4,6)P_4 in both vector-transfected and 3-kinase-transfected cells.* Therefore, the size of the receptor-dependent increase in [^{3}H]Ins(1,3,4,6)P_4 levels was independent of the specific radioactivity of [^{3}H]Ins(1,3,4,5,6)P_5. This observation also argues against any conversion of Ins(1,3,4,5,6)P_5 to Ins(1,3,4,6)P_4, at least in receptor-activated cells.

This is not to say that the Ins(1,3,4)P_3 6-kinase is exclusively responsible for the synthesis of Ins(1,3,4,6)P_4 and Ins(1,3,4,5,6)P_5 in all organisms. *Dictyostelium discoideum* (Stephens and Irvine, 1990), *Chlamydamonas eugametos* (Irvine *et al.*, 1992) and *Schizosaccharomyces pombe* (Hughes and Michell, 1993) are examples of organisms from which Ins(1,3,4)P_3 6-kinase is absent.† In *Dictyostelium* Ins(1,3,4,6)P_4 is still formed, but by the action of an Ins(3,4,6)P_3 1-kinase; this is one of the reactions that occurs during the stepwise phosphorylation of inositol to Ins(1,3,4,5,6)P_5 and InsP_6 (Stephens and Irvine, 1990). There are further examples in Nature of alternate pathways of Ins(1,3,4,5,6)P_5 synthesis: *Chlamydamonas* phosphorylates Ins(1,3,4,5)P_4 with a 6-kinase (Irvine *et al.*, 1992), while pea roots (Chattaway *et al.*, 1992) and *Saccharomyces cerevisiae* (Estevez *et al.*, 1994) phosphorylate Ins(1,4,5)P_3 first to Ins(1,4,5,6)P_4 and then to Ins(1,3,4,5,6)P_5. Some of these specialized enzyme reactions have their counterparts in animal cells. For example, the Ins(1,3,4,5)P_4 6-kinase is present in turkey erythrocytes (Irvine *et al.*, 1992), and mammalian cells contain an Ins(1,4,5,6)P_4 3-kinase (Section 3.2). There is no evidence that in these cases they contribute to *de novo* Ins(1,3,4,5,6)P_5 synthesis, so it will be interesting to determine both their functions and their

*There is an error in the time course presented in Figure 3 of the manuscript of Balla *et al.* (1994); it should have been indicated that there were no detectable levels of [^{3}H]Ins(1,3,4,6)P_4 at zero time in the transfected cells; cf. Figure 1 of the same paper. Thus, stimulation for 10 min with platelet-derived growth factor caused levels of [^{3}H]Ins(1,3,4,6)P_4 to increase from 0 to 6000 cpm/mg protein in kinase-transfected cells, and from 4000 to 10,000 in vector-transfected cells.

†Hughes and Michell (1993) pointed out that McIntosh and McIntosh (1990) were unable to detect Ins(1,3,4)P_3 6-kinase activity in *Xenopus* oocytes, but others have shown that these cells do indeed have this enzyme (Guse *et al.*, 1991).

evolutionary interrelationship with the similar enzyme activities of lower organisms.

Before leaving the subject of the Ins(1,3,4)P_3 6-kinase, we should discuss another peculiar characteristic worth mentioning; about 20% of the Ins(1,3,4)P_3 that this enzyme phosphorylates is converted to Ins(1,3,4,5)P_4 (Shears 1989; Hughes *et al.*, 1989, 1994; Abdullah *et al.*, 1992). The significance of this Ins(1,3,4)P_3 5-kinase activity is unknown; there is no evidence that it is sufficient to help sustain the levels of Ins(1,3,4,5)P_4, at least in stimulated cells (Hughes *et al.*, 1989).

There has also been some debate concerning the extent to which the Ins(1,3,4,6)P_4 5-kinase (Figure 1) contributes to Ins(1,3,4,5,6)P_5 synthesis *de novo*. Stephens and Downes (1990) compared the specific activities of individual phosphates of inositol phosphates extracted from cells labeled briefly with [^{32}P]-P_i. The underlying logic of these experiments is explained above; essentially, the greater the specific activity of a phosphate group, the later it would be added in a synthetic sequence. Stephens and Downes (1990) found that the 1-phosphate of Ins(1,3,4,5,6)P_5 had the highest specific activity, followed by the 5-phosphate. Thus, their proposed synthetic sequence was Ins(3,4,6)P_3 → Ins(3,4,5,6)P_4 → Ins(1,3,4,5,6)P_5. The Ins(3,4,6)P_3 was suggested to be produced by dephosphorylation of Ins(1,3,4,6)P_4; such a reaction has been observed in cell-free systems (Stephens *et al.*, 1989). This proposal downgrades the Ins(1,3,4,6)P_4 5-kinase to having, at best, an ancillary role in *de novo* Ins(1,3,4,5,6)P_5 synthesis.

It is, of course, entirely possible that the pathway of *de novo* Ins(1,3,4,5,6)P_5 synthesis in avian erythrocytes is uniquely different from that in mammalian cells. However, it was emphasized (Stephens and Downes, 1990; Downes and MacPhee, 1990) that the delineation of a *de novo* synthetic pathway using their methodology could be compromized by substrate cycles. Indeed, evidence for a substrate cycle between Ins(1,3,4,5,6)P_5 and Ins(3,4,5,6)P_4 was subsequently obtained (Menniti *et al.*, 1990; Oliver *et al.*, 1992). If this cycle were to be sufficiently active in avian erythrocytes, the 1-phosphate could have acquired a greater specific activity in the experiments of Stephens and Downes (1990), even though the 5-phosphate was the last to be added (to Ins(1,3,4,6)P_4) during Ins(1,3,4,5,6)P_5 synthesis *de novo*. Thus, the data obtained from avian erythrocytes (Stephens and Downes, 1990) can be considered not to conflict with there being a major role for the Ins(1,3,4,6)P_4 5-kinase in the *de novo* pathway of Ins(1,3,4,5,6)P_5 synthesis in this and many other animal cell types.

3. Ins(3,4,5,6)P_4 AND Ins(1,4,5,6)P_4

3.1. Identification

The metabolic pathway in Figure 1 depicts Ins(1,3,4,5,6)P_5 as being the precursor of both Ins(3,4,5,6)P_4 and Ins(1,4,5,6)P_4 through the actions of two "dead-end" substrate cycles. The evolution of this picture has been a rather complex process: Johnson and Tate (1969) provided the first description of Ins(3,4,5,6)P_4 and/or Ins(1,4,5,6)P_4 being a physiologically relevant constituent of an animal cell (the chicken erythrocyte), but the authors could not resolve which of the two enantiomers was present using nonchiral analytic procedures. This technical difficulty has contributed to some of the confusion in the literature on the occurrence of these isomers: in the process of trying to decide which enantiomer predominated in avian erythrocytes, Mayr and Dietrich (1987) reasoned that Ins(1,4,5,6)P_4, being closer to Ins(1,4,5)P_3 in structure, was a more likely configuration than Ins(3,4,5,6)P_4. This was the wrong choice, and an unfortunate one; the avian erythrocyte is the commercial source of this material, and so for some years "Ins(3,4,5,6)P_4" was incorrectly marketed as being Ins(1,4,5,6)P_4. Early examples of this confusion may be found in the papers by Mayr (1988), Donié *et al.* (1990), Palczewski *et al.* (1991), and Theibert *et al.* (1992), and the problem still occasionally arises in more recent reports, long after the suppliers reversed the nomenclature (see Figure 7 in the paper by Newton *et al.*, 1994).

Actually the situation is even more complex: although commercial sources of "Ins(3,4,5,6)P_4" isolated from avian erythrocytes are indeed predominantly that isomer, they are 10–30% contaminated with Ins(1,4,5,6)P_4 (S. B. Shears and M. Hirata, unpublished data). Indeed, evidence for the presence in chick erythrocytes of both of these isomers in about these proportions was published some years ago (Stephens *et al.*, 1988b). Even the first report of chemically synthesized racemically pure Ins(3,4,5,6)P_4 has created its own confusion (Pietrusiewicz *et al.*, 1992): it later turned out that the material was in fact Ins(1,4,5,6)P_4 (Pietrusiewicz *et al.*, 1994). Against this confusing background, it was clearly an important step forward when Stephens *et al.*, 1988b) devised a stereospecific, enzymatically based method for distinguishing Ins(3,4,5,6)P_4 from Ins(1,4,5,6)P_4. Since then, both enantiomers have been identified in various proportions in many cell types (Sections 3.2 and 3.3).

3.2. Synthesis and Metabolism of Ins(1,4,5,6)P_4

Ins(1,4,5,6)P_4 has been identified in a number of cells (Stephens *et al.*, 1988; Menniti *et al.*, 1990; Mattingly *et al.*, 1991; Barker *et al.*, 1992) but there is still some difficulty in understanding what controls its synthesis. The 3-phosphatase that produces Ins(1,4,5,6)P_4 from Ins(1,3,4,5,6)P_5 (Figure 1) was originally identified—and named—from its ability to also remove the 3-phosphate from Ins(1,3,4,5)P_4 (Doughney *et al.*, 1988; Nogimori *et al.*, 1991). Since then, several additional substrates of this enzyme have been identified, and it has been renamed as a multiple inositol polyphosphate phosphatase ("MIPP"; Craxton *et al.*, 1995). InsP_6 is the substrate with highest known affinity *in vitro* (0.3 nM!

Nogimori *et al.*, 1991). This enzyme is arguably the most enigmatic of those that metabolize the higher inositol polyphosphates: in liver at least, much of the cell's complement of MIPP is restricted to the interior of the endoplasmic reticulum (Ali *et al.*, 1993). To date, we have been unable to demonstrate the uptake of *any* inositol phosphate into isolated microsomes (Ali *et al.*, 1993). Perhaps we should therefore investigate this problem from the point of view of determining if there are circumstances under which MIPP might be released from the endoplasmic reticulum (Section 4.3).

Rat erythrocytes contain some MIPP in their plasma membrane (Craxton *et al.*, 1995); the Ins(1,3,4,5)P_4 3-phosphatase activity on the inner, cytosolic face of human erythrocytes (Doughney *et al.*, 1988; Estrada-Garcia *et al.*, 1991) is probably a closely related enzyme. There may be a uniquely good reason for MIPP being retained by erythrocytes (Section 4.3), so the plasma membrane may not be a significant location for MIPP in other cell types.

An Ins(1,3,4,5,6)P_5/Ins(1,3,4,5)P_4 phosphatase activity was briefly reported to be present on the surface of NIH3T3 cells (Carpenter *et al.*, 1989). Although these data have never been fully published, it was recently reiterated that such an enzyme may be a physiologically relevant inactivator of putative extracellular actions of Ins(1,3,4,5,6)P_5, and possibly also InsP_6 (Sasakawa *et al.*, 1995). However, there is an alternative interpretation of these workers' conclusions. The determination by Carpenter *et al.* (1989) that "intact" 3T3 cells could metabolize Ins(1,3,4,5,6)P_5 and Ins(1,3,4,5)P_4 was instead very likely due to the unavoidable presence in the culture of a small proportion of "damaged" cells. Indeed, how else can we explain why these same cells also dephosphorylated Ins(1,4,5)P_3 (albeit more slowly) even though Ins(1,4,5)P_3 5-phosphatase is entirely intracellular (Shears *et al.*, 1987b; 1988)? Carpenter *et al.* (1989) went on to state the the rate of Ins(1,3,4,5,6)P_5 metabolism was "unchanged" when the cells were permeabilized with digitonin. Clearly the observation that Ins(1,3,4,5,6)P_5 and Ins(1,3,4,5)P_4 were both dephosphorylated indicates that a MIPP-like enzyme (see above) was responsible. Digitonin treatment is often used with the specific purpose of permeabilizing the plasma membrane (e.g. Stuart *et al.*, 1994) and not membranes of intracellular organelles, such as endoplasmic reticulum, where MIPP resides (Ali *et al.*, 1993). With this hindsight, it can now be appreciated why Ins(1,3,4,5)P_4/Ins(1,3,4,5,6)P_5 dephosphorylation might not increase when the plasma membrane is permeabilized, without our having to invoke an extracellular location for this enzyme activity. It should also be mentioned that it is extremely difficult to quantify latency of MIPP in cell populations, because permeabilization techniques will release endogenous InsP_6, which will potently inhibit MIPP. In any case, the idea that Ins(1,3,4,5,6)P_5 and InsP_6 can be physiologically important extracellular signals has fallen out of favor (Section 4.1).

Ins(1,4,5,6)P_4 can be rephosphorylated back to Ins(1,3,4,5,6)P_5 by 3-

kinase (Oliver *et al.*, 1992; Craxton *et al.*, 1995). This kinase has been partly purified from rat liver and has been found to also phosphorylate Ins(1,2,4,5,6)P_5 but not Ins(1,4,5)P_3 (Craxton *et al.*, 1995; A. Craxton, M. Hirata, and S. B. Shears, unpublished data). As it happens, it has been known for some years that Ins(1,2,4,5,6)P_5 3-kinase activity is quite widespread; it has been observed in *Dictyostelium* (Stephens and Irvine, 1990), rat brain (Stephens *et al.*, 1991) and Jurkat T-lymphocytes (Guse and Emmrich, 1991). Some information on whether Ins(1,2,4,5,6)P_5 or Ins(1,4,5,6)P_4 is the preferred substrate of the 3-kinase *in vivo* was provided by Oliver *et al.*, (1992) in a study with the AR4-2J pancreatoma cell line. Since the enzymology exists for both substrates to participate in substrate cycles (InsP_6 $\leftrightarrows$ Ins (1,2,4,5,6)P_5 and Ins(1,3,4,5,6)P_5 $\leftrightarrows$ Ins(1,4,5,6)P_4), intact cells were treated with antimycin A to deplete cellular ATP levels and thereby perturb the phosphorylation pathways within the substrate cycles, without affecting dephosphorylation. Following the addition of antimycin A, the relative rates of appearance of Ins(1,2,4,5,6)P_5 and Ins(1,4,5,6)P_4 was taken to be indicative of the ongoing activities of both substrate cycles. It turned out that a 2-hr treatment of AR4-2J cells with antimycin A did not affect the levels of either InsP_6 or Ins(1,2,4,5,6)P_5 (Oliver *et al.*, 1992). In contrast, levels of Ins(1,4,5,6)P_4 increased fourfold within 1 hr (Oliver *et al.*, 1992). This result gives a strong indication that under normal cellular conditions the activity of the Ins(1,3,4,5,6)P_5/Ins(1,4,5,6)P_4 substrate cycle is significant. Moreover, these data confirm that some of the cell's Ins(1,3,4,5,6)P_5 does have access to MIPP *in vivo*.

There are some tentative, but sometimes inconsistent, indications that Ins(1,4,5,6)P_4 turnover within this substrate cycle might be a target of receptor activation. For example, the 3-kinase in extracts of Jurkat cells was activated by Ca^{2+}, at least when Ins(1,2,4,5,6)P_5 was the substrate (Guse and Emmrich, 1991). A Ca^{2+}-mediated stimulation of rat hepatic Ins(1,4,5,6)P_4 3-kinase has also been observed, but only by about 20% (A. Craxton and S. Shears, unpublished data). It is difficult to check if Ca^{2+} mobilization in intact cells is accompanied by a corresponding decrease in Ins(1,4,5,6)P_4 levels, because standard HPLC analyses cannot separate this compound from its enantiomer, Ins(3,4,5,6)P_4 (and, to further complicate matters, Ins(3,4,5,6)P_4 concentrations increase in response to activation of phospholipase C, see Section 3.4). However, there is one study where a chiral analysis clearly showed that levels of Ins(1,4,5,6)P_4 did not decrease upon receptor activation (Barker *et al.*, 1992). In fact, quite the opposite result was obtained. Following stimulation of WRK1 cells with vasopressin, levels of Ins(1,4,5,6)P_4 rose about twofold (Barker *et al.* 1992). Actually the latter result makes some sense, because the 3-kinase seems likely to be inhibited by the elevations in Ins(1,3,4,6)P_4 levels that accompany activation of phospholipase C (Craxton *et al.*, 1994). Increased concentrations of

Ins(1,4,5,6)P_4 have also been found in src-transformed fibroblasts (Mattingly *et al.*, 1991).

One final intriguing aspect to Ins(1,4,5,6)P_4 metabolism has arisen from a study with an engineered *Dictyostelium* mutant from which all phospholipase C activity was rendered nonfunctional (Drayer *et al.*, 1994). Surprisingly, the mutant expressed a normal phenotype. Moreover, despite the inactivation of this route of Ins(1,4,5)P_3 formation, its steady-state levels in the mutant were very similar to those in wild-type cells (Drayer *et al.*, 1994). This observation led to an unprecedented search for an alternative, phospholipase-independent source of Ins(1,4,5)P_3. It was already known that *Dictyostelium* synthesises Ins(1,3,4,5,6)P_5 by direct phosphorylation of inositol, independently of inositol lipid turnover (Stephens and Irvine, 1990). It was subsequently established that *Dictyostelium* can dephosphorylate Ins(1,3,4,5,6)P_5 to Ins(1,4,5)P_3, in part via Ins(1,4,5,6)P_4 (Van Dijken *et al.*, 1995). This appears to explain how these mutant cells maintain cellular Ins(1,4,5)P_3 levels. Van Dijken *et al.* (1995) made one further unexpected observation: purified rat hepatic MIPP can perform the very same reactions, i.e. Ins(1,3,4,5,6)$P_5 \rightarrow$ Ins(1,4,5,6)$P_4 \rightarrow$ Ins(1,4,5)P_3. It is not yet known if such a reaction sequence is significant in intact animal cells, and *Dictyostelium* remains the only organism in which it has been demonstrated that inositol can be directly phosphorylated to Ins(1,3,4,5,6)P_5. Nevertheless, a precedent has now been established for Ins(1,4,5)P_3 to be formed essentially independently of phospholipase C, at least in one cell type. It will be fascinating to see if this pathway turns out to be a more widespread addition to the metabolic versatility of Ins(1,3,4,5,6)P_5.

3.3. Ins(1,4,5,6)P_4 and Pleckstrin Homology Domains

The concentration of Ins(1,4,5,6)P_4 inside "nonstimulated" mammalian cells is probably around 1 μM (i.e., about the same as Ins(3,4,5,6)P_4, Vajanaphanich *et al.*, 1995). At these levels, Ins(1,4,5,6)P_4 has not been reported to be particularly significant as either a substitute or an antagonist of the metabolism of Ins(1,4,5)P_3, despite the obvious structural similarities between the two polyphosphates. For example, Ins(1,4,5,6)P_4 is not a substrate for the Ins(1,4,5)P_3 3-kinase (Craxton *et al.*, 1994). Neither does Ins(1,4,5,6)P_4 appear to have much impact upon Ca^{2+} homeostasis: First, it does not antagonize Ins(1,4,5)P_3-mediated Ca^{2+} release (Cullen *et al.*, 1990; Gawler *et al.*, 1991; DeLisle *et al.*, 1994). Second, it has only a relatively feeble ability to mobilize Ca^{2+} from rat brain microsomes that seems very unlikely to have any significance *in vivo* (the potency is more than 100-fold lower than that of Ins(1,4,5)P_3 (DeLisle *et al.*, 1994; Lu *et al.*, 1994). With regards to the putative Ins(1,3,4,5)P_4-receptor that gates Ca^{2+} entry across the plasma membrane,

Ins(1,4,5,6)P_4 binds at least 15-fold less avidly than Ins(1,3,4,5)P_4, so it is the latter which is justifiably considered the more relevant ligand *in vivo* (Cullen *et al.*, 1995a).

However, there is one class of polyphosphate-binding sites where the affinities for both Ins(1,4,5)P_3 and Ins(1,4,5,6)P_4 are not only very similar, but they also exceed the affinities for a range of other inositol polyphosphates. This observation was first made with a somewhat enigmatic 130 kD protein (p130) (Kanematsu *et al.*, 1992, 1995; Yoshida *et al.*, 1994). The cloning of p130 has further revealed it to have substantial homology to the δ-isoform of phospholipase C, although neither the native nor the cloned protein expressed any phospholipase C activity (Kanematsu *et al.*, 1995). It is intriguing that the Ins(1,4,5,6)P_4/Ins(1,4,5)P_3-binding site of p130 lies within the pleckstrin homology (PH) domain (Kanematsu *et al.*, 1995). This is the most promising line of evidence that there might be a physiologically relevant function for Ins(1,4,5,6)P_4. A number of proteins that mediate signal transduction contain a PH domain, which has been proposed to promote protein–protein interactions (Gibson *et al.*, 1994). Another idea, following the demonstration of the high affinity of some PH domains for PtdIns(4,5)P_2 (Harlan *et al.*, 1994) is that the motif can target certain proteins to the plasma membrane (where PtdIns(4,5)P_2 is localized). Ins(1,4,5)P_3, at least, seems able to competitively inhibit the interaction of PtdIns(4,5)P_2 with a noncatalytic, PH domain of phospholipase C-δ1 that is thought to guide the enzyme towards its substrate (Cifuentes *et al.*, 1994, 1995). The work of Hirata and colleagues (Kanematsu *et al.*, 1992, 1995; Yoshida *et al.*, 1994) strengthens the possibility that there are PH domains where Ins(1,4,5,6)P_4 might also be a physiologically relevant ligand. The potential for these soluble inositol phosphates to inhibit the ability of the PH domain to mediate protein/phospholipid interactions could be the basis of a quite widespread effect upon a form of protein traffic, i.e., the transfer of proteins between the plasma membrane and the cytosol.

3.4. Synthesis and Metabolism of Ins(3,4,5,6)P_4

In contrast to the limited progress in understanding the significance of Ins(1,4,5,6)P_4 (Section 3.3), the studies of its enantiomer, Ins(3,4,5,6)P_4, have been more illuminating. Levels of Ins(3,4,5,6)P_4 in unstimulated cells are approximately 1 μM (Pittet *et al.*, 1989; Vajanaphanich *et al.*, 1994). A number of groups have shown that levels of Ins(3,4,5,6)P_4 rise severalfold in response to activation of phospholipase C (Stephens *et al.*, 1988b; Balla *et al.*, 1989; Menniti *et al.*, 1990; Barker *et al.*, 1992). Menniti *et al.* (1990) were the first to provide significant evidence that Ins(3,4,5,6)P_4 was formed directly from Ins(1,3,4,5,6)P_5. This hypothesis was later consolidated by the demonstration of an Ins(1,3,4,5,6)P_5 1-phosphatase (Oliver *et al.*, 1992). Since an Ins(3,4,5,6)P_4

1-kinase had already been found in a number of cell types (Stephens *et al.*, 1988c; Balla *et al.*, 1989; McConnell *et al.*, 1991), it seems that Ins(3,4,5,6)P_4 and Ins(1,3,4,5,6)P_5 are participants in a 1-kinase/1-phosphatase substrate cycle (Figure 1).

The development of this concept was accompanied by the observation that the increase in cellular levels of Ins(3,4,5,6)P_4 was sometimes paralleled by a decrease in levels of Ins(1,3,4,5,6)P_5 (Menniti *et al.*, 1990). Therefore, the receptor-dependent increase in the metabolic flux from Ins(1,3,4,5,6)P_5 to Ins(3,4,5,6)P_4 was clearly not a mass action effect. On the contrary, levels of Ins(3,4,5,6)P_4 were proposed (Menniti *et al.*, 1990) to be elevated directly through the intervention of receptor-regulated processes in the activity of the Ins(3,4,5,6)P_4 1-kinase/Ins(1,3,4,5,6)P_5 1-phosphatase substrate cycle. Inhibition of the Ins(3,4,5,6)P_4 1-kinase by Ins(1,3,4)P_3 (Craxton *et al.*, 1994) may provide one means of elevating cellular Ins(3,4,5,6)P_4 levels. This would certainly help explain why persistent activation of any isoform of phospholipase C seems inevitably to be accompanied by elevations in levels of Ins(3,4,5,6)P_4 (see Menniti *et al.*, 1993b). Yet the demonstration that the 1-phosphatase activity in crude cell homogenates requires ATP (Oliver *et al.*, 1992) is suggestive of additional regulatory processes such as protein phosphorylation. Furthermore, in T_{84} colonic epithelial cells, activation of protein kinase C may increase the rate of dephosphorylation of Ins(1,3,4,5,6)P_5 to Ins(3,4,5,6)P_4 (Traynor-Kaplan *et al.*, 1994). However, in other cell types there is no evidence for this pathway being regulated by protein kinase C (Menniti *et al.*, 1990; Glennon and Shears, 1993).

In suggesting what might be the most important physiological consequence of a receptor-regulated breakdown of Ins(1,3,4,5,6)P_5 to Ins(3,4,5,6)P_4, Menniti *et al.*, (1990). pointed out that the severalfold changes in the relative small pool of Ins(3,4,5,6)P_4 greatly exceeded the slight impact on the relatively large pool of Ins(1,3,4,5,6)P_5. Therefore, it seemed most profitable to search for a physiological effect of Ins(3,4,5,6)P_4.

The metabolic considerations described above have been key elements in the elevation of Ins(3,4,5,6)P_4 to the status of a putative intracellular signal (Menniti *et al.*, 1993b). Clues to the characteristics of a target cellular process might lie in the time course by which agonists elevate Ins(3,4,5,6)P_4 (Pittet *et al.*, 1989; Balla *et al.*, 1989; Menniti *et al.*, 1990; Barker *et al.*, 1992): In general, levels of Ins(3,4,5,6)P_4 increase relatively slowly, particularly in comparison to the large changes in levels of Ins(1,4,5)P_3 and Ins(1,3,4,5)P_4 that occur within seconds of receptor activation. Ins(3,4,5,6)P_4 levels also progressively increase for up to 1 hr of cell stimulation. Thus, if Ins(3,4,5,6)P_4 were to regulate a physiological process, such regulation might be expected to develop relatively slowly after the application of an extracellular agonist, but would nevertheless be sustained. There has also been a determination of the rate at which various inositol phosphates returned to basal levels in AR4-2J cells after termination of the stimulation by bombesin (F. S. Menniti, J. W. Putney, Jr., and S. B. Shears, unpublished data). Levels of Ins(1,4,5)P_3 and Ins(1,3,4,5)P_4 returned to baseline relatively quickly ($t_{1/2} < 1$ min); levels of Ins(1,3,4,6)P_4 decayed with a $t_{1/2}$ of approximately 10 min. In these experiments, Ins(3,4,5,6)P_4 was the inositol phosphate that most slowly returned to the prestimulated baseline ($t_{1/2}$ approximately 50 min). Thus, it was anticipated that the regulation of a physiological activity by Ins(3,4,5,6)P_4 would reverse relatively slowly after displacement of the agonist from its receptor.

3.5. Evidence for Physiological Effects of Ins(3,4,5,6)P_4

3.5.1. Inhibition of Ins(1,3,4)P_3 6-kinase

Ins(3,4,5,6)P_4 is a potent, stereospecific inhibitor of Ins(1,3,4)P_3 6-kinase (Hughes *et al.*, 1989, 1994; Vajanaphanich *et al.*, 1994). The relevance of this inhibition is unclear; it is just one of several constraints placed upon the activity of the Ins(1,3,4)P_3 kinase, at least *in vitro*. Even in the absence of this particular restriction, Ins(1,3,4)P_3 phosphorylation would still be limited by product inhibition and the inherently low capacity of the enzyme (Section 2).

It may seem a remarkable coincidence that not only does Ins(3,4,5,6)P_4 potently inhibit the Ins(1,3,4)P_3 kinase, but Ins(1,3,4)P_3 also potently inhibits the Ins(3,4,5,6)P_4 kinase (Craxton *et al.*, 1994). However, a comparison of the structure of Ins(3,4,5,6)P_4 with "inverted" Ins(1,3,4)P_3 reveals a number of similarities (Figure 2). Both "ends" of each molecule may then be considered to be structurally equivalent and to have the same ring pucker. These similarities may be what permits reciprocal access of inhibitor to the substrate-binding site of each enzyme. The differences between these two molecules is embodied only in the orientation and composition of the substituents on the 2- and 5-carbons (Figure 2), although this is sufficient to ensure that the binding of each inhibitor is non-productive (Z. Tan, P. Enyedi, and S. B. Shears, unpublished data). It seems likely that we will find some important similarities in the active sites of these two enzymes.

3.5.2. Ins(3,4,5,6)P_4 and Ca^{2+} Mobilization

A racemic mixture of Ins(3,4,5,6)P_4 and Ins(1,4,5,6)P_4 can bind weakly to the Ins(1,4,5)P_3 receptor (Gawler *et al.*, 1991; Challiss *et al.*, 1991). However, such binding occurs without the release of significant amounts of Ca^{2+}, and it does not antagonize Ca^{2+}-mobilization due to Ins(1,4,5)P_3 (Cullen *et al.*, 1990; Gawler *et al.*, 1991; DeLisle *et al.*, 1994).

"Ins(3,4,5,6)P_4", at 10 μM, has been reported to promote Ca^{2+} entry into

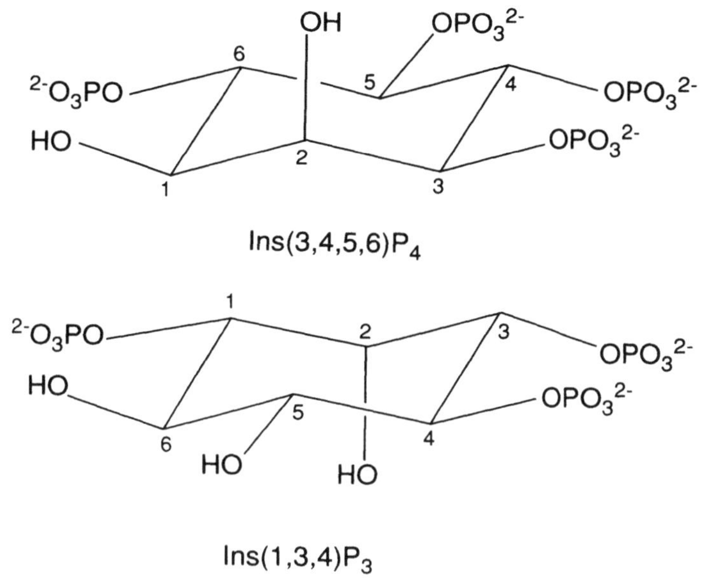

FIGURE 2. Comparison of the structures of Ins(3,4,5,6)P_4 and "inverted" Ins(1,3,4)P_3.

cells following its micoinjection into NIH-3T3 fibroblasts, (Hashii *et al.*, 1994a,b). Actually, these effects were obtained in response to the injection of a racemic mixture of Ins(3,4,5,6)P_4 plus Ins(1,4,5,6)P_4, so either enantiomer could have been responsible. Indeed, when Ins(3,4,5,6)P_4 and Ins(1,4,5,6)P_4 were separately tested for their ability to mobilize Ca^{2+} from either rat brain microsomes (Lu *et al.*, 1994) or *Xenopus* oocytes (DeLisle *et al.*, 1994), Ins(1,4,5,6)P_4 was found to be a weak agonist (100-fold less potent than Ins(1,4,5)P_3) and Ins(3,4,5,6)P_4 was entirely without effect.

Another problem with the experiments of Hashii *et al.* (1994a,b) is that 10 μM of either Ins(1,3,4,5)P_4, Ins(1,3,4,6)P_4 or Ins(1,2,5,6)P_4 also enhanced Ca^{2+} entry. There was also very little difference between InsP_4 isomers in the size of the cellular Ca^{2+} signal that they each initiated. Is it really possible that the effects of these various inositol phosphates arise from cross-reactivity with a genuine receptor, such as the plasma-membrane "Ins(1,3,4,5)P_4-receptor" that might gate Ca^{2+} entry (Irvine, 1992; Cullen *et al.*, 1995a,b)?

Unfortunately, the best candidate protein for this putative receptor binds Ins(3,4,5,6)P_4 130-fold more weakly than Ins(1,3,4,5)P_4 (Cullen *et al.*, 1995a). Other proteins for which Ins(1,3,4,5)P_4 is a strong ligand also bind Ins(3,4,5,6)P_4 and Ins(1,4,5,6)P_4 20- to 200-fold less avidly (Donié *et al.*, 1990; Challiss *et al.*, 1991; Theibert *et al.*, 1992). What seems to be missing from the data of Hashii *et al.* (1994a,b) are factors that would normally be expected to impart some specificity to a physiological effect of inositol phosphates. There fore, we should be cautious in interpreting what their results mean in relation to the mechanism of Ca^{2+} entry in NIH 3T3 cells.

3.5.3. Ins(3,4,5,6)P_4 and Ca^{2+}-Dependent Cl^- Secretion

The regulation of transcellular Cl^- fluxes by Ca^{2+} and by cyclic AMP is pivotal to the control of salt and fluid secretion, pH balance, neurotransmission, osmoregulation, and volume-dependent metabolic effects (Korn *et al.*, 1991; Häussinger and Lang, 1992; Nauntofte, 1992; Petersen, 1992). The actions of Ca^{2+} can be mediated on Cl^- channels through Ca^{2+}/calmodulin-dependent kinase II (Worrell and Frizzell, 1991). However, net Cl^- secretion also requires the coordinated activity of other ion-transporting proteins: the electrochemical driving force behind apical Cl^- efflux can be increased by Ca^{2+}-stimulated K^+ efflux across the basolateral membrane. Additional Cl^- is subsequently transported into the cell through the basolateral membrane by a Na^+–K^+–$2Cl^-$ cotransporter (Figure 3; Nauntofte, 1992; Petersen, 1992).

Some years ago, Morris *et al.* (1990) described a study of Ca^{2+}-activated Cl^- conductances in a colonic epithelial cell line, HT-29. These authors obtained whole-cell current recordings using the perforated patch technique. A significant technical point is that this methodology prevents the "rundown" of receptor responses that can result from loss of cytosolic activators during whole-cell patch-clamp experiments (Horn and Marty, 1988; Marty and Zimmerberg, 1989). Morris *et al.* (1990) observed that there was a rapid fall in the size of the Cl^- current following particularly large agonist-mediated rises in intracellular [Ca^{2+}], but not when similar changes in [Ca^{2+}] were induced by ionophores. They postulated that activation of phospholipase C promoted a feedback inhibitory process to protect cell volume and ionic composition from the consequences of large Cl^- conductance changes. It as the pursuit of this topic that ultimately uncovered evidence that Ins(3,4,5,6)P_4 may play an important role.

Support for the possibility of a phospholipase-C-linked inhibitory regulator of Ca^{2+}-dependent Cl^- secretion was obtained from experiments with T_{84} cells, another colonic epithelial cell line (Kachintorn *et al.*, 1993). In the latter study, Ca^{2+}-dependent Cl^- secretion across epithelial layers was measured in Ussing chambers. Cytosolic [Ca^{2+}] was elevated either by stimulating T_{84} cells with a phospholipase-C-linked agonist such as carbachol or by adding thapsigargin to inhibit the Ca^{2+}-ATPase of endoplasmic reticulum. It emerged that carbachol elicited a biphasic effect. This agonist initially stimulated Cl^- secretion. However, carbachol strongly inhibited the ability of a subsequent elevation in cytosolic [Ca^{2+}] to also activate Cl^- secretion, irrespective of whether this later mobilization of Ca^{2+} was brought about by histamine or thapsigargin (Kachintorn *et al.*, 1993; Vajanaphanich *et al.*, 1994). Importantly, carbachol did not alter the ability of either thapsigargin or histamine to raise cytosolic Ca^{2+}

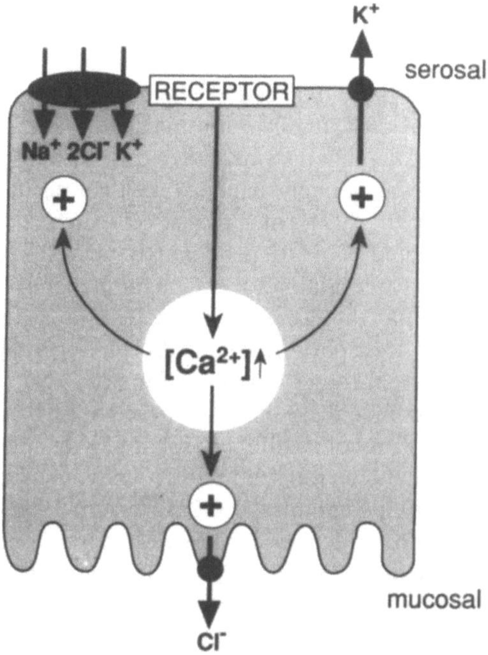

FIGURE 3. Calcium-stimulated Cl^- transport across T_{84} epithelial cells. The figure focuses on possible sites of action of Ca^{2+} during its stimulation of Cl^- transport across T_{84} epithelial cells. The model is applicable to other cell types, but net Cl^- secretion requires the participation of additional ion-transport systems that are not shown here. See Nauntofte (1992) and Petersen (1992) for further details.

(Dharmsathaphorn *et al.*, 1989; Kachintorn *et al.*, 1993). Even 20 min after carbachol is washed off T_{84} cells, the Cl^- secretory response to a second dose of this agonist is blunted by 50% (McEwan *et al.*, 1994.)

Attention has focused on the possibility that an inositol phosphate might mediate inhibition of Ca^{2+}-dependent Cl^- secretion. The time course for the development of the carbachol-dependent inhibition was paralleled by increases in several inositol phosphates, including Ins(3,4,5,6)P_4 (Vajanaphanich *et al.*, 1994). More strikingly, when carbachol's effects were reversed with atropine, the inhibition of Cl^- secretion persisted for more than 15 min; at this time, Ins(3,4,5,6)P_4 was the only inositol phosphate whose concentration remained elevated above basal levels (Vajanaphanich *et al.*, 1994). Thus, levels of Ins(3,4,5,6)P_4 decayed relatively slowly after removal of the agonist from the receptor, just as in the experiments with AR4-2J cells (Section 3.4).

This correlation between levels of Ins(3,4,5,6)P_4 and the inhibition of Ca^{2+}-dependent Cl^- secretion was pursued more directly using a novel cell-permeant analogue of Ins(3,4,5,6)P_4. This compound has the hydroxyl groups hidden by butyrates, and the phosphates are masked by acetoxymethyl esters (Vajanaphanich *et al.*, 1994). This enables the analogue to diffuse across the plasma membrane and into the cell, whereupon esterases liberate free Ins(3,4,5,6)P_4. Treatment of T_{84} cells with this analogue did not affect cellular Ca^{2+} pools, nor did it alter agonist-mediated Ca^{2+} fluxes, but thapsigargin-dependent Cl^- secretion was strongly inhibited; cell-permeant Ins(1,4,5,6)P_4 had no effect (Vajanaphanic *et al.*, 1994). In addition, cell-permeant Ins(3,4,5,6)P_4 also inhibited histamine-activated, Ca^{2+}-dependent Cl^- secretion. The action of this InsP_4 can be considered to be a novel mechanism of heterologous desensitization. Inositol phosphates may therefore be considered to have two opposing effects on Ca^{2+}-dependent Cl^- transport: Ins(1,4,5)P_3 [and perhaps Ins(1,3,4,5)P_4] stimulate Cl^- secretion through mobilization of Ca^{2+} stores, whereas Ins(3,4,5,6)P_4 may antagonize these effects. These positive and negative actions add to the functional versatility of inositol phosphates in the regulation of the many cellular processes that depend on transcellular Cl^- movements (Korn *et al.*, 1991; Häussinger and Lang, 1992; Nauntofte, 1992; Petersen, 1992).

Some important future directions for this research will be to understand the mechanism of this effect of Ins(3,4,5,6)P_4 and to determine how widespread it is. To date, all we can say on this subject with some conviction is that the particular type of Ca^{2+}-activated Cl^- channel(s) in *Xenopus* oocytes is unaffected by Ins(3,4,5,6)P_4 (Sasakawa *et al.*, 1994). Perhaps Ins(3,4,5,6)P_4 can control an ion transporter through an intermediary. Two regulatory proteins that impact on Ca^{2+}-dependent Cl^- channels are Ca^{2+}/calmodulin-dependent kinase II and annexin IV (Worrell and Frizzell, 1991; Chan *et al.*, 1994); could these be targets for Ins(3,4,5,6)P_4?

Several ion-transporting processes contribute to Ca^{2+}-dependent Cl^- secretion (Figure 3). It is possible that the activities of any or all of these ion carriers might be directly affected by Ins(3,4,5,6)P_4, which may therefore have far-reaching physiological consequences beyond down-regulation of Cl^- secretion. A K^+ conductance in macrophages has been reported by Wu *et al.* (1991) to be modified by "Ins(3,4,5,6)P_4" (actually the mixture of Ins(3,4,5,6)P_4 plus Ins(1,4,5,6)P_4 that is commercially available; see Section 3.1). However, the importance of this effect is difficult to determine, because Wu *et al.* (1991) also observed that the K^+ conductance was similarly affected by every other inositol phosphate that they tested: Ins(1,4)P_2, Ins(4,5)P_2, Ins(1,4,5)P_3, Ins(1,3,4)P_3, Ins(1,3,4,5)P_4. This lack of specificity should, at the very least, alert us to the possibility of misleading results as we search for specific, physiologically relevant effects of inositol phosphates on ion transporters.

As well as regulating the levels of inositol phosphates, agonists that act through phospholipase C also activate protein kinase C (Berridge and Irvine, 1989). There are circumstances in which protein kinase C might itself contribute to an agonist-dependent inhibition of Ca^{2+}-dependent Cl^- secretion in T_{84} cells.

For example, activation of protein kinase C by phorbol esters can attenuate agonist-dependent Ca^{2+} release (Kachintorn *et al.*, 1992). Protein kinase C may also act downstream of Ca^{2+} to inhibit both thapsigargin-mediated and ionomycin-mediated Cl^- secretion (Kachintorn *et al.*, 1992; Traynor-Kaplan *et al.*, 1994). However, the latter effect may at least partly be ultimately caused by Ins(3,4,5,6)P_4, levels of which are increased by treatment of T_{84} cells with phorbol esters (Traynor-Kaplan *et al.*, 1994). Furthermore, inhibition of protein kinase C by staurosporine did not prevent carbachol from attenuating thapsigargin-mediated Cl^- secretion (Traynor-Kaplan *et al.*, 1994). Thus, carbachol can act independently of protein kinase C to raise levels of Ins(3,4,5,6)P_4 and inhibit Cl^- secretion.

4. METABOLISM AND FUNCTIONS OF InsP_6

4.1. Early Thoughts on InsP_6

Until recently it was work with plants and microorganisms that had generated most of our information on InsP_6, where it is considered to be a store of both inositol and phosphate. Indeed, 50–80% of plant seed total phosphorus may be in the form of InsP_6–cation complexes, deposited as globular inclusions in membrane-bound storage bodies (for reviews, see Gibson and Ullah, 1990; Raboy, 1990). Biswas *et al.* (1978) proposed that, in plants at least, the 2-phosphate of InsP_6 could be transferred directly to ADP through the reversibility of the Ins(1,3,4,5,6)P_5 2-kinase; this idea was recently revived by the determination that the value of the equilibrium constant for this enzyme in soybean seeds was 14 (Phillippy *et al.*, 1994). Another location for high amounts of InsP_6 is the two apical cells of the 28-cell mesozoan dispersal larvae, *Dicyema typus;* it has been suggested that the ensuing negative buoyancy that these InsP_6 deposits impart may usefully keep this organism close to the sea bottom (Lapan, 1975).

Initially, InsP_6 was erroneously thought to be the major inositol polyphosphate in erythrocytes of birds and turtles (Rapoport, 1940; Rapoport and Guest, 1941); it was nearly 30 years later that the compound in these cells was found to be Ins(1,3,4,5,6)P_5 (Johnson and Tate, 1969), which is recognized to have an important regulatory role in the reduction of hemoglobin's affinity for oxygen birds, air-breathing fish, and possibly turtles (Issacks and Harkness, 1980). This specialized action of Ins(1,3,4,5,6)P_5 is another example of the functional versatility of inositol polyphosphates.

Although the first reports of InsP_6 being present in animals was incorrect (see above), it subsequently emerged that InsP_6 really is widely distributed throughout the animal kingdom (Heslop *et al.*, 1985; Vallejo *et al.*, 1987; Michell *et al.*, 1988). There were early reports that exogenous InsP_6 perturbed cellular Ca^{2+} fluxes (Nicoletti *et al.*, 1989; Sortino *et al.*, 1990). However, these data have been criticized as being experimental artifacts that arise from solubility problems and the ion-chelating prowess of InsP_6 (Mitchell *et al.*, 1991). Hanley and colleagues attributed another extracellular activity to InsP_6 when they proposed that it was a neurotransmitter (Vallejo *et al.*, 1987). These workers came to this conclusion after injecting InsP_6 into the nucleus tractus solitarius of rat brain, whereupon there was a decrease in both arterial pressure and heart rate. Since then, some independent workers have confirmed these observations (Barraco *et al.*, 1989), but others have been unable to (Brooks and Spyer, 1989). In any case, the ion-chelating properties of InsP_6 now seem likely to be responsible for those "neurotransmitter"-like effects of InsP_6 that have been observed (Smit and Dürmüller, 1990; Sun *et al.*, 1992). The increasing general acceptance of the latter explanation recently prompted Hanley to agree that Ins(1,3,4,5,6)P_5 and InsP_6 are ". . . unlikely to function physiologically in neuronal communication" (Sasakawa *et al.*, 1995). Nevertheless, he and his colleagues still leave open the possibility that the release of these polyphosphates from damaged cells may contribute to neuronal pathology (Sasakawa *et al.*, 1995).

Our appreciation of the nature of InsP_6 metabolism in animal cells has also changed considerably over the years. Research into this topic can be traced back nearly 90 years to a report that this polyphosphate was dephosphorylated by calf liver (McCollum and Hart, 1908). However, most of the early studies focused upon the digestive tract. Intestinal epithelial cells possess a battery of relatively nonspecific phosphatases that can digest InsP_6 to release dietary inositol for absorption (Moore and Veum, 1983; Rao and Ramakrishnan, 1985). Some of these enzyme activities seem to be located at the cell surface (Rubiera *et al.*, 1988; Vucenik and Shamsuddin, 1994). As for intracellular InsP_6 turnover, in general the slow rate of incorporation of exogenous [^{3}H]inositol into cellular InsP_6 generated the misleading impression that this polyphosphate was metabolically sluggish (e.g., Michell *et al.*, 1988). The perception of slow InsP_6 turnover inside cells led to the extrapolation to the animal kingdom of the idea that it might be a store of inositol and phosphate (Irvine, 1988; Berridge and Irvine, 1989). Another proposed role for InsP_6 that does not depend on a high turnover is that of an antioxidant (Graf *et al.*, 1987). Hawkins *et al.* (1993) have extended this hypothesis, by demonstrating that the remarkable affinity of InsP_6 for iron totally inhibited this metal's ability to catalyze the formation of hydroxyl radicals. These authors therefore suggested that a physiological function of InsP_6 was to transport iron within the cell in a form that protects against the potentially lethal consequences of free radical formation (Hawkins *et al.*, 1993). These are all entirely feasible roles for InsP_6—a phosphate store, an antioxidant and a chelator of iron—and if correct this would, by itself, be testimony to the flexibility with which nature has utilized this polyphosphate. Nevertheless, what has finally invigorated an interest in InsP_6 having additional effect that are much more dynamic are the demonstrations that this polyphosphate turns over rapidly

(Sections 4.3 and 4.4) and the uncovering of a family of InsP_6-binding proteins (Section 4.5).

Partly because of the crystalline deposits of InsP_6 in plants, it has been argued that, in animal cells too, InsP_6 is likely to be only partly soluble in the physiological ionic milieu (Irvine 1988; Stuart *et al.*, 1994). This concern has raised the possibility that some cellular InsP_6 may be compartmentalized. Stuart *et al.* (1994) recently studied InsP_5 and InsP_6 compartmentalization directly. These workers specifically permeabilized the plasma membrane of [^{3}H]inositol-labeled cells using several different techniques, and they subsequently monitored the rate of efflux of inositol phosphates in relation to cytosolic cellular makers. In the presence of cation chelators, and provided pores were created that were large enough for [^{3}H]InsP_5 and [^{3}H]InsP_6 to pass through, these polyphosphates were lost from the cell at the same rate as other intracellular markers (Stuart *et al.*, 1994). This result indicated that there could not be a substantial intravesicular pool of inositol phosphates. However, one caveat should be mentioned. Permeabilization, by whatever technique, introduces an enormous dilution of the cell interior. The amount of any InsP_6 that might be enclosed in a vesicle could be expected to be in equilibrium with the free cytosolic concentration of InsP_6. An instantaneous dilution of cytosolic InsP_6 could then perturb this equilibrium and thereby lead to a loss of this polyphosphate from its putative vesicular compartment; such a redistribution of InsP_6 would, however, need to be very rapid to be accommodated by the data of Stuart *et al.* (1994).

Even if cellular InsP_6 is not enclosed within a vesicular compartment, it might still not all be free in the cytosol. When Stuart *et al.* (1994) omitted the cation chelators from their media, InsP_5 and InsP_6 were then released from the permeabilized cells more slowly. This result is consistent with the observation (Poyner *et al.*, 1993) that, *in vitro*, cellular membranes have the capacity to bind significant amounts of InsP_5 and InsP_6 in a cation-dependent manner. If this occurs *in vivo*, it is likely to have a profound effect on the physicochemical nature of the interface between the membrane and the cytosol. This potential for cellular membranes to be "wallpapered" with InsP_5 and InsP_6 creates considerable uncertainty with respect to their free, cytosolic concentration *in vivo* and, by extrapolation, that of the diphosphoinositol polyphosphates (Section 4.3).

Even though the above discussion has shown that there are fundamental questions concerning InsP_6 that we still cannot answer—and more problems are outlined below—we can at least conclude this section of the review with the reassurance that we have corrected some mistakes and misconceptions that surrounded the early opinions concerning the presence of InsP_6 in animal cells.

4.2. Synthesis of InsP_6

Mammalian cells contain several InsP_5 kinase activities: Ins(1,3,4,5,6)P_5 2-kinase, Ins(1,2,3,4,6)P_5 5-kinase, Ins(1,2,4,5,6)P_5 3-kinase, and Ins(2,3,4,5,6)P_5

1-kinase (Ji *et al.*, 1989; Guse and Emmrich, 1991: Stephens *et al.*, 1991; Craxton *et al.*, 1994). These InsP_5 substrates have all been found inside animal cells, although Ins(1,3,4,5,6)P_5 almost always accounts for 95% or more of the total InsP_5 complement (Phillippy and Bland, 1988; Mayr, 1988; Menniti *et al.*, 1990; Mattingly *et al.*, 1991; Nogimori *et al.*, 1991; Stephens *et al.*, 1991; Wong *et al.*, 1992; Glennon and Shears, 1993). Each of the InsP_5 isomers that can be phosphorylated by these various kinase activities can also be formed by nonspecific dephosphorylation of preformed InsP_6 (Section 4.3). Therefore, a fundamental problem is to discriminate the InsP_5 isomer(s) that is a precursor in the *de novo* pathway of InsP_6 synthesis from those other InsP_5 isomers that participate in substrate cycles with InsP_6. Aside from that difficulty, there is the problem of determining the actual significance of this puzzling array of substrate cycles. There is also a mystery surrounding the nature of the presumed defect in InsP_5 metabolism in those cells that contain abnormally high levels of Ins(1,2,4,5,6)P_5 and/or Ins(2,3,4,5,6)P_5, i.e., T5-1 B lymphocytes (McConnell *et al.*, 1991) and some clones of Jurkat T lymphocytes (Guse and Emmrich, 1991)—other clones of Jurkat T cells contain the more usual, low levels of these particular InsP_5 isomers (da Silva *et al.*, 1994; Ali *et al.*, 1995b).

The *de novo* pathway of InsP_6 synthesis has been delineated in *Dictyostelium* (Ins → Ins3P → Ins(3,6)P_2 → Ins(3,4,6)P_3 → Ins(1,3,4,6)P_4 → Ins(1,3,4,5,6)P_5 → InsP_6, Stephens and Irvine, 1990). However, there is no evidence that this route occurs in any other organism. Because the reactions that lead to *de novo* synthesis of Ins(1,3,4,5,6)P_5 have been identified (Section 2), and because it is known that Ins(1,3,4,5,6)P_5 can be phosphorylated to InsP_6 by animal cells (Ji *et al.*, 1989; Stephens *et al.*, 1991), there is some justification for the pathway of InsP_6 synthesis that is outlined in Figure 1.

4.3. Dephosphorylation of InsP_6

The minor InsP_5 isomers of animal cells [i.e., all except Ins(1,3,4,5,6)P_5] are probably formed by nonspecific dephosphorylation of InsP_6 by a multiple inositol polyphosphate phosphatase (MIPP) (Nogimori *et al.*, 1991; Craxton *et al.*, 1995). *In vitro* at least, MIPP is sufficiently nonspecific that it can dephosphorylate InsP_6 and several InsP_5 isomers at least as far as InsP_3 (Nogimori *et al.*, 1991; Ali *et al.*, 1995b). It is possible that a MIPP-like enzyme is responsible for the dephosphorylation of InsP_6 by cell-free extracts of the mammary tumor WRK-1 cell line, one of the downstream metabolites of which is Ins(1,2,3)P_3 (Barker *et al.*, 1995). Some other isomers of inositol phosphates that contain a 2-phosphate have also been found to be present inside mammalian cells, i.e, Ins(1,2,3,4)P_4 and/or Ins(1,2,3,6)P_4 (Menniti *et al.*, 1990), Ins(1,2,6)P_3 and/or Ins(2,3,4)P_3 (McConnell *et al.*, 1992), and Ins(1,2)P_2/Ins(2,3)P_2 (Barker *et al.*, 1995). These are probably additional metabolites of InsP_6. It would be informative from an evolutionary standpoint to compare the relationship of MIPP to the phytases isolated from wheat and *Paramecium*, which can also dephosphorylate InsP_6 to Ins(1,2,3)P_3 (Tomlinson and Ballou, 1962; Lim and Tate, 1973; Freund *et al.*, 1992).

Levels of Ins(1,2,3)P_3 were found to increase transiently at the late G_1 phase of the cell cycle (Barker *et al.*, 1995), which is indicative of increased InsP_6 metabolism. Changes in levels of Ins(1,3,4,5,6)P_5 and InsP_6 have also been observed to accompany cell growth and differentiation in some case (French *et al.*, 1991; Guse *et al.*, 1993; Balla *et al.*, 1994). The possible significance of these effects is discussed in Section 4.5. A MIPP is a candidate for mediating these long-term changes in levels of inositol polyphosphates. One possible mechanism for altering the levels of Ins(1,3,4,5,6)P_5 and InsP_6 could be by regulation of their uptake into endoplasmic reticulum, where MIPP is located (Ali *et al.*, 1993). An alternative possibility is that the intracellular distribution of MIPP might be controlled. This could be somewhat facilitated by MIPP apparently being a soluble, luminal enzyme (Ali *et al.*, 1993; Craxton *et al.*, 1995). Such proteins are normally exported from the endoplasmic reticulum in the bulk flow, only to be retrieved by a vesicular compartment that recognizes a four-amino-acid retention signal at the protein's carboxy terminus (Pelham, 1989). Perhaps there is a mechanism for regulating the extent to which MIPP can escape this retrieval process and gain access to its substrates.

MIPP is also present in plasma membranes of mature mammalian erythrocytes, even though these cells do not synthesize inositol phosphates (Doughney *et al.*, 1988; Estrada-Garcia *et al.*, 1991; Craxton *et al.*, 1995). There may be a good reason for MIPP activity outliving the ability of these cells to produce the enzyme's substrates: Ins(1,3,4,5,6)P_5 and possibly InsP_6 seem likely to accumulate during erythropoiesis through the pathways described in Figure 1 because erythropoietic cells express receptor-regulated phospholipase C activity (Ren *et al.*, 1994). However, if inositol polyphosphates were to be retained by the mature erythrocyte, they would then bind tightly to hemoglobin and impair the normal regulation of its affinity for oxygen by 2,3-diphosphoglycerate (Arnone and Perutz, 1974). The persistence of MIPP as the developing erythrocyte attains maturity may be part of the process by which the removal of any residual Ins(1,3,4,5,6)P_5 and InsP_6 can be assured.

4.4. Synthesis and Metabolism of Diphosphoinositol Polyphosphates

Diphosphoinositol polyphosphates ("inositol pyrophosphates") recently added yet another new dimension to the structural versatility of inositol derivatives. These compounds were first characterized in *Entamoeba histolytica* (Martin *et al.*, 1993), *Dictyostelium* (Stephens *et al.*, 1993) and AR4-2J pancreatoma cells (Menniti *et al.*, 1993a), although several preceding studies had foretold the presence of metabolites of inositol that had greater polarity than InsP_6 (Europe-Finner *et al.*, 1989; Stephens *et al.*, 1991; Wong *et al.*, 1992; Oliver *et al.*, 1992). Two groups of InsP_6 derivatives have received particular attention. These are now well-known colloquially as "InsP_7" and "InsP_8" (e.g., see Hughes and Michell, 1993), but their more formal, albeit less readily articulated nomenclatures are, respectively, diphosphoinositol pentakisphosphate (PP-InsP_5) and bis-diphosphoinositol tetrakisphosphate ([PP]$_2$-InsP_4).* In the case of the PP-InsP_5 extracted from *Entamoeba histolytica,* NMR analysis has indicated that the diphosphate is attached to the 5-position (Martin *et al.*, 1993). A particularly rigorous structural characterization of the compounds isolated from *Dictyostelium* indicated that a different isomer of PP-InsP_5 was present. Here, the diphosphate was in the 1/3-position (Stephens *et al.*, 1993). The two diphosphates on [PP]$_2$-InsP_4 were determined to be attached to the 1/3-position and the 4/6 position (Stephens *et al.*, 1993). The major PP-InsP_5 made by a mammalian tissue—rabbit brain—has now recently been found to co-chromatograph on HPLC with the PP-InsP_5 from *Dictyostelium* (G. Mayr, M. Bembenek, and S. Shears, unpublished data). Although such data indicate that the same isomer of PP-InsP_5 is present in both organisms, there is still evidence for there being further structural diversity to this group of compounds *in vitro,* since both rat liver and *Saccharomyces* possess the enzymatic capacity to synthesize a second isomer of PP-InsP_5 (Ali *et al.*, 1995a; Shears *et al.*, 1995). Ins(1,3,4,5,6)P_5 is also converted to yet another diphosphate derivative (PP-InsP_4) (Menniti *et al.*, 1993a; Glennon and Shears, 1993).

An approximate estimate of the total levels of diphosphoinositol polyphosphates in cells can be obtained by labeling them *in situ* with [^{3}H]inositol, because PP-InsP_5 and [PP]$_2$-InsP_4 seem to be in isotopic equilibrium with InsP_6 (Menniti *et al.*, 1993a; Glennon and Shears, 1993). The levels of PP-[^{3}H]InsP_5 and [PP]$_2$-[^{3}H]InsP_4 are 0.5 to 5% of the level of [^{3}H]InsP_6 (Stephens *et al.*, 1993; Menniti *et al.*, 1993a; Glennon and Shears, 1993; Shears *et al.*, 1995). Estimates of the total cellular concentration of InsP_6 range from 5–100 μM (Szwergold *et al.*, 1987; French *et al.*, 1991; Bunce *et al.*, 1993). This puts the levels of PP-InsP_5 and [PP]$_2$-InsP_4 in the low micromolar range at most; the estimated levels of PP-InsP_4, based on a comparison with Ins(1,3,4,5,6)P_5, are in the same range (Menniti *et al.*, 1993a; Glennon and Shears, 1993).

We do not know if the equivocal debate over compartmentation of InsP_6 (Section 4.1) might also be relevant to the diphosphoinositol polyphosphates. However, interest in the latter compounds rises above uncertainty over their intracellular distribution. For example, there is considerable fascination with

*The original nomenclature that Menniti *et al.* (1993a) used for the diphosphoinositol polyphosphates was "InsP_5P" and "InsP_6P"; these were later changed to PP-InsP_4 and PP-InsP_5, respectively. "InsP_6X," the compound that Menniti *et al.* (1993a) were unable to characterize, was later shown to be [PP]$_2$-InsP_4 (sometimes written as PP-InsP_4-PP, Shears *et al.*, 1995).

their remarkably high turnover through substrate cycles (Figure 1). The latter information was obtained by exploiting the ability of F^- to inhibit the phosphatase(s) that attack PP-InsP_5 and $[PP]_2$-InsP_4 (Menniti *et al.*, 1993a; Shears *et al.*, 1995). By blocking the dephosphorylation of PP-InsP_5 and $[PP]_2$-InsP_4 *in situ*, and exposing their ongoing rate of synthesis, F^- caused the levels of these compounds to increase manyfold within minutes (Menniti *et al.*, 1993a; Glennon and Shears, 1993; Shears *et al.*, 1995). There was also an informative, corresponding drop in the levels of the Ins(1,3,4,5,6)P_5 and InsP_6 precursors that yielded new information on the rapid turnover of Ins(1,3,4,5,6)P_5 and InsP_6. It became clear that, every hour, the equivalent of 50% of the cellular pools of these two compounds was being cycled through the diphosphoinositol polyphosphates (Menniti *et al.*, 1993a; Shears *et al.*, 1995). One proposed explanation for this high turnover is that the diphosphate moiety can support a transphosphorylation reaction (Stephens *et al.*, 1993), or perhaps it presents the cell with the means of quickly regulating the local concentrations of these particular compounds. Understanding the reason for this ongoing metabolism is surely pivotal to the determination of their contributions to any physiological process.

Another interesting feature of the turnover of diphosphoinositol polyphosphates is its sensitivity to perturbation of cellular Ca^{2+} pools by thapsigargin. The latter specifically inhibits the Ca^{2+}-ATPase of endoplasmic reticulum, which depletes intraluminal Ca^{2+} stores, elevates cytosolic $[Ca^{2+}]$, and thereby promotes Ca^{2+} entry into the cell (see Putney, 1992). Thapsigargin also reduces steady-state levels of the diphosphoinositol polyphosphates by up to 50%, apparently by inhibiting their rates of synthesis (Glennon and Shears, 1993). A similar response was obtained when cells were treated with a structurally unrelated inhibitor of Ca^{2+}-ATPase inhibitor, cyclopiazonic acid (N. Ali and S. B. Shears, unpublished data). Therefore, it seems likely that the effect common to both the thapsigargin and cyclopiazonic acid, namely, changes in the status of cellular Ca^{2+} pools, is responsible for the inhibition of synthesis of the diphosphoinositol polyphosphates. This effect of an intracellular signal such a Ca^{2+} is further testimony to the importance of these polyphosphates, but its relevance has not yet become clear: the response is slow to develop (20–30 min, N. Ali and S. B. Shears, unpublished data). Moreover, mobilization of Ca^{2+} through the activation of extracellular receptors has, at best, only a marginal effect on levels of the diphosphoinositol polyphosphates (Menniti *et al.*, 1993a; Glennon and Shears, 1993; N. Ali and S. B. Shears, unpublished data)

4.5. Proteins that Bind InsP_6 with High Affinity: A Functionally Related Family?

A few years ago, it was found that InsP_6 bound with high affinity to the clathrin assembly protein AP-2 (Beck and Keen, 1991; Timerman *et al.*, 1992;

Voglmaier *et al.*, 1992). More proteins have since been found to share this characteristic, including coatomer (Fleischer *et al.*, 1994; Ali *et al.*, 1995a), AP-3 (Norris *et al.*, 1995; Ye *et al.*, 1995), and bovine retinal arrestin (Palczewski *et al.*, 1991). The absolute binding parameters *in vitro* were out of necessity determined under nonphysiological conditions; the absolute affinities *in vivo* may be affected by the ionic composition of the physiological milieu and other cellular factors. Thus, we should be cautious in extrapolating these absolute affinities to the intracellular environment. Nevertheless, with the exception of arrestin (see below), the K_d values for binding of InsP_6 were determined to be in the low nanomolar range. Another consistent feature of this binding is its specificity: The less phosphorylated the inositol polyphosphate ligand, the less avidly it binds (i.e., the rank order is InsP_6 > Ins(1,3,4,5,6)P_5 > Ins(1,3,4,5)P_4 > Ins(1,4,5)P_3; Beck and Keen, 1991; Palczewski *et al.*, 1991; Fleischer *et al.*, 1994; Norris *et al.*, 1995; Ali *et al.*, 1995a; Ye *et al.*, 1995). Ins(1,4,5)P_3 and Ins(1,3,4,5)P_4 are not only the weaker ligands, but they are also the least abundant of this group of compounds, so their binding seems unlikely to be generally physiologically significant. Whenever the relative affinities of InsP_6, PP-InsP_5 and [PP]$_2$-InsP_4 have been compared, i.e., for AP-2 (S. Fleischer and S. Shears, unpublished data), coatomer (Fleischer *et al.*, 1994; Ali *et al.*, 1995a) and AP-3 (Ye *et al.*, 1995), there has generally been less than a tenfold difference between these three polyphosphates. Because InsP_6 is present in such large excess this again seems likely to be the more physiologically relevant ligand.

It seems more than a coincidence that three distinct InsP_6-binding proteins, namely, AP-2, AP-3, and coatomer, all participate in the control of vesicle traffic: AP-2 regulates receptor-mediated endocytosis (Rothman, 1994); AP-3 is a synaptosomal-specific protein that helps control the synaptic vesicle lifecycle (Bennett and Scheller, 1994; Rothman, 1994). Coatomer is believed to regulate vesicular traffic though Golgi stacks and may also participate in the retrograde traffic from the intermediate vesicular compartment between the Golgi and endoplasmic reticulum (Pelham, 1989; Rothman, 1994).

At the very least, it will be interesting to search for polyphosphate-binding motifs in these various proteins. It is also of interest that vesicle trafficking could be the cellular process that proves a profitable focal point in our search for the physiological activities of InsP_6. This would bring the study of this molecule into a fundamental problem in cell biology, namely, understanding how vesicular transport targets proteins toward specific cellular sites. This is an essential component of secretion, synaptic transmission, accumulation of exogenous macromolecules, organelle biogenesis, and maintenance of cell polarity. If inositol polyphosphates contributed to the regulation of all these important processes, this would truly represent a remarkably versatile range of physiological effects.

Evidence that inositol polyphosphates affect vesicle traffic goes beyond the correlate of their binding to participating proteins. For example, the assembly of a clathrin coat around the endocytic vesicle is promoted by AP-2 and AP-3; the addition of low micromolar concentrations of InsP_6 inhibits the ability of these proteins to promote clathrin assembly (Beck and Keen, 1991; Norris *et al.*, 1995; Ye *et al.*, 1995). Thus, InsP_6 may be a constitutive inhibitor of endocytosis. It is not a new idea that some vesicle trafficking events are, in default, kept 'off.' For example, regulated exocytosis has been proposed to be constitutively restrained by so-called "fusion "clamps", which then need to be released by specific intracellular signals (see Rothman, 1994). It has been believed that "fusion clamps" would probably need to be proteins so as to impart specificity (Rothman, 1994). Perhaps we should now consider that inositol polyphosphates might be "clamps" for some aspects of vesicle trafficking. The degree of inhibition would then be related to its relative sensitivity to the prevailing free concentration of InsP_6. Note that total cellular levels of this polyphosphate (15–100 μM; Swergold *et al.*, 1987; Pittet *et al.*, 1989; Bunce *et al.*, 1993; Mountford *et al.*, 1994) are certainly sufficient to constitutively "clamp" the functions of AP-2 and AP-3 (Beck and Keen, 1991; Norris *et al.*, 1995; Ye *et al.*, 1995). However, there is some uncertainty as to the free, soluble concentration of InsP_6, not in the least because of the substantial potential of cellular membranes to bind this polyphosphate (Poyner *et al.*, 1993; see also Section 4.1). Furthermore, just as we must exercise some caution in a quantitative extrapolation to intact cells of InsP_6-binding parameters determined *in vitro* (see above), so we must be equally judicious with the absolute values we have determined for the potency with which InsP_6 inhibits clathrin assembly.

These complications prevent us from developing this hypothesis in a quantitative fashion. However, it is notable that AP-2 (Pauloin and Thurieau, 1993) and AP-3 (Murphy *et al.*, 1993) are both subject to phosphorylation; AP-3 is also glycosylated (Murphy *et al.*, 1993). Such observations provide enough latitude for speculating that the sensitivity of these adaptor proteins to InsP_6 might be controlled by covalent modification. These putative regulatory processes would then be protein-specific, just as Rothman (1994) perceived when the "fusion-clamp" hypothesis was initially proposed. Adaptor-protein function and vesicle traffic could also be regulated by changes in the free, soluble concentrations of inositol polyphosphates. Quite substantial changes in total levels of InsP_6 occur during the progression through the cell cycle or upon cellular differentiation (French *et al.*, 1991; Guse *et al.*, 1993; Balla *et al.*, 1994); perhaps this may be brought about in part by a redistribution of cellular MIPP (Section 4.2). If so, these changes in InsP_6 levels may have some relatively long-term regulatory importance. However, in general the total cellular concentration of InsP_6 does not respond quickly and substantially to cell stimulii. The 50% stimulus-dependent increase in the levels of InsP_6 in N1E-115 neuroblastoma cells (Sasakawa *et al.*, 1992) represents a rare—and the largest—example of a substantial acute change. It is possible that there may be greater but localized changes in polyphosphate levels that are beyond the detection of our current techniques. Such an event could, for example, be brought about by regional changes in $[Ca^{2+}]$. This cation promotes InsP_6 binding to cellular membranes (Poyner *et al.*, 1993) and this could form the basis of a process by which the free InsP_6 concentration is regulated.

The idea that InsP_6 regulates vesicle traffic has arisen from experiments with isolated proteins. Therefore, it is of particular interest that data in support of this hypothesis can be seen in two studies of the consequences of injecting InsP_6 into intact cells (Llinás *et al.*, 1994; Sasakawa *et al.*, 1994). For example, consider endocytosis, mediated by clathrin assembly proteins such as AP-2, which is a mechanism by which certain ligand-receptor complexes can be internalized, thereby down-regulating the response (e.g., Sorkin and Waters, 1993). If there were any direct evidence that InsP_6 could attenuate down-regulation, this could be taken as supportive of the idea that this polyphosphate genuinely inhibits clathrin assembly *in vivo*. Sasakawa *et al.* (1994) have published data which speak directly to this idea. They expressed a functional rat substance P receptor in *Xenopus laevis* oocytes, and confirmed that stimulation with agonist led to a loss of receptor response to a subsequent, second stimulation with substance P. It was striking that microinjection of InsP_6 specifically protected against this loss of receptor response (Sasakawa *et al.*, 1994). The time-course of these experiments ($>$10 min) was within the timeframe during which the substance P receptor undergoes substantial ligand-dependent internalization in other systems (Garland *et al.*, 1994; Mantyh *et al.*, 1995). Therefore, these data are entirely consistent with InsP_6 being a physiologically relevant inhibitor of endocytosis.

It should be emphasized that Sasakawa *et al.* (1994) interpreted their data somewhat differently; they believed that InsP_6 had attenuated receptor desensitization, and not receptor internalization. In part, the logic behind this proposal came from an earlier demonstration that InsP_6 binds with low (micromolar) affinity to visual arrestin, which perturbs its binding to rhodopsin, an action that acts to prevent desensitization to a subsequent light-induced signal (Palczewski *et al.*, 1991, 1992). The arrestin superfamily includes β-arrestin, which participates in desensitization of β_2-adrenergic (Pippig *et al.*, 1993) and m2-muscarinic (Gurevich *et al.*, 1993) G-protein coupled receptors. Thus, Sasakawa *et al.* (1994) proposed that InsP_6 acted by perturbing a hypothetical interaction between the substance P receptor and a putative *Xenopus* form of arrestin. This hypothesis makes further assumptions since it is not known if the ability of retinal arrestin to bind InsP_6 is shared by any other members of the arrestin superfamily. Even the physiological significance of InsP_6 binding to visual arrestin is uncertain, not in the least because retinal cells may not even synthesize InsP_6; squid retina, for example, does not phosphorylate Ins(1,4,5)P_3 (Szuts *et al.*, 1986; Wood *et al.*, 1990) and in animal cells *de novo* InsP_6 synthesis has never been shown to occur independently of Ins(1,4,5)P_3 phosphorylation (although such a pathway exists in *Dictyostelium* and could occur in higher organisms, see Section 3.2). At this point, the evidence is stronger that InsP_6 attenuates internalization rather than desensitization.

Experiments with a different cell type also support the idea that inositol polyphosphate can inhibit vesicle traffic (Llinás *et al.*, 1994). The focus of this study was synaptic transmission. Specifically, the release of neurotransmitter from the giant synapse of the stellate ganglion of squid was impaired by microinjecting inositol polyphosphates into the cells (Llinás *et al.*, 1994). These authors could not determine the volume of the injection fluid or the amount of inositol polyphosphate injected, and so they did not determine the rank order of efficacy for the various inositol phosphates that were tested. Nevertheless, InsP_6, Ins(1,3,4,5,6)P_5 and Ins(1,3,4,5)P_4 inhibited synaptic transmission, whereas Ins(1,4,5)P_3 did not (Llinás *et al.*, 1994).

It was proposed (Fukuda *et al.*, 1994) that these effects of polyphosphates upon synaptic transmission were actually based on an interaction with synaptotagmin. Neuronal members of this family of proteins regulate release of neurotransmitter through the control of fusion of the synaptic vesicle with the plasma membrane (Südhof, 1995). However, it is just as feasible that the life-cycle of the synaptic vesicle was perturbed at the endocytic, retrieval stage through an interaction of InsP_6 with AP-2 and/or AP-3.

Although synaptotagmin may not have been the primary target of InsP_6 in the experiments of Llinás *et al.* (1994), this protein does bind InsP_6. However, synaptotagmin differs from AP-2, AP-3, and coatomer in that Ins(1,3,4,5)P_4 binds more strongly than InsP_6 (Fukuda *et al.*, 1994; Niinobe *et al.*, 1994). The polyphosphate binding site of synaptotagmin lies within the so-called C2B domain, a lysine-rich sequence that has also been implicated in mediating protein/phospholipid and protein/protein interactions (Fukud *et al.*, 1994). As pointed out by Fukuda *et al.* (1994), it is possible that by binding to C2B domains and competing with PtdIns(4,5)P_2, inositol phosphates may regulate the association of proteins with cellular membranes. Another possible inositol polyphosphate /C2B interaction is exemplified by the Ins(1,3,4,5)P_4-binding protein (GAP1^{IP4BP}) that has GTPase activity towards Ras and Rap (Cullen *et al.*, 1995c). Here again, there is competition between PtdIns(4,5)P_2 and Ins(1,3,4,5)P_4 (Cullen *et al.*, 1995c), but the physiological functions ascribed to these two proteins (neurotransmitter release for synaptotagmin, Ca^{2+} signalling for GAP1^{IP4BP}) are too different to permit us to generalize about the physiological interactions of inositol polyphosphate with C2B domains. There is also a difference in specificity: Ins(1,3,4,5,6)P_5 is a physiological candidate ligand for synaptotagmin, which it binds slightly more avidly than Ins(1,3,4,5)P_4 (Niinobe *et al.*, 1994). GAP1^{IP4BP} prefers Ins(1,3,4,5)P_4 over Ins(1,3,4,5,6)P_5 by a factor of 14 (Cullen *et al.*, 1995a,b,c), and of course Ins(1,3,4,5,6)P_5 plays no role in

Ca^{2+} signalling. Still, these observations underscore the theme in this review of the versatility of inositol polyphosphates.

5. CONCLUDING STATEMENT

The demonstration that receptor-generated Ins(1,4,5)P_3 mobilized cellular Ca^{2+} stores (Streb *et al.*, 1983) was the *raison d'être* for studies into Ins(1,4,5)P_3 metabolism (Storey *et al.*, 1984). In contrast, the early research into the pathways of InsP_5 and InsP_6 metabolism did not have the benefit of any sound, physiologically relevant focal points (Shears, 1992). It is hoped that this review has convinced the reader that there is now good evidence that a diverse range of fundamental cellular processes may be regulated by these highly charged compounds. Better still, perhaps this review will encourage a broader-based and more concerted effort to understand the molecular mechanisms that underlie the actions of these inositol polyphosphates.

6. REFERENCES

Abdullah, M., Hughes, P. J., Craxton, A., Gigg, R., Desai, T., Marecek, J. F., Prestwich, G. D., and Shears, S. B. 1992, Purification and characterization of inositol-1,3,4-trisphosphate 5/6-kinase from rat liver using an inositol hexakisphosphate affinity column, *J. Biol. Chem.* **267:**22340–22345.

Ali, N., Craxton, A., and Shears, S. B., 1993, Hepatic Ins(1,3,4,5)P_4 3-phosphatase is compartmentalized inside endoplasmic reticulum. *J. Biol. Chem.* **268:**6161–6167.

Ali, N., Duden, R., Bembenek, M. E., and Shears, S. B., 1995a, The interaction of coatomer with inositol polyphosphates is conserved in *Saccharomyces cerevisiae, Biochem. J.* **310:**279–284.

Ali, N., Craxton, A., Sumner, M., and Shears, S. B., 1995b, Effects of aluminium upon the hepatic inositol polyphosphate phosphatase, *Biochem. J.* **305:**557–561.

Arnone, A., and Perutz, M. F., 1974, Structure of inositol hexaphosphate-human deoxyhaemoglobin complex, *Nature* **249:**34–36.

Balla, T., Guillemette, G., Baukal, A. J., and Catt, K. J., 1987, Metabolism of inositol 1,3,4-trisphosphate to a new tetrakisphosphate isomer in angiotensin-stimulated adrenal glomerulosa cells, *J. Biol. Chem.* **242:**9952–9955.

Balla, T., Baukal, A. J., Hunyady, L., and Catt, K. J., 1989, Agonist-induced regulation of inositol tetrakisphosphate isomers and inositol pentakisphosphate in adrenal gloerulosa cells, *J. Biol. Chem.* **264:**13605–13611.

Balla, T., Sim, S. S., Baukal, A. J., Rhee, S. G., and Catt, K. J., 1994, Inositol polyphosphates are not increase by overexpression of Ins(1,4,5)P_3 3-kinase but show cell-cycle dependent changes in growth factor-stimulated fibroblasts, *Mol. Biol. Cell.* **5:**17–27.

Barker, C. J., Wong, N. S., Maccallum, S. M., Hunt, P. A., Michell, R. H., and Kirk, C. J., 1992, The interrelationships of the inositol phosphates formed in vasopressin-stimulated WRK-1 mammary tumour cells, *Biochem. J.* **286:**469–474.

Barker, C. J., French, P. J., Moore, A. J., Nilsson, T., Berggren, P.-O., Bunce, C. M., Kirk, C. J., and Michell, R. H., 1995, Inositol 1,2,3-trisphosphate and inositol 1,2- and/or 2,3-bisphosphate are normal constituents of mammalian cells, *Biochem. J.* **306:**557–564.

Barraco, R. A., Phillis, J. W., and Simpson, L. L., 1989, Cardiorespiratory effects of inositol hexakisphosphate following microinjections into the nucleus tractus solitarii, *Eur. J. Pharmacol.* **173:**75–84.

Beck, K. A., and Keen, J. H., 1991, Interaction of phosphoinositide cycle intermediates with the plasma membrane-associated clathrin assembly protein AP-2, *J. Biol. Chem.* **266:**4442–4447.

Bennett, M. K., and Scheller, R. H., 1994, A molecular description of synaptic vesicle trafficking, *Annu. Rev. Biochem.* **63:**63–100.

Berridge, M. J., and Irvine, R. F. 1989, Inositol phosphates and cell signalling, *Nature* **341:**197–205.

Biswas, S., Maity, I. B., Chakrabarti, S., and Biswas, B. B., 1978, Purification and characterization of *myo*-inositol hexaphosphate–adenosine diphosphate phosphotransferase from *Phaseolus aureus, Arch. Biochem. Biophys.* **185:**557–566.

Brooks, P. A., and Spyer, K. M., 1990, Investigation of inositol hexakisphosphate actions in rat nucleus tractus solitarium *in vitro, Neurosci. Lett.* **105:**120–124.

Bunce, C. M., French, P. J., Allen, P., Mountford, J. C., Moor, B., Greaves, M. F., Michell, R. H., and Brown, G., 1993, Comparison of the levels of inositol metabolites in transformed haemopoietic cells and their normal counterparts, *Biochem. J.* **289:**667–673.

Carpenter, D., Hanley, M. R., Hawkins, P. T., Jackson, T. R., Stephens, L. R., and Vallejo, M., 1989, The metabolism and functions of inositol pentakisphosphate and inositol hexakisphosphate, *Biochem. Soc. Trans.* **17:**3–5.

Cifuentes, M. E., Honkanen, L., and Rebecchi, M. J., 1993, Proteolytic fragments of phosphoinositide-specific phospholipase C-δ1. Catalytic and membrane binding properties. *J. Biol. Chem.* **268:**11586–11593.

Challiss, R.A.J., Willcocks, A. L., Mulloy, B., Potter, B.V.L., and Nahorski, S. R., 1991, Characterization of inositol 1,4,5-trisphosphate and inositol 1,3,4,5-tetrakisphosphate binding sites in rat cerebellum, *Biochem. J.* **274:**861–867.

Chan, H. C., Kaetzell, M. A., Gotter, A. L., Dedman, J. R., and Nelson, D. J., 1994, Annexin IV inhibits calmodulin-dependent protein kinase II-activated chloride conductance. A novel mechanism for ion channel regulation, *J. Biol. Chem.* **269:**32462–32468.

Chattaway, J. A., Drobak, B. K., Watkins, P.A.C., Dawson, A. P., Letcher, A. J., Stephens, L. R., and Irvine R. F., 1992, An inositol 1,4,5-trisphosphate 6-kinase in pea roots, *Planta* **187:**542–545.

Cifuentes, M. E., Delaney, T., and Rebecchi, M. J., 1994, D-*myo*-inositol 1,4,5-trisphosphate inhibits binding of phospholipase C-δ1 to bilayer membranes, *J. Biol. Chem.* **269:**1945–1948.

Cornish-Bowden, A., 1991, Failure of channelling to maintain low concentrations of metabolic intermediates. *Eur. J. Biochem.* **195:**103–108.

Craxton, A., Erneux, C., and Shears, S. B., 1994, Inositol 1,4,5,6-tetrakisphosphate is phosphorylated in rat liver by a 3-kinase that is distinct from inositol 1,4,5-trisphosphate 3-kinase, *J. Biol. Chem.* **269:**4337–4342.

Craxton, A., Ali, N., and Shears, S. B., 1995, Comparison of the activities of a multiple inositol polyphosphate phosphatase obtained from several sources: A search for heterogeneity in this enzyme. *Biochem. J.* **305:**557–561.

Cullen, P. J., Irvine, R. F., and Dawson, A. P., 1990, Synergistic control of Ca^{2+} mobilization in permeabilized mouse L1210 lymphoma cells by inositol 2,4,5-trisphosphate and inositol 1,3,4,5-tetrakisphosphate, *Biochem. J.* **271:**549–553.

Cullen, P. J., Chung, S.-K., Chang, Y.-T., Dawson, A. P., and Irvine, R. F., 1995a, Specificity of the purified inositol (1,3,4,5) tetrakisphosphate-binding protein from porcine platelets, *FEBS Lett.* **358:**240–242.

Cullen, P. J., Dawson, A. P., and Irvine, R. F., 1995b, Purification and characterization of an inositol 1,3,4,5-tetrakisphosphate binding protein from pig platelets: Possible identification of a novel non-neuronal Ins(1,3,4,5)P_4 receptor, *Biochem. J.* **305:**139–143.

Cullen, P. J., Hsuan, J. J., Truong, O., Letcher, A. J., Jackson, T. R., Dawson, A. P., and Irvine, R. F. 1995c, Identification of a specific Ins(1,3,4,5)P_4-binding protein as a member of the GAP1 family, *Nature* **376:**527–530.

da Silva, C. P., Emmrich, F., and Guse, A. H., 1994, Adriamycin inhibits 1,4,5-trisphosphate 3-kinase activity *in vitro* and blocks formation of inositol 1,3,4,5-tetrakisphosphate in stimulated jurkat T-lymphocytes. Does inositol 1,3,4,5-tetrakisphosphate play a role in Ca^{2+} entry? *J. Biol. Chem.* **269:**12521–12526.

DeLisle, S., Radenberg, T., Wintermantel, M. R., Tietz, C., Parys, J. B., Pittet, D., Welsh, M. J., and Mayr, G., 1994, Second messenger specificity of the novel trisphosphate receptor: reappraisal based on novel inositol phosphates, *Am. J. Physiol.* **266:**C429–C436.

Dharmsathaphorn, K., Cohn, J., and Beuerlein, G., 1989, Multiple calcium-mediated effector mechanisms regulate chloride secretory responses in T_{84} cells, *Am. J. Physiol.* **256:**C1224–C1230.

Dingwell, C., and Laskey, R., 1992, The nuclear membrane, *Science* **258:**942–947.

Donié, F., Hülser, E., and Reiser, G., 1990, High-affinity inositol 1,3,4,5-tetrakisphosphate receptor from cerebellum: Solubilization, partila purification and characterization, *FEBS Lett.* **268:**194–198.

Doughney, C., McPherson, M. A., and Dormer, R. L., 1988, Metabolism of inositol 1,3,4,5-tetrakisphosphate by human erythrocyte membranes. A new mechanism for the formation of inositol 1,4,5-trisphosphate, *Biochem. J.* **251:**927–929.

Downes, C. P., and MacPhee, C. H., 1990, *myo*-Inositol metabolites as cellular signals, *Eur. J. Biochem.* **195:**1–18.

Drayer, A. L. Van Der Kaay, J., Mayr, G. W., and Van Haastert, P.J.M., 1994, Role of phospholipase C in *Dictyostelium:* Formation of inositol 1,4,5-trisphosphate and normal development in cells lacking phospholipase C activity, *EMBO J.*, **13:**1601–1609.

Estevez, F., Pulford, D., Stark, M.J.R., Carter, A. N., and Downes, C. P., 1994, Inositol trisphosphate metabolism in *Saccharomyces cerevisiae:* Identification, purification and properties of inositol 1,4,5-trisphosphate 6-kinase, *Biochem. J.* **302:**709–716.

Estrada-Garcia, T., Craxton, A., Kirk, C. J., and Michell, R. H., 1991, A salt-activated inositol 1,3,4,5-tetrakisphosphate 3-phosphatase at the inner surface of the human erythrocyte membrane, *Proc. R. Soc. Lond. [Biol.]* **244:**63–68.

Europe-Finner, G. N., Gammon, B., Wood, C. A., and Newell, P. C., 1989, Inositol tris- and polyphosphate formation during chemotaxis of *Dictyostelium, J. Cell. Sci.* **93:**585–592.

Falck, J. R., Reddy, K. K., Ye, J., Saady, M., Mioskowski, C., Shears, S. B., Tan, Z., and Safrany, S., 1995, Synthesis and structure of cellular mediators: inositol polyphosphate diphosphates, *JACS,* submitted.

Fleischer, B., Zie, J., Mayrleitner, M., Shears, S. B., Palmer, D. J., and Fleischer, S., 1994, Golgi coatomer binds, and forms K^+-selective channels gated by, inositol polyphosphates, *J. Biol. Chem.* **269:**17826–17832.

French, P. J., Bunce, C. M., Stephens, L. R., Lord, J. M., McConnell, F. M., Brown, G., Creba, J. A., and Michell, R. H., 1991, Changes in the levels of inositol lipids and phosphates during the differentiation of HL60 promyelocytic cells towards neutrophils or monocytes, *Proc. R. Soc. Lond. [Biol.]* **245:**193–201.

Freund, W.-D., Mayr, G. W., Tietz, C., and Schultz, J. E., 1992, Metabolism of inositol phosphates in the protozoan *Paramecium.* Characterization of a novel inositol-hexakisphosphate-dephosphorylating enzyme, *Eur. J. Biochem.* 207:359–367.

Fukuda, M., Aruga, J., Niinobe, M., Aimoto, S., and Mikoshiba, K., 1994, Inositol-1,3,4,5-tetrakisphosphate binding to the C2B domain of IP4BP/synaptotagmin II, *J. Biol. Chem.* **269:**29206–29211.

Garland, A. M., Grady, E. F., Payan, D. G., Vigna, S. R., and Bunnett, N. B., 1994, Agonist-induced internalization of the substance P (NK1) receptor expressed in epithelial cells, *Biochem. J.* **303:**177–186.

Gawler, D. J., Potter, B.V.L., Gigg, R., and Nahorski, S. R., 1991, Interactions between inositol tris- and tetrakis-phosphates. Effects on intracellular Ca^{2+} mobilization in SH-SY5Y cells, *Biochem. J.* **276:**163–167.

Gibson, D. M., and Ullah, A.B.J., 1990, Phytases and their action on phytic acid, in *Inositol Metabolism in Plants* (D. J. Morré, W. F. Boss, and F. A. Loewus, eds.), pp. 77–92, Wiley-Liss, Inc., New York.

Glennon, M. C., and Shears, S. B., 1993, Turnover of inositol pentakisphosphates, inositol hexakisphosphate and diphosphoinositol polyphosphates in primary cultured hepatocytes, *Biochem. J.* **293:**583–590.

Graf, E., Empson, K. L., and Eaton, J. W., 1987, Phytic acid. A natural antioxidant, *J. Biol. Chem.* **262:**11647–11650.

Gurevich, V. V., Richardson, R. M., Kim, C. M., Hosey, M. M., and Benovic, J. L., 1993, Binding of wild-type and chimeric arrestins to the m2 muscarinic cholinergic receptor, *J. Biol. Chem.* **268:**16879–16882.

Guse, A. H., and Emmrich, F., 1991, T-cell receptor-mediated metabolism of inositol polyphosphates in Jurkat T-lymphocytes. Identification of a D=*myo*-inositol 1;2,3,4,6-pentakisphosphate 2-phosphomonoesterase activity, a D-*myo*-inositol 1,3,4,5,6-pentakisphosphate 1/3-phosphate activity and a D/L-*myo*-inositol 1,2,4,5,6-pentakisphosphate 1/3-kinase activity, *J. Biol. Chem.* **266:**24498–24502.

Guse, A. H., Gercken, G., Boysen, H., Schwarz, J. R., and Meyerhof, W., 1991, Inositol tetrakisphosphates as second messenegers induce Ca^{2+}-dependent chloride currents in *Xenopus laevis* oocytes, *Biochem. Biophys, Res. Commun.* **179:**641–647.

Guse, A. H., Greiner, E., Emmrich, F., and Brand, K., 1993, Mass changes of inositol 1,3,4,5,6-pentakisphosphate and inositol hexakisphosphate during cell cycle progression in rat thymocytes, *J. Biol. Chem.* **268:**7129–7133.

Hansen, C. A., Von Dahl, S., Huddell, B., and Williamson, J. R., 1988, Characterization of inositol 1,3,4-trisphosphate phosphorylation in rat liver, *FEBS Lett.* **236:**53–56.

Harlen, J. E., Hajduk, P. J., Yoon, H. S., and Fesik, S. W., 1991, Plekstrin homology domains bind to phosphatidylinositol 4,5-bisphosphate, *Nature* **371:**168–170.

Hashii, M., Hirata, M., Ozaki, S., Nozawa, Y., and Higashida, H., 1994a, Ca^{2+} influx evoked by inositol-3,4,5,6-tetrakisphosphate in *ras*-transformed NIH/3T3 fibroblasts, *FEBS Lett.* **340:**-276–280.

Hashii, M., Hirata, M., Ozaki, S., Nozawa, Y., and Higashida, H., 1994b, Ca^{2+} influx gated by inositol-3,4,5,6-tetrakisphosphate in NIH/3T3 fibroblasts, *Biochem. Biophys, Res. Commun.* **200:**1300–1306.

Häussinger D., and Lang, F., 1992, Cell volume and hormone action, *Trends. Pharmacol. Sci.* **13:**371–373.

Hawkins, P. T., Poyner, D. R., Jackson, T. R., Letcher, A. J., Lander, D. A., and Irvien, R. F., 1993, Inhibition of iron-catalysed hydroxyl radical formation by inositol polyphosphates: A possible physiological function for *myo*-inositol hexakisphosphate, *Biochem. J.* **294:**929–934.

Heslop, J. P., Irvine, R. F., Tashjian, A. H., and Berridge, M. J., 1985, Inositol tetrakis- and pentakisphosphates in GH_4C_1 cells, *J. Biol.* **199:**395–402.

Hildebrandt, J.-P., and Shuttleworth, T. J., 1992, Calcium-sensitivity of inositol 1,4,5-trisphosphate metabolism in exocrine cells from the avian salt gland, *Biochem. J.* **282:**703–710.

Horn, R., and Marty, A., 1988, Muscarinic activation of ionic currents measured by a new whole-cell recording method, *J. Gen. Physiol.* **92:**145–159.

Hughes, A. R., and Putney, J. W., Jr., 1989, Source of ^{3}H-labeled inositol bis- and monophosphates in agonist-stimulated rat parotid acinar cells, *J. Biol. Chem.* **264:**9400–9407.

Hughes, P. J., and Michell, R. H., 1993, Novel inositol containing phospholipids and phosphates: Their synthesis and possible new roles in cellular signalling, *Curr. Opin. Neurobiol.* **3:**383–400.

Hughes, P. J., Hughes, A. R., Putney, J. W., Jr., and Shears, S. B., 1989, The regulation of the phosphorylation of inositol 1,3,4-trisphosphate in cell-free preparations and its relevance to the formation of inositol 1,3,4,6-tetrakisphosphate in agonist-stimulated rat parotid cells, *J. Biol. Chem.* **264:**19871–19878.

Hughes, P. J., Kirk, C. J., and Michell, R. H., 1994, Inhibition of porcine brain inositol 1,3,4-trisphosphate kinase by inositol polyphosphates, other inositol phosphates, polyanions and polycations, *Biochim. Biophys. Acta* **1223:**57–70.

Irvine, R. F., 1988, Inositol phosphates: Proliferation, metabolism and function, *Phil. Trans. R. Soc. London. [Biol.]* **320:**281–298.

Irvine, R. F., 1991, Inositol tetrakisphosphate as a second messenger: Confusions, contradictions and a potential resolution, *BioEssays* **13:**419–427.

Irvine, R. F., 1992, Is inositol tetrakisphosphate the second messenger that controls Ca^{2+} entry into cells? *Adv. Second Messenger Phosphoprotein Res.* **26:**161–185.

Irvine, R. F., Änggård, E. E., Letcher, A. J., and Downes, C. P., 1985, Metabolism of inositol 1,4,5-trisphosphate and inositol 1,3,4-trisphosphate in rat parotid acinar cells, *Biochem. J.* **229:**505–511.

Irvine, R. F., Letcher, A. J., Stephens, L. R., and Musgrave, A., 1992, Inositol polyphosphate metabolism and inositol lipids in a green alga, *Chlamydomonas eugametos, Biochem. J.* **281:**261–266.

Isaacks, R. E., and Harkness, D. R., 1980, Erythrocyte organic phosphates and hemoglobin function in birds, reptiles and fishes, *Am. Zool.* **20:**115–129.

Ji, H., Sandberg, K., Baukal, A. J., and Catt, K. J., 1989. Metabolism of inositol pentakisphosphate to inositol hexakisphosphate in *Xenopus laevis* oocytes, *J. Biol. Chem.* **264:**20185–20188.

Johnson, L. F., and Tate, M. E., 1969, Structure of "phytic acids," *Can. J. Chem.* **47:**63–73.

Kachintorn, U., Vongkovit, P., Vajanaphanich, M., Dinh, S., Barrett, K. E., and Dharmsathaphorn, K., 1992, Dual effects of a phorbol ester on calcium-dependent chloride secretion by T84 epithelial cells, *Am. J. Physiol.* **262:**C15–C22.

Kachintorn, U., Vajanaphanich, M., Barrett, K. E., and Traynor-Kaplan, A. E., 1993, Elevation of inositol tetrakisphosphate parallels inhibition of Ca^{2+}-dependent Cl^-secretion in T84 cells, *Am. J. Physiol.* **264:**C671–C676.

Kanematsu, T., Misumi, Y., Watanabe, Y., Ozaki, S., Koga, T., Iwanaga, S., Ikehara, Y., and Hirata, M., 1995, New Ins(1,4,6)P_3- binding protein similar to phospholipase C-δ1, *Biochem. J.*, in press.

Kanematsu, T., Takeya, H., Watanabe, Y., Ozaki, S., Yoshida, M., Koga, T., Iwanaga, S., and Hirata, M., 1992, Putative inositol 1,4,5-trisphosphate binding proteins in rat brain cytosol, *J. Biol. Chem.* **267:**6518–6525.

King, W. G., Downes, C. P., Prestwich, G. D., and Rittenhouse, S. E., 1990, Ca^{2+}-stimulatable and protein kinase C-inhibitable accumulation of inositol 1,3,4,6-tetrakisphosphate in human platelets, *Biochem. J.* **270:**125–131.

Koppler, P., Matter, N., and Malviya, A. N., 1993, Evidence for stereospecific inositol 1,3,4,5-tetrakisphosphate binding sites on rat liver nucleii. Delineating inositol 1,3,4,5-tetrakisphosphate interaction in nuclear calcium signalling processes, *J. Biol. Chem.* **268:**26248–26252.

Korn, S. J., Bolden, A., and Horn, R., 1991, Control of action potentials and Ca^{2+} influx by the Ca^{2+}-dependent chloride current in mouse pituitary cells, *J. Physiol.* **439:**423–437.

Lapan, E. A., 1975, Magnesium inositol hexaphosphate deposits in mesozoan dispersal larvae, *Exp. Cell. Res.* **94**:277–282.

Lim, P. E., and Tate, M. E., 1973, The phytases. II. Properties of phytase fractions F_1 and F_2 from wheat bran and the *myo*-inositol phosphates produced by fraction F_2. *Biochim. Biophys. Acta* **302**:316–328.

Llinás, R., Sugimori, M., Lang, E. J., Morita, M., Fukuda, M., Niinobe, M., and Mikoshiba, K., 1994, The inositol high-polyphosphate series blocks synaptic transmission by preventing vesicular fusion: A squid giant synapse study, *Proc. Natl. Acad. Sci. U.S.A.* **91**:12990–12993.

Lu, P.-J., Gou, D.-M., Shieh, W.-R., and Chen, C.-S., 1994, Molecular interactions of endogenous D-myo-inositol phosphates with the intracellular D-myo-Inositol 1,4,5-triphosphate recognition site, *Biochem. J.* **33**:11586–11597.

Mantyh, P. W., Allen, C. J., Ghilardi, J. R., Rogers, S. D., Mantyh, C. R., Liu, H., Basbaum, A. I., Vigna, S. R., and Maggio, J. E., 1995, Rapid endocytosis of a G protein coupled receptor: Substance P evoked internalization of its receptor in the rat striatum in vivo, *Proc. Natl. Acad. Sci. U.S.A.* **92**:2622–2626.

Martin, J. B., Bakker-Grunwald, T., and Klein, G., 1993, ^{31}P-NMR analysis of *Entamoeba histolytica*. Occurrence of high amounts of two inositol phosphates, *Eur. J. Biochem.* **214**:711–718.

Marty, A., and Zimmerberg, J., 1989, Diffusion into the patch-clamp recording pipette of a factor necessary for the muscarinic current response, *Cell. Signal.* **1**:259–268.

Mattingly, R. R., Stephens, L. R., Irvine, R. F., and Garrison, J. C., 1991, Effects of transformation with the v-*src* oncogene on inositol phosphate metabolism in rat-1 fibroblasts: D-*myo*-Inositol 1,4,5,6-tetrakisphosphate is increased in v-*src*-transformed rat-1 fibroblasts and can be synthesized from D-*myo*-inositol 1,3,4-trisphosphate in cytosolic extracts, *J. Biol. Chem.* **266**:15144–15153.

Mayr, G. W., 1988, A novel metal-dye detection system permits picomolar-range h.p.l.c. analysis of inositol polyphosphates from non-radioactively labelled cell or tissue specimens, *Biochem. J.* **254**:585–591.

Mayr, G. W., and Dietrich, W., 1987, The only inositol tetrakisphosphate in avian erythrocytes is the isomer lacking phosphate at position 3: A NMR study, *FEBS Lett.* **213**:278–282.

McCollom, E. V., and Hart, E. B., 1908, On the occurrence of a phytin-splitting enzyme in animal tissues, *J. Biol. Chem.* **4**:497–500.

McConnell, F. M., Stephens, L. R., and Shears, S. B., 1991, Multiple isomers of inositol pentakisphosphates in Epstein–Barr virus transformed (T5-1) B-lymphocytes. Identification of inositol (1,3,4,5,6) pentakisphosphate, D-inositol (1,2,4,5,6) pentakisphosphate and L-inositol (1,2,4,5,6) pentakisphosphate, *Biochem. J.* **280**:323–329.

McConnell, F. M., Shears, S. B., Lane, P.J.L., Scheibel, M. S., and Clark, E. A., 1992, Relationships between the degree of cross-linking of surface immunoglobulin and the associated inositol 1,4,5-trisphosphate and Ca^{2+} signals in human B cells, *Biochem. J.* **284**:447–455.

McEwan, G.T.A., Hirst, B. H., and Simmons, N. L., 1994, Carbachol stimulates Cl^- secretion via activation of two distinct apical Cl^- pathways in cultured human T_{84} intestinal epithelial monolayers, *Biochim. Biophys. Acta* **1220**:241–247.

McIntosh, R. P., and McIntosh, J.E.A., 1990, Formation of inositol pentakisphosphate by ovarian follicles of *Xenopus laevis* oocytes from metabolism of inositol (1,4,5) trisphosphate and inositol (1,3,4,5) tetrakisphosphate and from receptor activation, *Biochem. Biophys. Res. Commun.* **166**:380–386.

Menniti, F. S., Oliver, K. G., Nogimori, K., Obie, J. F., Shears, S. B., and Putney, J. W., Jr., 1990, Origins of *myo*-inositol tetrakisphosphates in agonist-stimulated rat pancreatoma cells. Stimulation by bombesin of *myo*-inositol 1,3,4,5,6-pentakisphosphate breakdown to *myo*-inositol 3,4,5,6-tetrakisphosphate, *J. Biol. Chem.* **265**:11167–11176.

Menniti, F. S., Miller, R. N., Putney, J. W., Jr., and Shears, S. B., 1993a, Turnover of inositol polyphosphate pyrophosphates in pancreatoma cells, *J. Biol. Chem.* **268**:3850–3856.

Menniti, F. S., Oliver, K. G., Putney, J. W., Jr., and Shears, S. B., 1993b, Inositol phosphates and cell signalling: New view of InsP_5 and InsP_6, *Trends Biochem. Sci.* **18:**53–56.

Michell, R. H., King, C. E., Piper, C. E., Stephens, L. R., Bunce, C. M., Guy, G. R., and Brown, G., 1988, Inositol lipids and phosphates in erythrocytes and HL60 cells, *Soc. Gen. Physiol. Ser.* **43:**345–355.

Mitchell, R., MacEwan, D., Dougan, L., Johnson, M., and Thomson, F., 1991, Does inositol hexakisphosphate induce calcium entry into GH3 cells, *Biochem. Soc. Trans.* **19:**116s.

Moore, R. J., and Veum, T. L., 1983, Adaptive increase in phytate digestibility by phosphorus-deprived rats and the relationship of intestinal phytase (EC 3.1.3.8) and alkaline phosphatase (EC 3.1.3.1) to phytate utilization, *Br. J. Nutr.* **49:**145–152.

Morris, A. P., Kirk, K. L., and Frizzell, R. A., 1990, Simultaneous analysis of cell Ca^{2+} and Ca^{2+}-stimulated chloride conductance in colonic epithelial cells (HT-29), *Cell Regul.* **1:**951–963.

Mountford, J. C., Bunce, C. M., French, P. J., Michell, R. H., and Brown, G., 1994, Intracellular concentrations of inositol, glycerophosphoinositol, and inositol pentakisphosphate increase during haematopoietic cell differentiation, *Biochem. Biophys. Acta.* **1222:**101–108.

Murphy, J.-E., Hanover, J. A., Froehlich, M., DuBois, G., and Keen, J. H., 1994, Clathrin assembly protein AP-3 is phosphorylated and glycosylated on the 50 kD structural domain, *J. Biol. Chem.* **269:**21346–21352.

Nauntofte, B., 1992, Regulation of electrolyte and fluid secretion in salivary acinar cells, *Am. J. Physiol.* **263:**G823–G837.

Newton, C. L., Mignery, G. A., and Südhof, T. C., 1994, Co-expression in vertebrate tissues and cell lines of multiple inositol 1,4,5-trisphosphate (InsP_3) receptors with distinct affinities for InsP_3, *J. Biol. Chem.* **269:**28613–28619.

Nicoleti, F., Bruno, V., Fiore, L., Cavallaro, S., and Canonico, P . L., 1989, Inositol hexakisphosphate (phytic acid) enhances Ca^{2+} influx and D-[^{3}H]aspartate release in cultured cerebellar neurons, *J. Neurochem.* **53:**1026–1030.

Nogimori, K., Hughes, P. J., Glennon, M. C., Hodgson, M. E., Putney, J. W., Jr., and Shears, S. B., 1991, Purification of an inositol 1,3,4,5-tetrakisphosphate 3-phosphatase from rat liver and the evaluation of its substrate specificity, *J. Biol. Chem.* **266:**16499–16506.

Norris, F. A., Ungewickell, E., and Majerus, P. W., 1995, Inositol hexakisphosphate binds to clathrin assembly protein 3 (AP-3/AP180) and inhibits clathrin cage assembly *in vitro, J. Biol. Chem.***270:**214–217.

Oliver, K. G., Putney, J. W., Jr., Obie, J. F., and Shears, S. B., 1992, The interconversion of inositol 1,3,4,5,6-pentakisphosphate and inositol tetrakisphosphates in AR4-2J cells, *J. Biol. Chem.* **267:**21528–21534.

Palczewski, K., Pulvermuller, A., Buczylko, J., Gutmann, C., and Hofmann, K. P., 1991, Binding of inositol phosphates to arrestin, *FEBS Lett.* **295:**195–199.

Palczewski, K., Rispoli, G., and Detwiler, P. B., 1992, The influence of arrestin (48K protein) and rhodopsin kinase upon visual transduction, *Neuron* **8:**117–126.

Palmer, S., Hughes, K. T., Lee, D.D.Y., and Wakelam, M.J.O., 1988, Development of a novel, Ins(1,4,5)P_3-specific, binding assay. Its use to determine the intracellular concentration of Ins(1,4,5)P_3 in unstimulated and vasopressin-stimulated rat hepatocytes, *Cell Signal.* **1:**147–156.

Pauloin, A., and Thurieau, C., 1993, The 50 kDa protein subunit of assembly polypeptide (AP) AP-2 adaptor from clathrin-coated vesicles is phosphorylated on theronine-156 by AP-1 and a soluble AP50 kinase which co-purifies with the assembly polypeptides, *Biochem. J.* **296:**409–415.

Pelham, H.R.B., 1989, Control of protein exit from the endoplasmic reticulum, *Annu. Rev. Cell Biol.* **5:**1–23.

Petersen, O. H., 1992, Stimulus–secretion coupling: Cytoplasmic calcium signals and the control of ion channels in exocrine acinar cells, *J. Physiol.* **448:**1–51.

Phillippy, B. Q., and Bland, J. M., 1988, Gradient ion chromatography of inositol phosphates, *Anal. Biochem.* **175:**162–168.

Phillippy, B. Q., Ulah, A.H.J., and Ehrlich, K.C., 1994, Purification and some properties of inositol 1,3,4,5,6-pentakisphosphate 2-kinase from immature soybean seeds, *J. Biol. Chem.* **269:** 28393–28399.

Pietrusiewicz, K. M., Salamornczyk, G. M., Bruzik, K. S., and Wieczorek, W., 1992, The synthesis of homochiral inositol phosphates from *myo*-inositol, *Tetrahedron* **48:**5523–5542.

Pietrusiewicz, K. M., Salamornczyk, G. M., Bruzik, K. S., and Wieczorek, W., 1994, Corrigendum, *Tetrahedro* **50:**573.

Pippig, S., Andexinger, S., Daniel, K., Puzicha, M., Caron, M. G., Lefkowitz, R. J., and Lohse, M. J., 1993, Overexpression of β-arrestin and β-adrenergic receptor kinase augment desensitization of β2-adreneric receptors. *J. Biol. Chem.* **268:**3201–3208.

Pittet, D., Schlegel, W., Lew, D. P., Monod, A., and Mayr, G. W., 1989, Mass changes in inositol tetrakis- and pentakisphosphate isomers induced by chemotactic peptide stimulation in HL-60 cells, *J. Biol. Chem.* **264:**18489–18493.

Poyner, D. R., Cooke, F., Hanley, M. R., Reynolds, D.J.M., and Hawkins, P. T., 1993, Characterization of metal-ion induced [^{3}H]inositol hexakisphosphate binding to rat cerebella membranes, *J. Biol. Chem.* **268:**1032–1038.

Putney, J. W., Jr., 1992, Inositol phosphates and calcium entry, *Adv. Second Messenger Phosphoprotein Res.* **26:**143–160.

Raboy, V., 1990, Biochemistry and genetics of phytic acid synthesis, in *Inositol Metabolism in Plants* (D. J. Morré, W. F. Boss, and F. A. Loewus, eds.), pp. 55–76, Wiley-Liss, New York.

Rao, R. K., and Ramakrishnan, C. V., 1985, Studies on inositolphosphatase in rat small intestine, *Enzyme* **33:**205–215.

Rapoport, S., 1940, Phytic acid in avian erythrocytes, *J. Biol. Chem.* **135:**403–406.

Rapoport, S., and Guest, G. M., 1941, Distribution of acid-soluble phosphorus in blood cells of various vertebrates, *J. Biol. Chem.* **138:**269–282.

Reaves, B., and Banting, G. (1994) Vacuolar ATPase inactivation blocks recycling to the trans-Golgi network from the plasma membrane, *FEBS Lett.* **345:**61–66.

Ren, H.-Y., Komatsu, N., Shimizu, R., Okada, K., and Miura, Y., 1994, Erythropoietin induces tyrosine phosphorylation and activation of phospholipase C-γ1 in a human erythropoietin-dependent cell line, *J. Biol. Chem.* **269:**19633–19638.

Rothman, J. E., 1994, Mechanisms of intracellular protein transport, *Nature* **372:**55–63.

Rubiera, C., Velasco, G., Michell, R. H., Lazo, P. S., and Shears, S. B., 1988, 1D-*myo*-Inositol 1,4,5-trisphosphate dephosphorylation by rat enterocytes involves an intracellular 5-phosphatase and non-specific phosphatase activity at the cell surface, *Biochem. J.* **255:**131–137.

Sasakawa, N., Nakaki, T., Kashima, R., Kanba, S., and Kato, R., 1992, Stimulus-induced accumulation of inositol tetrakis- pentakis- and hexakisphosphate in N1E-115 neuroblastoma cells, *J. Neurochem.* **58:**2116–2123.

Sasakawa, N., Ferguson, J. E., Sharif, M., and Hanley, M. R., 1994, Attenuation of agonist-induced desensitization of the rat substance P receptor by microinjection of inositol pentakis- and hexakisphosphates in *Xenopus laevis* oocytes, *Mol. Pharmacol.* **46:**380–385.

Sasakawa, N., Sharif, M., and Hanley, M. R., 1995, Metabolism and biological activities of inositol pentakisphosphate and inositol hexakisphosophate, *Biochem. Pharmacol.* **50:**137–146.

Shears, S. B., 1989, The pathway of *myo*-inositol 1,3,4-trisphosphate phosphorylation in liver. Identification of *myo*-inositol 1,3,4-trisphosphate 6-kinase, *myo*-inositol 1,3,4-trisphosphate 5-kinase and *myo*-inositol 1,3,4,6-tetrakisphosphate 6-kinase, *J. Biol. Chem.* **264:**19879–19886.

Shears, S. B., 1992, Metabolism of inositol phosphates, *Adv. Second Messenger Phosphoprotein Res.* **26:**63–92.

Shears, S. B., Parry, J. B., Tang, E.K.Y., Irvine, R. F., Michell, R. H., and Kirk, C. J., 1987, Metabolism of D-*myo*-inositol 1,3,4,5-tetrakisphosphate by rat liver, including the synthesis of a novel isomer of *myo*-inositol tetrakisphosphate, *Biochem. J.* **246:**139–147.

Shears, S. B., Ali, N., Craxton, A., and Bembenek, M. E., 1995, Synthesis and metabolism of bis-diphosphoinositol tetrakisphosphate *in vitro* and *in vivo, J. Biol. Chem.* **270:**10489–10497.

Smith, S. E., and Dürmüller, N., 1990, Inositol hexakisphosphate is convulsant in mice and rats in the nanaolar range, *Eur. J. Pharmacol.* **191:**337–343.

Sorkin, A., and Waters, C. M., 1993, Endocytosis of growth factor receptors, *BioEssays* **15:**375–382.

Sortino, M. A., Nicoletti, F., and Cannnnonico, P. O., 1990, Inositol hexakisphosphate stimulates $^{45}Ca^{2+}$ uptake in anterior pituitary cells in culture, *Eur. J. Pharmacol.,* **189:**115–118.

Srere, P. A., 1987, Complexes of sequential metabolic enzymes, *Ann. Rev. Biochem.* **56:**89–24.

Stephens, L. R., and Downes, C. P., 1990, Product–precursor relationships amongst inositol polyphosphates. Incorporation of $[^{32}P]P_i$ into *myo*-inositol 1,3,4,5-tetrakisphosphate, *myo*-inositol 1,3,4,6-tetrakisphosphate, *myo*-inositol 3,4,5,6-tetrakisphosphate and *myo*-inositol 1,3,4,5,6-pentakisphosphate, *Biochem. J.* **265:**435–452.

Stephens, L. R., and Irvine, R. F., 1990, Stepwise phosphorylation of *myo*-inositol leading to *myo*-inositol hexakisphosphate in *Dictyostelium, Nature* **346:**580–583.

Stephens, L. R., Hawkins, P. T., Barker, C. J., and Downes, C. P., 1988a, Synthesis of *myo*-inositol 1,3,4,5,6-pentakisphosphate from inositol phosphates generated by receptor activation, *Biochem. J.* **253:**721–733.

Stephens, L. R., Hawkins, P. T., Carter, N., Chahwala, S. B., Morris, A. J., Whetton, A. D., and Downes, C. P., 1988b, L-*MYO*-Inositol 1,4,5,6-tetrakisphosphate is present in both mammalian and avaian cells, *Biochem. J.* **249:**271–282.

Stephens, L. R., Hawkins, P. T., Morris, A. P., and Downes, C. P., 1988c, L-*myo*-Inositol 1,4,5,6-tetrakisphosphate (3-hydroxy)kinase, *Biochem. J.* **249:**283–292.

Stephens, L. R., Hawkins, P. T., and Downes, C. P., 1989, An analysis of *myo*-$[^3H]$inositol trisphosphate found in *myo*-$[^3H]$inositol prelabelled avian erythrocytes, *Biochem. J.* **262:**727–737.

Stephens, L. R., Hawkins, P. T., Stanley, A. F., Moore, T., Poyner, D. R., Morris, P. J., Hanley, M. R., Kay, R. R., and Irvine, R. F., 1991, *myo*-Inositol pentakisphosphates. Structure, biological occurrence and phosphorylation to *myo*-inositol hexakisphosphate, *Biochem. J.* **275:**485–499.

Stephens, L. R., Radenberg, T., Thiel, U., Vogel, G., Khoo, K.-H., Dell, A., Jackson, T. R., Hawkins, P. T., and Mayr, G. W., 1993, The detection, purification, structural characterization and metabolism of diphosphoinositol pentakisphosphate(s) and bisdisphosphoinositol tetrakisphosphate(s), *J. Biol. Chem.* **268:**4009–4015.

Stewart, S. J., Kelley, L. L., and Powers, F. S., 1987, Production of inositol pentakisphosphate in a human T-lymphocyte cell-line, *Biochem. Biophys. Res. Commun.* **145:**902–985.

Storey, D. J., Shears, S. B., Kirk, C. J., and Michell, R. H., 1984, Stepwise enzymatic dephosphorylation of inositol 1,4,5-trisphosphate to inositol in liver, *Nature* **312:**374–376.

Streb, H., Irvine, R. F., Berridge, M. J., and Schultz, I., 1983, Release of C from a non-mitochondrial store in pancreatic acinar cells by inositol 1,4,5-trisphosphate, *Nature* **306:**67–69.

Stuart, J. A., Anderson, K. L., French, P. J., Kirk, R. H., and Michell, R. H., 1994, The intracellular distribution of inositol polyphosphates in HL60 promyeloid cells, *Biochem. J.* **303:**517–525.

Südhof, T. C., 1995, The synaptic vesicle cycle: A cascade of protein–protein interactions, *Nature* **375:**645–653.

Sun, M., Wahlestedt, C., and Reis, D. J., 1992, Inositol hexakisphosphate excites rat medullary sympathoexcitatory neurones *in vivo, Eur. J. Pharmacol.* **215:**9–16.

Szwergold, B. S., Graham, R. A., and Brown, T. R., 1987, Observation of inositol pentakis- and hexakis-phosphates in mammalian tissues by ^{31}P-NMR, *Biochem. Biophys. Res. Commun.* **149:**874–881.

Szuts, E. Z., Wood, S. F., Reid, M. S., and Fein, A., 1986, Light stimulates the rapid formation of inositol trisphosphate in squid retina, *Biochem. J.* **240:**929–932.

Tarver, A. P ., King, W. G., and Rittenhouse, S. E., 1987, Inositol 1,4,5-trisphosphate and inositol 1,2-cyclic 4,5-trisphosphate are minor components of total mass of inositol trisphosphate in thrombin-stimulated platelets, *J. Biol. Chem.* **262:**17268–17271.

Theibert, A. B., Estevez, V. A., Mourey, R. J., Maracek, J. F., Barrow, R. K., Prestwich, G. D., and Snyder, S. H ., 1992, Photoaffinity labeling and characterization of isolated inositol 1,3,4,5-tetrakisphosphate and inositol hexakisphosphate-binding proteins, *J. Biol. Chem.* **267:**9071–9079.

Timerman A. P., Mayrleitner, M. M., Lukas, T. J., Chadwick, C. C., Saito, A., Watterson, D. M., Schindler, H., and Fleischer, S., 1992, Inositol polyphosphate receptor and clathrin assembly protein AP-2 are related proteins that form potassium-sensitive ion channels in planar lipid layers, *Proc. Natl. Acad. Sci. U.S.A.* **89:**8976–8980.

Tomlinson, R. V., and Ballou, C. E., 1962, Myoinositol polyphosphate intermediates in the dephosphorylation of phytic acid by phytase, *Biochemistry* **1:**166–171.

Traynor-Kaplan, A. E., Buranawuti, T., Vajanaphanich, M., and Barrett, K. E., 1994, Protein kinase C activity does not mediate the inhibitory effects of carbachol on chloride secretion by T84 cells, *Am. J. Physiol.* **267:**C1224–C1230.

Vajanaphanich, M., Schultz, C., Rudolf, M. T., Wasserman, M., Enyedi, P., Craxton, A., Shears, S. B., Tsien, R. Y., Barrett, K. E., and Traynor-Kaplan, A., 1994, Long-term uncoupling of chloride secretion from intracellular calcium by Ins(3,4,5,6)P_4, *Nature* **371:**711–714.

Vallejo, M., Jackson, T., Lightman, S., and Hanley, M. R., 1987, Occurrence and extracellular actions of inositol pentakis- and hexakisphosphate in mammalian brain, *Nature* **330:**656–658.

Van Dijken, P., de Hass, J.-R., Craxton, A., Erneux, C., Shears, S. B., and Van Hasstert, 1995, A novel, phospholipase C independent, pathway of Ins(1,4,5)P_3 synthesis in *Dictyostelium* and rat liver, *J. Biol. Chem.*, **270:**29724–29731.

Voglmaier, S. M., Keen, J. H., Murphy, J.-E., Ferris, C. D., Prestwich, G. D., Snyder, S. S., and Theibert, A. B., 1992, Inositol hexakisphosphate receptor identified as the clathrin assembly protein AP-2, *Biochem. Biophys. Res. Commun.* **187:**158–163.

Vucenik, I., and Shamsuddin, A. M., 1994, [^{3}H]Inositol hexaphosphate (phytic acid) is rapidly absorbed and metabolized by murine and human malignant cells *in vitro, J. Nutr.* **124:**861–868.

Wilson, D. W., Lewis, M. J., and Pelham, H.R.B., 1993, pH-dependent binding of KDEL to its receptor *in vitro, J. Biol. Chem.* **268:**7465–7468.

Wong, N. S., Barker, C. J., Morris, A. J., Craxton, A., Kirk, C. J., and Michell, R. H., 1992, The inositol phosphates in WRK1 rat mammary tumour cells, *Biochem. J.* **286:**459–468.

Wood, S. F., Szuts, E. Z., and Fein, A., 1989, Inositol trisphosphate production in squid photoreceptors. Activation by light, aluminum fluoride and guanine nucleotides, *J. Biol. Chem.* **264:**12970–12976.

Worrell, R. T., and Frizzell, R. A., 1991, CaMKII mediates stimulation of chloride conductance by calcium in T84 cells, *Am. J. Physiol.* **260:**C877–C882.

Wu, J. T., Gong, Q., Chou, R. H., and Weiland, S. J., 1991, Ca^{2+}-insensitive modulation of a K^+ conductance by inositol polyphosphates, *J. Biol. Chem.* **266:**14893–14895.

Xie, J., and Fleischer, B., 1994, Isolation and characterization of a novel inositol hexakisphosphate binding protein from mammalian cell cytosol, *Biochemistry* **33:**7908–7916.

Ye, W., Ali, N., Bembenek, M. E., Shears, S. B., and Lafer, E. M., 1995, Inhibition of clathrin assembly by high-affinity binding of specific inositol polyphosphates to the synapse-specific clathrin assembly protein AP-3, *J. Biol. Chem.* **270:**1564–1568.

Yoshida, M., Kanematsu, T., Watanabe, Y., Koga, Y., Ozaki, S., Iwanaga, S., and Hirata, M., 1994, D-myo-inositol 1,4,5-trisphosphate binding in rat brain membranes, *J. Biochem.* **115:**973–980.

Zhang, J. Z., Davletov, B. A., Südhof, T. C., and Anderson, R.G.W., 1994, Synaptotagmin I is a high affinity receptor for clathrin AP-2: Implications for membrane recycling, *Cell* **78:**751-760.

Chapter 8

Inositol Phosphates and Their Metabolism in Plants

Pushpalatha P. N. Murthy

1. INTRODUCTION

The discovery that plasma membrane phosphoinositides mediate cellular responses to external signals has led to tremendous interest in the structure and metabolism of phosphoinositides and inositol phosphates. The role of inositol-1,4,5-trisphosphate [$Ins(1,4,5)P_3$] as a mediator of receptor-initiated changes in intracellular calcium is well characterized (Berridge, 1993; Berridge and Irvine, 1989). The role of $Ins(1,3,4,5)P_4$ (the abbreviations used are defined in Section 2) in regulating cellular calcium entry at the plasma membrane (Berridge, 1993; Berridge and Irvine, 1989) is evolving rapidly. Neurotransmitter functions for $Ins(1,3,4,5,6)P_5$ and $InsP_6$ (Vallejo *et al.*, 1988) have been proposed. The ability of $Ins(1,3,4,5,6,)P_5$ to modulate the affinity of hemoglobin for oxygen is widely accepted (Isaacks and Harkness, 1980). The possibility that other inositol phosphates found in cells may perform cellular functions has heightened interest in the structure and metabolism of inositol phosphates (Menniti *et al.*, 1993b). *In vivo,* inositol phosphates are interrelated by a complex web of reactions (Majerus *et al.*, 1988), and a rapid dynamic equilibrium exists among the different inositol

Pushpalatha P. N. Murthy Chemistry Department, Michigan Technological University, Houghton, Michigan 49931.

Subcellular Biochemistry, Volume 26: myo-Inositol Phosphates, Phosphoinositides, and Signal Transduction, edited by B. B. Biswas and Susweta Biswas. Plenum Press, New York, 1996.

phosphates. A number of recent reviews (Drøbak, 1992, 1993; Coté and Crain, 1993; Hetherington and Drøbak, 1992; Rincon and Boss, 1990) and other chapters in this book contain information on the structure and metabolism of phosphoinositides in plant cells. This review describes the pathways and enzymes involved in the metabolism of inositol phosphates in plant cells.

2. NOMENCLATURE OF *myo*-INOSITOL AND ITS PHOSPHATES

Cyclitols are cyclic carbohydrates containing more than two hydroxyl groups on ring atoms. These compounds contain special elements of absolute and relative configuration that are not clearly indicated by the general methods of nomenclature. Therefore, special rules for specifying structures have been proposed (IUB, 1989; IUPAC, 1976). Significant confusion exists in the nomenclature of inositol because of a number of changes recommended by the IUPAC Commission on the Nomenclature of Organic Chemistry (1976) and the International Union of Biochemistry Nomenclature Committee (1989) over the past 25 years. The inherent stereochemical complexity of inositols and their derivatives and the resulting problems in nomenclature have been expertly discussed in a number of publications (Parthasarathy and Eisenberg, 1986, 1990; Cosgrove, 1980a) that will prove illuminating to the interested reader.

To completely define *myo*-inositol, a meso compound, or its derivatives, which may be chiral or meso, both the absolute configuration (D or L) and the positional numbers have to be specified. The numbering system currently in practice for *myo*-inositol is shown in Figure 1. According to IUPAC recommendation I-4 (1976), locants (positional numbers), and thus the direction of numbering, are assigned to the carbon atoms of the ring with reference to the steric relations and the nature of substituents attached to the ring. In *myo*-inositol or its derivatives, substituents above the plane of the ring constitute one set, and those lying below the plane a second set. Lowest locants are then assigned to one set, according to criteria described in recommendation I-4 (IUPAC, 1976). In the case of *myo*-inositol, locants 1, 2, 3, and 5 are assigned to the four hydroxyl groups that are on the same side of the ring (substituents are more numerous on this side of the ring than the other) without specifying the starting point. The starting point could be as shown in **(I)**, D-numbering or **(II)**, L-numbering. The lowest possible locant is then assigned to the substituted carbon. Therefore, in substituted inositol phosphates **(III)** and **(IV)**, the starting points could be either of the enantiotopic carbons, as shown. The configuration, D or L, is specified by the lowest numbered asymmetric carbon, C-1 (IUPAC, 1976), as follows: If the molecule is viewed in the vertical (Fischer–Tollens) projection with C-1 at top and C-2 and C-3 on the front edge of the ring, the configuration is assigned D **(V)** if the hydroxyl group or other substituent at the lowest-numbered chiral center

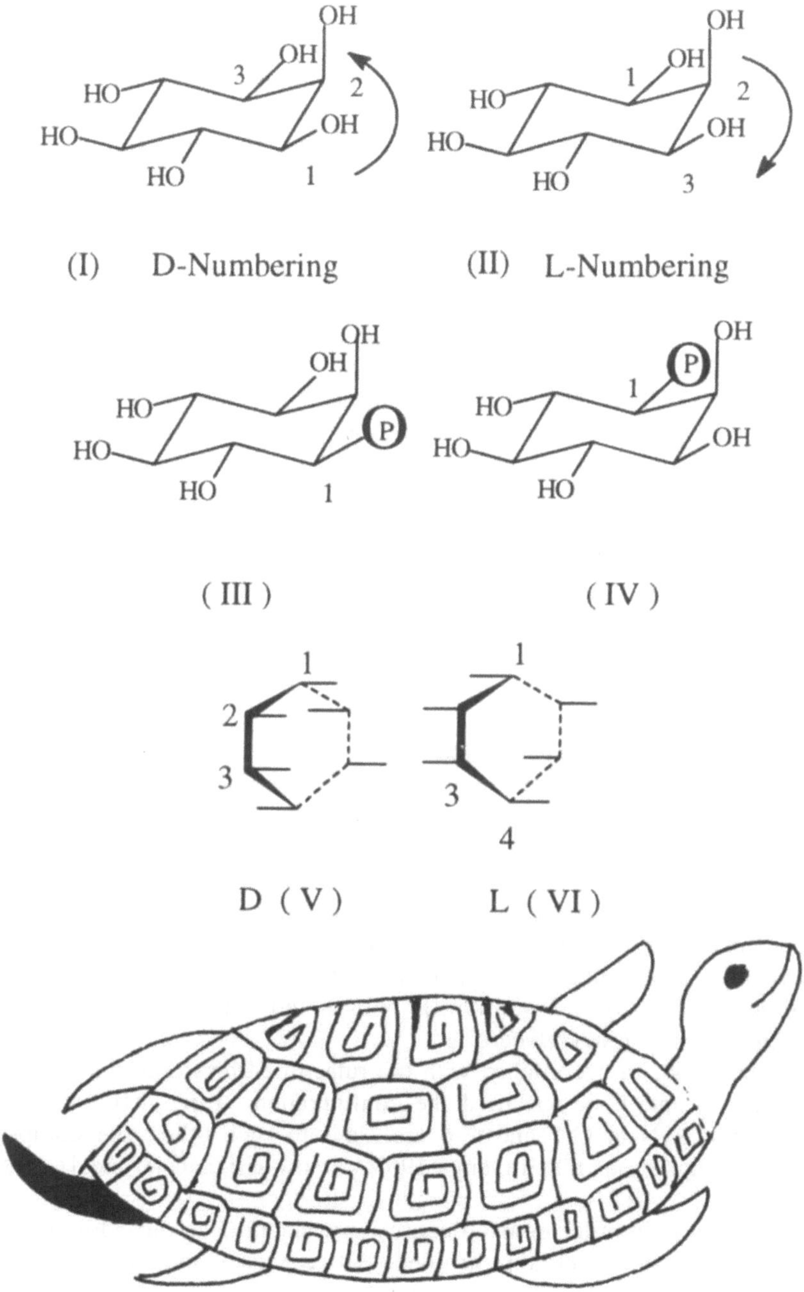

FIGURE 1. Convention for numbering *myo*-inositol and its derivatives. See text for explanation of turtle.

projects to the right and L (**VI**) if it projects to the left. More conveniently, a horizontal projection can be used as follows: when the formula is drawn so that the substituent on the lowest-numbered asymmetric carbon atom is above the plane of the ring, and the numbering is clockwise, the compound is L (**II**); if counterclockwise, it is D (**I**). The "1" in the name indicates that C-1 is the chiral center used to define configuration. According to this rule, enantiomeric compounds **III** and **IV** would be 1D-*myo*-inositol-1-monophosphate and 1L-*myo*-inositol-1-monophosphate, respectively, because the substituent-bearing carbon must be assigned the lowest possible locant. The 1D and 1L prefixes highlight enantiomeric relationships. For meso compounds such as unsubstituted *myo*-inositol or 1,3- or 4,6-disubstituted derivatives, numbering that will lead to the L designation is applied. Therefore, between 1968 and 1986, if the absolute configuration was unspecified, it was taken to imply the L designation.

Strict adherence to the IUPAC rules sometimes obscures metabolic relationships; so, in 1988, the Nomenclature Committee of the International Union of Biochemists (IUB) (1989) recommended that the lowest-locant rule be relaxed and stereospecific numbering be allowed so that structure **III** could be indicated either by designation 1D-*myo*-inositol-1-monophosphate or by 1L-*myo*-inositol-3-monophosphate so long as the prefix 1D or 1L is specified. It is left to the author to use either of the alternative designations, whichever is convenient or more appropriate to the discussion. It was further recommended (IUB, 1989) that the symbol Ins be taken to mean *myo*-inositol with numbering in the 1D configuration, unless 1L is explicitly added. This recommendation was largely a response to the enormous interest in phosphatidylinositol lipids in which the phosphatidyl group is esterified on the D-1 hydroxyl group.

As an aid to remembering the numbering of *myo*-inositol, Agranoff (1978) suggested the image of a turtle (Figure 1). The head of the turtle represents the axial hydroxyl at the 2-position, while the four limbs and the tail represent the five equatorial hydroxyl groups. The right front limb of the turtle is the D-1 position, and, proceeding counterclockwise, the head is the D-2, etc. The left front limb is the L-1 or D-3 position. In phosphoinositides, the right front limb, D-1, is esterified to diacylglycerol.

Before 1968, in accordance with the rules of carbohydrate nomenclature, the highest locant, C-6, specified the configuration, D or L (IUPAC, 1976). Because hydroxyls at C-1 and C-6 are *trans* to each other, the compounds assigned configurations D and L circa 1968 are now assigned 1L and 1D, respectively.

In summary, in order to avoid confusion when reading older literature, the following points should be kept in mind. Between 1968 and 1986, meso compounds were numbered in the L designation, and therefore, if no chirality is specified, assuming the L-designation is a safer choice. In the literature published before 1968, compounds of the L designation are of the current D designation, and vice versa. For example, before 1968, compound **III** would have been labeled L-*myo*-inositol-1-monophosphate (carbon-6 used to assign configuration). After 1968, it would be labeled 1D-*myo*-inositol-1-monophosphate or *myo*-inositol-3-monophosphate (unspecified configuration is taken to mean L). Now, **III** may be labeled Ins(1)P_1, which is the numbering system followed in this chapter.

3. CENTRAL ROLE OF *myo*-INOSITOL IN PLANT METABOLISM

In plant tissues, Ins is converted to a large number of compounds (Coté and Crain, 1993; Drøbak, 1992; Loewus *et al.*, 1990). Figure 2 depicts some of the reactions that involve Ins. The metabolism of Ins can be divided into the following three groups:

1. Biosynthesis of Ins from glucose-6-phosphate (pathway 1) via Ins(3)P_1.
2. Incorporation of the Ins moiety into a variety of compounds such as phosphoinositides and PI-glycans (pathway 2), sphingolipids (pathway 3), inositol phosphates (pathway 4), galactinol (pathway 5), methyl ethers (pathway 6), and indole esters (pathway 7).
3. Reactions on the inositol ring that transform the Ins moiety into other sugars, such as glucuronic acid (pathway 8), and inositol isomers such as chiro-, scyllo-, neo-, and muco- (pathway 9) via methyl ethers (Hoffmann-Ostenhof and Pittner, 1982).

Figure 2 presents a more complete picture of inositol metabolism in plant cells, and the discussion presented in this chapter should be viewed within this larger frame. This chapter includes those aspects of inositol metabolism that relate to inositol phosphates. Other aspects of inositol metabolism have been reviewed elsewhere (Loewus *et al.*, 1990), including other chapters of this book.

Inositol is biosynthesized from glucose-6-phosphate by *myo*-inositol-1L-phosphate synthase. This enzyme is present in plants, animals, and fungi. The enzymatic reaction has been intensively studied, and many details of the isomerization reaction are known. A good discussion of the historical development is presented in several publications (Loewus, 1990a; Loewus and Loewus, 1983; Hoffmann-Ostenhof and Pittner, 1982; Pasternak, 1965). As early as 1900, 42 years before the configuration of Ins was established by Fisher and Posternak, Masquenne postulated that Ins could be the product of head-to-tail cyclization of a hexose molecule. In 1944, Fischer pointed out the similarities in the stereochemical configurations of C-1, C-2, C-3, and C-4 of glucose and D1, D4, D5, and D6 of *myo*-inositol, respectively (Figure 3), and further developed the idea. However, it was not until 1962 that more conclusive evidence was obtained by

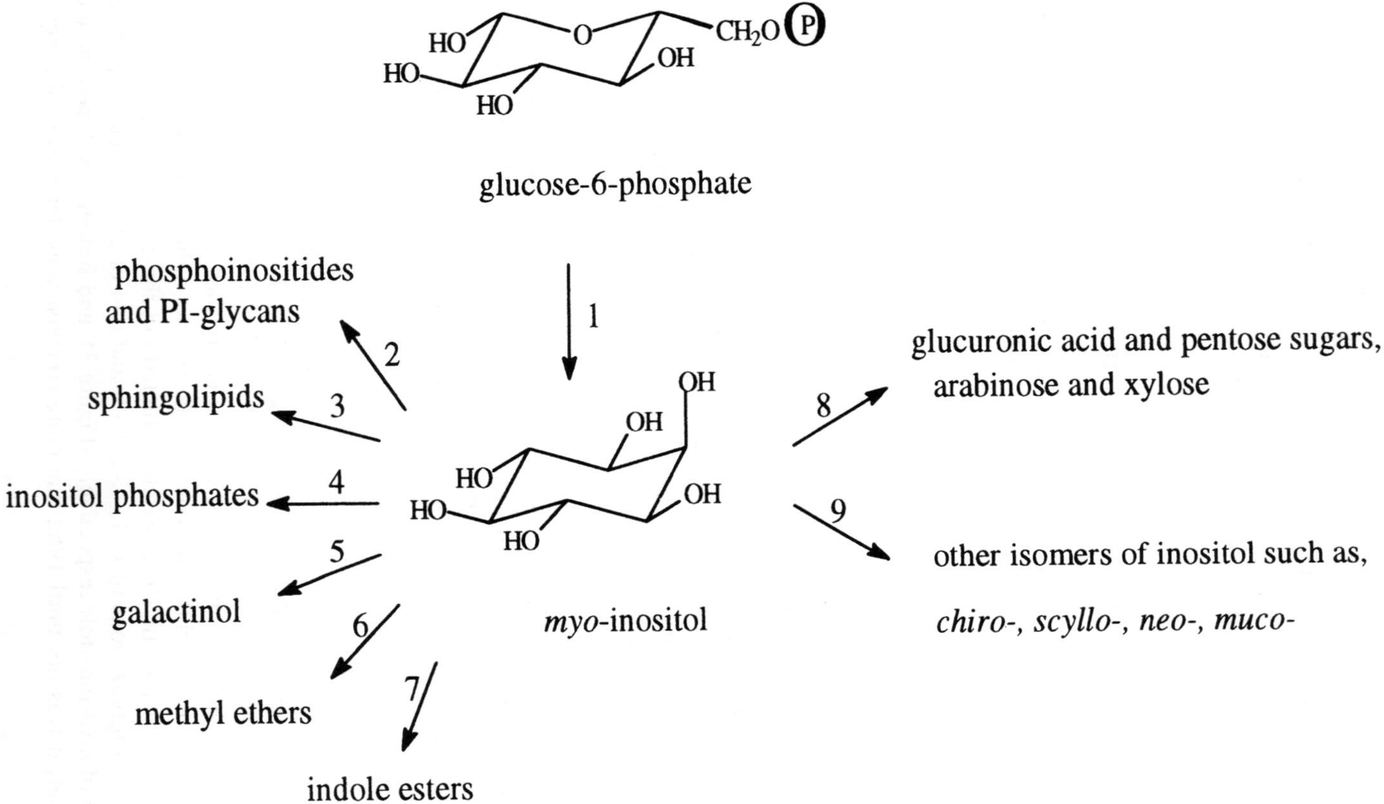

FIGURE 2. Schematic representation of *myo*-inositol metabolism in plant cells.

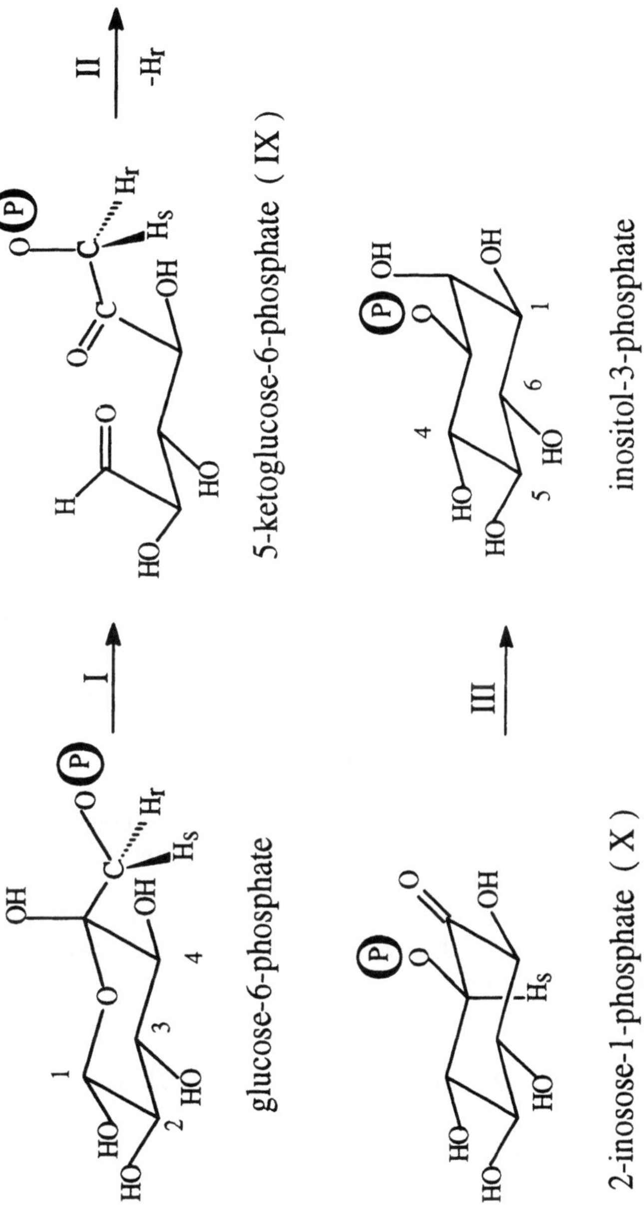

FIGURE 3. Conversion of D-glucose-1-phosphate to *myo*-inositol-3-phosphate.

Loewus and Kelly (1962), who showed that D-[1-^{14}C]glucose was converted to *myo*-[4-^{14}C]inositol in parsley leaves. Three steps are involved in the conversion of D-glucose to *myo*-inositol (Figure 3):

$$\text{D-glucose} \rightarrow \text{D-glucose-6-phosphate} \rightarrow \text{Ins(3)P}_1 \rightarrow \text{Ins}$$

The conversion of glucose-6-phosphate to *myo*-inositol-3-phosphate is the first committed step of inositol metabolism. This reaction is catalyzed by the enzyme 1L-*myo*-inositol phosphate synthase. The enzyme has been extensively studied, and many details of the mechanism are known. The first step of the reaction involves the oxidation of the C-5 carbon (and reduction of NAD^+) to yield 5-ketoglucose-6-phosphate (**IX**). The pro-R hydrogen adjacent to the keto group of **IX** is removed, and the enolate anion undergoes cyclization to yield *myo*-2-inosose-1-phosphate (**X**). Reduction of the keto group by addition of the hydride of the *si* face of carbonyl generates *myo*-inositol-3-phosphate, the final product of the reaction.

This reaction is the only biosynthetic route to *myo*-inositol. It is widely accepted that the 1L-*myo*-inositol phosphate synthase from plants, mammals, and yeast all function in the same manner. The molecular weight of the native enzyme in *Saccharomyces cerevisiae* is approximately 240,000 (Dean-Johnson and Henry, 1989); SDS electrophoresis of the purified enzyme showed the presence of a single polypeptide of molecular weight 62,000. Biosynthesis of the enzyme is regulated by exogenous inositol and unliked regulatory genes. A gene (INO1) corresponding to the enzyme has been isolated from *Saccharomyces cerevisiae*. The largest open reading frame of 553 amino acids predicts a peptide of molecular weight 62,842. The nucleotide sequence of the entire INO1 gene and surrounding regions has been established, and the translational start point and amino terminus have been determined. Information regarding the structure–activity domains and the relative locations of these domains remains to be established.

The action of Ins monophosphatase, a Mg^{2+}-dependent enzyme, hydrolyzes Ins(3)P_1 to inositol (Loewus and Loewus, 1982). Section 4.1 presents further details on this enzyme.

4. METABOLISM OF *myo*-INOSITOL PHOSPHATES

4.1. Inositol Monophosphates

Of the six possible InsP_1 derivatives, four isomers, Ins(1)P_1, Ins(2)P_1, Ins(3)P_1, and Ins(4)P_1, have been detected in plants. As discussed above, the formation of Ins(3)P_1 by 1L-*myo*-inositol phosphate synthase is the first committed step of inositol metabolism. Interestingly, $Ins(3)P_1$ is also produced by an Ins kinase, an ATP- and Mg^{2+}-dependent enzyme. The fact that the only known enzyme capable of directly phosphorylating Ins should produce the same isomer that is produced by 1L-*myo*-inositol phosphate synthase is intriguing.

Phospholipase-C-catalyzed hydrolysis of PI is an alternate route by which inositol monophosphate is produced. Inositol-phospholipid-specific phospholipase C activity has been detected in a number of plant tissues (reviewed in Drøbak, 1992). Both soluble and plasma-membrane-bound forms of the enzymes have been detected. More recently, multiple forms of PI-PLC have been isolated from the cytosol and plasma membrane of oat roots (Huang *et al.*, 1994). Both $Ins(1)P_1$ and $cIns(1,2)P_1$ were produced by the enzyme from celery stalk (Irvine *et al.*, 1980). Like the mammalian enzymes, PI-PLC from plants hydrolyzes PIP and PIP_2 to produce inositol bis- and trisphosphates. The membrane-bound form from wheat roots exhibits preference for PIP and PIP_2 (Melin *et al.*, 1992).

Both $Ins(1)P_1$ and $Ins(4)P_1$ have been detected in enzyme-catalyzed hydrolysis of $Ins(1,4)P_2$, which was derived by hydrolysis of $Ins(1,4,5)P_3$ (Martinoia *et al.*, 1993; Drøbak *et al.*, 1991; Joseph *et al.*, 1989; Memon *et al.*, 1989); $Ins(2)P_1$ has been identified as the last intermediate of acid-phytase-catalyzed hydrolysis of phytic acid (Cosgrove, 1980c; Lim and Tate, 1973; Tomlinson and Ballou, 1962). To my knowledge, $Ins(5)P_1$ and $Ins(6)P_1$ have not been reported in plant tissues thus far.

Inositol monophosphatases from plants and animals are unusual enzymes in that both $Ins(1)P_1$ and its enantiomer, $Ins(3)P_1$, are hydrolyzed with equal efficiency (Loewus and Loewus, 1982; Ackermann *et al.*, 1987). In addition, the enzyme from plants is able to hydrolyze $Ins(2)P_1$ at 50% the rate of $Ins(1)P_1$ (Loewus and Loewus, 1982), whereas the inositol monophosphatase from mammalian sources is inhibited by $Ins(2)P_1$ (Ackermann *et al.*, 1987). Inositol monophosphatase from lily pollen (Loewus and Loewus, 1982) does not hydrolyze *p*-nitrophenylphosphate or glucose-6-phosphate, differentiating it from nonspecific phosphatases present in cells.

Lithium ions, a drug prescribed in the form of citrate or carbonate salts to treat manic–depressive illness, inhibits the action of inositol monophosphatases from mammalian sources (Hallcher and Sherman, 1980). This inhibitory effect has been extremely valuable in investigating the transient production of $InsP_3$ on receptor activation in mammalian cells (Berridge and Irvine, 1989). The inhibitory action of Li^+ on inositol monophosphatases from plants depends on the tissue and subcellular localization. A partially purified inositol monophosphatase from lily pollen is susceptible to inhibition by Li^+, but at higher concentrations (approximately 50% inhibition at 50 mM Li^+ compared with 0.8 mM Li^+ required to inhibit the enzyme from mammalian sources by 50%). The $InsP_2$ hydrolysis catalyzed by the soluble fractions from carrot suspension culture cells was inhibited by Li^+ to a small extent (25% by 50 mM Li^+) (Memon *et al.*,

1989). Lithium (100 mM) also inhibited Ins(1,4)P_2 hydrolysis by a soluble fraction from tobacco culture cells but had no effect on Ins(1)P_1 dephosphorylation (Joseph *et al.*, 1989). It is likely that the enzyme preparations used in these studies contained phosphatases from multiple subcellular compartments with varying susceptibility to Li^+, and therefore, the observed effect of Li^+ was the net result of the action of multiple phosphatases. This point was borne out by more recent study in which mesophyll cells from barley leaves were fractionated into cytosolic, tonoplast, and soluble vacuolar components, and the effects of Li^+ on inositol phosphatase in these fractions were investigated (Martinoia *et al.*, 1993). Lithium (20 mM) and Ca^{2+} (100 μM) strongly inhibited the hydrolysis of Ins(1,4,5)P_3 to InsP_1 by soluble vacuolar fractions but had only a slight inhibitory effect (10% to 20%) on InsP_3 hydrolysis by cytosolic and tonoplast fractions.

4.2. Inositol Bisphosphates

Of the 15 possible InsP_2 isomers, the following have been identified in plant tissue: Ins(1,2)P_2, Ins(1,4)P_2, and Ins(4,5)P_2. The Ins(1,2)P_2 is formed as an intermediate in acid-phytase-catalyzed hydrolysis of phytic acid (reviewed in Cosgrove, 1980c); Ins(1,4)P_2 and Ins(4,5)P_2 have been identified as hydrolysis products of Ins(1,4,5)P_3 in a number of tissues (see Section 4,3); Ins(1,4,5)P_2 has not been detected in any of the studies investigating Ins(1,4,5)P_3 metabolism; Ins(1,4)P_2 is also produced by the action of PI-PLC on phosphatidylinositol-4-monophosphate. A brief report by Igaue *et al.* (1982) stated that, on the basis of proton-decoupled ^{31}P-NMR, the IP_2 isomers Ins(1,3)P_2 and Ins(2,6)P_2 were characterized, and they propose that these isomers are intermediates in the biosynthesis of phytic acid.

4.3. Inositol Trisphosphates

Of the 20 possible InsP_3 isomers, the following have been identified in plant tissues: Ins(1,2,3)P_3 Ins(1,2,6)P_3, and Ins(1,4,5)P_3. The Ins(1,2,3)P_3 and Ins(1,2,6)P_3 are formed as intermediates in the acid-phytase-catalyzed hydrolysis of phytic acid (reviewed in Cosgrove, 1980c). Both of these triphosphates are further hydrolyzed by acid phytases; however, alkaline phytase from lily pollen and *Typha latifolia* yield Ins(1,2,3)P_3 as the final product (Barrientos *et al.*, 1994, Loewus *et al.*, 1990). The action of PI-PLC on phosphatidylinositol-4,5-bisphosphate results in the formation of Ins(1,4,5)P_3 (Huang *et al.*, 1994). To my knowledge, no information on IP_2 kinases has been published so far.

Interest in Ins(1,4,5)P_3 as a second messenger has led to investigations of its metabolism in a number of tissues (Martinoia *et al.*, 1993; Drøbak *et al.*, 1991;

Joseph *et al.*, 1989; Memon *et al.*, 1989). These investigations have indicated that Ins(1,4,5)P_3 is rapidly hydrolyzed by plant extracts to inositol bisphosphates, Ins(1,4)P_2 and Ins(4,5)P_2, inositol monophosphates, Ins(1)P_1 and Ins(4)P_1, and finally to Ins. Failure to detect Ins(1,5)P_2 in these studies indicates that plant cells contain enzymes that selectively remove phosphates from the 1- and 5-positions of Ins(1,4,5)P_3 but that phosphate from the 4-position is not removed. The similarities end there. Ratios of the intermediate Ins monophosphates and bisphosphate varied with tissue, as did pH optima and sensitivity to metal ions such as Mg^{2+}, Li^{+}, and Ca^{2+}.

$$\mathrm{Ins(1,4,5)P_3} \rightarrow \mathrm{Ins(1,4)P_2} + \mathrm{Ins(4,5)P_2} \rightarrow$$
$$\mathrm{Ins(4)P_1} + \mathrm{Ins(1)P_1} \rightarrow \mathrm{Ins}$$

The effect of calcium ions on phosphatase activity varied with enzyme source. In the soluble fraction from carrot cells grown in suspension culture, $CaCl_2$ (100 μM) enhanced IP_3 dephosphorylation but inhibited IP_2 dephosphorylation (Memon *et al.*, 1989). With a soluble fraction of *Nicotiana tobacam* cells, the inclusion of $CaCl_2$ stimulated the hydrolysis of all inositol phosphate substrates: I(1,4,5)P_3, I(1,4)P_2, I(4,5)P_2, and I(4)P_1 (Joseph *et al.*, 1989). When soluble fraction from pea roots was the source of enzyme, the presence of $CaCl_2$ (400 nM) enhanced the formation of I(4,5)P_2 but did not affect the formation of I(1,4)P_2 (Drøbak *et al.*, 1991). In barley mesophyll cells, $CaCl_2$ (100 μM) inhibited activity of soluble vacuolar fractions but did not affect the activity of cytosolic and tonoplast fractions (Martinoia *et al.*, 1993).

Additional confirmation of the varied biochemical properties of different phosphatases is evident in the effect of Li^{+} ions on the enzyme activity. Lithium (50 mM) had no effect on IP_3 dephosphorylation by soluble fraction from carrot cells but inhibited IP_2 dephosphorylation by 25% (Memon *et al.*, 1989). Lithium (100 mM) inhibited I(1,4)P_2 hydrolysis by a soluble fraction from tobacco cells by 15% but had no effect on I(1)P_1 dephosphorylation (Joseph *et al.*, 1989). When interpreting these results, it is useful to keep in mind that the enzymes in these investigations quite possibly contained multiple phosphatases from different subcellular compartments, and therefore, the variation in biochemical properties and product distribution that was observed could be the net result of the action of multiple phosphatases. Martinoia *et al.*, (1993) fractionated mesophyll cells from barley leaves into cytosolic, tonoplast, and soluble vacuolar components and investigated the metabolic fate of Ins(1,4,5)P_3 in these fractions. Although the same products, Ins(1,4)P_2 and Ins(4,5)P_2 and $InsP_1$ of undetermined regiochemistry, were observed, large differences in the degradation rates of initial trisphosphates and the intermediate bisphosphates and monophosphates were detected. Not surprisingly, vacuolar phosphatases hydrolyzed $InsP_3$ several times faster than cytosolic fractions, which was faster than tonoplast fractions.

The hydrolyzing ability of cytosolic fractions was strictly Mg^{2+} dependent, whereas vacuolar and tonoplast phosphatases were inhibited by Mg^{2+}. The phosphatases in tonoplast produced putative Ins(4,5)P_2 under acidic conditions and Ins(1,4)P_2 at a higher pH. Cytosolic and vacuole soluble phosphatases exhibit similar but less pronounced pH dependence. Lithium (20 mM) and Ca^{2+} (100 μM) strongly inhibited the activity of the soluble vacuolar fraction but had only slight inhibitory effect (10–20%) on $InsP_3$ hydrolysis by cytosolic and tonoplast fractions. This study reiterates the fact that results from unfractionated cellular homogenates should be interpreted with caution.

4.4. Inositol Tetrakisphosphates

Of the 15 tetrakisphosphates possible, the following isomers have been identified in plant tissues: Ins(1,2,3,6)P_4, Ins(1,2,5,6)P_4, D- or L-Ins(1,2,3,4)P_4, and Ins(1,4,5,6)P_4. The first three isomers have been identified as intermediates in the phytase-catalyzed hydrolysis of phytic acid (Cosgrove, 1980c; Barrientos *et al.*, 1994). The $InsP_4$ isomer Ins(1,4,5,6)P_4 was produced by the action of kinases on Ins(1,4,5)P_3 *in vivo*. In addition, unpublished work reported in a recent review (Drøbak, 1993) mentions the formation of small amounts of D/L-Ins(1,2,3,4)P_4 from Ins(1,4,5)P_3, which suggests the action of a 5-phosphatase and a 2-kinase.

4.5. Inositol Pentakisphosphates

Of the six possible $InsP_5$ isomers, Ins(1,2,3,4,6)P_5, Ins(1,2,3,5,6)P_5, and Ins(1,3,4,5,6)P_5 have been identified. All three isomers have been detected in phytase-catalyzed hydrolysis of phytic acid. No report on $InsP_4$ kinases has appeared as yet, and as the biosynthesis of phytic acid has not been established, the structures of the $InsP_4$ intermediates involved in this pathway are unknown.

4.6. Inositol Hexakisphosphate (Phytic Acid)

As is evident from the above discussion, current understanding of the metabolism of inositol phosphates in plant tissues results largely from long-standing interest in phytic acid and more recent interest in Ins(1,4,5)P_3. Phytic acid is a major constituent of seeds and pollen (1–5% of dry weight) (Loewus, 1990b; Raboy, 1990). In mature seeds and lily pollen, phytic acid is localized in membrane-bound, phytate-rich granules. The hydrolysis of phytic acid to Ins by the action of phytases has been extensively investigated, and the specificity of hydrolysis has been elucidated for a number of phytases (Cosgrove, 1980c). However, despite numerous attempts in whole plants, plant organs, subcellular organelles, and cell culture, the biosynthesis of phytic acid is poorly understood

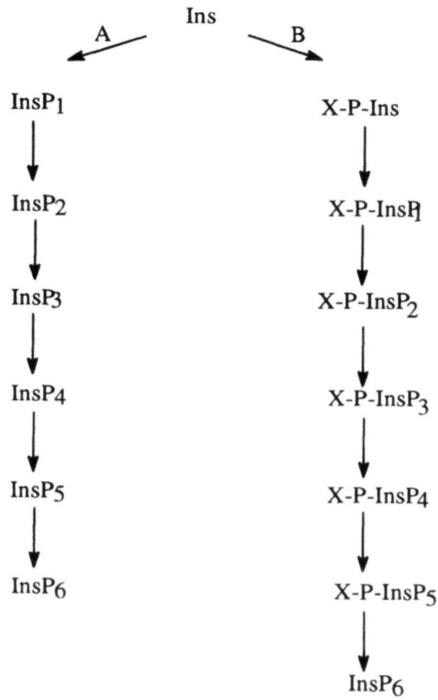

FIGURE 4. Hypothetical pathways for the biosynthesis of phytic acid.

(Scott and Loewus, 1986). The hypotheses that have been proposed can be divided into two categories (Figure 4). One school of thought is that biosynthesis of phytic acid proceeds by sequential phosphorylation of inositol by one or more kinases (pathway A). The second hypothesis is that the biosynthesis involves sequential phosphorylation, either partially or completely, of an inositol derivative such as X-P-Ins, which is then hydrolyzed to phytic acid (pathway B).

The question of whether phytic acid is biosynthesized from Ins or Ins(3)P_1 or both is unresolved (Scott and Loewus, 1986): InsP_1, the first intermediate in pathway A, could be Ins(3)P_1, the product of either Ins(3)P_1 synthase or Ins kinase, or one of the other six monophosphates. Although Ins kinases that yield other isomers of InsP_1 have not yet been identified, other isomers of InsP_1 such as Ins(1)P_1, Ins(4)P_1, and Ins(2)P_1 are produced in plant cells by the action of phytases and phospholipase C. Irrespective of the initial precursor, the formation of intermediary inositol phosphates (InsP_2 to InsP_5) by sequential phosphorylation is supported by a number of studies in duckweed, germinating mung bean, and rice suspension cultures. When incubated with [2-^{14}C]Ins, *Lemna gibba* G3 produced all intermediary forms, InsP_1 to InsP_5 whereas in *Wolfialla floridana*, another duckweed, only small amounts of InsP_4 and InsP_5 were detected (Roberts and Loewus, 1968). Bollman *et al.* (1980) isolated three enzyme fractions from *Lemna gibba* that, together, could convert Ins to $InsP_6$. One fraction converted Ins to Ins(3)P_1, a second to $InsP_3$, and a third to $InsP_6$.

A number of experiments conducted with developing mung bean by Biswas *et al.* provided additional information regarding the biosynthesis of phytic acid. Consistent with the operation of a sequential pathway A, Ins phosphates $InsP_1$ to $InsP_6$ were detected during mung bean development. As development progressed, $InsP_6$ was the predominant Ins phosphate (Majumdar and Biswas, 1973; Majumdar *et al.*, 1972). In germinating mung bean, a partially purified protein fraction catalyzed the conversion of Ins(3)P_1 to $InsP_5$, but not to $InsP_6$. However, surprisingly, Ins(2)P_1 was converted to $InsP_6$ (Chakrabarti and Biswas, 1981). The enzyme preparation showed higher affinity for the highly phosphorylated forms in the order $IP_1 < IP_2 < IP_3 < IP_4 < IP_5$. As two biosynthetic routes for the formation of Ins(3)P_1 were already known, the authors considered Ins(3)P_1 to be the most likely monophosphate intermediate. To complete the conversion to $InsP_6$, the authors proposed the action of a phytic acid–adenosine diphosphate phosphotransferase that was identified in germinating mung beans (Biswas *et al.*, 1978).

Further evidence came from investigations using suspension culture rice cells after incorporation of ^{32}Pi and [^{3}H]inositol (Igaue *et al.*, 1980). Inositol phosphate content at 3 hr, 27 hr, and 4 days was determined by measuring phosphorus content, ^{32}P content, and ^{3}H content in separated inositol phosphates. The presence of $InsP_1$, $InsP_2$, $InsP_3$, $InsP_4$ (two isomers), $InsP_5$ (three isomers), and $InsP_6$ were observed. A brief report stated that the structures of inositol phosphates were determined by proton-decoupled ^{31}P-NMR, and the data were interpreted to indicate the operation of two pathways, one starting with Ins(3)P_1 and a second starting with Ins(2)P_1 below (Igaue *et al.*, 1982). However, ^{31}P chemical shifts dispersed over a narrow range, resulting in significant overlap of resonances (Barrientos *et al.*, 1994; Costello *et al.*, 1976; Emsley and Niazi, 1981), and additional details of structural assignment would be useful.

$$\text{Ins(3)P}_1 \rightarrow \text{Ins(1,3)P}_2 \rightarrow \text{Ins(1,3,5)P}_3 \rightarrow \text{Ins(1,3,5,6)P}_4 \rightarrow \text{Ins(1,2,3,5,6)P}_5 + \text{Ins(1,3,4,5,6)P}_5 \rightarrow \text{InsP}_6$$

$$\text{Ins(2)P}_1 \rightarrow \text{Ins(2,6)P}_2 \rightarrow \text{Ins(2,5,6)P}_3 \rightarrow \text{Ins(2,3,5,6)P}_4 \rightarrow \text{Ins(2,3,4,5,6)P}_5 \rightarrow \text{InsP}_6$$

Recent work by Stephens and Irvine (1990) indicates that, in *Dictyostelium* the biosynthesis of phytic acid proceeds by the formation of the following intermediates: Ins(3)P_1, Ins(3,6)P_2, Ins(3,4,6)P_3, Ins(1,3,4,6)P_4, and Ins(1,3,4,5,6)P_5.

These results provide strong evidence for the sequential phosphorylation of Ins. The presence of Ins(2)P_1 and its conversion to phytic acid by both rice cells and germinating mung bean is very intriguing. No evidence for an Ins kinase that phosphorylates at the 2-position is available so far. However, Ins(2)P_1 is an intermediate in the phytase-catalyzed hydrolysis of phytic acid. Whether these steps are catalyzed by one kinase or multiple kinases is unknown. Evidence by Greenwood and Bewley (1984) suggests that phytic acid biosynthesis may occur in cisternal endoplasmic reticulum, and the product subsequently deposited in phytin granules (Loewus *et al.*, 1990; Scott and Loewus, 1986).

Evidence in support of pathway B has been more indirect. Absence of detectable levels of inositol phosphates during seed development when $InsP_6$ biosynthesis is occurring has been interpreted as evidence supporting pathway 2 (Asada *et al.*, 1969). The presence of phospholipase C that yields $Ins(1,4,5)P_3$ from PIP_2 and kinases that phosphorylate $Ins(1,4,5)P_3$ to Ins(1,3,4,5) and to $InsP_5$ and $InsP_6$ have been invoked to suggest a pathway for $InsP_6$ biosynthesis (Drøbak, 1992; Raboy, 1990). Just as the presence of enzymes or intermediates does not provide conclusive proof of a biosynthetic pathway, the absence of detectable levels cannot be confidently interpreted as evidence for or against biosynthetic routes. Further work will clarify this question.

The primary enzymes responsible for the hydrolysis of phytic acid are phytases. Phytases are a special class of phosphatases that catalyze the sequential hydrolysis of phytic acid to $Ins(2)P_1$ and finally to Ins. Phytases occur in a variety of organisms including plants, animals, and fungi (Cosgrove, 1980b; Gibson and Ullah, 1990). The sequence of hydrolysis by acid phytases was investigated extensively in the 1960s and 1970s, and a detailed review of this work is presented in the book by Cosgrove (1980c). Based on the specificity of initial hydrolysis, two classes of acid phytases are recognized by the IUPAC and IUB (1975), the 6-phytase found in plants and the 3-phytase found in fungi. More recently, the presence of an alkaline phytase (Baldi *et al.*, 1988; Scott and Loewus, 1986) that yields $Ins(1,2,3)P_3$ as a final product (Barrientos *et al.*, 1994) was discovered in lily pollen. Subsequently, the presence of alkaline phytase in a variety of legumes was shown (Scott, 1991). The presence of multiple enzymes in wheat bran phytase and lily pollen with differing specificity, pH optima, and biochemical properties seems to suggest that the hydrolysis of phytic acid may be under the control of multiple phytases (Baldi *et al.*, 1988; Lin *et al.*, 1987). The physiological relevance of the multiple enzymes and products remains to be established.

The sequence of hydrolysis of phosphate groups has been investigated in plants (wheat bran phytase, mung bean, and lily pollen) and microorganisms (*Aspergillus*, yeast, *E. coli*). Specificity of hydrolysis by some plant phytases is shown in Figure 5. In all phytases studied thus far (Cosgrove, 1980c; Barrientos *et al.*, 1994), hydrolysis occurs adjacent to the free hydroxyl group, and therefore, the initial position of hydrolysis is a major determinant of subsequent points of hydrolysis. Attack adjacent to a free hydroxyl group could result from the greater nucleophilicity of the hydroxyl oxygen compared to the ester bond oxygen (Cosgrove, 1980c).

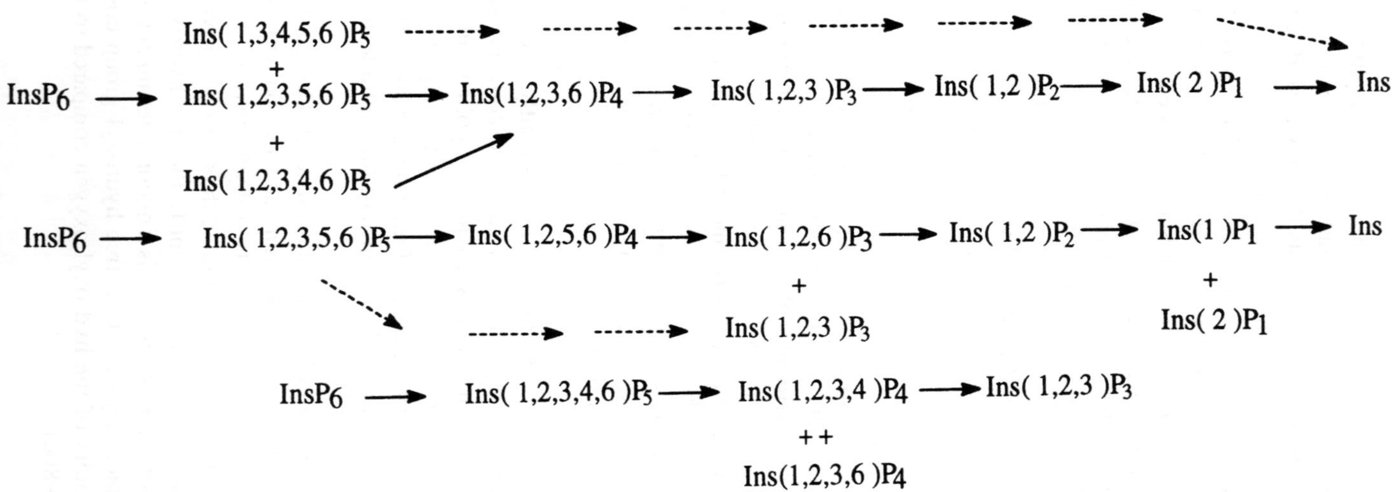

FIGURE 5. Hydrolysis of phytic acid by phytases. Top: Hydrolysis of phytic acid by wheat bran phytase fraction F2. Middle: Hydrolysis of phytic acid by wheat bran phytase fraction F1. Bottom: Hydrolysis of phytic acid by alkaline phytase from lily pollen.

Previous investigations of phytases have relied heavily on indirect methods such as degradation of inositol phosphates to acyclic sugars by a sequence of reactions (periodate cleavage, reduction, and dephosphorylation) and subsequent identification of the final product by a variety of chromatographic techniques (Lim and Tate, 1973). These methods are extremely elaborate and time-consuming. More recently, in our investigation of the specificity of alkaline phytase (Barrientos *et al.*, 1994), the structures of the intermediates and final product were established by two-dimensional NMR techniques (total correlation spectroscopy and ^{1}H–^{31}P correlation spectroscopy), which are substantially faster, provide more conclusive structural information, and obviate the need to separate isomers.

Phytic acid is ubiquitous in cereal grains and legumes, which make up the bulk of animal feed. Reduction of phytic acid is desirable because numerous studies have led to the conclusion that phytic acid–mineral ion complex formed in the digestive tract is not readily absorbed into the bloodstream (Maga, 1982). This results in reduced bioavailability of both phosphorus and minerals to the animal. Therefore, the addition of phytase as a food supplement has been attractively viewed by agricultural companies. A gene from *Aspergillus niger* coding for extracellular glycosylated phytase was isolated (Pen *et al.*, 1993; van Hartingsveldt *et al.*, 1993) and engineered into transgenic tobacco seeds. The enzyme was expressed as 1% of the soluble protein in mature seeds. In addition, supplementation of chicken feed with transgenic seeds led to improved growth rate (Pen *et al.*, 1993).

5. SYNTHETIC ANALOGUES OF INOSITOL PHOSPHATES

The second messenger role of Ins(1,4,5)P_3 and the rapid phosphorylations and dephosphorylations that these compounds undergo have made them attractive targets for drug design. Numerous research efforts to synthesize and investigate the biological activity of inositol phosphate analogues have been undertaken. These research efforts have been triggered by the belief that such chemicals will be useful in further delineating the biological role of InsP_3 and in the hope that these analogues may find applications as pharmaceuticals. Toward this end, synthetic methods for the stereoselective formation of inositol phosphates and analogues of inositol phosphates and separation methods for the resolution of stereoisomers have been developed. The substantial body of knowledge that has resulted from these synthetic endeavors have been discussed in two recent books (Billington, 1993; Reitz *et al.*, 1991). The rationale of analogue synthesis and the information gleaned from some analogues will be briefly presented here. Synthetic analogues have been aimed at the following targets: the Ins(1,4,5)P_3 receptor, with the rationale that the analogue may compete at the

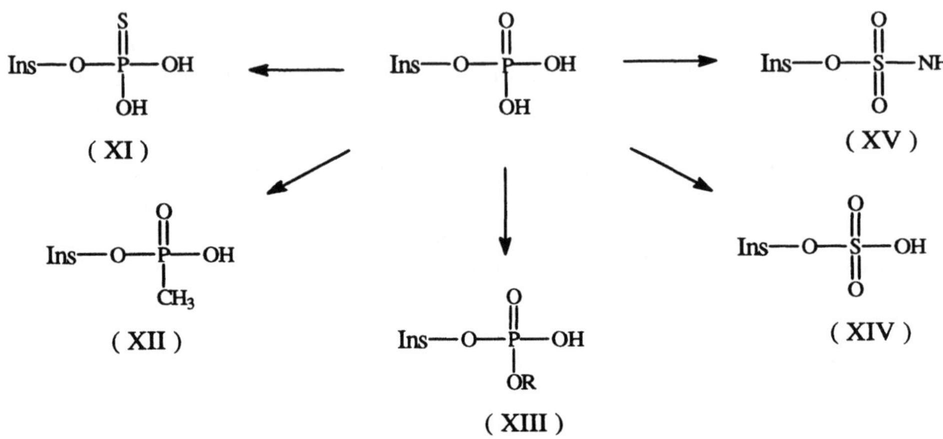

FIGURE 6. Analogues of inositol phosphates.

receptor binding site and function as an agonist or antagonist; and inositol phosphatases and kinases, with the rationale that the analogue may be stable to the action of these enzymes and thereby result in a longer half-life. In addition, analogues for affinity chromatography and cell-permeable caged molecules for intracellular calcium studies have been reported (Billington, 1993; Reitz *et al.*, 1991).

On the basis of chemical structure, the synthetic analogues can be divided into two broad categories, compounds that contain modifications of the phosphate group (Figure 6) and compounds in which the hydroxyl groups have been modified or replaced (Figure 7). Modifications of the phosphate group, or replacement by a phosphate isosteric group, include (Figure 6) replacement by phosphorothioate (**XI**) (to reduce hydrolysis by phosphatases), methylphosphonate (**XII**), or phosphate esters (**XIII**). In addition, replacement of the phosphate group by sulfate (**XIV**) or sulfamoylate (**XV**) has been reported. Modifications of the hydroxyl groups (Figure 7) include replacement by hydrogen (**XVI**), fluorine (**XVII**), an ether linkage (**XVIII**), an alkyl group (**XIX**), or inversion of stereochemistry (**XX**) at any of the six carbons of the ring.

A substantial amount of synthetic effort has been aimed at synthesizing analogues of Ins(1,4,5)P_3 because of its role as a regulator of cellular calcium, and of inositol monophosphatase because of its pivotal role in the recycling of inositol phosphates in animal cells. To illustrate the utility of synthetic analogues, information obtained from a few Ins(1,4,5)P_3 analogues will be discussed. The 1,4,5-trisphosphorothioate analogue of Ins(1,4,5)P_3 binds to the IP_3 receptor (Taylor *et al.*, 1988; Willcocks *et al.*, 1988) and releases calcium from intracellular stores (Taylor *et al.*, 1989). It is only three times less potent than Ins(1,4,5)P_3 (Taylor *et al.*, 1988). As expected, it is not hydrolyzed by the

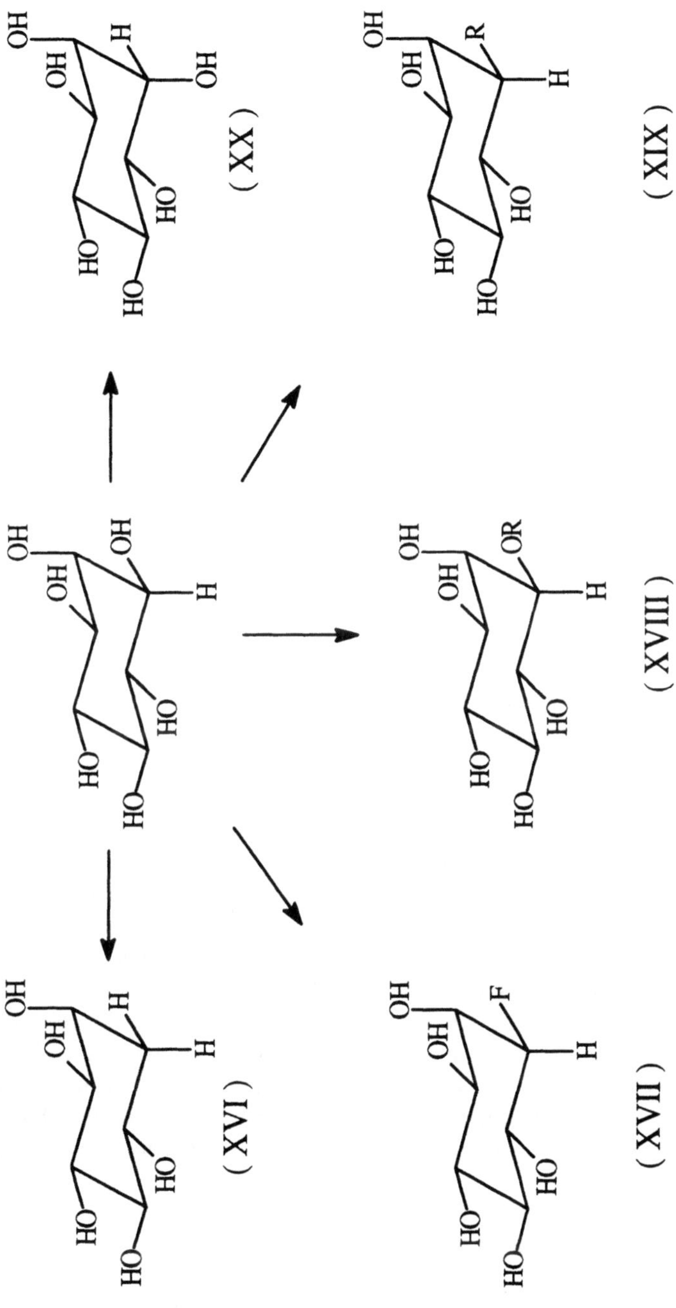

FIGURE 7. Analogues at the D-1 position of *myo*-inositol.

5-phosphatase that acts on Ins(1,4,5)P_3 and consequently gives rise to long-lived calcium increases (Taylor *et al.*, 1988; Willcocks *et al.*, 1988). It is a potent inhibitor of 5-phosphatase (Cooke *et al.*, 1989a) but not an inhibitor of Ins(1,4,5)P_3 3-kinase (Taylor *et al.*, 1989; Safrany *et al.*, 1991; Wojcikiewicz *et al.*, 1990).

In contrast to the trisphosphorothioate, the monophosphorothioate analogue that contains phosphorothioate at the 5-position and phosphates at the 1- and 4-positions binds to the Ins(1,4,5)P_3 receptor, releases calcium, inhibits the 5-phosphatase, and binds to the 3-kinase (Safrany *et al.*, 1991; Cooke *et al.*, 1989a). Similarly, the analogue with the phosphorothioate at the 1-position (**XXI**) also releases calcium from permeabilized cells. In addition, **XXI** has been used as a base for the attachment of a photoaffinity probe, which was used to label three IP_3-binding proteins (Schäfer *et al.*, 1990; Henne *et al.*, 1988).

A methylphosphonate, **XXII,** is the only analogue known to act as a antagonist of Ins(1,4,5)P_3-stimulated calcium release (Dreef *et al.*, 1991a,b) and therefore promises to provide some fascinating details of IP_3 action. The fluorinated analogue (**XXIII**) showed no biological activity (Dreef *et al.*, 1991b). A number of Ins(1,4,5)P_3 analogues in which the phosphate group at the 1-position was in the form of a diester (of glycerol, short-chain aldehyde and acids, and quaternary ammonium salts) all act as agonists at the Ins(1,4,5)P_3 receptor, suggesting that substantial modification at the 1-position is tolerated (Henne *et al.*, 1988). The immobilized analogue (**XXIV**) has been successfully used to purify the Ins(1,4,5)P_3 receptor by affinity chromatography (Prestwich, 1991; Lampe and Potter, 1990).

To convert inositol phosphates to a cell-permeable analogue, "caged" analogues of Ins(1,4,5)P_3 in which the phosphate diester contains a photolabile group have been synthesized (Billington, 1993; Adams *et al.*, 1988; Gurney and Lester, 1987; Walker *et al.*, 1987; Tsien and Zucker, 1986). Once inside the cell, active IP_3 can be released by photolytic deprotection. The use of "caged" molecules in stomatal guard cells provided strong evidence that Ins(1,4,5)P_3 plays a role in calcium regulation in plant cells (Gilroy *et al.*, 1990).

The use of the synthetic analogues in *in vitro* assays has contributed significantly to our understanding of the functional groups required for substrate recognition and catalytic activity (reviewed in Billington, 1993). Despite the significant progress in synthesizing analogues that compete at the active site in *in vitro* assays, the synthesis of a cell-permeable analogue that functions in intact cells has not been accomplished. However, the enormous strides that have recently been made suggest that a cell-permeable analogue may not be far away.

6. INOSITOL PYROPHOSPHATES

More recently, molecules that contain a pyrophosphate group have been detected in *Dictyostelium discoideum,* a slime mold (Stephens *et al.*, 1993), and

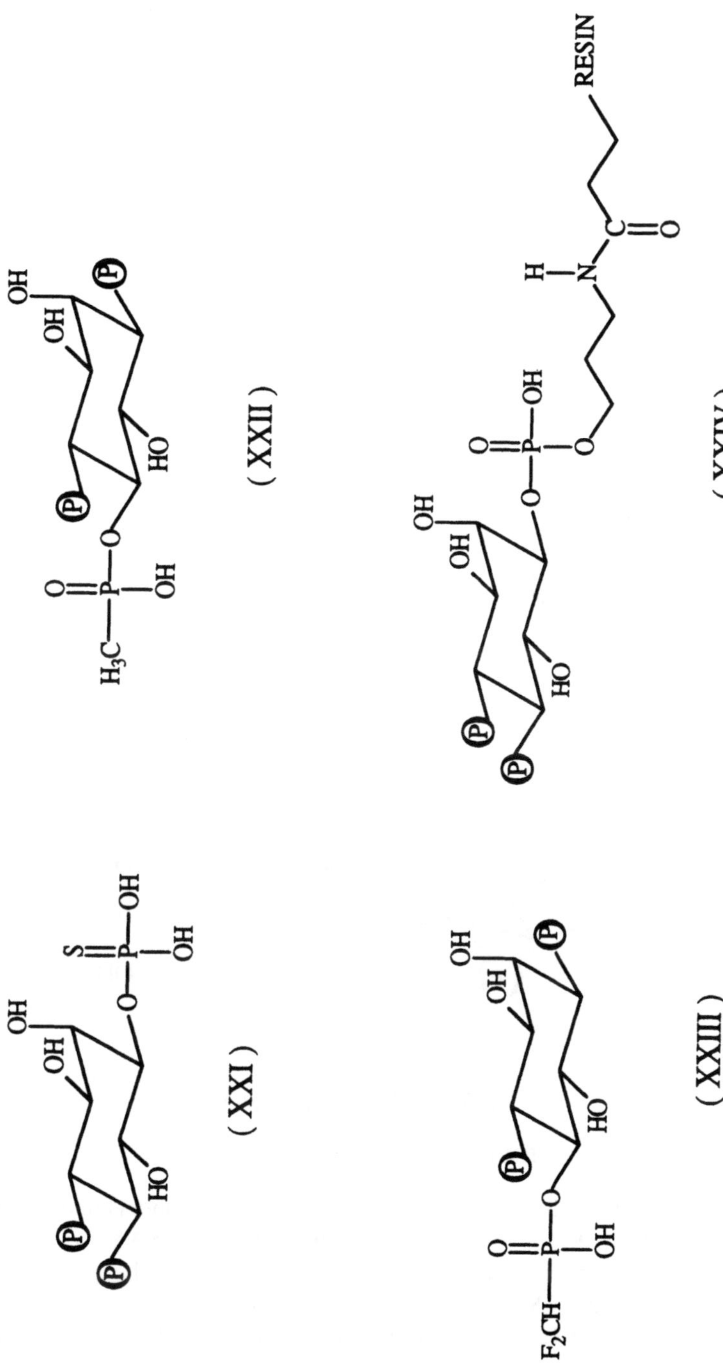

FIGURE 8. Analogues of inositol(1,4,5)trisphosphate.

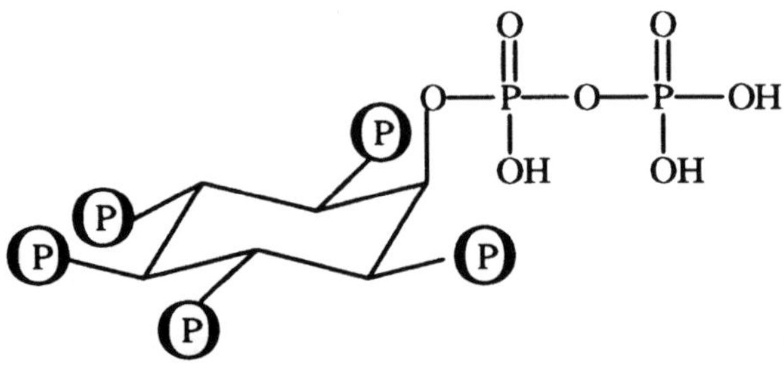

(XXV)

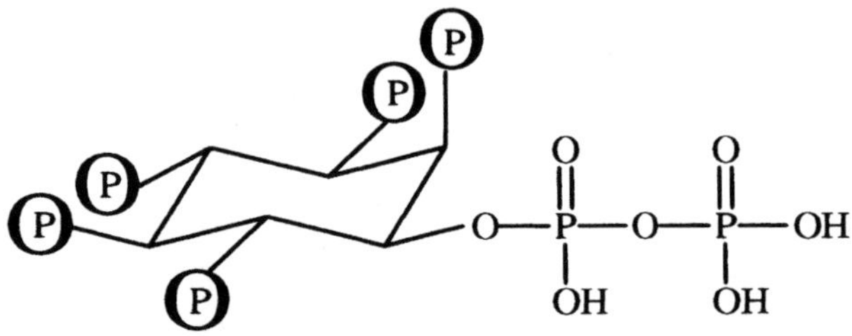

(XXVI) PP-InsP5

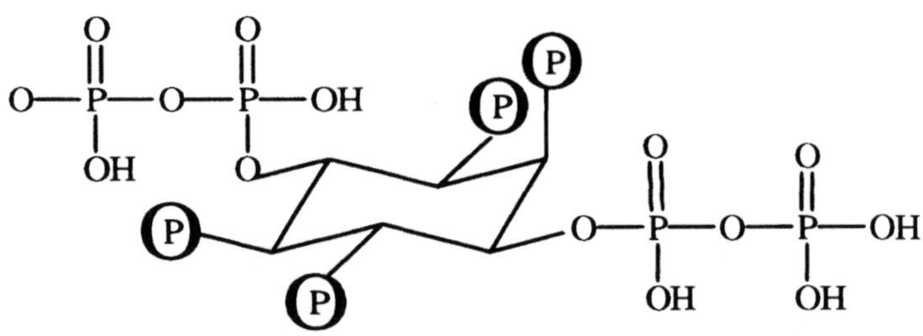

(XXVII) bis PP-InsP4

FIGURE 9. Inositol pyrophosphates.

mammalian cells (Menniti *et al.*, 1993a,b). The pyrophosphorylated phytic acid is indicated as PPInsP$_5$ or InsP$_6$P (Figure 9). Although **XXV** was originally discovered as an artifact of workup (Mayr *et al.*, 1992), the pyrophosphate-containing compounds **XXVI,** where the pyrophosphate group is at D-1, and a second molecule (**XXVII**) with pyrophosphates at D-1 and D-2 or D-1 and D-4 have been found to occur naturally (Stephens *et al.*, 1993). The endogenous concentrations of the pyrophosphate-containing compounds were found to be in the range 0.05–0.25 mM in *Dictyostelium discoideum.*

Fast metabolic turnover of these compounds was suggested when ARG-2J cells, which contain low concentrations of $PPInsP_4$ and $PPInsP_5$, accumulated significantly higher concentrations in the presence of 1–10 mM fluoride, an inhibitor of pyrophosphatase that degrades $InsP_5P$ and $InsP_6P$ (Menniti *et al.*, 1993). [^{3}H]Inositol-labeled experiments showed that 50% of the Ins(1,3,4,5,6)P_5 pool and 20% of the $InsP_6$ pool in AR4-2J cells is cycled through pyrophosphate every hour! A more complete discussion of these molecules is presented in another chapter of this book.

7. METABOLIC PROCESSES INVOLVING INOSITOL PHOSPHATES

7.1. Signal Transduction

In recent years numerous published reports suggest that signal-induced hydrolysis of phosphoinositides to yield diacylglycerol and Ins(1,4,5)P_3 occurs in plant cells. The signals include phytohormones, light-induced leaf movements, stomatal closure of guard cells, plant–pathogen interactions, and gravity. For a detailed discussion of this vast area, the reader should consult the many extensive and critical reviews that have appeared in the recent past (Drøbak, 1993; Coté and Crain, 1993; Hetherington and Drøbak, 1992; Rincon and Boss, 1990), including Coté and colleagues in Chapter 11 of this volume.

7.2. Stress-Related Processes

Abiotic stresses such as drought, low temperature, and high salinity are environmental factors that result in the accumulation of low-molecular-weight metabolites such as mannitol (Tarczynski *et al.*, 1993), proline, and glycine in both eukaryotes and prokaryotes. It was only recently discovered that high concentrations of cyclitols such as pinitol (**XXVIII**), the 3-methyl ether of *chiro*-inositol, and ononitol (**XXIX**), the 4-methyl ether of *myo*-inositol (Figure 10), are found in high concentrations in many osmotolerant plants. The role of cyclitol methyl ethers in environmental stress response has provided additional impetus for developments in the field of inositol methyl ethers and inositol isomers. Pinitol is thought to be biosynthesized from *myo*-inositol in a two-step sequence involving methylation followed by epimerization (Loewus and Dickinson, 1982; Hoffman-Ostenhoff and Pittner, 1982). With the aim of increasing endogenous concentrations of cyclitols and thereby conferring stress tolerance, a gene, *imt1*, which encodes an inositol methyltransferase, was isolated from a

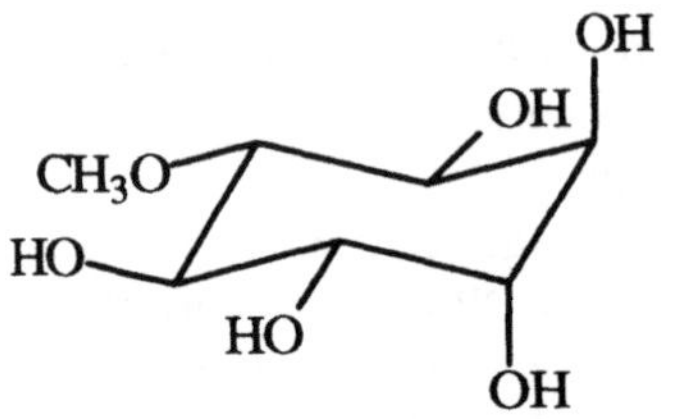

pinitol (XXVIII)

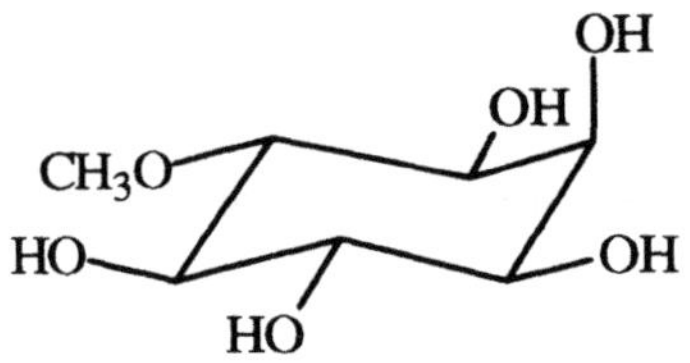

ononitol (XXIX) **FIGURE 10.** Methyl ethers of *myo*- and *chiro*-inositol.

halophyte and expressed in tobacco (Vernon *et al.*, 1993). Increased interest in understanding the factors that determine stress tolerance will result in better understanding of cyclitol biosynthesis.

8. CONCLUSIONS

The role of inositol phosphates in regulating cellular calcium in plant cells is not yet well understood. That inositol tris- and tetrakisphosphates play a critical role in controlling both internal stores and entry from external stores in animal cells is well accepted (Berridge, 1993; Berridge and Irvine, 1989). Observations that Ins(1,4,5)P_3 can release Ca^{2+} from tonoplast vesicles (Schumaker and Sze, 1987) and increase cytoplasmic calcium concentration (Gilroy *et al.*, 1990) provide strong evidence that inositol phosphates may play an active role in calcium regulation in plant cells. The question of whether structural isomers of $InsP_3$ or other inositol phosphates such as $InsP_4$, $InsP_5$, and $InsP_6$ play a role in calcium movement across plant cellular membranes such as plasma membrane, endoplasmic reticulum, and nucleus has yet to be addressed. Intracellular compartmentalization of calcium pools also raises questions regarding spatial and temporal regulation of calcium concentration and the role that inositol phosphates play in maintaining or changing intracellular calcium equilibrium. The role of inositol phosphates in regulating short-term and long-term physiological processes is yet to be addressed.

The recent discovery of inositol pyrophosphates, combined with the information that these compounds turn over at a rapid rate, necessitating the commitment of significant amounts of cellular energy, suggests that these compounds may play an active role in physiological processes (Menniti *et al.*, 1993a,b; Stephens *et al.*, 1993). Rapid turnover of energy-rich inositol pyrophosphates indicates a significant commitment of cellular energy. The question of whether such compounds are present in plant cells and the role that these compounds play in cells will be an area of active research.

Environmental stress-induced changes in cyclitol metabolism have provided much-needed impetus for a more complete investigation of cyclitol metabolism (Vernon *et al.*, 1993). Whether phosphorylated forms of other isomers of inositol are produced and the role of such compounds as metabolic intermediates need to be mapped out.

9. REFERENCES

Ackermann, K. E., Gish, M. P., Honchar, M. P., and Sherman, W. R., 1987, Evidence that inositol-1-phosphate in brain of lithium-treated rats results mainly from phosphatidylinositol metabolism, *Biochem. J.* **242:**517–524.

Adams, S. R., Kao, J.P.Y., Grynkiewicz, G., Minta, A., and Tsien, R. Y., 1988, Biologically useful chelators that release Ca^{+2} upon illumination, *J. Am. Chem. Soc.* **110:**3212–3220.

Agranoff, B. W., 1978, Cyclitol confusion, *Trends Biochem. Sci.* **3:**N283–N285.

Asada, K., Tanaka, K., and Kasai, Z., 1969, Formation of phytic acid in cereal grains, *Ann. N.Y. Acad. Sci.* **165:**801–814.

Baldi, B. G., Scott, J. J., Everard, J. D., and Loewus, F. A., 1988, Localization of constitutive phytases in lily pollen and properties of the pH 8 form, *Plant Sci.* **56:**137–147.

Barrientos, L., Scott, J. J., and Murthy, P.P.N., 1994, Specificity of hydrolysis of phytic acid by alkaline phytase from lily pollen, *Plant Physiol.* **106:**1489–1495.

Berridge, M. J., 1993, Inositol trisphosphate and calcium signalling, *Nature* **361:**315–325.

Berridge, M. J., and Irvine, R. F., 1989, Inositol trisphosphate and diaclyglycerol: Two interactive second messengers, *Nature* **341:**197–205.

Billington, D. C., 1993, *The Inositol Phosphates, Chemical Synthesis and Biological Significance*, VCH Publishers, New York, pp. 87–138.

Biswas, S., Maity, I. B., Chakrabarti, S., and Biswas, B. B., Purification and characterization of *myo*-inositol hexaphosphate–adenosine diphosphate phosphotransferase from *Phaeolus aureus*, *Arch. Biochem. Biophys.* **185:**557–566.

Bollman, O., Strother, S., and Hoffman-Ostenhof, O., 1980, The enzymes involved in the synthesis of phytic acid in *Lemna gibba* (studies on the biosynthesis of cyclitols, XL), *Mol. Cell. Biochem.* **30:**171–175.

Chakrabarti, S., and Biswas, B. B., 1981, Two forms of phosphoinositol kinase from germinating mung bean seeds, *Phytochemistry* **20:**1815–1817.

Cooke, A. M., Nahorski, S. R., and Potter, B.V.L., 1989a, *Myo*-inositol 1,4,5-trisphosphorothioate is a potent competitive inhibitor of human erythrocyte 5-phosphatase, *FEBS Lett.* **242:**373–377.

Cooke, A. M., Noble, N. J., Gigg, R., Willcoks, A. L., Strupish, J., Nahorski, S. R., and Potter, B.V.L., 1989b, Synthesis of *myo*-inositol 1,4-bisphosphate-5-phosphorothioate, *Biochem. Soc. Trans.* **16:**992.

Cosgrove, D. J., 1980a, *Inositol Phosphates, Their Chemistry, Biochemistry and Physiology,* Elsevier, Amsterdam.

Cosgrove, D. J., 1980b, Phytase, in *Inositol Phosphates, Their Chemistry, Biochemistry, and Physiology,* pp. 85–98, Elsevier, Amsterdam.

Cosgrove, D. J., 1980c, Intermediates in the dephosphorylation of P_6-inositols by phytase enzymes, in *Inositol Phosphates, Their Chemistry, Biochemistry, and Physiology,* pp. 99–105, Elsevier, Amsterdam.

Costello, A.J.R., Glonek, T., and Myers, T. C., 1976, ^{31}P nuclear magnetic resonance—pH titrations of *myo*-inositol hexaphosphate, *Carbohydrate Res.* **46:**159–171.

Coté, G. G., and Crain, R. C., 1993, Biochemistry of phosphoinositides, *Annu. Rev. Plant Physiol. Plant Mol. Biol.* **44:**333–356.

Dean-Johnson, M., and Henry, S. A., 1989, Biosynthesis of inositol in yeast: Primary structure of *myo*-inositol-1-phosphate synthase (EC 5.5.1.4) and functional analysis of its structural gene, the *INO1* locus, *J. Biol. Chem.* **264:**1274–1283.

Dreef, C. E., Mayr, G. W., Jansze, J.-P., Roelen, H.C.P.F., Van der Marel, G. A., and van Boom, J. H., 1991a, An expeditious synthesis of biologically important *myo*-inositol phosphorothioates, *Bioorg. Med. Chem. Lett.* **1:**239–242.

Dreef, C. E., Schiebier, W., van der Marel, G. A., and van Boom, 1991b, Synthesis of 5-Phosphorate analogs of *myo*-inositol 1,4,5-trisphosphate: Possible intracellular calcium antagonists, *Tetrahedron Lett.* **32:**6021–6024.

Drøbak, B. K., 1992, The plant phosphoinositide system, *Biochem. J.* **288:**697–712.

Drøbak, B. K., 1993, Plant phosphoinositides and intracellular signaling, *Plant Physiol.* **102:**705–709.

Drøbak, B. K., Watkins, P.A.C., Chattaway, J. A., Roberts, K., and Dawson, A. P., 1991, Metabolism of inositol(1,4,5)trisphosphate by a soluble enzyme fraction from pea (*Pisum sativum*) roots, *Plant Physiol.* **95:**412–419.

Emsley, J., and Niazi, S., 1981, The structure of *myo*-inositol hexaphosphate in solution: ^{31}P N.M.R. investigation, *Phosphorus Sulfur* **10:**401–408.

Gibson, D. M., and Ullah, A.B.J., 1990, Phytases and their action on phytic acid, in *Inositol Metabolism in Plants,* (D. J. Morré, W. F. Boss, and F. A. Loewus, eds.), pp. 77–92, Wiley-Liss, New York.

Gilroy, S., Read, N. D., and Trewavas, A. J., 1990, Elevation of cytoplasmic calcium by caged calcium or caged inositol trisphosphate initiates stomatal closure, *Nature* **346:**769–771.

Greenwood, J. S., and Bewley, J. D., 1984, Subcellular distribution of phytin in the endosperm of developing castor bean: A possibility for its synthesis in the cytoplasm prior to deposition within protein bodies, *Planta* **160:**113–120.

Gurney, A. M., and Lester, H. A., 1987, Light-flash physiology with synthetic photosensitive compounds, *Physiol. Rev.* **67:**583–617.

Hallcher, L. M., and Sherman, W. R., 1980, The effects of lithium ion and other agents on the activity of *myo* inositol-1-phosphatase from bovine brain, *J. Biol. Chem.* **255:**10896–10901.

Henne, V., Mayr, G. W., Grawowski, B., Koppitz, B., and Soeling, H.-D., 1988, Semisynthetic derivatives of inositol 1,4,5-trisphosphate substituted at the 1-phosphate group. Effect on calcium release from permeabilized guinea pig parotid acinar cells and comparison with binding to aldolase A, *Eur. J. Biochem.* **194:**95–101.

Hetherington, A. M., and Drøbak, B. K., 1992, Inositol-containing lipids in higher plants, *Prog. Lipid Res.* **31:**53–63.

Hoffmann-Ostenhof, O., and Pittner, F., 1982, Biosynthesis of *myo*-inositol and its isomers, *Can. J. Chem.* **60:**1863–1871.

Huang, C.-H., Yan, A.K.F., Crain, R. C., and Coté, G. C., 1994, Phosphoinositide specific phospholipase C in oat roots, *Plant Physiol.* **105:**105.

Igaue, I., Shimizu, M., and Miyauchi, S., 1980, Formation of a series of *myo*-inositol phosphates during growth of rice plant cells in suspension culture, *Plant Cell Physiol.* **21:**351–356.

Igaue, I., Miyauchi, S., and Saito, K., 1982, Formation of *myo*-inositol phosphates in a rice cell suspension culture, in *Proceedings of the 5th International Congress of Plant Tissue and Cell Culture,* (A. Fujwara, ed.) pp. 265–266, Mauruzen, Tokyo.

Irvine, R. F., Letcher, A. J., and Rawson, R.M.C., 1980, Phosphatidylinositol phosphodiesterase in higher plants, *Biochem. J.* **192:**279–283.

Isaacks, R. E., and Harkness, D. R., 1980, Erythrocyte organic phosphates and hemoglobin function in birds, reptiles and fishes, *Am. Zool.* **20:**115–129.

IUB Nomenclature Committee, 1989, Numbering of atoms in *myo*-inositol, *Biochem. J.* **258:**1–2.

IUPAC Commission on the Nomenclature of Organic Chemistry and IUPAC-IUB Commission on Biochemical Nomenclature, 1976, Nomenclature of cyclitols, *Biochem. J.* **153:**23–31.

IUPAC-IUB Enzyme Nomenclature Recommendation, 1975, Supplement 1: Corrections and additions, *Biochim. Biophys. Acta* **429:**1–2.

Joseph, S. K., Esch, T., and Bonner, W. D., Jr., 1989, Hydrolysis of inositol phosphates by plant cell extracts, *Biochem. J.* **264:**851–856.

Lampe, D., and Potter, B.V.L., 1990, Synthesis of *myo*-inositol-1-phosphorothioate 4,5-bisphosphate: Preparation of a fluorescently labeled *myo*-inositol 1,4,5-trisphosphate analog, *J. Chem. Soc. Chem. Commun.* 1500–1501.

Lim, P. E., and Tate, M. E., 1973, The Phytases II. Properties of phytase fraction F_1 and F_2 from wheat bran and *myo*-inositol phosphates produced by fraction F_2, *Biochim. Biophys. Acta* **302:**316–328.

Lin, J.-J., Dickinson, D. B., and Ho, T.-H. D., 1987, Phytic acid metabolism in lily (*Lilium longiflorum* Thumb.) pollen, *Plant Physiol.* **83:**408–413.

Loewus, F. A., 1990a, Inositol biosynthesis, in *Inositol Metabolism in Plants* (D. J. Morré, W. F. Boss, and F. A. Loewus, eds.), Wiley-Liss, New York, pp. 13–19.

Loewus, F. A., 1990b, Structure and occurrence of inositol in plants, in *Inositol Metabolism in Plants* (D. J. Morré, W. F. Boss, and F. A. Loewus, eds.), Wiley-Liss, New York, pp. 1–11.

Loewus, F., and Dickinson, D. B., 1982, Cyclitols, in *Encyclopedia of Plant Physiology:* Vol. 13A, *Plant Carbohydrates I: Intracellular Carbohydrates* (F. A. Loewus, and W. Tanner, eds.), pp. 193–206, Springer-Verlag, Berlin.

Loewus, F. A., and Kelly, S., 1962, Conversion of glucose to inositol in parsley leaves, *Biochem. Biophys. Res. Commun.* **7:**204–208.

Loewus, F. A., and Loewus, M. W., 1983, *myo*-Inositol: Its biosynthesis and metabolism, *Annu. Rev. Plant Physiol.* **34:**137–161.

Loewus, F. A., Everard, J. D., and Young, K. A., 1990, Inositol metabolism: Precursor role and breakdown, in *Inositol Metabolism in Plants* (D. J. Morré, W. F. Boss, and F. A. Loewus, eds.), pp. 21–45, Wiley-Liss, New York.

Loewus, M. W., and Loewus, F. A., 1982, *myo*-Inositol-1-phosphatase from the pollen of *Lilium longiflorum* Thumb., *Plant Physiol.* **70:**765–770.

Maga, J. A., 1982, Phytate: Its chemistry, occurrence, food interactions, nutritional significance, and methods of analysis, *J. Agric. Food Chem.* **30:**1–9.

Majerus, P. W., Connolly, T. M., Bansal, V. S., Inhorn, R. C., Ross, T. S., and Lips, D. L., 1988, Inositol phosphates: Synthesis and degradation, *J. Biol. Chem.* **263:**3051–3054.

Majumdar, A.N.L., and Biswas, B. B., 1973, Further characterization of phosphoinositol kinase isolated from germinating mung bean seeds, *Phytochemistry* **12:**315–319.

Majumdar, A.N.L., Mandal, N. C., and Biswas, B. B., 1972, Phosphoinositol kinase from germinating mung bean seeds, *Phytochemistry* **11:**503–508.

Martinoia, E., Locher, R., and Vogt, E., 1993, Inositol trisphosphate metabolism in subcellular fractions of barley (*Hordeum vulgare* L.) mesophyll cells, *Plant Physiol.* **102:**101–105.

Mayr, G. W., Radenberg, T., Theil, U., Vogel, G., and Stephens, L. R., 1992, Phosphoinositol disphosphates: Non-enzymic formation *in vitro* and occurrence *in vivo* in the cellular slime mold *Dictyostelium, Carbohydrate Res.* **234:**247–262.

Melin, P. M., Pical, C., Tergil, B., and Sommarin, M., 1992, Phosphoinositide phospholipase C in wheat root plasma membranes. partial purification and characterization, *Biochim. Biophys. Acta* **1123:**163–169.

Memon, A. R., Rincon, M., and Boss, W. F., 1989, Inositol trisphosphate metabolism in carrot (*Daucus carota* L.) cells, *Plant Physiol.* **91:**477–480.

Menniti, F. S., Miller, R. N., Putney, J. W., Jr., and Shears, S. B., 1993a, Turnover of inositol polyphosphate pyrophosphates in pancreatoma cells, *J. Biol. Chem.* **268:**3850–3856.

Menniti, F. S., Oliver, K. G., Putney, J. W., Jr., and Shears, S. B., 1993b, Inositol phosphates and cell signaling: New views of $InsP_5$ and $InsP_6$, *Trends Biochem. Sci.* **18:**53–56.

Parthasarathy, R., and Eisenberg, F., Jr., 1986, The inositol phospholipids: A stereochemical view of biological activity, *Biochem. J.* **235:**313–322.

Parthasarathy, R., and Eisenberg, F., Jr., 1990, Biochemistry, stereochemistry, and nomenclature of the inositol phosphates, in *Inositol Phosphates and Derivatives: Synthesis, Biochemistry, and Therapeutic Potential* (A. B. Reitz, ed.), pp. 1–19, ACS Symposium Series 463, American Chemical Society, Washington, DC.

Pasternak, T., 1965, *The Cyclitols,* pp. 341–351, Holden-Day, San Francisco.

Pen, J., Verwoerd, T. C., van Paridon, P. A., Beudeker, R. F., van den Elzen, P.J.M., Geerse, K. van der Klis, J. D., Versteegh, H.A.J., van Ooyen, A.J.J., and Hoekema, A., 1993, Phytase-containing transgenic seeds as a novel feed additive for improved phosphorus utilization, *Bio/Technology* **11:**811–814.

Prestwich, G. D., Marecek, J., Mourey, R. J., Theibert, A. B., Ferris, C. D., Danoff, S. K., and Snyder, S. H., 1991, Tethered IP_3, synthesis and biochemical application of the 1-D-(3-aminopropyl)ester of inositol (1,4,5)trisphosphate, *J. Am. Chem. Soc.* **113:**1822.

Raboy, V., 1990, Biochemistry and genetics of phytic acid synthesis, in *Inositol Metabolism in Plants* (D. J. Morré, W. F. Boss, and F. A. Loewus, eds.), pp. 55–76, Wiley-Liss, New York.

Reitz, A. B., 1991, *Inositol Phosphates and Derivatives, ACS Symposium Series 463*. American Chemical Society, Washington, DC.

Rincon, M., and Boss, W. F., 1990, Second messenger role of phosphoinositides, in *Inositol Metabolism in Plants* (D. J. Morré, W. F. Boss, and F. A. Loewus, eds.), pp. 173–200, Wiley-Liss, New York.

Roberts, R. M., and Loewus, F., 1968, Inositol metabolism in plants. VI. Conversion of *myo*-inositol to phytic acid in *Wolffiella floridana, Plant Physiol.* **43:**1710–1716.

Safrany, S. T., Wojcikiewicz, R.J.H., Strupish, J., McBain, J., Cooke, A., Potter, B.V.L., and Nahorski, S. R., 1991, Synthetic phosphorothioate-containing analogs of inositol 1,4,5-trisphosphate mobilize intracellular calcium stores and interact differentially with inositol 1,4,5-trisphosphate 5-phosphatase and 3-kinase, *Mol. Pharmacol.* **39:**754–761.

Schäfer, R., Nehls-Sahabandu, M., Grabowski, B., Dehlinger-Kremer, M., and Mayr, G. W., 1990, Synthesis and application of photoaffinity analogues of inositol 1,4,5-trisphosphate selectively substituted at the 1-phosphate group, *Biochem. J.* **272:**817–825.

Schumaker, K. S., and Sze, H., 1987, Inositol 1,4,5-trisphosphate releases Ca^{++} from vacuolar membrane vesicles of oat roots, *J. Biol. Chem.* **262:**3944–3946.

Scott, J. J., 1991, Alkaline phytase activity in nonionic detergent extracts of legume seeds, *Plant Physiol.* **95:**1298–1301.

Scott, J. J., and Loewus, F. A., 1986, Phytate metabolism in plants, in *Phytic Acid: Chemistry and Applications* (E. Graf, ed.), pp. 23–42, Pilatus Press, Minneapolis.

Stephens, L. R., and Irvine, R. F., 1990, Stepwise phosphorylation of *myo*=inositol leading to *myo*-inositol hexakisphosphate in *Dictyostelium, Nature* **346:**580–583.

Stephens, L. R., Radenberg, T., Thiel, U., Vogel, G., Khoo, K.-H., Dell, A., Jackson, T. R., Hawkins, P. T., and Mayr, G. W., 1993, The detection, purification, structural characterization, and metabolism of diphosphoinositol pentakisphosphate(s) and bisdiphosphoinositol tetrakisphosphate(s), *J. Biol. Chem.* **268:**4009–4015.

Tarczynski, M. C., Jensen, R. G., and Bohnert, H. J., 1993, Stress protection of transgenic tobacco by production of the osmolyte mannitol, *Science* **259:**508–510.

Taylor, C. W., Berridge, M. J., Cooke, A. M., and Potter, B.V.L., 1988, DL-*myo*-Inositol 1,4,5-trisphosphorothioate mobilizes intracellular calcium in Swiss 3T3 cells and *Xenopus* oocytes, *Biochem. Biophys. Res. Commun.* **150:**626–632.

Taylor, C. W., Berridge, M. J., Cooke, A. M., and Potter, B.V.L., 1989, Inositol 1,4,5-trisphosphorothioate, a stable analog of inositol trisphosphate which mobilizes intracellular calcium, *Biochem. J.* **259:**645–650.

Tomlinson, R. V., and Ballou, C. E., 1962, *myo*-Inositol polyphosphate intermediates in the dephosphorylation of phytic acid by phytase, *Biochemistry* **1:**166–171.

Tsien, R. Y., and Zucker, R. S., 1986, Control of cytoplasmic calcium with photolabile tetracarboxylate 2-nitrobenzhydrol chelators, *Biophys. J.* **50:**843–853.

Vallejo, M., Jackson, T., Lightman, S., and Hanley, M. R., 1988, Occurrence and extracellular actions of inositol pentakis and hexakis-phosphate in mammalian brain, *Nature* **330:**656–658.

van Hartingsveldt, W., van Zeijl, C.M.J., Harteveld, G. M., Gouka, R. J., Suykerbuyk, M.E.G., Luiten, R.G.M., van Paridon, P. A., Selten, G.C.M., Veenstra, A. E., van Gorcom, R.F.M., and van den Hondel, C.A.M.J.J., 1993, Cloning, characterization and overexpression of the phytase-encoding gene (*phyA*) of *Aspergillus niger, Gene* **127:**87–94.

Vernon, D. M., Tarczynski, M. C., Jensen, R. G., and Bohnert, H. J., 1993, Cyclitol production in transgenic tobacco, *Plant J.* **4:**199–205.

Walker, J. W., Somlyo, A. V., Goldman, Y. E., Somlyo, A. P., and Trentham, D. R., 1987, Kinetics of smooth and skeletal muscle activation by laser pulse photolysis of caged inositol 1,4,5-trisphosphate, *Nature* **327:**249–252.

Willcocks, A. L., Potter, B.V.L., Cooke, A. M., and Nahorski, S. R., 1988, *Myo*-inositol (1,4,5) trisphosphorothiorate binds to specific [^{3}H]inositol (1,4,5) trisphosphate sites in rat cerebellum and is resistant to 5-phosphatase, *Eur. J. Pharmacol.* **155:**181–183.

Wojckiewicz, R.J.H., Cooke, A. M., Potter, B.V.L., and Nahorski, S. R., 1990, Inhibition of inositol 1,4,5-trisphosphate metabolism in permeabilized SH-5Y5Y human neuroblastoma cells by a phosphorothioate-containing analog of inositol 1,4,5-trisphosphate, *Eur. J. Biochem.* **192:**459–467.

Chapter 9

Genetics of *myo*-Inositol Phosphate Synthesis and Accumulation

Victor Raboy and Paolo Gerbasi

1. INTRODUCTION

Two cellular functions for the *myo*-inositol (Ins) polyphosphate esters (Ins bis-, tris-, tetrakis-, pentakis-, and hexakisphosphate or phytic acid) are prominent in the literature. These are as metabolites in phosphate and mineral storage or regulation pathways and as metabolites in signal transduction pathways (see below). A third discrete function involves the modulation of hemoglobin oxygen affinity in the avian erythrocyte (Isaacks and Harkness, 1980). Recently Ins hexa-, hepta-, and octaphosphates containing one or two pyrophosphate moieties have been identified in a number of cell types (Mayr *et al.*, 1992; Menniti *et al.*, 1993). They are probably synthesized from Ins pentakis- and hexakisphosphate as part of a kinase–pyrophosphatase cycle. This cycle may represent an additional discrete role for Ins phosphate metabolism as a component of "high-energy" phosphate bond metabolism.

Although it was identified first as an abundant organic phosphate in the plant seed and subsequently as a common constituent of other plant tissues, particularly those with clear storage functions (reviewed in Cosgrove, 1980;

Victor Raboy and Paolo Gerbasi USDA-ARS Range Weeds and Cereals Research Unit, Montana State University, Bozeman, Montana 59717.

Subcellular Biochemistry, Volume 26: myo-Inositol Phosphates, Phosphoinositides, and Signal Transduction, edited by B. B. Biswas and Susweta Biswas. Plenum Press, New York, 1996.

Raboy, 1990), phytic acid may in fact be ubiquitous in eukaryotic tissues (Heslop *et al.*, 1985; Loewus *et al.*, 1990; Stephens *et al.*, 1991; Hawkins *et al.*, 1993). Analyses conducted to date indicate that phytic acid and Ins pentakisphosphate represent the bulk of total Ins phosphates observed in mammalian cells (Berridge and Irvine, 1989; Heslop *et al.*, 1985; Menniti *et al.*, 1993). The storage and subsequent redistribution of nutrients play an important role in plant biology. Of the major classes of nutrient stores, phosphate and mineral storage and redistribution often involve the synthesis of phytic acid, the subsequent deposition of phytic acid as a mixed "phytin" salt of mineral cations, followed by disassembly of phytin deposits.

Several lines of evidence (reviewed in Raboy, 1990) indicate that phytic acid metabolism in plant tissues may play a role in the regulation of phosphate and mineral concentration. In this context, the "storage" function may simply represent one component of the "regulation" function. Phytic acid is a strong chelator of mineral cations and a common counterion in their salts (Lott, 1984). It is this functional property that is taken advantage of in its role in mineral storage and regulation. Two examples serve to illustrate the different ways phytic acid as mineral chelator may be put to use. In the first example, an elegant study of cereal grain development revealed that differences in the time course of the formation of potassium and magnesium phytin deposits reflects time-course differences in major storage metabolite pathways which require potassium and magnesium (Ogawa *et al.*, 1979). The second example is the more recent interest in phytic acid as a putative anti-oxidant and anti-colon cancer agent; phytic acid is an effective chelator of iron, and thus possibly an inhibitor of iron-catalyzed hydroxyl radical formation (Graf and Eaton, 1993; Hawkins *et al.*, 1993).

The role that phosphatidylinositol (PtdIns) phosphate-derived Ins polyphosphates play in signal transduction has grown from the initial or "classical" case of D-Ins 1,4,5-trisphosphate (D-Ins 1,4,5-P3) and calcium mobilization, to include an ever-growing number of signal transduction pathways that involve a large number of the possible isomers of any given Ins polyphosphate, and, in one important pathway, specific PtdIns polyphosphates (Whitman *et al.*, 1988; Berridge and Irvine, 1989; Cantley *et al.*, 1991; Majerus, 1992; Majerus *et al.*, 1990; Stephens *et al.*, 1993a). There is growing evidence that the PtdIns/Ins phosphate metabolic machinery exists in plant cells, functions in signal transduction, and is at least in one case essential for viability (Drøbak, 1992; Trewavas and Gilroy, 1991; Welters *et al.*, 1994).

In addition to any basic interest in understanding fundamental processes, two very practical rationales motivate studies in these areas. First, heritable or acquired disruptions in PtdIns/Ins phosphate metabolism may be the cause of many human pathologies (for an example, see Attree *et al.*, 1992). Second, seed-derived dietary phytic acid is perceived to play a negative role in nonruminant mineral nutrition (McCance and Widdowson, 1935; Erdman, 1981) but also possibly a positive role as an antioxidant and anticancer agent (Graf and Eaton, 1993).

Here we review the status and contribution of Mendelian and molecular genetics to our understanding of the metabolic pathways and possible roles of these metabolites in animal, microbial, and plant models. These studies often begin with the isolation and characterization of heritable mutants. Continuing advances in the physiology and biochemistry of Ins polyphosphate metabolism has provided a sound framework for advances in Mendelian and molecular genetics. These studies have contributed substantially to our understanding of the role of these metabolites in microbial and animal signal transduction pathways. The contribution of such genetic resources to plant systems, or to phosphate and mineral storage and regulation pathways in any system, is, in relative terms, undeveloped.

2. *myo*-INOSITOL AND L-*myo*-INOSITOL 1-PHOSPHATE: SYNTHESIS/REGULATION

Pathways involving Ins ultimately begin with the synthesis of Ins, which, proceeding from glucose-6-phosphate, is catalyzed by two enzymes (reviewed in Loewus, 1990). The sole *de novo* source of the Ins ring is via the activity of 1L-*myo*-inositol-1-phosphate synthase (EC 5.5.1.4, hereafter "synthase"), which converts D-glucose-6-phosphate to L-Ins(1)P. The L-Ins(1)P phosphatase (EC 3.1.3.25) activity then yields free Ins, a component of numerous pathways. *Inositol-less* or Ins auxotroph mutants, first isolated in *Neurospora crassa* (Beadle, 1944; Beadle and Tatum, 1945), could presumably result from disruptions in genes encoding the synthase or a phosphatase functionally essential for Ins production or in other loci essential for the regulation or expression of synthase or phosphatase genes. A strategy for selecting auxotrophs in mammalian cells was described and yielded Chinese hamster ovary cell lines that included Ins auxotrophs (Kao and Puck, 1968). Yeast (*Saccharomyces cerevisiae*) Ins auxotroph mutants (Culbertson and Henry, 1975) were determined to represent ten loci (complimentation groups), mutants of which lack synthase. Sixty-eight percent were determined to be alleles of the *ino1* locus (Donahue and Henry, 1981a), which was identified as the structural gene encoding the synthase (Donahue and Henry, 1981b). This enzymatic activity is highly regulated in yeast, at the level of *ino1* gene expression, by both Ins concentration and a number of unlinked regulatory loci that may repress or derepress *INO1* expression (reviewed in Carman and Henry, 1989; Hirsch and Henry, 1986; Hosaka *et al.*, 1992; Hoshizaki *et al.*, 1990; White *et al.*, 1991; Nikoloff *et al.*, 1992). This regulation is pivotal to lipid metabolism in yeast. The *INO1* gene was subsequently cloned via complementation (Klig and Henry, 1984), and following sequence analysis, the primary structure of the synthase was determined (Dean-Johnson and Henry, 1989). Recent studies have progressed to analyses of 5′ *cis*-regulatory sequences of the structural gene (Lopes *et al.*, 1991, 1993).

A plant gene with homology to this synthase was found to be rapidly up-regulated during abscisic-acid-induced morphogenic response in *Spirodela polyrrhiza* (Smart and Fleming, 1993). A sequence of a cDNA encoding an *Arabidopsis thaliana* synthase was reported (Dean Johnson, 1994), and Ins concentration regulates the expression of the corresponding gene (Dean-Johnson and Sussex, 1995). This cDNA was isolated via complimentation in yeast, and similarities between the yeast, *S. polyrrhiza* and *A. thaliana* sequences indicated a high degree of conservation in synthase genes (Dean Johnson, 1994).

The first proposed function for the L-Ins(1)P phosphatase was in Ins production, and initial studies indicated that the phosphatase copurified with the synthase (reviewed in Loewus, 1990). Originally identified as an L-Ins 1-phosphatase, this enzyme was subsequently found to hydrolyze both D- and L-Ins(1)P enantiomers as well as the D- and L(4)P enantiomers, and to be noncompetitively inhibited by Li^+ ions (Hallcher and Sherman, 1980; Gee *et al.*, 1988). Thus, this enzyme can contribute to the Ins pool via cleavage of the synthase product, or products of PtdIns and PtdIns phosphate breakdown.

A study of *Neurospora crassa* Ins auxotrophs grown in the presence of [^{14}C]glucose indicated incorporation of label into Ins(1)P (Williams, 1971). It was proposed that such mutants are deficient in Ins(1)P phosphatase. A cDNA encoding bovine brain Ins(1)P phosphatase was isolated and revealed the protein to be a homodimer of 30-kDa subunits, in good agreement with estimates derived from purified proteins (Diehl *et al.*, 1990). Subsequently, cDNA homologues were isolated from human and rat (McAllister *et al.*, 1992) and from *Xenopus laevis*, the African clawed frog (Wreggett, 1992). A very high degree of primary sequence conservation was first noted among mammalian homologues, and subsequently between these and the *X. laevis* homologue. This observation places this sequence among those highly conserved during the evolution of eukaryotes. The inhibition of this type of phosphatase, regardless of source, by lithium ions provides a plausible explanation for the efficacy of lithium in the treatment of psychiatric disorders and for other neural and developmental effects of lithium (Berridge *et al.*, 1989). However, to date, no heritable disorder or mutant phenotype has been unequivocally shown to be specifically related to a disruption in this type of phosphatase gene. We will return to this subject during the discussion of the Ins polyphosphate phosphatases below.

The L-Ins 1-kinase (EC 2.7.1.64) activity also yields the same product, L-Ins(1)P, as does the synthase (Dietz and Albersheim, 1965; English *et al.*, 1966). One possible function for this enzyme is in the synthetic pathway to phytic acid (Biswas *et al.*, 1978; Bollmann *et al.*, 1980; Stephens *et al.*, 1990; discussed further in Section 9). Another role may be, along with the synthase and phosphatase, in the maintenance or regulation of L-Ins(1)P levels (Loewus, 1990). To date, no mutants, naturally occurring or otherwise, have been shown to represent heritable disruptions in a gene encoding an L-Ins 1-kinase, nor has a sequence encoding an L-Ins 1-kinase been identified. An analysis of such a mutant or clone might help resolve questions concerning this kinase's physiological role.

3. PHOSPHATIDYLINOSITOL SYNTHESIS AND TRANSPORT

The first step in the PtdIns phosphate pathway is the synthesis of PtdIns, catalyzed by CDP-diacylglycerol–inositol 3-phosphatidyltransferase. A yeast mutant was shown to represent a lesion in a gene encoding this enzyme and was selected as an auxotroph requiring high levels (100 μg/ml) of exogenous Ins (Nikawa and Yamashita, 1982). The selection strategy was based on the assumption that yeast Ins synthesis mutants, although auxotrophic, can grow on low concentrations of exogenous Ins (2 μg/ml) and thus could be excluded from this screen. Like the Ins synthesis mutants mentioned above, auxotrophy indicates the essentiality of this function. The gene encoding this enzyme, designated *PIS*, was cloned via genetic complementation (Nikawa and Yamashita, 1984).

Once synthesized, targeted PtdIns transport between membranes and compartments is mediated by PtdIns transport proteins. A series of studies have clearly established a role for cytosolic PtdIns transfer proteins in secretory processes. The protein encoded by the yeast *sec14* gene is essential for Golgi complex protein transport (Bankaitis *et al.*, 1986). The phenotype of *sec14* mutants includes an exaggerated Golgi apparatus and a block in protein secretion (Novick *et al.*, 1980; Bankaitis *et al.*, 1989). A gene encoding a yeast PtdIns transport protein was isolated (*pit1;* Aitken *et al.*, 1990) and found to be identical to *sec14* (Bankaitis *et al.*, 1990). Antibodies raised against a bovine brain PtdIns transfer protein permitted a rat brain PtdIns transfer protein cDNA to be isolated and, when used as a probe of genomic DNA blots, indicated that the rat genome probably contains more than one PtdIns transfer protein gene (Dickeson *et al.*, 1989). The rat gene could complement some but not all yeast *sec14* mutants (Skinner *et al.*, 1993). The partial complementation of *sec14* nulls was correlated with the inability of the rat protein to associate stably with the yeast Golgi membrane. It was concluded that PtdIns transfer activity is not the sole essential activity of the PtdIns transfer protein encoded by *sec14*. In mammalian cells, regulated secretion requires in part an ATP-requiring priming event, which in turn involves priming-specific factors called PEP proteins. One such protein, PEP3, is a PtdIns transfer protein, and the yeast *sec14* protein could substitute for PEP3 in priming (Hay and Martin, 1993).

These studies indicate that PtdIns transfer proteins play a role in both constitutive and regulated secretory pathways and that this function is conserved between yeast and mammalian cells. It is interesting that although these studies illustrate the extant of functional interchangeability of yeast and mammalian PtdIns transfer proteins, their corresponding coding sequences share little homology. However, the rat and human (Dickeson *et al.*, 1994) genes are 98.9% homologous, making this PtdIns transfer protein one of the most highly conserved between two species. This high degree of conservation extended to 5′- and 3′-untranslated regions, indicating conservation of regulatory as well as structural functions. The human PtdIns transfer protein gene (*PIPTN*) was mapped to 17p13.3 (Fitzgibbon *et al.*, 1994).

The *Drosophila retinal degeneration B* (*rdgB*) gene was found to encode a membrane-associated PtdIns transfer protein that may be regulated by light-induced changes in intracellular calcium (Vihtelic *et al.*, 1993). The *rdgB* mutants were initially isolated as part of a genetic studies of the *Drosophila* nervous system (Hotta and Benzer, 1970). Disruptions in *rdgB* result in abnormal photoreceptor response and light-enhanced retinal degeneration (Harris and Stark, 1977; Stark *et al.*, 1983). The isolation of *rdgB* alleles in a screen for olfaction mutants revealed the importance of the *rdgB* gene product in olfactory signal transduction (Woodward *et al.*, 1992). The 160-kDa rdgB protein contains membrane-spanning domains and a calcium-binding domain; its amino-terminal 281 amino acid residues are more than 40% identical to the rat PtdIns transfer protein (Vihtelic *et al.*, 1993) and possesses PtdIns transfer activity *in vitro*. We will return to the subject of *rdgB* below, in the discussion of the *Drosophila norpA* gene, which encodes a PtdIns-specific phospholipase C. An allele of *norpA* suppresses the retinal degeneration conditioned by an *rdgB* allele, providing a genetic model of the functional interrelationship of these two loci in the PtdIns phosphate pathway.

4. PHOSPHATIDYLINOSITOL KINASES AND PHOSPHATE KINASES

Once synthesized, PtdIns is a substrate for a D-PtdIns 4-kinase, which produces D-PtdIns(4)P, which in turn is a substrate for a D-PtdIns(4)P 5-kinase. A yeast 4-kinase genomic clone has been obtained via PCR using primers based on a peptide sequence derived from purified enzyme preparations (Flanagan *et al.*, 1993). The gene encoding this enzyme was designated *pik1* and was similar to mammalian and yeast D-PtdIns 3-kinases discussed below. A *pik1* null construct was prepared by replacing two-thirds of the *pik1* sequence with a fragment containing the *Leu2* gene and used to transform wild-type cells. Sporulation of resulting *pik1Δ1::Leu2/Pik1* heterozygotes revealed that spore viability segregated 2:2 in every tetrad examined and that all viable spores were Leu^-, indicating the essentiality of the *pik1* gene. In contrast, a 3-kinase null (*vps32*, see below) survives at 30°C.

Although the "conventional" Ins phosphate second messenger pathway proceeds via the synthesis of D-PtdIns(4)P, subsequent studies have clearly demonstrated a second pathway involving one or more of the D-PtdIns 3-phosphates (Whitman *et al.*, 1988; Stephens *et al.*, 1993a). These compounds are not substrates for phospholipase C, which suggests that they function directly as second messengers (Lips *et al.*, 1989; Serunian *et al.*, 1989). The PtdIns 3-kinase activity has been found to be associated with many protein tyrosine kinases and has a functional role in multiprotein signal transduction complexes involved in transformation and mitogenesis (reviewed by Cantley *et al.*, 1991; Downes and Carter, 1991). Association of receptor proteins with the 3-kinase requires phosphorylation of the appropriate receptor protein tyrosine residue (Talmage *et al.*, 1989). Studies to date indicate that PtdIns 3-kinase activity is catalyzed by a heterodimeric enzyme consisting of an 85-kDa subunit that has no intrinsic kinase activity and a 110-kDa subunit with kinase activity. Sequences encoding examples of both subunits have been isolated, and genetic and molecular evidence indicates that at least two closely related forms of both subunits may exist in mammalian genomes (Escobedo *et al.*, 1991; Otsu *et al.*, 1991; Skolnik *et al.*, 1991; Hiles *et al.*, 1992). The 85-kDa subunit appears to function in the formation of multiprotein signal transduction complexes containing 3-kinase activity. It contains Src homology 2 (SH2) domains, which bind to phosphorylated tyrosine residues in other proteins and thus functions in protein–protein interaction. It also contains a domain found in members of a family of rac- and rho-GTPase-activating proteins (GAPs) (Otsu *et al.*, 1991; Yoakim *et al.*, 1992, 1994). The protein encoded by the human OCRL gene, an Ins polyphosphate 5-phosphatase (see below), also contains a similar GAP domain (Baldwin and Zhang, 1993). This finding provides evidence that genes in PtdIns phosphate and Ins phosphate metabolism contain sequences encoding similar regulatory domains.

The C-terminal half-sequence of the 110-kDa gene shows significant homology to the yeast *vps34* gene (Herman and Emr, 1990; Hiles *et al.*, 1992). The *vps34* was isolated in a screen for yeast vacuolar protein sorting (*vps*) mutants (Bankaitis *et al.*, 1986; Robinson *et al.*, 1988) and was subsequently shown to encode a PtdIns 3-kinase (Schu *et al.*, 1993). Disruption in *vps34* resulted in severe defects in sorting of proteins to what is otherwise a wild-type vacuole, as well as defects in other membrane structures within the cytoplasm. It was proposed that the *vps34* protein may be a component of "a large intracellular multiprotein structure" that functions in sorting pathways (Herman and Emr, 1990). Taken together, these data support the hypothesis that a PtdIns 3-kinase is involved in protein targeting, vesicular transport, and vesicle-mediated responses. The yeast *tor2* gene was also found to have significant homology to the C-terminal half of the bovine 110-kDa gene and the yeast *vps34* gene (Kunz *et al.*,

1993). The *tor2* mutants were isolated in a screen for resistance to the immunosuppressive drug rapamycin, which arrests the yeast cell cycle in the G_1 phase. Double mutants that are *tor2* and *tor1* (a homologue of *tor2*) similarly disrupt the cell division cycle. However, although *tor2* alone is lethal, it does not by itself disrupt the cell division cycle. These data suggest that the *tor2* PtdIns kinase, possibly a 3-kinase, is essential for yeast viability, and that although at least one function is critical to the cell division cycle, the *tor2* 3-kinase has at least one other function, not related to the cell division cycle.

An *Arabidopsis thaliana* PtdIns 3-kinase cDNA, referred to as *At*VPS34, was isolated using PCR with primers based on conserved sequences of bovine and yeast PtdIns 3- and 4-kinases (Welters *et al.*, 1994). The derived amino acid sequence has a greater degree of homology to the yeast *vps34* 3-kinase than to the mammalian 110 kDa 3-kinase (40% and 25%, respectively) and contains two conserved domains: a C-terminal catalytic domain containing kinase and ATP-binding motifs and an N-terminal "calcium-dependent lipid-binding" or CaLB domain (Clark *et al.*, 1991; reviewed in Stephens *et al.*, 1993a). The CaLB domain provides a mechanism for targeting the protein to the membrane or its substrate and is found in the 110 kDa 3-kinase but not the *Vps34* 3-kinase. While the *Arabidopsis* cDNA did not rescue a yeast *Vps34* deletion mutant, a chimeric yeast *Arabidopsis* cDNA containing the *Arabidopsis* C-terminal catalytic domain did. *At*VPS34 encodes an active PtdIns 3-kinase. Plants transformed with an antisense construct displayed severely inhibited growth and development. Thus at least one PtdIns 3-kinase is essential to plant viability, as in animals and yeast.

Insulin reception involves the formation of binary and ternary signaling complexes that contain a PtdIns 3-kinase, the insulin receptor which contains both tyrosine phosphorylation motifs and an intrinsic tyrosine kinase activity, and a protein called IRS-1, which also contains multiple tyrosine phosphorylation motifs (Backer *et al.*, 1993). Tyrosine-phosphorylation of the receptor results in the subsequent phosphorylation of IRS-1, which then modulates 3-kinase activity. A substantial fraction of *in vivo* 3-kinase activity may be caused by the formation of binary complexes with phosphorylated IRS-1. Ternary complex formation is probably mediated in large part via the association of IRS-1 with the 3-kinase, specifically via interaction with the SH2 domain in the 3-kinase 85-kDa subunit described above. The IRS-1 also contains a "pleckstrin homology" (PH) domain that functions in targeting proteins to membranes and is found in a PtdIns-specific phospholipase C (see Section 5, below).

5. PHOSPHATIDYLINOSITOL-SPECIFIC PHOSPHOLIPASE C AND 1,2-CYCLIC INOSITOL PHOSPHATES

Phosphatidylinositol phosphate-specific phospholipase C (PLC) plays a central role in the initially described PtdIns signal transduction pathway. Much progress has been made in the study of PLC isozymes and the genomic sequences that encode them (Rhee *et al.*, 1989). The relatively high degree of observed PLC sequence heterogeneity reflects the evolution of relatively specific, defined functions for a given isozyme, which contrasts with the high degree of sequence homogeneity observed in other cases.

Disruptions in the *Drosophila no receptor potential A* (*norpA*) locus reduce or eliminate photoreceptor potential in an allele-specific manner, and severe alleles render the fly completely blind (Hotta and Benzer, 1970; Pak *et al.*, 1970). The absence of PtdIns-specific PLC in *norpA* eyes provided a clue to the lesion in signal transduction conditioned by this mutant (Yoshioka *et al.*, 1985). Genetic and molecular evidence, including the isolation and characterization of a transposon-tagged *norpA* allele, indicated that a clone containing sequences homologous to a bovine PLC represented the *norpA* gene (Bloomquist *et al.*, 1988). Subsequently, peptide sequences obtained from purified *Drosophila* eye PtdIns-specific PLC were found to be part of the deduced amino acid sequence of *norpA* cDNAs (Toyoshima *et al.*, 1990). These studies provide direct evidence for the essential role in invertebrate phototransduction of a PtdIns-specific PLC represented by the *norpA* gene.

A weak or "hypomorphic" allele of *norpA* suppresses the mutant phenotype of a hypomorphic allele of the *rdgB* gene discussed above, which encodes a PtdIns transfer protein (Harris and Stark, 1977; Stark *et al.*, 1983). Differences in tissue-specific expression of these genes argue against the hypothesis that this suppression reflects the direct interaction of the proteins encoded by these loci (Vihtelic *et al.*, 1993). An alternative hypothesis is that retinal degeneration is the result of changes in PtdIns metabolite levels. In this alternative hypothesis, the reduced level of PtdIns transport in an *rdgB* hypomorph is compensated by the reduced catabolism of PtdIns in the background of a *norpA* hypomorph, thus maintaining levels of PtdIns that do not lead to degeneration (Vihtelic *et al.*, 1993). In either case, these studies clearly illustrate the role of both functions, and their interaction, in phototransduction in *Drosophila*.

Activation of PLC was shown to be a critical component of the mitogenic signaling pathway in Chinese hamster lung fibroblast cells (CCL39) induced with the mitogen α-thrombin (Paris *et al.*, 1987). Because α-thrombin also activates a Na^+/H^+ antiporter in CCL39 cells, it was reasoned (Rath *et al.*, 1989) that mutants defective in α-thrombin response would survive an H^+ suicide selection developed for the isolation of antiporter mutants (Pouysségur *et al.*, 1984). Two α-thrombin nongrowing CCL39 mutants were isolated, D1-6b and D1-9b, and were shown to have defects in PLC function (Rath *et al.*, 1990; Leonis and Silbert, 1993). Wild-type CCL39 cells were shown to contain at least two chromatographically distinct PLCs corresponding to PLC_δ and PLC_γ. Biochemical and immunologic studies indicated that both D1-6b and D1-9b cells were lacking in PLC_δ (Rath *et al.*, 1990; Leonis and Silbert, 1993). Genetic tests indicated that both mutants were codominant and allelic. However, these two mutants differed in a number of ways with respect to both phenotype and agonist-induced phosphoinositide turnover. Transfection of D1-6b with a rat brain PLC_{δ} cDNA corrected the *in vitro* biochemical defect but not the *in vivo* effects having to do with agonist induction of PtdIns turnover or Na^{+}/H^{+} antiporter activation. This may result from the inability of the heterologous protein to interact properly with hamster-specific signaling machinery. Although these studies provide genetic evidence for a mitogenesis-specific role for this PLC_{δ}, a more detailed understanding of its function in various cellular processes will proceed from the molecular characterization of each mutant's lesion (Leonis and Silbert, 1993).

A gene encoding a yeast homologue of the mammalian PLC_{δ}, *PLC1*, was found to be important to cell growth (Yoko-o *et al.*, 1993). Neither the CCL39 nor the yeast PLC_{δ} contains sequences necessary for regulation by protein tyrosine kinases or by G proteins. However, the amino-terminal portion of $PLC_{\delta 1}$ (which has no intrinsic catalytic activity) contains a "pleckstrin homology" (PH) domain that binds to PtdIns(4,5)P_2 (Harlan *et al.*, 1994). The PH domain targets the $PLC_{\delta 1}$ to PtdIns(4,5)P_2-containing membranes, and this binding is necessary for catalytic activity. The PH domain is found in diverse cytosolic proteins that are components of several signal transduction pathways and was so named because it was first isolated as part of the platelet protein pleckstrin (Tyers *et al.*, 1988; Mayer *et al.*, 1993; Musacchio *et al.*, 1993). An analysis of a rat brain $PLC_{\delta 1}$ revealed that the amino terminus also binds D-Ins(1,4,5)P_3, which acts as a noncompetitive inhibitor (Yagisawa *et al.*, 1994). The binding sequence was found to have little homology to sequences in other proteins that bind D-Ins(1,4,5)P_3. A negative-feedback model for the regulation of this enzyme by one of its products was proposed and awaits testing. In this model: (1) the PH domain targets the PLC to the PtdIns(4,5)P_2-containing membrane via binding to PtdIns(4,5)P_2; (2) PLC catalysis of PtdIns(4,5)P_2 both reduces the PtdIns(4,5)P_2 content of the membrane and transiently increases the level of Ins(1,4,5)P_3, which competes for binding to the PH domain; (3) the net effect is release and therefore inactivation of the PLC.

PLC activity produces Ins phosphates and Ins-1,2-cyclic phosphates (Dawson *et al.*, 1971). On stimulation of pancreatic minilobules with carbamylcholine, D-Ins(1,4,5)P_3 peaks rapidly (at 10 sec) and then rapidly drops to a steady-state level approximately two-thirds that of the initial peak, which is maintained for at least 20 min. In contrast, Ins-1,2-cyclic phosphates rise slowly to levels approaching the D-Ins(1,4,5)P_3 steady-state level (Dixon and Hokin, 1987). Assuming approximately equivalent potencies in eliciting calcium release, it was hypothesized that D-Ins(1,4,5)P_3 production functions in short-term response, whereas the 1,2-cyclic phosphates may play a role in more sustained cellular responses (such as cell growth). Ins-1,2-cyclic 2-phosphohydrolase converts the D-1,2-cyclic phosphate to the D-1-phosphate (Dawson and Clarke,

1972, 1973). Amino acid sequence analysis of a purified cyclic phosphohydrolase was found to be nearly identical to that predicted from the cDNA of human lipocortin III (Ross *et al.*, 1990). Cells transfected with the lipocortin III cDNA contained elevated cyclic phosphohydrolase activity, reduced 1,2-cyclic phosphates, and had an altered phenotype with respect to growth (a lower cellular density at confluence; Ross *et al.*, 1991b). This supports the above hypothetical physiological function of the cyclic phosphates produced by phospholipase C.

The intracellular pathogen *Listeria monocytogenes* secretes a PtdIns-specific PLC encoded by the monocistronic *plcA* gene (Camilli *et al.*, 1991). Disruption of this gene via transposon insertion in the mutant DP-L1054 resulted in a loss of secreted PLC activity and a greatly reduced virulence (the ability of the mutant bacterium to propagate in host tissue). Whether the secreted enzyme acts on host or pathogen substrates is unknown, and therefore its specific role in pathogenicity is unknown.

6. INOSITOL PHOSPHATE RECEPTORS AND KINASES

D-Ins(1,4,5)P_3, the "classical" elicitor of calcium release, does so following binding to specific receptors (reviewed in Majerus, 1992). In a study of cerebellar ataxic mouse mutants, a membrane glycoprotein, P_{400}, was found to be enriched in Purkinje cells from normal mice as compared with Purkinje cells in mutant mice. Following purification of the protein, a cDNA was obtained, and the protein it encoded was demonstrated to have D-Ins(1,4,5)P_3 binding properties (Furuichi *et al.*, 1989). In an alternative approach, a Purkinje cell D-Ins(1,4,5)P_3 receptor cDNA was cloned by subtracting a wild-type mouse cerebellar cDNA library from one prepared from Purkinje cell development (*pcd*) mutants (Mignery *et al.*, 1989). This D-Ins(1,4,5)P_3 receptor, subsequently termed type 1, was found to be partially homologous to the skeletal muscle ryanodine receptor and, like it, functions as a gated calcium channel. Following mutagenesis studies (Mignery *et al.*, 1990), a domain model was proposed in which the receptor contains an amino-terminal D-Ins(1,4,5)P_3 binding domain, a carboxy-terminal calcium channel domain, and an intervening "coupling" domain that regulates the relationship between the two distal binding domains and contains a cyclic-AMP-dependent phosphorylation site. Several lines of evidence indicated that multiple D-Ins(1,4,5)P_3 receptors serve multiple transduction functions. This was supported by the isolation and characterization of a second rat brain D-Ins(1,4,5)P_3 receptor, termed type 2, which was found to have 69% sequence homology with the type 1 receptor (Südhof *et al.*, 1991). Although the overall design of the two receptors is similar, comparison of sequences indicated that the D-Ins(1,4,5)P_3 binding sites were more similar than were the calcium channel domains, which in turn were more similar than the coupling domains. This suggest that the two receptors have distinct functions and that variation in calcium gating and regulation or coupling sequences evolved to satisfy these functions.

A D-Ins(1,4,5)P_3 receptor cDNA was cloned from a *Drosophila* head cDNA library using the mouse receptor clone as a probe (Yoshikawa *et al.*, 1992). The *Drosophila* receptor is expressed throughout fly development. In the adult, its expression appears highest in appendages having photo- and olfactory reception and in skeletal muscle tissues. This indicates that signal transduction mechanisms in *Drosophila* skeletal muscles are similar to that in vertebrate smooth muscles, which contain a high level of Ins(1,3,4)P_3 receptor mRNA.

D-Ins(1,4,5)P_3, but not D-Ins(1,2-cyclic,4,5)P_3, is also the substrate for a D-Ins(1,4,5)P_3 3-kinase. The D-Ins(1,4,5)P_3 3-kinase is apparently widely distributed and, as in most of the above cases, subject to a variety of regulatory mechanisms and patterns of expression. A rat brain cDNA library was screened with a mixture of ten monoclonal antibodies to a 53-kDa D-Ins(1,4,5)P_3 3-kinase isolated from rat brain (Choi *et al.*, 1990). One of the two longest inserts obtained contained an open reading frame shown to encode a 3-kinase. Interestingly, although data-base sequence comparisons found no overall homology with any other protein, several short regions of homology were found with the first D-Ins(1,4,5)P_3 receptor protein described above. Expression of the cDNA in COS cells resulted in a 150-fold increase in 3-kinase activity, supporting the conclusion that the cDNA in question encoded the 3-kinase. The rat brain D-Ins(1,4,5)P_3 3-kinase was also transfected into NIH 3T3 and CCL39 fibroblast cells (Balla *et al.*, 1991). In such cells, the conversion of D-Ins(1,4,5)P_3 to D-Ins(1,3,4,5)P_4 was greatly increased on agonist stimulation. The resulting decrease in D-Ins(1,3,4)P_3 was correlated with an attenuation in cytoplasmic calcium increase and calcium response, and there was no evidence that the increased D-Ins(1,3,4,5)P_4 levels increased calcium release. These data suggest that in this case the function of the 3-kinase is to modulate D-Ins(1,4,5)P_3 levels rather than produce D-Ins(1,3,4,5)P_4 as a messenger.

Receptors for other Ins phosphates, including D-Ins(1,3,4,5)P_4 and $InsP_6$ (phytic acid) have been identified (reviewed by Irvine, 1992). The presence of high-affinity D-Ins(1,3,4,5)P_4 receptors adds support for other studies indicating a direct role of D-Ins(1,3,4,5)P_4 as a second messenger (Irvine, 1992). Thus, evidence to date seems to indicate that D-Ins(1,4,5)P_3 3-kinase functions to produce the D-Ins(1,3,4,5)P_4 as second messenger or to modulate the level of the D-Ins(1,4,5)P_3 messenger, depending on the specific case. An $InsP_6$ receptor was purified, and a partial amino acid sequence was obtained (Theibert *et al.*, 1991; Voglmaier *et al.*, 1992). Degenerate PCR primers based on this sequence were used to amplify the corresponding sequence from a rat total cerebellar RNA. The $InsP_6$ receptor was nearly identical to subunits of a mouse AP-2 protein, which functions in the assembly of clathrin into coated vesicles. AP-2 may be involved in receptor recycling in the plasma membrane and is enriched in the brain. Although the $InsP_6$ receptor has greatest affinity for Ins P_6, it also binds D-Ins(1,4,5)P_3, $InsP_4$ isomers, and D-PtdIns(4,5)P_2, so its actual physiologically relevant effector is unclear and may not be $InsP_6$.

7. INOSITOL POLYPHOSPHATE PHOSPHATASES OTHER THAN PHYTASES

Inositol phosphate phosphatases other than phytases (phytic acid-specific phosphohydrolases, see below) can be thought of a falling into three classes: the Ins monophosphatases discussed above, whose function initially was seen as being the production or maintenance of Ins level; the Ins polyphosphate 1-phosphatases; and the Ins polyphosphate phosphatases other than the 1-phosphatases, such as Ins polyphosphate 3- and 5-phosphatases (for a review see Majerus, 1992). While both PtdIns 3- and 5-phosphate phosphatase activities are critical to their respective pathways, their molecular identity remained obscure until it was demonstrated that certain Ins 3- and 5-phosphate phosphatases also hydrolyze PtdIns 3- and 5-phosphate monoesters (Caldwell *et al.*, 1991; Matzaris *et al.*, 1994; Zhang *et al.*, 1995). Several lines of evidence indicate that genes encoding either Ins phosphate 1- or 5-phosphatases are primary candidates for those gene disruptions that are the cause of heritable human psychiatric and brain development disorders (Hallcher and Sherman, 1980; Berridge *et al.*, 1989; Attree *et al.*, 1992). In addition to the Ins monophosphate phosphatases, Ins polyphosphate 1-phosphatase is also noncompetitively inhibited by lithium. Substrates for this enzyme are D-Ins(1,4)P_2 and D-Ins(1,3,4)P_3 but not other D-1-phosphate-containing Ins polyphosphates. A bovine cDNA encoding this function was obtained and encoded a 44-kDa protein (York and Majerus, 1990). Sequence comparisons indicated no significant similarity to other known proteins, with the exception of one limited region of similarity to the bovine Ins monophosphate phosphatase. Subsequently, a human homologue with 84% amino acid identity to the bovine gene was cloned and localized to chromosome 2 band q32 (York *et al.*, 1993). Three regions of amino acid sequence are totally conserved in the bovine and human homologues: 25 residues at a metal-binding site; 28 residues at a putative substrate-binding site; and 35 residues at a potential nuclear localization site.

A number of studies have shown that enzymes in PtdIns metabolism are present in the nucleus and play a role in nuclear function (reviewed in Divecha *et al.*, 1993). Recent studies have shown that nuclear PtdIns phosphates are degraded during the S phase of synchronized HeLa cell cycle and that the Ins polyphosphate 1-phosphatase described above appears to be primarily a nuclear protein (York and Majerus, 1994; York *et al.*, 1994). Its overexpression inhibited

DNA synthesis, possibly via breakdown of Ins(1,4)P_2, a putative nuclear signal that increases the affinity of DNA polymerase α for template/primer and increases activity tenfold (Syliva *et al.*, 1988).

Multiple types of Ins polyphosphate 5-phosphatase, both soluble and particulate, have been identified to date, each with different properties (reviewed in Majerus, 1992). Biochemical evidence indicates that a lesion in a membrane-bound D-Ins(1,4,5)P_3 5-phosphatase may be the cause of both human and swine malignant hyperpyrexia (MH), a life-threatening complication of anesthesia (Foster *et al.*, 1989). Malignant hyperpyrexia is associated with an underlying muscle disease. MH-susceptible muscle contains an elevated myoplasmic calcium ion concentration and is hypercontractile. Although myoplasmic 5-phosphatase activity was similar in MH-susceptible and control muscle tissues, sarcoplasmic-reticulum-bound 5-phosphatase activity was substantially reduced in MH-susceptible muscle.

A cDNA encoding a 75-kDa type II 5-phosphatase was isolated (Ross *et al.*, 1991a) via a novel approach in which human placental "sublibraries" were first amplified and then screened for those whose protein lysates contained fusion protein(s) that bound to [^{3}H]D-Ins(1,3,4,5)P_4 in solution. These sublibraries were further divided until a single cDNA was isolated, which turned out to represent only a portion of the 5-phosphatase gene. Screening of additional libraries provided the sequence necessary to obtain the composite sequence representing a complete 5-phosphatase gene. The gene encoding this 5-phosphatase, *INPP5B*, was mapped to the human chromosome band 1p34 (Jänne *et al.*, 1994).

The cloning and subsequent mapping of sequences encoding Ins mono and polyphosphate phosphatases provide the genetic resources necessary in studies of the putative relationship between disruptions in Ins phosphate metabolism and certain inherited psychiatric disorders. As in the case of the Ins monophosphate phosphatases, to date genetic evidence for a linkage between a disruption in an Ins polyphosphate 1-phosphatase and a heritable disorder is lacking. However, genetic and molecular evidence indicates that lesions in an Ins polyphosphate 5-phosphatase gene may be the cause of the inherited human disorder Lowe's oculocerebrorenal syndrome (OCRL; Attree *et al.*, 1992), which is an X-linked developmental disorder with a pleiotropic phenotype affecting the lens, brain, and kidney. It has been mapped to Xq25-q26, and two unrelated women with OCRL were found to have X:autosome translocation break points at that map location. The translocation break point in one of these individuals was found to interrupt a gene, *OCRL-1*, with 53% amino-acid identity and 71% similarity to the cDNA sequence for the soluble 75-kDa human platelet Ins polyphosphate 5-phosphatase type II (Ross *et al.*, 1991a). Differences in predicted amino-acid content and molecular mass, however, indicate that *OCRL1* may represent the gene encoding a brain 100-kDa Ins polyphosphate 5-phosphatase. Analyses of additional OCRL and normal individuals supported the relationship between disruption in the phosphatase gene and the OCRL syndrome.

A recent study demonstrated that the *OCRL1*-encoded protein hydrolyzed the lipid PtdIns 4,5-P2 (to yield PtdIns 4-P1) 590- and 1080-fold better than it hydrolyzed the water-soluble substrates Ins 1,4,5-P3 and Ins 1,3,4,5-P4, respectively (Zhang *et al.*, 1995). This and other evidence supported the conclusion that this enzyme is primarily a lipid phosphatase, and that the assorted OCRL pathologies are due to perturbed regulation of PtdIns 4,5-P2 levels. The 75 kDa Type II Ins polyphosphate 5-phosphatase described above is also a PtdIns polyphosphate 5-phosphatase (Matzaris *et al.*, 1994). However, not all Ins polyphosphate 5-phosphatases hydrolyze lipid substrates, and not all PtdIns polyphosphate 5-phosphatases hydrolyze Ins phosphate substrates. Further studies should lead to an understanding of both the physiological and evolutionary significance of these interesting findings. An Ins polyphosphate 3-phosphatase also catalyzes PtdIns 3-phosphate hydrolysis (Caldwell *et al.*, 1991). In this later case, two forms of the enzyme were isolated from rat brain: a homo-dimer containing subunits of M_r 65,000, referred to as Type I; and a hetero-dimer containing the 65,000 subunit and a M_r 78,000 subunit, termed Type II. The Type I enzyme hydrolyzed the soluble Ins polyphosphate more efficiently than the Type II enzyme, and conversely the Type II enzyme hydrolyzed the lipid substrate more efficiently than the Type I enzyme. It was concluded that the 78,000 subunit determined the relative substrate specificities of the two Types.

Most recently, a cDNA encoding a 43-kDa membrane-associated 5-phosphatase was cloned from a human placental cDNA library using degenerate oligonucleotides based on peptide sequences derived from the purified protein (Laxminarayan *et al.*, 1994). If conservative amino acid replacements are allowed, the COOH-terminal segment of the 43-kDa sequence is 70% and 67% similar to the corresponding domains in the OCRL and 75-kDa 5-phosphatases, respectively. The COOH terminus also contains a "CAAX" motif, which is found in many proteins involved in signal transduction and, following prenylation, determines membrane localization required for biological function (Glomset *et al.*, 1990).

8. PHYTASES

Phytases, phytic-acid-specific phosphohydrolases, are widespread and have been isolated from plant, microbial, and animal sources (reviewed in Cosgrove, 1980; Gibson and Ullah, 1990). Phytases are classified according to the initial phosphate removed: 6-phytases (EC 3.1.3.26) initially remove the 6-phosphate and are typical of plant phytases; and 3-phytases (EC 3.1.3.8) initially remove the 3-phosphate and are typical of microbial phytases. There have been numerous studies of the biochemistry of seed and pollen phytases. As one example, studies have indicated that in lily (*Lilium longiflorum*) pollen there are at least two constitutively expressed phytases and one developmentally induced phytase, each with a different pattern of expression, localization, and regulation (Baldi *et al.*, 1988; reviewed in Loewus *et al.*, 1990). Understanding the regulation of phytase expression and activity is integral to understanding phytic acid metabolism during the development and germination of seed and pollen. For example, seed phytases may be regulated by inorganic P concentration in a way that is analogous to the regulation of fungal phytases discussed below (Bianchetti and Sartirana, 1967; Sartirana and Bianchetti, 1967). Despite the interest in these phytases from the perspective of plant biology, to date no plant phytase mutants have been described, nor have any plant phytase genes or sequences been cloned or reported.

The regulation of microbial phytases is similar to the regulation of the more general class of microbial acid phosphatases; high levels of inorganic P repress their synthesis while limiting levels of inorganic P induce their synthesis (reviewed by Gibson and Ullah, 1990). Typically these phytases are extracellular and function in inorganic P supply. Sequences encoding two extracellular phytases have recently been cloned from *Aspergillus niger* (Mullaney *et al.*, 1991; Ehrlich *et al.*, 1993; Pen *et al.*, 1993; Piddington *et al.*, 1993; van Hartingsveldt *et al.*, 1993). An applied rationale motivated these efforts. These extracellular phytases of fungal origin are valued as an industrial protein to be used as an additive in poultry and swine feeds or in other applications where grain- or legume-derived phytic acid presents a problem (Pen *et al.*, 1993). Design of constructs for their overexpression (Pen *et al.*, 1993) will contribute to efforts to produce the quantities necessary for these industrial uses.

9. PHYTIC ACID

The synthesis of phytic acid may proceed via the sequential phosphorylation of L-Ins(1)P (Biswas *et al.*, 1978; Stephens and Irvine, 1990; Stephens *et al.*, 1991). If so, what is the source of L-Ins(1)P? A pathway in *Dictyostelium* was described that began with the synthesis of L-Ins(1)P catalyzed by an L-Ins 1-kinase (or D-Ins 3-kinase; Stevens and Irvine, 1990; Stephens *et al.*, 1990), but the possible contribution of L-Ins(1)P synthase in the production of L-Ins(1)P was not determined. In fact, the relative roles, in spatial or temporal terms, of the synthase versus the kinase in pathways to phytic acid have not been determined unequivocally in any system. The isolation of plant genes encoding the synthase (Dean-Johnson, 1994; Dean-Johnson and Sussex, 1995; Smart and Fleming; 1993), should provide an important part of the genetic resources necessary to attack this question. The cloning of L-Ins 1-kinase genes would also contribute greatly to studies of this question.

Starting with L-Ins(1)P, most studies indicate that sequential phosphorylation to phytic acid is catalyzed by two or more Ins phosphate or polyphosphate kinases or phosphotransferases (Biswas *et al.*, 1978; Stephens and Irvine, 1990; Phillippy *et al.*, 1994). In essentially all described pathways the final step, the conversion of Ins pentakisphosphate to phytic acid, is catalyzed by an Ins 1,3,4,5,6-pentakisphosphate 2-kinase. However, the pathway from L-Ins(1)P to Ins 1,3,4,5,6-pentakisphosphate may differ greatly in different organisms. For example, studies of phytic acid synthesis in the developing and germinating mung bean seed described a single "phosphoinositol kinase" that converts L-Ins(1)P to Ins 1,3,4,5,6-pentakisphosphate in a sequential fashion (reviewed in Biswas *et al.*, 1978). In these studies the identity of the positional Ins polyphosphate isomer intermediates was not determined. In contrast, in the *Dictyostelium* pathway to phytic acid the conversion of L-Ins(1)P (or D-Ins(3)P) to Ins 1,3,4,5,6-pentakisphosphate may be catalyzed by a series of four putative kinases, with the intermediates being L-Ins 1,4-P2 (or D-Ins 3,6-P2), L-Ins 1,4,6-P3 (or D-Ins 3,4,6-P3), and D/L-Ins 1,3,4,6-P4 (Stephens and Irvine, 1990). Thus in this case the pathway to phytic acid may involve positional isomers of Ins tris- or tetrakisphosphate different from those known to function as "second messengers," such as D-Ins 1,4,5-P3 or D-Ins 1,3,4,5-P4. Phytic acid synthesis may also proceed initially via PtdIns polyphosphate-derived intermediates. For example, studies of Ins phosphate metabolism in *Saccharomyces* and in peas (*Pisum sativum*) roots identified a D-Ins 1,4,5-P3 6-kinase, and proposed that its product is an intermediate in Ins pentakis- or hexakisphosphate production (Chattaway *et al.*, 1992; Estevez *et al.*, 1994). From this brief discussion it is clear that there are many open questions concerning pathways to phytic acid, and that these pathways in the developing and germinating seed, a subject of decades-long interest, are still not well understood. The isolation of phytic-acid-pathway-specific mutants, and the cloning of genes in this pathway, such as genes encoding the various kinases proposed to play a role, would greatly contribute to its dissection.

Numerous studies of phytic acid metabolism in plant tissues clearly indicate a role in phosphate, Ins and mineral storage and phosphate homeostasis (reviewed in Raboy, 1990). It had been proposed more than three decades ago that phytic acid may also serve as a phosphate donor for ATP regeneration in the initial phases of seed germination, but this hypothesis was viewed unfavorably based on the low phosphoryl bond energy of the phosphomonoesters in phytic acid (reviewed in Cosgrove, 1980). A recent study of a soybean Ins 1,3,4,5,6-pentakisphosphate 2-kinase demonstrated that the reaction it catalyzes is reversible and can lead to ATP regeneration if the concentrations of substrate and product are favorable, as might be the case during the initial stages of seed germination (Phillippy *et al.*, 1994). This confirms the earlier studies of an analogous enzyme and pathway during mung bean development and germination (Biswas *et al.*, 1978). Interest in the possible role of Ins hexakisphosphate as metabolite in "high-energy" phosphate bond pathways was recently invigorated by the observation of pyrophosphoryl-containing Ins polyphosphates in a variety of animal and microbial cells (Mayr *et al.*, 1992; Menniti *et al.*, 1993; Stephens *et al.*, 1993b). As an example of this, two pyrophosphoryl-containing Ins phosphates were observed in *Dictyostellium*: 1) a D/L-1-diphosphoinositol pentakisphosphate containing monophosphate esters at the 2, 3, 4, 5, and 6 positions, and a pyrophosphate at the 1 position, and thus an Ins heptaphosphate; 2) a D/L-bis-(1-4)-disphosphoinositol tetrakisphosphate containing monophosphate esters at the 2, 3, 5, and 6 positions and pyrophosphates at the 1 and 4 positions, and thus an Ins octaphosphate. The phosphomonoester groups have the predicted bond energies of between 3.3 and 4 kcal mol^{-1}. However, the diphosphate groups have bond energies of 6.6 kcal mol^{-1}, similar to that in ADP. This indicates that these molecules may act as phosphate donors in phosphotransferase reactions. It thus may turn out that the earlier proposals, based on studies of seed development and germination, were in fact closer to the truth than thought.

We have been studying the genetics of phytic acid accumulation in seeds (Raboy *et al.*, 1990, 1994). Phytic acid is the most abundant phosphorus (P)-containing compound in seeds, typically representing from 60% to 80% of seed total P (reviewed in Cosgrove, 1980; Raboy, 1990). During seed development, phytic acid most typically accumulates as mixed "phytate" salts of several mineral cations, including K, Mg, Ca, Mn, Zn, Fe, and other cations (reviewed in Lott, 1984). These salts are often deposited in storage microbodies (protein bodies) in discrete inclusions referred to as globoids (reviewed in Pernollet, 1978). During germination, phytate salts are disassembled, supplying the germinating seed and growing seedling with P, *myo*-inositol, and minerals (Dmitrieva and Sobolev, 1984).

In historical terms, perhaps the greatest interest in seed phytic acid has been in its putative role as an antinutrient in the diets of nonruminant animals (reviewed in McCance and Widdowson, 1935; Erdman, 1981). As a result, there have been numerous studies of the variation in seed phytic acid content among lines and cultivars of crop species (reviewed in Raboy, 1990). These studies typically find that: (1) variation in seed phytic acid P is highly and positively correlated with variation in seed total P, with correlation coefficients typically equal to or greater than 0.95; (2) nearly all observed variation in seed total P can be accounted for by variation in phytic acid P, with little change in other seed P components; (3) variation in seed phytic acid P is a function of variation in seed total P, which in turn is a function of both genetic and environmental factors that influence plant P status. These studies reported no known heritable variants in which this close and positive relationship between seed total P and phytic acid P was altered. As an example, the only single-gene effect on seed phytic acid that is Mendelian or qualitative in nature, reported before the work described below, was that of the allelic variants of the soybean (*Glycine max* L. Merr.) *Np* locus, which regulate the uptake of phosphate at the root–rhizosphere interface (Raboy and Dickinson, 1984).

Recent studies have shown that the close and positive relationship between seed total P and phytic acid P is altered in two nonlethal maize (*Zea mays* L.) *low phytic acid* (*lpa*) mutants termed *low* phytic acid 1 (Lpa1) and Lpa2 (Raboy *et al.*, 1994). Kernels homozygous for an *Lpa* mutant are, to the naked eye, wild-type in phenotype and contain wild-type levels of total P, but contain reductions in kernel phytic acid P ranging from 33% (Lpa2) to 66% (Lpa1). The first two *Lpa* mutants, *Lpa1* and *Lpa2*, were isolated by screening a population of mutants using a high-voltage paper electrophoresis (HVPE) assay for kernel phytic acid (Figure 1). At maturity, a wild-type kernel typical of the stocks used in these studies contains about 4.5 mg/g phytic acid P, representing about 75% to 80% of kernel total P, and from 0.1 to 0.5 mg/g inorganic P. If one extracts a "wild-type" maize kernel in 10 volumes of 0.4 M HCl, and tests 10 μl with HVPE, one observes an abundant compound representing about 3.5 μg P, which comigrates with standard phytic acid (Figure 1A, lanes 1 through 5). At this level of sensitivity one observes a trace of inorganic P and no lower Ins phosphates. The HVPE phenotype of *Lpa1* is about a 66% reduction in phytic acid P accompanied by what appears to be a molar-equivalent increase in inorganic P (Figure 1, lanes 6 through 10). No increases in lower Ins phosphates are observed at this level of sensitivity. The HVPE phenotype of *Lpa2* is about a 33% reduction in phytic acid P, accompanied by two novel P-containing spots with mobilities similar to that of Ins pentakisphosphate and intermediate between Ins pentakis- and tetrakisphosphate and a molar-equivalent increase in inorganic P (Figure 1, lanes 11 through 15). Quantitative analyses confirmed that homozygosity for either *Lpa1* or *Lpa2* (1) has little or no effect on kernal total P, and (2) the reductions in phytic acid P appear to be essentially completely accounted for by increases in inorganic P.

Both these and the earlier studies of defective kernel or *DEK* mutants (Raboy *et al.*, 1990) indicate that reductions in phytic acid P are typically accompanied by molar-equivalent increases in kernel inorganic P. Because the only obvious effect of these mutants is in phytic acid accumulation, we hypothesize that the functions disrupted in mutants of this class are proximal to the phytic acid pathways. Thus, these mutants may provide valuable tools in attacking some of the outstanding questions concerning phytic acid metabolism in plant tissues. One conclusion we can draw from the initial results is that the phytic acid level typical of wild-type seeds, and the close relationship between seed phytic acid P and total P in wild-type seeds, are not absolute requirements for seed development and function.

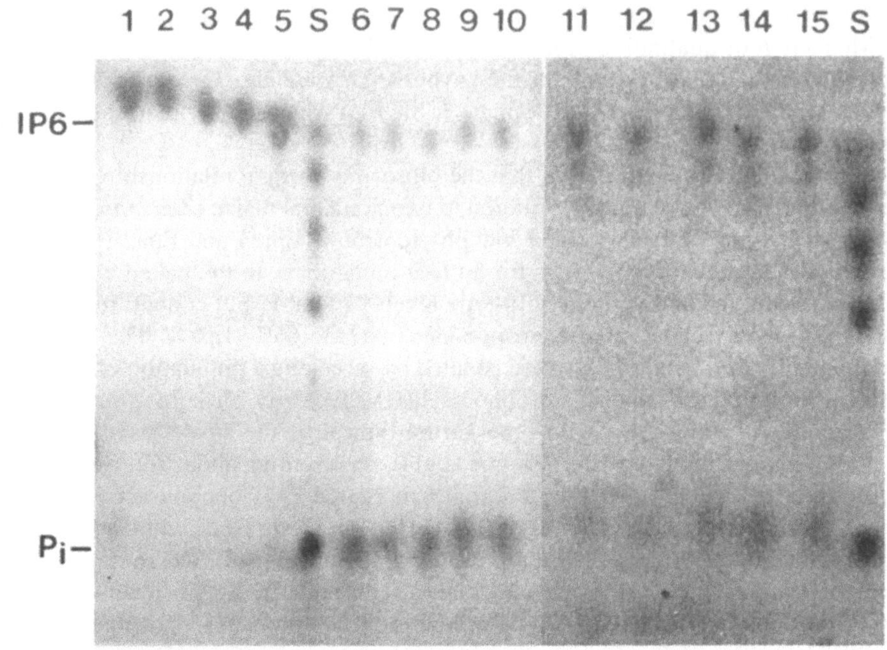

FIGURE 1. The high-voltage paper electrophoresis (HVPE) phenotype of maize *low phytic acid* (*lpa*) mutants. Individual kernels were crushed and extracted in 10 v/w 0.4 M HCl. Ten microliters of extract was applied to the origin of a Whatman No. 1 sheet and fractionated electrophoretically (25 V/cm, 2 hr) in 0.1 M Na oxalate. The chromatogram was dried and stained for phosphate. Standards (lane S) contain a mixture of phytic acid (IP6) and, in descending order, *myo*-inositol pentakis-, tetrakis-, tris-, and bisphosphate (sometimes not visible to the naked eye), and inorganic P (P_i). Lanes 1 through 5 are single-kernel tests of wild-type kernels; lanes 6 through 10 show tests of homozygous *lpa1* kernels, lanes 11 through 15 tests of homozygous *lpa2* kernels.

10. SUMMARY: CURRENT STATUS OF INOSITOL AND PHOSPHATIDYLINOSITOL PHOSPHATE GENETICS

This review indicates that for signal transduction pathways in animal and microbial models, mutants disrupting examples of many if not most functions have been isolated and studied. These studies have often provided definitive proofs of functional essentiality, including identifying the specific lesion causing a heritable human disorder. This is also leading in certain cases both to a relatively detailed knowledge of a given function's structure and regulation and to a detailed understanding of the interaction of distinct functions. Outstanding examples of this include the interaction of alleles at the *Drosophila rdgB* and *norpA* loci (Stark *et al.*, 1983) and the "modular domains" found in one or more of the proteins in the PtdIns phosphate/Ins phosphate signal transduction pathways as well as in proteins in other signal transduction pathways that serve targeting or regulatory functions (Mayer *et al.*, 1993; Musacchio *et al.*, 1993). These mutants have also proven a valuable resource in the molecular cloning of genes encoding these functions.

It is clear that in the evolution of complex, multicellular organisms there has been a proliferation of these pathways. Thus, in many cases the initial isolation of a gene encoding a particular function has led to the isolation of multiple variants of that function. Currently this field is progressing rapidly toward an elucidation of the genetic basis for this proliferation. One interesting possibility is that the PtdIns phosphate signal transduction pathway first evolved as a function in the nucleus or nuclear membrane and subsequently was duplicated in the plasma membrane for cytosolic signal transduction (Divecha *et al.*, 1993).

Similar progress in the genetics of PtdIns phosphate signal transduction pathways in plant systems such as the recent isolation of an *Arabidopsis* PtdIns 3-kinase (Welters *et al.*, 1994), would be of significant value in attacking some of the questions concerning these pathways in plants. The genetics of Ins phosphate pathways associated with phosphate and mineral storage and regulation, or in "high energy" bond metabolism in plant or nonplant models, also is not as advanced. However, continuing studies of phytic acid metabolism in models such as maize, *Arabidopsis, Dictyostellium,* and *Saccharomyces* should ultimately rectify this deficit.

11. REFERENCES

Aitken, J. F., van Heusdan, G.P.H., Temkin, M., and Dowhan, W., 1990, The gene encoding the phosphatidylinositol transfer protein is essential for cell growth, *J. Biol. Chem.* **265:**4711–4717.

Attree, O., Olivos, I. M., Okabe, I., Bailey, L. C., Nelson, D. L., Lewis, R. A., McInnes, R. R., and Nussbaum, R. L., 1992, The Lowe's oculocerebrorenal syndrome gene encodes a protein highly homologous to inositol polyphosphate-5-phosphatase, *Nature* **358:**239–242.

Backer, J. M., Myers, M. G., Jr., Sun, X.-J., Chin, D. J., Shoelson, S. E., Miralpeix, M., and White, M. F., 1993, Association of IRS-1 with the insulin receptor and the phosphatidylinositol 3′-kinase, *J. Biol. Chem.* **268:**8204–8212.

Baldi, B. G., Scott, J. J., Everard, J. D., and Loewus, F. A., 1988, Localization of constitutive phytases in lily pollen and properties of the pH 8 form, *Plant Sci.* **56:**137–147.

Baldwin, G. S., and Zhang, Q-X, 1993, Related GAP domains in inositol polyphosphate 5-phosphatase and the p85 subunit of phosphatidylinositol 3-kinase, *Trends Biochem. Sci.* **18:**378–380.

Balla, T., Sim, S. S., Iida, T., Choi, K. Y., Catt, K. J., and Rhee, S. G., 1991, Agonist-induced calcium signaling is impaired in fibroblasts overproducing inositol 1,3,4,5-tetrakisphosphate, *J. Biol. Chem.* **266:**24719–24726.

Bankaitis, V. A., Johnson, L. M., and Emr, S. D., 1986, Isolation of yeast mutants defective in protein targeting to the vacuole, *Proc. Natl. Acad. Sci. U.S.A.* **83:**9075–9079.

Bankaitis, V. A., Malehorn, D. E., Emr, S. D., and Greene, R., 1989, The *Saccharomyces cerevisiae Sec14* gene encodes a cytosolic factor that is required for transport of secretory proteins from the yeast Golgi complex, *J. Cell Biol.* **108:**1271–1281.

Bankaitis, V. A., Aitken, J. R., Cleves, A. E., and Dowhan, W., 1990, An essential role for a phospholipid transfer protein in yeast Golgi function, *Nature* **347**:561–562.

Beadle, G. W., 1944, An inositolless mutant strain of *Neurospora* and its use in bioassays, *J. Biol. Chem.* **156**:683–689.

Beadle, G. W., and Tatum, E. L., 1945, *Neurospora*. II. Methods of producing and detecting mutations concerned with nutritional requirements, *Am. J. Bot.* **32**:678–686.

Berridge, M. J., and Irvine, R. F., 1989, Inositol phosphates and cell signalling, *Nature* **341**:197–205.

Berridge, M. J., Downes, C. P., and Hanley, M. R., 1989, Neural and developmental actions of lithium: A unifying hypothesis, *Cell* **59**:411–419.

Bianchetti, R., and Sartirana, M. L., 1967, The mechanism of repression by inorganic phosphate of phytase synthesis in the germinating wheat embryo, *Biochim. Biophys. Acta* **145**:485–490.

Biswas, B. B., Biswas, S., Chakrabarti, S., and De, B. P., 1978, A novel metabolic cycle involving *myo*-inositol phosphates during formation and germination of seeds, in *Cyclitols and Phosphoinositides* (W. W. Wells and F. Eisenberg, Jr., eds.), pp. 57–68, Academic Press, New York.

Bloomquist, B. T., Shortridge, R. D., Schneuwly, S., Perdew, M., Montell, C., Steller, H., Rubin, G., and Pak, W. L., 1988, Isolation of a putative phospholipase C gene of *Drosophila, norpA*, and its role in phototransduction, *Cell* **54**:723–733.

Bollmann, O., Strother, S., and Hoffman-Ostenhoff, O., 1980, The enzymes involved in the synthesis of phytic acid in *Lemna gibba, Mol. Cell. Biol.* **30**:171–175.

Caldwell, K. K., Lips, D. L., Bansal, V. S., and Majerus, P. W., 1991, Isolation and characterization of two 3-phosphatases that hydrolyze both phosphatidylinositol 3-phosphate and inositol 1,3-bisphosphate, *J. Biol. Chem.* **266**:18378–18386.

Camilli, A., Goldfine, H., and Portnoy, D. A., 1991, *Listeria monocytogenes* mutants lacking phosphatidylinositol-specific phospholipase C are avirulent, *J. Exp. Med.* **173**:751–754.

Cantley, L. C., Auger, K. R., Carpenter, C., Duckworth, B., Graziani, A., Kapeller, R., and Soltoff, S., 1991, Oncogenes and signal transduction, *Cell* **64**:281–302.

Carman, G. M., and Henry, S. A., 1989, Phospholipid biosynthesis in yeast, *Annu. Rev. Biochem.* **58**:635–669.

Chattaway, J. A., Drøbak, B. K., Watkins, P.A.C., Dawson, A. P., Letcher, A. J., Stephens, L. R., and Irvine, R. F., 1992, An inositol 1,4,5-trisphosphate-6-kinase activity in pea roots, *Planta* **187**:542–545.

Choi, K. Y., Kim, H. K., Lee, S. Y., Moon, K. H., Sim, S. S., Kim, J. W., Chung, H. K., and Rhee, S. G., 1990, Molecular cloning and expression of a complementary DNA for inositol 1,4,5-trisphosphate 3-kinase, *Science* **248**:64–66.

Clark, J. D., Lin, L-L., Kriz, R. W., Ramesha, C. S., Sultzman, L. A., Lin, L. Y., Milona, N., and Knopf, J. L., 1991, A novel arachidonic acid-selective cytosolic PLA_2 contains a Ca^{2+}-dependent translocation domain with homology to PKC and GAP, *Cell* **65**:1043–1051.

Cosgrove, D. J., 1980, *Inositol Phosphates: Their Chemistry, Biochemistry and Physiology,* Elsevier Scientific, Amsterdam.

Culbertson, M. R., and Henry, S. A., 1975, Inositol-requiring mutants of *Saccharomyces cerevisiae, Genetics* **80**:23–40.

Dawson, R.M.C., and Clarke, N., 1972, D-*myo*-Inositol 1:2-cyclic phosphate 2-phosphohydrolase, *Biochem. J.* **127**:113–118.

Dawson, R.M.C., and Clarke, N. G., 1973, A comparison of D-inositol 1:2-cyclic phosphate 2-phosphohydrolase with other phosphodiesterases of kidney, *Biochem. J.* **134**:59–67.

Dawson, R.M.C., Freinkel, N., Jungalwala, F. B., and Clarke, N., 1971, The enzymatic formation of *myo*-inositol 1:2-cyclic phosphate from phosphatidylinositol, *Biochem. J.* **122**:605–607.

Dean Johnson, M., 1994, The *Arabidopsis thaliana myo*-inositol 1-phosphate synthase (EC 5.5.1.4), *Plant Physiol.* **105**:1023–1024.

Dean-Johnson, M., and Henry, S. A., 1989, Biosynthesis of inositol in yeast, Primary structure of *myo*-inositol-1-phosphate synthase (EC 5.5.1.4) and functional analysis of its structural gene, the *INO1* locus, *J. Biol. Chem.* **264:**1274–1283.

Dean Johnson, M., and Sussex, I. M., 1995, 1L-*myo*-inositol 1-phosphate synthase from *Arabidopsis thaliana, Plant Physiol.* **107:**613–619.

Dickeson, S. K., Lim, C. N., Schuyler, G. T., Dalton, T. P., Helmkamp, G. M., Jr., Yarbrough, L. R., 1989, Isolation and sequence of cDNA clones encoding rat phosphatidylinositol transfer protein, *J. Biol. Chem.* **264:**16557–16564.

Dickeson, S. K., Helmkamp, G. M., Jr., and Yarbrough, L. R., 1994, Sequence of a human cDNA encoding phosphatidylinositol transfer protein and occurrence of a related sequence in widely divergent eukaryotes, *Gene* **142:**301–305.

Diehl, R. E., Whiting, P., Potter, J., Gee, N., Ragan, C. I., Linemeyer, D., Schoepfer, R., Bennet, C., and Dixon, R.A.F. 1990, Cloning and expression of bovine brain inositol monophosphatase, *J. Biol. Chem.* **265:**5946–5949.

Dietz, M., and Albersheim, P., 1965, The enzymatic phosphorylation of *myo*-inositol. *Biochem. Biophys. Res. Commun.* **19:**598–603.

Divecha, N., Banfić, H., and Irvine, R. F., 1993, Inositides and the nucleus and inositides in the nucleus, *Cell* **74:**405–407.

Dixon, J. F., and Hokin, F. E., 1987, Inositol 1,2-cyclic 4,5-trisphosphate concentration relative to inositol 1,4,5-trisphosphate in pancreatic minilobules on stimulation with carbamylcholine in the absence of lithium: Possible role as a second messenger in long- but not short-term responses, *J. Biol. Chem.* **262:**13892–13895.

Dmitrieva, M. I., and Sobolev, A. M., 1984, Mobilization of phytin in castor bean during germination, *Soviet Plant Physiol.* **31:**1028–1035.

Donahue, T. F., and Henry, S. A., 1981a, Inositol mutants of *Saccharomyces cerevisiae:* Mapping the *ino1* locus and characterizing alleles of the *ino1, ino2,* and *ino4* loci, *Genetics* **98:**491–503.

Donahue, T. F., and Henry, S. A., 1981b, *myo*-Inositol-1-phosphate synthase, Characteristics of the enzyme and identification of its structural gene in yeast, *J. Biol. Chem.* **256:**7077–7085.

Downes, C. P., and Carter, A. N., 1991, Phosphoinositide 3-kinase: A new effector in signal transduction? *Cellular Signalling* **3:**501–513.

Drøbak, B. K., 1992, The plant phosphoinositide system, *Biochem. J.* **288:**697–712.

Ehrlich, K. C., Montalbano, B. G., Mullaney, E. J., Dischinger, H. C., Jr., and Ullah, A.H.J., 1993, Identification and cloning of a second phytase gene (*phyB*) from *Aspergillus niger (ficuum), Biochem. Biophys. Res. Commun.* **195:**53–57.

English, P. D., Dietz, M., and Albersheim, P., 1966, Myoinositol kinase: Partial purification and identification of product, *Science* **151:**198–199.

Erdman, J. W., 1981, Bioavailability of trace minerals from cereals and legumes, *Cereal Chem.* **58:**21–26.

Escobedo, J. A., Navankasattusas, S., Kavanaugh, W. M., Milfay, D., Fried, V. A., and Williams, L. T., 1991, cDNA cloning of a novel 85 kd protein that has SH2 domains and regulates binding of PI_3-kinase to the PDGF β-receptor, *Cell* **65:**75–82.

Estevez, F., Pulford, D., Stark, M.J.R., Crater, A. N., and Downes, C. P., 1994, Inositol trisphosphate metabolism in *Saccharomyces cerevisiae*: Identification, purification and properties of inositol 1,4,5-trisphosphate 6-kinase, *Biochem. J.* **302:**709–716.

Fitzgibbon, J., Pilz, A., Gayther, S., Appukuttan, B., Dulai, K. S., Dalhanty, J. D. A., Helmkamp, G. M., Jr., Yarbrough, L. R., and Hunt, D. M., 1994, Localization of the gene encoding human phosphatidylinositol transfer protein (PITPN) to 17p 13.3: A gene showing homology to the *Drosophila* retinal degeneration B gene (*rdgB*), *Cytogenet. Cell. Genet.* **67:**205–207.

Flanagan, C. A., Schnieders, E. A., Emerick, A. W., Kunisawa, R., Admon, A., and Thorner, J.,

1993, Phosphatidylinositol 4-kinase: Gene structure and requirement for yeast cell viability, *Science* **262:**1444–1448.

Foster, P. S., Gesini, E., Claudianos, C., Hopkinson, K. C., and Denborough, M. A., 1989, Inositol 1,4,5-trisphosphate phosphatase deficiency and malignant hyperpyrexia in swine, *Lancet* **8655:**124–127.

Furuichi, T., Yoshikawa, S., Miyawaki, A., Wada, K., Maeda, N., and Mikoshiba, K., 1989, Primary structure and functional expression of the inositol 1,4,5-trisphosphate-binding protein P_{400}, *Nature* **342:**32–38.

Gee, N. S., Ragan, C. I., Watling, K. J., Aspley, S., Jackson, R. G., Reid, G. G., Gani, D., and Shute, J. K., 1988, The purification and properties of *myo*-inositol monophosphatase from bovine brain, *Biochem. J.* **249:**883–889.

Gibson, D. M., and Ullah, A.B.J., 1990, Phytases and their action on phytic acid, in *Inositol Metabolism in Plants* (D. J. Morre, W. F. Boss, and F. A. Loewus, eds.), pp. 77–92, Wiley-Liss, New York.

Glomset, J. A., Gelb, M. H., and Farnsworth, C. S., 1990, Prenyl proteins in eukaryotic cells: A new type of membrane anchor, *Trends Biochem. Sci.* **15:**139–142.

Graf, E., and Eaton, J. W., 1993, Suppression of colonic cancer by dietary phytic acid. *Nutr. Cancer* **19:**11–19.

Hallcher, L. M., and Sherman, W. R., 1980, The effects of lithium ion and other agents on the activity of *myo*-inositol-1-phosphatase from bovine brain, *J. Biol. Chem.* **255:**10896–10901.

Harlan, J. E., Hajduk, P. J., Yoon, H. S., and Fesik, S. W., 1994, Peckstrin homology domains bind to phosphatidylinositol-4,5-bisphosphate, *Nature* **371:**168–170.

Harris, W. A., and Stark, W. S., 1977, Hereditary retinal degeneration in *Drosophila melanogaster.* A mutant defect associated with the phototransduction process, *J. Gen. Physiol.* **69:**261–291.

Hawkins, P. T., Poyner, D. R., Jackson, T. R., Letcher, A. J., Lander, D. A., and Irvine, R. F., 1993, Inhibition of iron-catalysed hydroxyl radical formation by inositol polyphosphates: A possible physiological function for *myo*-inositol hecakisphosphate, *Biochem. J.* **294:**929–934.

Hay, J. C., and Martin, T.F.J., 1993, Phosphatidylinositol transfer protein required for ATP-dependent priming of Ca^{2+}-activated secretion, *Nature* **366:**572–575.

Herman, P. K., and Emr, S. D., 1990, Characterization for *VPS34*, a gene required for vacuolar protein sorting and vacuole segregation in *Saccharomyces cerevisiae, Mol. Cell. Biol.* **10:**6742–6754.

Herman, P. K., Stack, J. H., and Emr, S. D., 1991, A genetic and structural analysis of the yeast Vps15 protein kinase: Evidence for a direct role of Vps15p in vacuolar protein delivery, *EMBO J.* **10:**4049–4060.

Heslop, J. P., Irvine, R. F., Tashjian, A. H., and Berridge, M. J., 1985, Inositol tetrakis- and pentakisphosphates in GH_4 cells, *J. Exp. Biol.* **119:**395–401.

Hiles, I. D., Otsu, M., Volinia, S., Fry, M. J., Gout, I., Dhand, R., Panayotou, G., Ruiz-Larrea, F., Thompson, A., Totty, N. F., Hsuan, J. J., Courtneidge, S. A., Parker, P. J., and Waterfield, M. D., 1992, Phosphatidylinositol 3-kinase: Structure and expression of the 110 kd catalytic subunit, *Cell* **70:**419–429.

Hirsch, J. P., and Henry, S. A., 1986, Expression of the *Saccharomyces cerevisiae* inositol-1-phosphate synthase (*INO1*) gene is regulated by factors that affect phospholipid synthesis, *Mol. Cell. Biol.* **6:**3320–3328.

Hosaka, K., Nikawa, J., Kodaki, T., and Yamashita, S., 1992, A dominant mutation that alters the regulation of *INO1* Expression in *Saccharomyces cerevisiae, J. Biochem.* **111:**352–358.

Hoshizaki, D. K., Hill, J. E., and Henry, S. A., 1990, The *S. cerevisiae Ino4* gene encodes a small, highly basic protein required for derepression of phospholipid biosynthetic enzymes, *J. Biol. Chem.* **265:**4736–4745.

Hotta, Y., and Benzer, S., 1970, Genetic dissection of the *Drosophila* nervous system by means of mosaics, *Proc. Natl. Acad. Sci. U.S.A.* **67:**1156–1163.

Irvine, R. F., 1992, Inositol phosphates and Ca^{2+} entry: Toward a proliferation or a simplification? *FASEB J.* **6:**3085–3091.

Isaaks, R. E., and Harkness, D. R., 1980, Erythrocyte organic phosphates and hemoglobin function in birds, reptiles, and fishes, *Amer. Zool.* **20:**115–129.

Jänne, P. A., Dutra, A. S., Dracopoli, N. C., Charnas, L. R., Puck, J. M., and Nussbaum, R. L., 1994, Localization of the 75-kDa inositol polyphosphate-5-phosphatase (INPP5B) to human chromosome band 1p34, *Cytogenet. Cell Genet.* **66:**164–166.

Kao, F.-T., and Puck, T. T., 1968, Genetics of somatic mammalian cells, VII. Induction and isolation of nutritional mutants in Chinese hamster cells, *Proc. Natl. Acad. Sci. U.S.A.* **60:**1275–1281.

Klig, L. S., and Henry, S. A., 1984, Isolation of the yeast *INO1* gene: Located on an autonomously replicating plasmid, the gene is fully regulated, *Proc. Natl. Acad. Sci. U.S.A.* **81:**3816–3820.

Kunz, J., Henriquez, R., Schneider, U., Deuter-Reinhard, M., Movva, N. R., and Hall, M. N., 1993, Target of rapamycin in yeast, TOR2, is an essential phosphatidylinositol kinase homolog required for G_1 progression, *Cell* **73:**585–596.

Laxminarayan, K. M., Chan, B. K., Tetaz, T., Bird, P. I., and Mitchell, C. A., 1994, Characterization of a cDNA encoding the 43-kDa membrane-associated inositol-polyphosphate 5-phosphatase, *J. Biol. Chem.* **269:**17305–17310.

Leonis, M. A., and Silbert, D. F., 1993, Characterization of a second hamster lung fibroblast mutant with defects in phosphatidylinositol-specific phospholipase C, *J. Biol. Chem.* **268:**9416–9424.

Lips, D. L., Majerus, P. W., Gorga, F. R., Young, A. T., and Benjamin, T. L., 1989, Phosphatidylinositol 3-phosphate is present in normal and transformed fibroblasts and is resistant to hydrolysis by bovine brain phospholipase C II, *J. Biol. Chem.* **264:**8759–8763.

Loewus, F. A., 1990, Inositol biosynthesis, in *Inositol Metabolism in Plants* (D. J. Morre, W. F. Boss, and F. A. Loewus, eds.), pp. 13–19, Wiley-Liss, New York.

Loewus, F. A., Everard, J. D., and Young, K. A., 1990, Inositol metabolism: Precursor role and breakdown, in *Inositol Metabolism in Plants* (D. J. Morre, W. F. Boss, and F. A. Loewus, eds.), pp. 21–45, Wiley-Liss, New York.

Lopes, J. M., Hirsch, J. P., Chorgo, P. A., Schulze, K. L., and Henry, S. A., 1991, Analysis of sequences involved in the *Ino1* promoter that are involved in its regulation by phospholipid precursors, *Nucleic Acids Res.* **19:**1687–1693.

Lopes, J. M., Schulze, K. L., Yates, J. W., Hirsch, J. P., and Henry, S. A., 1993, The *Ino1* promoter of *Saccharomyces cerevisiae* includes an upstream repressor sequence (URS1) common to a diverse set of yeast genes, *J. Bacteriol.* **175:**4235–4238.

Lott, J.N.A., 1984, Accumulation of seed reserves of phosphorus and other minerals, in *Seed Physiology,* (D. R. Murray, ed.), pp. 139–166, Academic Press, New York.

Majerus, P. W., 1992, Inositol phosphate biochemistry, *Annu. Rev. Biochem.* **61:**225–250.

Majerus, P. W., Ross, T. S., Cunningham, T. W., Caldwell, K. K., Jefferson, A. B., and Bansal, V. S., 1990, Recent insights in phosphatidylinositol signaling, *Cell* **63:**459–465.

Matzaris, M., Jackson, S. P., Laxminarayan, K. M., Speed, C. J., and Mitchell, C. A., 1994, Identification and characterization of the phosphatidylinositol-(4,5)-bisphosphate 5-phosphatase in human platelets, *J. Biol. Chem.* **269:**3397–3402.

Mayer, B. J., Ren, R., Clark, K. L., and Baltimore, D., 1993, A putative modular domain present in diverse signaling proteins, *Cell* **73:**629–630.

Mayr, G. W., Radenberg, T., Thiel, U., Vogel, G., and Stephens, L. R., 1992, Phosphoinositol diphosphates: Non-enzymic formation *in vitro* and occurrence *in vivo* in the cellular slime mold *Dictyostelium, Carbohydrate Res.* **234:**247–262.

McAllister, G., Whiting, P., Hammond, E. A., Knowles, M. R., Atack, J. R., Bailey, F. J.,

Maigetter, R., and Ragan, C. I., 1992, cDNA cloning of human and rat brain *myo*-inositol monophosphatase, *Biochem. J.* **284:**749–754.

McCance, R. A., and Widdowson, E. M., 1935, Phytic acid in human nutrition. *Biochem. J.* **29:**42694–42699.

Menniti, F. S., Miller, R. N., Putney, J. W. Jr., and Shears, S. B., 1993, Turnover of inositol polyphosphate pyrophosphates in pancreatoma cells, *J. Biol. Chem.* **268:**3850–3856.

Mignery, G. A., Südhof, T. C., Takei, K., and Camilli, P. D., 1989, Putative receptor of inositol 1,4,5-trisphosphate similar to ryanodine receptor, *Nature* **342:**192–195.

Mignery, G. A., Newton, C. L., Archer, B. T. III, and Südhof, T. C., 1990, Structure and expression of the rat inositol 1,4,5-trisphosphate receptor, *J. Biol. Chem.* **265:**12679–12685.

Mullaney, E. J., Gibson, D. M., and Ullah, A.H.J., 1991, Positive identification of a lambda gt11 clone containing a region of fungal phytase gene by immunoprobe and sequence verification, *Appl. Microbiol. Biotechnol.* **35:**611–614.

Musacchio, A., Gibson, T., Rice, P., Thompson, J., and Saraste, M., 1993, The PH domain: A common piece in the structural patchwork of signalling proteins, *Trends Biochem. Sci.* **18:**343–348.

Nikawa, J., and Yamashita, S., 1982, Yeast mutant defective in synthesis of phosphatidylinositol, *Eur. J. Biochem.* **125:**445–451.

Nikawa, J., and Yamashita, S., 1984, Molecular cloning of the gene encoding CDPdiacylglycerol-inositol 3-phosphatidyl transferase in *Saccharomyces cerevisiae*, *Eur. J. Biochem.* **143:**251–256.

Nikoloff, D. M., McGraw, P., and Henry, S. A., 1992, The *Ino1* gene of *Saccharomyces cerevisiae* encodes a helilx–loop–helix protein that is required for activation of phospholipid biosynthesis, *Nucleic Acids Res.* **20:**3253.

Novick, P., Field, C., and Schekman, R., 1980, Identification of 23 complementation groups required for post-translational events in the yeast secretory pathway, *Cell* **21:**205–215.

Ogawa, M., Tanaka, K., and Kasai, Z., 1979, Accumulation of phosphorus, magnesium and potassium in developing rice grains followed by electron microprobe X-ray analysis focusing on the aleurone layer, *Plant Cell Physiol.* **20:**19–27.

Otsu, M., Hiles, I., Gout, I., Fry, M. J., Ruiz-Larrea, F., Panayotou, G., Thompson, A., Dhand, R., Hsuan, J., Totty, N., Smith, A. D., Morgan, S. J., Courtneidge, S. A., Parker, P. J., and Waterfield, M. D., 1991, Characterization of two 85 kd proteins that associate with receptor tyrosine kinases, middle-T/pp60$^{c\text{-}src}$ complexes, and PI_3-kinase, *Cell* **65:**91–104.

Pak, W. L., Grossfield, J., and Arnold, K., 1970, Mutants of the visual pathway of *Drosophila melanogastor, Nature* **227:**518–520.

Paris, S., Chambard, J.-C., and Pouysségur, J., 1987, Coupling between phosphoinositide breakdown and early mitogenic events in fibroblasts, *J. Biol. Chem.* **262:**1977–1983.

Pen, J., Verwoerd, T. C., van Paridon, P. A., Beudeker, R. F., van den Elzen, P.J.M., Geerse, K., van der Klis, J. D., Versteegh, H.A.J., van Ooyen, A.J.J., and Hoekema, A., 1993, Phytase-containing transgenic seeds as a novel feed additive for improved phosphorus utilization, *Bio/technology* **11:**811–814.

Pernollet, J.-C., 1978, Protein bodies of seeds: Ultrastructure, biochemistry, biosynthesis and degradation, *Phytochemistry* **17:**1473–1480.

Phillippy, B. Q., Ullah, A.H.J., and Ehrlich, K. C., 1994, Purification and some properties of inositol 1,3,4,5,6-pentakisphosphate 2-kinase from immature soybean seeds, *J. Biol. Chem.* **269:**28393–28399.

Piddington, C. S., Houston, C. S., Paloheimo, M., Cantrell, M., Miettinen-Oinonen, A., Nevalainen, H., and Rambosek, J., 1993, The cloning and sequencing of the genes encoding phytase (*phy*) and pH 25-optimum acid phosphatase (*aph*) from *Aspergillus niger* var. *awamori, Gene* **133:**55–62.

Pouysségur, J., Sardet, C., Franchi, A., L'Allemain, G., and Paris, S., 1984, A specific mutation abolishing Na^+/H^+ antiport activity in hamster fibroblasts precludes growth at neutral and acidic pH, *Proc. Natl. Acad. Sci. USA* **81:**4833–4837.

Raboy, V., 1990, The biochemistry and genetics of phytic acid synthesis, in *Inositol Metabolism in Plants* (D. J. Moore, W. Boss, and F. A. Loewus, eds.), pp. 52–73, Alan R. Liss, New York.

Raboy, V., and Dickinson, D. B., 1984, Effect of phosphorus and zinc nutrition on soybean seed phytic acid and zinc, *Plant Physiol.* **75:**1094–1098.

Raboy, V., Dickinson, D. B., and Neuffer, M. G., 1990, A survey of maize mutants for variation in phytic acid, *Maydica* **35:**383–390.

Raboy, V., Young, K., and Gerbasi, P., 1994, Maize *low phytic acid* (Lpa) mutants, Abstracts: 4th International Congress of Plant Molecular Biology: Abs. No. 1827.

Rath, H. M., Doyle, G. A. R., and Silbert, D. F., 1989, Hamster fibroblasts defective in thrombin-induced mitogenesis, *J. Biol. Chem.* **264:**13387–13390.

Rath, H. M., Fee, J. A., Rhee, S. G., and Silbert, D. F., 1990, Characterization of phosphatidylinositol-specific phospholipase C defects associated with thrombin-induced mitogenesis, *J. Biol. Chem.* **265:**3080–3087.

Rhee, S. G., Suh, P.-G., Ryu, S.-H., and Lee, S. Y., 1989, Studies of inositol phospholipid-specific phospholipase C, *Science* **244:**546–550.

Robinson, J. S., Klionsky, D. J., Banta, L. M., and Emr, S. D., 1988, Protein sorting in *Saccharomyces cerevisiae*: Isolation of mutants defective in the delivery and processing of multiple vacuolar hydrolases, *Mol. Cell. Biol.* **8:**4936–4948.

Ross, T. S., Tait, J. F., and Majerus, P. W., 1990, Identity of inositol 1,2-cyclic phosphate 2-phosphoydrolase with lipocortin III, *Science* **248:**605–607.

Ross, T. S., Jefferson, A. B., Mitchell, C. A., and Majerus, P. W., 1991a, Cloning and expression of human 75-kDa inositol polyphosphate-5-phosphatase, *J. Biol. Chem.* **266:**20283–20289.

Ross, T. S., Whiteley, B., Graham, R. A., and Majerus, P. W., 1991b, Cyclic hydrolase-transfected 3T3 cells have low levels of inositol 1,2-cyclic phosphate and reach confluence at low density, *J. Biol. Chem.* **266:**9086–9092.

Sartirana, M. A., and Bianchetti, R., 1967, The effects of phosphate on the development of phytase in the wheat embryo, *Physiol. Plant.* **20:**1066–1075.

Schu, P. V., Takegawa, K., Fry, M. J., Stack, J. H., Waterfield, M. D., and Emr, S. D., 1993, Phosphatidylinositol 3-kinase encoded by yeast *VPS*34 gene essential for protein sorting, *Science* **260:**88–91.

Serunian, L. A., Haber, M. T., Fukui, T., Kim, J. W., Rhee, S. G., Lowenstein, J. M., and Cantley, L. C., 1989, Polyphosphoinositides produced by the phosphatidylinositol 3-kinase are poor substrates for phospholipase C from rat liver and bovine brain, *J. Biol. Chem.* **264:**17809–17815.

Skinner, H. B., Alb, J. G., Jr., Whitters, E. A., Helmkamp, G. M., Jr., and Bankaitis, V. A., 1993, Phospholipid transfer activity is relevant to but not sufficient for the essential function of the yeast *Sec14* gene product, *EMBO J.* **12:**4775–4784.

Skolnik, E. Y., Margolis, B., Mohammadi, M., Lowenstein, E., Fischer, R., Drepps, A., Ullrich, A., and Schlessinger, J., 1991, Cloning of PI_3 kinase-associated p85 utilizing a novel method for expression/cloning of target proteins for receptor tyrosine kinases, *Cell* **65:**83–90.

Smart, C. C., and Fleming, A. J., 1993, A plant gene with homology to D-*myo*-inositol-3-phosphate synthase is rapidly and spatially up-regulated during an abscisic acid induced morphogenic response in *Spirodela polyrrhiza, Plant J.* **4:**279–293.

Stack, J. H., Herman, P. K., Schu, P. V., and Emr, S. D., 1993, A membrane-associated complex containing the Vps15 protein kinase and the Vps34 PI 3-kinase is essential for protein sorting to the yeast lysosome-like vacuole, *EMBO J.* **12:**2195–2204.

Stark, W. S., Chen, D.-M., Johnson, M. A., and Frayer, K. L., 1983, The *rdgB* gene of *Drosophila*: Retinal degeneration in different alleles and inhibition by *norpA*, *J. Insect Physiol.* **29:**123–131.

Stephens, L. R., and Irvine, R. F., 1990, Stepwise phosphorylation of *myo*-inositol leading to *myo*-inositol hexakisphosphate in *Dictyostellium. Nature* **346:**580–583.

Stephens, L. R., Kay, R. R., and Irvine, R. F., 1990, A *myo*-inositol D-3 hydroxykinase activity in *Dictyostelium, Biochem. J.* **272:**201–210.

Stephens, L. R., Hawkins, P. T., Stanley, A. F., Moore, T., Poyner, D. R., Morris, P. J., Hanley, M. R., Kay, R. R., and Irvine, R. F., 1991, *myo*-Inositol pentakisphosphates: Structure, biological occurrence and phosphorylation to *myo*-inositol hexakisphosphate, *Biochem. J.* **275:**485–499.

Stephens, L. R., Jackson, T. R., and Hawkins, P. T., 1993a, Agonist-stimulated synthesis of phosphatidylinositol(3,4,5)-trisphosphate: a new intracellular signalling system?, *Biochim. Biophys. Acta* **1179:**27–75.

Stephens, L., Radenberg, T., Thiel, U., Vogel, G., Khoo, K.-H., Dell, A., Jackson, T. R., Hawkins, P. T., and Mayr, G. W., 1993, The detection, purification, structural characterization, and metabolism of diphosphoinositol pentakisphosphate(s) and bisdiphosphoinositol tetrakisphosphate(s), *J. Biol. Chem.* **268:**4009–4015.

Südhof, T. C., Newton, C. L., Archer, B. T., III, Ushkaryov, Y. A., and Mignery, G. A., 1991, Structure of a novel $InsP_3$ receptor, *EMBO J.* **10:**3199–3206.

Sylvia, V., Curtin, G., Norman, J., Stec, J., and Busbee, D., 1988, Activation of a low specific activity form of DNA polymerase α by inositol-1,4-bisphosphate, *Cell* **54:**651–658.

Talmage, D. A., Freund, R., Young, A. T., Dahl, J., Dawe, C. J., and Benjamin, T. L., 1989, Phosphorylation of middle T by $pp60^{c\text{-}src}$: A switch for binding of phosphatidylinositol 3-kinase and optimal tumorigenesis, *Cell* **59:**55–65.

Theibert, A. B., Estevez, V. A., Ferris, C. D., Danoff, S. K., Barrow, R. K., Prestwich, G. D., and Snyder, S. H., 1991, Inositol 1,3,4,5-tetrakisphosphate and inositol hexakisphosphate receptor proteins: Isolation and characterization from rat brain, *Proc. Natl. Acad. Sci. U.S.A.* **88:**3165–3169.

Toyoshima, S., Matsumoto, N., Wang, P., Inoue, H., Yoshioka, T., Hotta, Y., and Osawa, T., 1990, Purification and partial amino acid sequences of phosphoinositide-specific phospholipase C of *Drosophila* eye, *J. Biol. Chem.* **265:**14842–14848.

Trewavas, A., and Gilroy, S., 1991, Signal transduction in plant cells, *Trends Genet.* **7:**356–361.

Tyers, M., Rachubinski, R. A., Stewart, M. I., Varrichio, A. M., Shorr, R.G.L., Haslam, R. J., and Harley, C. B., 1988, Molecular cloning and expression of the major protein kinase C substrate of platelets, *Nature* **333:**470–473.

van Hartingsveldt, W., van Zeijl, C.M.J., Harteveld, G. M., Gouka, R. J., Suykerbuyk, M.E.G., Luiten, R.G.M., van Paridon, P. A., Selten, G.C.M., Veenstra, A. E., van Gorcom, R.F.M., and van den Hondel, C.A.M.J.J., 1993, Cloning, characterization and overexpression of the phytase-encoding gene (*phyA*) of *Aspergillus niger, Gene* **127:**87–94.

Vihtelic, T. S., Goebl, M., Milligan, S., O'Tousa, J. E., and Hyde, D. R., 1993, Localization of *Drosophila retinal degeneration B,* a membrane-associated phosphatidylinositol transfer protein, *J. Cell Biol.* **122:**1013–1022.

Voglmaier, S. M., Keen, J. H., Murphy, J., Ferris, C. D., Prestwich, G. D., Snyder, S. H., and Theibert, A. B., 1992, Inositol hexakisphosphate receptor identified as the clathrin assembly protein AP-2, *Biochem. Biophys. Res. Commun.* **187:**158–163.

Welters, P., Takegawa, K., Emr, S. D., and Chrispeels, M. J., 1994, ATVPS34, a phosphatidylinositol 3-kinase of *Arabidopsis thaliana,* is an essential protein with homology to a calcium-dependent lipid binding domain, *Proc. Natl. Acad. Sci. U.S.A.* **91:**11398–11402.

White, M. J., Hirsch, J. P., and Henry, S. A., 1991, The *Opi1* gene of *Saccharomyces cerevisiae,* a negative regulator of phospholipid biosynthesis, encodes a protein containing polyglutamine tracts and a leucine zipper, *J. Biol. Chem.* **266:**863–872.

Whitman, M., Downes, C. P., Keeler, M., Keller, T., and Cantley, L., 1988, Type I phosphatidylinositol kinase makes a novel inositol phospholipid, phosphatidylinositol-3-phosphate, *Nature* **332:**644–646.

Williams, S. G., 1971, Biosynthesis of inositol by inositol-less mutants of *Neurospora crassa*, *Aust. J. Biol. Sci.* **24:**1181–1188.

Woodward, C., Alcorta, E., and Carlson, J., 1992, The *rdgB* gene of *Drosophila*: A link between vision and olfaction, *J. Neurogenetics* **8:**17–31.

Wreggett, K. A., 1992, Inositol monophosphatase is a highly conserved enzyme having localized structural similarity to both glycerol 3-phosphate dehydrogenase and haemoglobin, *Biochem. J.* **286:**147–152.

Yagisawa, H., Hirata, M., Kanematsu, T., Watanabe, Y., Ozaki, S., Sakuma, K., Tanaka, H., Yabuta, N., Kamata, H., Hirata, H., and Nojima, H., 1994, Expression and characterization of an inositol 1,4,5-trisphosphate binding domain of phosphatidylinositol-specific phospholipase C-δ1, *J. Biol. Chem.* **269:**20179–20188.

Yoakim, M. W., Hou, W., Liu, Y., Carpenter, C. L., Kapeller, R., and Schaffhausen, B. S., 1992, Interactions of polyomavirus middle T with the SH2 domains of the pp85 subunit of phosphatidylinositol 3-kinase, *J. Virol.* **66:**5485–5491.

Yoakim, M., Hou, W., Songyang, Z., Liu, Y., Cantley, L., and Schaffhausen, B., 1994, Genetic analysis of a phosphatidylinositol 3-kinase SH2 domain reveals determinants of specificity, *Mol. Cell. Biol.* **14:**5929–5938.

Yoko-o, T., Matsui, Y., Yagisawa, H., Nojima, H., Uno, I., and Toh-e, A., 1993, The putative phosphoinositide-specific phospholipase C gene, *PLC1*, of the yeast *Saccharomyces cerevisiae*, is important for cell growth, *Proc. Natl. Acad. Sci. U.S.A.* **90:**1804–1808.

York, J. D., and Majerus, P. W., 1990, Isolation and heterologous expression of a cDNA encoding bovine inositol polyphosphate 1-phosphatase, *Proc. Natl. Acad. Sci. U.S.A.* **87:**9548–9552.

York, J. D., and Majerus, P. W., 1994, Nuclear phosphatidylinositols decrease during S-phase of the cell cycle of HeLa cells, *J. Biol. Chem.* **269:**7847–7850.

York, J. D., Veile, R. A., Donis-Keller, H., and Majerus, P. W., 1993, Cloning, heterologous expression, and chromosomal localization of human inositol polyphosphate 1-phosphatase, *Proc. Natl. Acad. Sci. U.S.A.* **90:**5833–5837.

York, J. D., Saffitz, J. E., and Majerus, P. W., 1994, Inositol polyphosphate 1-phosphatase is present in the nucleus and inhibits DNA synthesis, *J. Biol. Chem.* **269:**19992–19999.

Yoshikawa, S., Tanimura, T., Miyawaki, A., Nakamura, M., Yuzaki, M., Furuichi, T., and Mikoshiba, K., 1992, Molecular cloning and characterization of the inositol 1,4,5-trisphosphate receptor in *Drosophila melanogaster*, *J. Biol. Chem.* **267:**16613–16619.

Yoshioka, T., Inoue, H., and Hotta, Y., 1985, Absence of phosphatidylinositol phosphodiesterase in the head of a *Drosophila* visual mutant, *norpA* (no receptor potential A), *J. Biochem.* **97:**1251–1254.

Zhang, X., Jefferson, A. B., Auethavekiat, V., and Majerus, P. W., 1995, The protein deficient in Lowe syndrome is a phosphatidylinositol-4,5-bisphosphate 5-phosphatase, *Proc. Natl. Acad. Sci. U.S.A.* **92:**4853–4856.

Chapter 10

Metabolism of *myo*-Inositol Phosphates and the Alternative Pathway in Generation of *myo*-Inositol Trisphosphate Involved in Calcium Mobilization in Plants

Susweta Biswas and B. B. Biswas

1. INTRODUCTION

The *myo*-inositol polyphosphates and their role in cellular metabolism and signal transduction processes in plants have now attracted the attention of scientists in growing numbers. Phytic acid, *myo*-inositol-1,2,3,4,5,6-hexakisphosphate has long been known as the storage form of phosphorus in seeds (see Cosgrove, 1980). The calcium and magnesium salt of phytic acid is also known as phytin. Of the total phosphorus, 50–80% has been found to be associated with phytin in different seeds. Although the composition of phytic acid has been known for more than 100 years, many problems concerning its metabolism and functions

Susweta Biswas Department of Biochemistry, Bose Institute, Calcutta 700 054, India.
B. B. Biswas Department of Biophysics, Molecular Biology and Genetics, University of Calcutta, Calcutta 700 009, India.

Subcellular Biochemistry, Volume 26: myo-Inositol Phosphates, Phosphoinositides, and Signal Transduction, edited by B. B. Biswas and Susweta Biswas. Plenum Press, New York, 1996.

are not fully solved (see Loewus and Loewus, 1983; B. Biswas *et al.*, 1984; Raboy, 1990; Drøbak, 1992). What is apparent is that plant cells representing different tissue types synthesize phytic acid both to sequester phosphorus and also to chelate different metallic cations such as Ca^{2+} and others. Several reviews have dealt with the occurrence, chemistry, and nutritional implications of phytates in legumes and cereals (see Reddy *et al.*, 1982; B. Biswas *et al.*, 1984). Phytate in general rapidly accumulates during the development of seeds and disappears during germination of seeds.

The main question arises as to whether *myo*-inositol phosphate(s) other than phytic acid does occur in plants and, if so, how these are derived and what specific role they can play in plant metabolic events. Similarly, the occurrence of pyrophosphate derivatives of $InsP_6$ might project their involvement in the energetics of cellular metabolism as well as in signaling (Stephens *et al.*, 1993). In this review, attention has specifically been focused first on the biosynthetic and degradative pathways of *myo*-inositol phosphates entailing (1) metabolic cycle in seed formation and germination, (2) phosphoinositide and $InsP_6$ pathway of generation of *myo*-inositol trisphosphate(s), and (3) their role in Ca^{2+} mobilization in plant cells. Second, we discuss the elucidation of signal transduction by *myo*-inositol trisphosphate(s) and some of the interesting signals in plants and their relevance in the global phenomenon of signal transduction pathways particular to the plant cell. Last, attempts have also been made as far as possible to project a coherent and emerging picture of the status of *myo*-inositol phosphates, phosphoinositides, and their involvement in the cascade of events in the process of responses to different signals perceived by plants. The review of the literature is not intended to be all inclusive. Vast literature has accumulated on this aspect, and the readers are advised to consult relevant reviews (Berridge and Irvine, 1989; Drøbak, 1992; Palme, 1992; Poovaiah and Reddy, 1993; Cote and Crain, 1993).

2. METABOLISM OF INOSITOL PHOSPHATES IN PLANTS

Because there are nine stereoisomers of inositol, the possible existence of phosphorylated derivatives of different inositols is expected. However, it is observed that the principal phosphocompounds present in the cell are the *myo*-inositol phosphates (see Cosgrove, 1980; B. Biswas *et al.*, 1984). It is currently believed that *myo*-inositol synthesis in the cell comprises three enzymatic steps: (1) glucose is converted to glucose-6-phosphate (Gluc-6-P) by hexokinase; (2) Gluc-6-P is reacted by the action of *myo*-inositol-1-P synthase; and (3) the *myo*-inositol-1-P is converted to *myo*-inositol by *myo*-inositol-1-P phosphatase (see Wells and Eisenberg, 1978; B. Biswas *et al.*, 1984; Loewus *et al.*, 1990).

2.1. Biosynthetic Pathways of *myo*-Inositol Phosphates

Because *myo*-inositol hexakisphosphate ($InsP_6$) is present abundantly in plant cells, particularly in almost all seeds, the pathway of its synthesis might comprise a number of steps catalyzed by different enzymes. There are three probable pathways for the synthesis of $InsP_6$ as proposed earlier (see Cosgrove, 1966). One of the mechanisms appears to involve kinase conversion of *myo*-inositol to L-*myo*-inositol-1-P, and in fact, such as enzyme has been reported (Loewus *et al.*, 1982). L-*myo*-Inositol-1-P is also synthesized from Gluc-6-P by the action of *myo*-inositol-1-P synthase (Chen and Charalampous, 1986).

The next question arises as to how further phosphorylation does occur. Are there different kinases that can specifically phosphorylate *myo*-inositol-1-P to higher homologues or a single enzyme that can catabolize the reaction up to $InsP_6$. Our laboratory has shown that *myo*-inositol-1-P can be further phosphorylated to Ins(1,3,4,5,6)P_5 by kinase(s); however, it can not phosphorylate the C_2 position of $InsP_5$ (S. Biswas *et al.*, 1978). That reaction appears to involve another enzyme, i.e., $InsP_6$–ADP-phosphotransferase, a kinase that can phosphorylate Ins(1,3,4,5,6)P_5 to Ins(1,2,3,4,5,6)P_6 in mung bean seeds (S. Biswas *et al.*, 1978). In contrast, Ins(2)P has been suggested to be the precursor for $InsP_6$ formation in rice (Igaue *et al.*, 1980), though how Ins(2)P is synthesized is not clear. When Ins(2)P was used as initial substrate for the enzyme phosphoinositol kinase(s), the synthesis of $InsP_6$ was demonstrated (S. Biswas *et al.*, 1978). Igaue *et al.* (1982) further analyzed the different *myo*-inositol phosphates and confirmed that different isomers, such as Ins(1,3)P_2, Ins(2,6)P_2, Ins(1,3,5)P_3, Ins(2,5,6)P_3, Ins(1,3,5,6)P_4, Ins(2,3,5,6)P_4, Ins(2,3,4,5,6)P_5, Ins(1,2,3,5,6)P_5, and Ins(1,3,4,5,6)P_5, were obtained.

This finding added further complication in the biosynthetic pathway of $InsP_6$. However, one point is clear: there might be two routes of phosphorylation starting with either Ins(3)P (L-*myo*-inositol-1-P) or Ins(2)P. The picture has apparently become more complicated by the concomitant dephosphorylation of different *myo*-inositol phosphates.

It has also been demonstrated that in germinating mung bean, D- and/or L-Ins(1,2,3,4,5)P_5 and D/L-Ins(1,2,4,5,6)P_5 are predominately present (Stephens *et al.*, 1991). Phillippy and Bland (1988) earlier reported that in addition to the $InsP_5$ reported by Stephens *et al.* (1991), D/L-Ins(1,2,3,4,6)P_5 and D/L-Ins(1,3,4,5,6)P_5 are also present. This, in fact, supported our data that D/L-Ins(1,3,4,5,6)P_5 is likely to be predominantly present in the case of mung bean as a result of the action of phospho-*myo*-inositol kinase and $InsP_6$–ADP-phosphotransferase-mediated synthesis of $InsP_6$ (S. Biswas *et al.*, 1978).

Stephens and Irvine (1990) established that in *Dictyostelium*, $InsP_6$ was formed as a result of stepwise phosphorylation of *myo*-inositol catalyzed by more than one kinase. The steps of phosphorylation are as follows. Starting from Ins(3)P (synthesized possibly by *myo*-inositol kinase), Ins(3,6)P_2, Ins(3,4,6)P_3, Ins(1,3,4,6)P_4, and Ins(1,3,4,5,6)P_5 were obtained. Interestingly Ins(1,3,4,5,6)P_5 might be phosphorylated further to $InsP_6$ by a phosphotransferase-type enzyme detected in mung bean and soybean (S. Biswas and Biswas, 1965; S. Biswas *et al.*, 1978; Phillippy *et al.*, 1994). However, multiple kinases were detected that could phosphorylate the 2, 3, or 5 positions of $InsP_5$ in *Dictyostelium* (Stephens *et al.*, 1991), because other $InsP_5$s such as Ins(1,2,4,5,6)P_5 and Ins(1,2,3,4,6)P_5 have also been detected. It has also been suggested that these $InsP_5$s might be formed by dephosphorylation only to be recycled to $InsP_6$ by phosphorylation (Stephens and Irvine, 1990).

If a comparison is made between the stepwise phosphorylation of *myo*-inositol to $InsP_6$ in *Dictyostelium* and other plants, a wide variation in the synthesized isomers of $InsP_2$ to $InsP_5$ is evident, and no definitive phosphorylation pattern emerges except that a common isomer of $InsP_5$, Ins(1,3,4,5,6)P_5, is detectable in those cases. Interestingly, in the case of mung bean, formation of $InsP_6$ from Ins(1,3,4,5,6)P_5 was demonstrated by the enzyme system $InsP_6$–ADP-phosphotransferase (S. Biswas *et al.*, 1978).

All these observations indicate the involvement of a number of kinases that can phosphorylate specific positions of *myo*-inositol monophosphate. Recently, in the case of the animal system, specific kinases have been reported (Craxton *et al.*, 1994). Implicit also is the fact that phosphoinositol kinase from mung bean has two different forms (Chakraborti and Biswas, 1981a), and the phosphotransferase system has two distinct activities, one dependent on ADP and the other on GDP (S. Biswas and Biswas, 1965; S. Biswas *et al.*, 1975; S. Biswas et al., 1978). However, nothing is known so far about their involvement in differential phosphorylation of specific positions of Ins(1)P. Boltman *et al.* (1980) reported three enzymes from *Lemna gibba* that could catalyze the synthesis of $InsP_6$. One of them had a *myo*-inositol kinase yielding Ins(1)P; the second converted Ins(1)P to $InsP_3$; the third converted $InsP_3$ to $InsP_6$.

The second pathway by which *myo*-inositol phosphates are synthesised might be through the synthesis of phosphoinositides. Two reactions for the incorporation of *myo*-inositol into phosphoinositides have been found to exist in plants (Moore, 1982). Phosphatidylinositol has long been known to be a common constituent of plant phospholipids (Harwood, 1980). One mechanism for the synthesis of phosphatidylinositol involves the enzyme CDP-diacylglycerol:inositol phosphatidyltransferase, and the other is carried out by phosphatidylinositol:inositol phosphatidyltransferase. Both reactions seem to occur in the endoplasmic reticulum of plant cells and also in Golgi bodies (Moore, 1990). However, of the two reactions, only the first is capable of net synthesis of phosphatidylinositol. Irvine *et al.* (1989) convincingly showed that only two phosphoinositides were present in pea leaves, and their structures have been

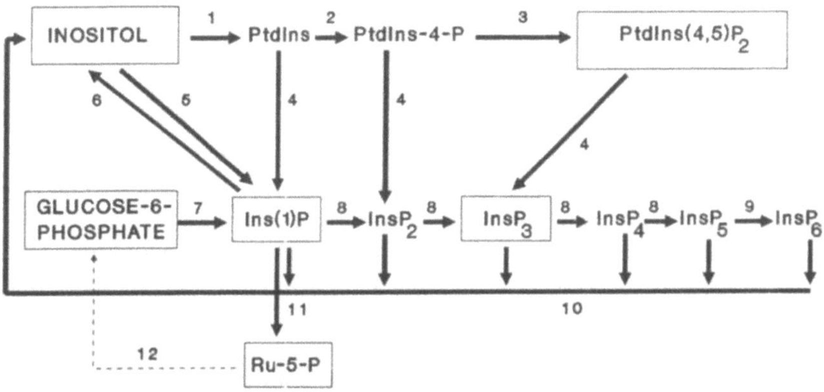

FIGURE 1. Pathway of synthesis and degradation of inositolphosphates and phosphoinositides in plants. 1, CDP-diacylglycerol:inositol phosphatidyl transferase; 2 and 3, phosphoinositide kinases; 4, phopholipase C; 5, *myo*inositol kinase; 6, Ins(1)P-phosphatase; 7, *Myo*inositol-1-P-synthase; 8, phosphoinositol kinase(s); 9, $InsP_6$-ADP-phosphotransferase; 10, Phytase(s); 11, Ins(1)P-dehydrogenase; 12, pentose shunt pathway enzymes.

found to be PtdIns(4)P and PtdIns(4,5)P_2; Ins(1,4)P_2 and Ins(1,4,5)P_3 are respectively derived through the action of phospholipase C. In fact, the presence of phospholipase C has also been reported.

Once *myo*-inositol bis- or trisphosphate is formed, this can be further phosphorylated by kinase-type enzymes as mentioned (Majerus *et al.*, 1988). The *myo*-inositol trisphosphate obtained from phosphoinositide, when incubated with phosphoinositol kinase, formed $InsP_5$ (Chakraborty and Biswas, 1981b). A recent study (Estevez *et al.*, 1994) confirmed this, finding that a D-Ins(1,4,5)P_3-6 kinase in yeast can lead to the synthesis of D-Ins(1,4,5,6)P_4 and eventually to $InsP_5$. A scheme outlining the steps involved and considering the experimental evidence thus far available as well as some of the lacunae in the biosynthesis of $InsP_6$ in plants is given in Figure 1. Obviously, this is far from an unequivocal and universal pathway but should be looked for if it exists. Because of complications of experiments involving simultaneous synthesis and degradation of $InsP_6$ *in vivo* as well as lack of proper identification of different isomers of *myo*-inositol phosphates from different plant systems, the pathway as proposed is still tentative.

2.2. Dephosphorylation of *myo*-Inositol Polyphosphates

Because the synthesis of *myo*-inositol polyphosphates is seemingly associated with the degradation of these compounds *in vivo*, it might be interesting to look into the mechanisms of degradation and the utilization of the products by the cell for growth and development. It is also tempting to consider that some of the degradation products of $InsP_6$ in plants, particularly $InsP_3$, might be involved in the signal transduction process. Enzymatic dephosphorylation of $InsP_6$ by phytase(s) has been reported to be a sequential and stepwise process. This aspect of dephosphorylation and the types of phytases so far reported have already been discussed (see Cosgrove, 1980; B. Biswas *et al.*, 1984; Loewus *et al.*, 1990; Drøbak, 1992). Tomlinson and Ballou (1962) reported that wheat bran phytase produces a series of optically active *myo*-inositol phosphates such as 1- and 2-monophosphates, Ins(1,2)P_2, Ins(1,2,3)P_3, L-Ins(1,2,6)P_3, and Ins(1,2,5,6)P_4. Lim and Tate (1973) indicated that wheat bran phytase F2 produces Ins(2)P, L-Ins(1,2)P_2, Ins(1,2,3)P_3, L-Ins(1,2,3,4)P_4, Ins(1,2,3,4,6)P_5, L-Ins(1,2,3,4,5)P_5, and Ins(1,3,4,5,6)P_5.

Recently, a new form of phytase from lily pollen has been reported (Scott and Loewus, 1986). It is an alkaline phytase having a pH optimum of 8, is activated by calcium ions, and dephosphorylates $InsP_6$ to $InsP_3$ as a final product (Loewus *et al.*, 1990). The $InsP_3$ has been identified as Ins(1,2,3)P_3 (see Chapter 8). It is proposed that initial hydrolysis occurs at the D-5 position of $InsP_6$ to yield Ins(1,2,3,4,6)P_5, followed by two subsequent dephosphorylations adjacent to the D-5 hydroxyl group to produce, finally, Ins(1,2,3)P_3 (Barrientos *et al.*, 1994). Mung bean phytase (pH 7.5) yields primarily Ins(2,4,5)P_3 from $InsP_6$ *in vitro,* as tentatively identified by HPLC (Maitra *et al.*, 1988). However, there is now growing evidence that some products of inositol phosphates play an important role in the biological functions of plant cells, ultimately including plant growth and development.

In vivo $InsP_6$ can be hydrolyzed by a number of phytases. The biological significances of different phytases and intracellular roles of the products produced are still not clear. A pertinent question is whether Ins(1,2,3)P_3 and Ins(2,4,5)P_3 can act as second messenger in the plant system as well as Ins(1,4,5)P_3. We discuss this aspect later.

2.3. Metabolic Cycle Involving Glucose-6-Phosphate and *myo*-Inositol Phosphates in Plants

From the viewpoint of seed physiology, research in biosynthesis and breakdown of $InsP_6$ has lagged far behind that on other major seed reserves such as starch and protein. Seed crops constitute a major renewable resource of our civilization. To fully understand the metabolic processes that control the major phosphate reserve in seed would be a major advance. In fact, a metabolic cycle emerged out of the work done in our laboratory (see B. Biswas *et al.*, 1978, 1984; S. Biswas and Biswas, 1994). Glucose-6-P is used during seed formation by *myo*-inositol-1-P synthetase for the production of Ins(1)P, which is subsequently phosphorylated to Ins(1,3,4,5,6)P_5 by phosphoinositol kinase(s) and ultimately to $InsP_6$ by another enzyme, i.e., $InsP_6$–ADP-phosphotransferase, already discussed (B. Biswas *et al.*, 1984). During germination, $InsP_6$ is hydrolyzed by the enzyme phytase, ultimately to *myo*-inositol, which is required for cell wall biosynthesis (see Loewus *et al.*, 1990). In the early phase of germination, Ins(1,3,4,5,6)P_5 is formed by the reversible reaction of $InsP_6$–ADP-phosphotransferase, and its subsequent hydrolysis by phytase yields to Ins(1)P before finally giving rise to *myo*-inositol. But Ins(1)P, in turn, can be converted to ribulose-5-P (Ru-5-P) by *myo*-inositol-1-phosphate dehydrogenase, giving feedback to the production of glucose-6-P during the early phase of germination through the pentose shunt pathway. The probable mechanism of combined dehydrogenation and decarboxylation of Ins(1)P leading to the formation of Ru-5-P has already been described (De and Biswas, 1979; Ghosh *et al.*, 1984).

The stoichiometry of this reaction suggests that 2 mol of NAD^+ is reduced for each mole of Ru-5-P produced. However, the fact that NAD^+-dependent dehydrogenation is an initial requirement for decarboxylation has been documented by the observation that omission of NAD^+ from the reaction mixture results in striking inhibition of decarboxylation. The presence of this enzyme provides a link among the metabolic pathways during germination of seeds. The significance of this pathway lies in the fact that it supplies the energy as ATP, reducing power as NADH, and pentose phosphate for nucleotide biosynthesis. It appears, then, that $InsP_6$ degradation during the early period of imbibition does play a significant role in seed germination.

3. CALCIUM MOBILIZATION AND ITS ROLE IN PLANT RESPONSES

The importance of ion fluxes in various metabolic processes such as uptake of nutrients by roots, stomatal opening, development of polarity, and the expansion of cells have been recognized (see Gilroy and Trewavas, 1994, and references therein). Circumstantial evidence suggests that this is true for most other aspects of plant development (Bowler and Chua, 1994). The concept that Ca^{2+} acts as a messenger is based on the fact that the signals, either chemical, physical, or mechanical, can induce changes in cytosolic Ca^{2+} (Bush, 1993). In fact, using Ca^{2+}-binding fluorescent dyes and microinjection of aequorin (a Ca^{2+}-binding protein giving fluorescence) or inserting an aequorin gene has shown that signal-induced changes in cytosolic Ca^{2+} occur in plants (Bush and Jones, 1990; Gilroy *et al.*, 1990; McAinsh *et al.*, 1992; Poovaiah and Reddy, 1987; Scheuerlein *et al.*, 1991).

Recently, Poovaiah and Reddy 1993) reviewed the subject of calcium and signal transduction in plants. Mechanical stimuli such as touch and wind and stress signals such as cold shock and fungal elicitors have been shown to induce a transient increase in cytosolic Ca^{2+} (Knight *et al.*, 1991, 1992; Sistrunk *et al.*, 1994). Studies also indicated that the source of Ca^{2+} is different for cold-shock and wind signals. Stomatal guard cells have now been used extensively to establish the messenger role of Ca^{2+} and signaling pathways in plants (Mansfield *et al.*, 1990; McAinsh *et al.*, 1992).

3.1. Calcium Channels in Plants

Recent studies in plants indicate that there are several types of Ca^{2+} channels in plasma membrane and vacuoles that are regulated by different mechanisms (Johannes *et al.*, 1991; Schroeder and Thuleau, 1991). There are two general classes of Ca^{2+} ion channels: (1) plasma membrane Ca^{2+} channels, which allow Ca^{2+} influx from cell wall space into cytosol, and (2) Ca^{2+} channels in microsomes and vacuoles, which can release Ca^{2+} from these stores. In addition, both Ca^{2+} influx and Ca^{2+} release occur within the same cell type. In higher plants little is known about the molecular components of Ca^{2+} channels in the plasma membrane. However, many of the channel blockers used in animal systems inhibit a variety of physiological responses (Sounders and Helper, 1983). The application of microelectrode techniques and drugs on rapid, sensitive responses in *Mimosa pudica* suggests that activation of voltage-dependent Ca^{2+} channels may be required for triggering cellular responses (Lijima and Sibaoka, 1985). It appears that the Ca^{2+}-modulated responses are mediated through Ca^{2+}-binding protein(s) or receptor(s).

3.2. Calmodulin: The Calcium Receptor

The cytosolic calcium receptors are a class of Ca^{2+}-binding proteins referred to as calcium-modulated proteins. Examples of Ca^{2+}-modulated proteins are troponin C, parvalbumin, S-100 protein, vitamin-D-dependent Ca^{2+}-binding protein, calmodulin, and various other proteins in a variety of cells (Roberts *et al.*, 1986). It has also been reported that plants possess several different calmodulin genes and proteins (see Roberts and Harmon, 1992). These isoforms might be involved in different functions in plant cells. Calmodulin is a ubiquitous, highly conserved low-molecular-weight protein that is present in all eukaryotic cells including those of plants, and we discuss only Ca^{2+}-modulated calmodulin-mediated responses in plants. Plant calmodulin, its gene structure and expression in transgenic plants, along with the calmodulin-dependent enzymes and other proteins have been discussed recently by Poovaiah and Reddy (1993). Only the salient features are mentioned in the context of (1) calcium- and calmodulin-dependent protein phosphorylations along with (2) Ca^{2+}-dependent and calmodulin-independent protein kinases.

3.3. Calcium- and Calmodulin-Dependent Protein Phosphorylations

Calcium-promoted phosphorylations of both soluble and membrane proteins in various plant tissues have been reported by several groups of workers (Hetherington and Trewavas, 1984; Veluthambi and Poovaiah, 1984; Ranjeva and Boudet, 1987; Poovaiah and Reddy, 1987; Anderson, 1989; Blowers and Trewavas, 1989). They showed that the amount of phosphate (^{32}P) incorporated into the membrane fraction increased when the free Ca^{2+} concentration in the medium increased above 0.3 μM. Addition of bovine calmodulin (20 μM) significantly increased the overall protein phosphorylation. Binding of Ca^{2+} to calmodulin results in activation of calmodulin, and the active form of calmodulin initiates a cascade of biochemical events by activating several calmodulin-dependent enzymes including protein kinases. Purification and characterization of Ca^{2+}-dependent protein kinase have been reported from several plant sources (Li *et al.*, 1991; Putnam-Evanans *et al.*, 1990; Polya *et al.*, 1990; Harper *et al.*, 1991).

3.4. Calcium- and Calmodulin-Independent Protein Kinases

A protein kinase that requires Ca^{2+} but not calmodulin or phospholipids for activity has been purified from soybean (Harper *et al.*, 1991). Although this is one of the best-characterized plant kinases, its substrate and function *in vivo* are still unknown. A variety of approaches to studying the presence of this type of kinase have been reported from a number of plants (Roberts and Harmon, 1992). Recently, two genes that encode Ca^{2+}-dependent protein kinases inducible by drought and high-salt stresses in *Arabidopsis thaliana* have been reported (Urao *et al.*, 1994). Protein kinase C (PKC) is a family of related protein kinases distinguished by their dependence on Ca^{2+}, phospholipid, and diacyl glycerol for activity but not on calmodulin; PKC also has a direct role in the cascades of protein phosphorylation and dephosphorylation in the process of signal transduction. This aspect is discussed under protein kinases and phosphatases later (Section 4.4).

3.5. Inositol-Trisphosphate-Activated Ca^{2+} Channels in Plants

The relationship between Ca^{2+} and the $InsP_3$ receptor in animal systems is complex in nature (Irvine, 1990; Taylor and Marshall, 1992). The question arises as to whether plant cells utilize $InsP_3$ as second messenger for Ca^{2+} release from the cellular as well as from the intracellular stores such as microsomes and/or vacuoles? Inositol trisphosphate has already been shown to stimulate Ca^{2+} efflux from higher plant vacuoles (Schumaker and Szi, 1987; Ranjeva *et al.*, 1988) and microsomes (Maitra *et al.*, 1988; Brosnan and Sanders, 1990). Patch-clamp studies of red beet have demonstrated that $InsP_3$ specifically activates Ca^{2+}

channels in the tonoplast (Alexandre *et al.*, 1990). Further, it has been shown that heparin, which binds to $InsP_3$ receptor in animal systems, inhibits microsomal $InsP_3$-mediated Ca^{2+} release (Brosnan and Sanders, 1990), but both nifedipine and ryanodine are ineffective. The $Ins(1,4,5)P_3$-induced Ca^{2+} release has also been shown to be accompanied by an influx of K^+ and sensitive to K^+ channel blockers (Canut *et al.*, 1989).

The effect of $Ins(1,4,5)P_3$ and K^+ fluxes in stomatal guard cells has been studied more extensively (Blatt *et al.*, 1990; Gilroy *et al.*, 1990). Further, Shimazaki *et al.* (1992) demonstrated that a variety of calmodulin antagonists (e.g., trifluoperazine) can inhibit H^+ extrusion by *Vicia faba* guard cell protoplasts and stomatal opening in *Commelina benghalensis*. Though regulation of opening and closing of stomata by the application of $Ins(1,4,5)P_3$, by Ca^{2+} release, and by efflux of K^+ and H^+ makes the mechanism more complicated, attempts are now being made to establish a direct link between $Ins(1,4,5)P_3$ action and modulation of K^+ channel activity (see Assman, 1993, and references therein). Coupling interactions of auxin at its receptor on the plasma membrane change the activity of both inward and outward K^+ channels, suggesting that auxin-binding protein initiates signal transduction (Helen and Goldsmith, 1993). Modulation of K^+ channels in *Vicia* stomatal guard cells by a peptide homologous to auxin-binding protein C terminus has strengthened this viewpoint (Thiel *et al.*, 1993).

3.6. Abscisic-Acid-Activated Ca^{2+} Channels in Plants

Abscisic acid (ABA) has been shown to increase Ca^{2+} levels in a variety of plant cells (Gehring *et al.*, 1990), indicating that Ca^{2+} mobilization may be a common property of ABA signaling. Several studies have suggested that Ca^{2+} influx plays an important role in initiating abscisic-acid-induced stomatal closure during water stress (Mansfield *et al.*, 1990). Abscisic acid inhibits stomatal opening and promotes stomatal closure and can trigger an elevation of cytosolic Ca^{2+} concentration in guard cells (McAinsh *et al.*, 1990). The ABA stimulation of Ca^{2+} uptake could cause membrane depolymerization, which is required for stomatal closure and sufficient for inhibition of stomatal opening. However, an obligate link between ABA application and Ca^{2+} uptake has not been established unequivocally, although Ca^{2+} channel blockers reduce the effectiveness of ABA in inhibiting stomatal opening.

Schroeder and Hagiwara (1990) showed that ABA-induced cytosolic Ca^{2+} elevations are variable, and Ca^{2+}-permeable channels in the plasma membrane contribute to rises in cytosolic Ca^{2+}. These ABA-activated ion channels were shown to be nonselective, allowing K^+ efflux along with Ca^{2+} influx. Data so far accumulated indicate that there is a rapid ABA-stimulated turnover of phosphoinositides as well as InsP because a doubling of $InsP_3$ levels after application of ABA in the presence of Li^+ has been recorded (see Assmann, 1993). Further, the picture of closure and opening of stomata is complicated because ABA, Ca^{2+}, and $InsP_3$ can promote closure, but auxin and DAG can help in opening of stomata. We do not discuss this aspect because it has been covered extensively (see Assmann, 1993). What is apparent is that ABA distinctly plays an important role in mobilizing Ca^{2+} as well as in turnover of phosphoinositides in plants as far as stomatal closing and opening are concerned. There is good evidence that in guard cells there is a repetitive increase in cytosolic Ca^{2+} in guard cells by ABA (Schroeder and Hagiwara, 1990).

3.7. Voltage-Dependent Ca^{2+} Channels in Plants

Studies on touch sensitivity responses in *Mimosa pudica* and *Dionaea muscipula* using inserted microelectrodes have suggested that activation of voltage-dependent Ca^{2+} channels may be required for triggering cellular responses (Hepler and Wayne, 1985; Lijima and Sibaoka, 1985). Voltage sensitivity of the H^+/Ca^{2+} antiport in higher plant tonoplasts suggests a role in vacuolar Ca^{2+} accumulation (Blackford, 1990; Pantoja *et al.*, 1992). However, the differences between plant and animal VDCC have yet to be established.

3.8. Calcium-Activated Ca^{2+} Channels

The bulk of the channels sensitive to Ca^{2+} are K^+ and Cl^- selective. Interestingly, it has been demonstrated that stretch-activated chloride, potassium, and Ca^{2+} channels coexist in plasma membranes of guard cells of *Vicia faba* (Cosgrove and Hedrich, 1991). However, Ca^{2+} can also affect the gating of Ca^{2+}-permeable channels that could mediate Ca^{2+} release from the vacuole into the cytoplasm see Johannes *et al.*, 1992). Thus, parallel pathways for intracellular Ca^{2+} release from the vacuoles of higher plants have been proposed (Johannes *et al.*, 1992).

The observed role of Ca^{2+} in plant signal transduction in conjunction with the existence of plant phosphoinositides, inositol phosphates, and $InsP_3$-sensitive Ca^{2+} stores has evoked considerable interest on the role of phosphoinositide metabolism in plants (see Lehle, 1990; Trewavas and Gilroy, 1991; Drobak, 1992; Coté and Crain, 1993). However, the link among signal perception, $InsP_3$ production, Ca^{2+} mobilization, and cell response is yet to be established unequivocally.

4. PHOSPHOINOSITIDES IN PLANTS

Current interest in the metabolism of phosphoinositides in plants has arisen from studies of animal systems. Evidence is now accumulating in favor of the

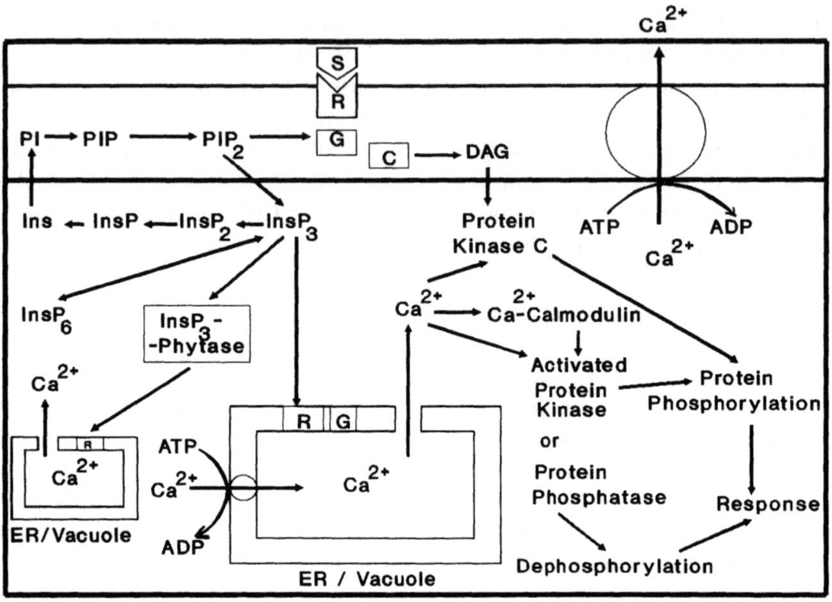

FIGURE 2. Proposed phosphoinositide and inositol hexakisphosphate signaling pathway eliciting intracellular Ca^{2+} release and subsequent phosphorylation and dephosphorylation of proteins leading to a particular response in plant cell. S, stimulus; R, receptor; G, G-protein; C, phospholipase C; DAG, diacylogycerol; PI, phosphatidyl inositol; PIP, phosphatidylinositol-4-phosphate; PIP_2, phosphatidylinositol 4,5,-bisphosphate; ER, endoplasmic reticulum; $InsP_3$, *myo*inositol(1,4,5)trisphosphate. In the $InsP_3$-phytase complex, $InsP_3$ represents other *myo*inositol trisphosphates as well. Phytase activity is developmentally or environmentally controlled.

existence of phosphoinositide turnover and the presence of various components such as $PtdInsP_2$, phospholipase C, G proteins, $InsP_3$ receptor, protein kinases, and phosphatases in plant systems. The different steps in phosphoinositide signaling operative in the case of plant systems are given in Figure 2. Very interesting reviews on the plant phosphoinositide system (Drøbak, 1992) and on the biochemistry of phosphoinositides (Coté and Crain, 1993, and references therein) are available.

The presence of PtdIns, PtdInsP, and $PtdInsP_2$ in plants was first demonstrated by Boss and Massel (1985). Irvine and colleagues (1989) demonstrated that phosphoinositides contain phosphatidylinositol-4-phosphate and phosphatidylinositol-4,5-bisphosphate. Subsequently, phosphoinositides have been reported from different plant systems (see Drøbak, 1992; Coté and Crain, 1993). These studies indicate the presence of a very small amount of $PtdInsP_2$ (0.5–1% of phosphoinositides).

Two approaches have been used to study the localization of phosphoinositides. One was to identify the sites of lipid synthesis, i.e., PtdIns and PtdInsP kinase activity (Sandelius and Sommarin, 1986, 1988). The other was to isolate membranes from prelabeled cells and identify the membranes enriched in phosphoinositides (Wheeler and Boss, 1987). Plasma membrane of carrot cells contains not only PtdInsP and $PtdInsP_2$ but also inositol lypophospholipids. The biosynthesis and subcellular localization of inositol lysophospholipids has been reviewed in details by Wheeler and Boss (1990). Both PtdIns(3)P and PtdIns(3,4)P_2 have also been reported in plants (Brearley and Honke, 1992; Irvine *et al.*, 1989). Association of phosphotidylinositol 4-kinase with the plant cytoskeleton has recently been reported (Xu *et al.*, 1992). The presence of scyllo-inositol containing phosphatidylinositol in plant cells has indicated the possibility of its participation in phosphoinositide and InsP metabolism (Kinnard *et al.*, 1995).

4.1. Phospholipase C in Plants

Phospholipase C (PLC) in plants was reported first by Irvine *et al.* (1980). Pfaffmann *et al.* (1987) reported PLC activity in plant stems, but some of the activity (10%) could not be solubilized but was associated with the membrane fraction. The PLC activity associated with the plasma membrane was found to hydrolyze all three phosphoinositides, liberating Ins(1)P, Ins(1,4)P_2, and Ins(1,4,5)P_3, although PtdInsP is the preferred substrate (Melin *et al.*, 1987). For optimal activity of this PLC, Ca^{2+} (micromolar) was needed. The presence of phosphoinositide-specific phospholipase C enzyme has subsequently been reported from different plant systems (see Drøbak, 1992). A phosphatidylinositol-specific phospholipase C (PI-PLC-δ) has been found to be induced by dehydration and salt stress in *Arabidopsis thaliana,* indicating a possible role during plant responses to various environmental stimuli (Hirayama *et al.*, 1995). Interestingly, McMurray and Irvine (1985) found that phospholipase C activity can be stimulated by GTP γ S over broad range of Ca^{2+} concentrations, suggesting an involvement of G proteins in the stimulation of PLC. Though G proteins have been implicated in physiological processes in plants (see Terryn *et al.*, 1993), regulation of plant phospholipase C is yet to be established unequivocally.

4.2. GTP-Binding Proteins in Plants

GTP-binding proteins or G proteins are important molecules participating in signal transduction pathways in animal cells (Gilman, 1987; Simon *et al.*, 1991; Gupta *et al.*, 1992). G proteins are composed of α, β, and γ subunits that are associated in an inactive GDP-bound form. Activation by a ligand of a cell surface receptor causes replacement of GDP by GTP. The GTP-α subunit complex separates from the βγ dimer and stimulates its downstream effector. The α subunit is thought to confer the specificity of the interaction with the receptor and the effector. However, there is accumulating evidence for a role of the $\beta\gamma$ dimer in determining the specificity of G protein function (Birnbaumer, 1992). Termination of the signal occurs when GTP bound by the α subunit of the G protein is hydrolyzed to GDP. The α subunit then reassociates with the $\beta\gamma$ complex. In yeast, heterotrimeric G proteins are involved in mating (Dietzel and Kurjan, 1987; Obara *et al.*, 1991) and sensing nutrition (Isshiki *et al.*, 1992). In the slime mold *Dictyostelium,* G proteins are needed for aggregation (Firtel *et al.*, 1989) and multicellular development (Hadwiger and Firtel, 1992). In *Drosophila,* G protein is involved in embryo development (Parks and Weischau, 1991). In fact, involvement of G proteins in different pathways in different organisms is quite extensive, and the purpose here is not to catalogue these. The pertinent question is how far G protein is involved in signal transduction pathways in plants. Evidence is accumulating for the presence of G proteins in plants (Drøbak *et al.*, 1988; Lam, 1993).

Only two genes coding for a G protein α subunit (Gα) have been isolated from *Arabidopsis* (Ma *et al.*, 1990;) and from tomato (Ma *et al.*, 1991). GPa1 is present through all stages of development and in all organs except the seeds. The level of GPa1 protein is higher in immature than mature organs. The complexity of the GPa1 localization pattern suggests that it might be involved in different signaling pathways during the different stages of development of *Arabidopsis* (Weiss *et al.*, 1993). G proteins have been isolated and characterized from different plant systems. In addition to G proteins, plants also have small GTP -binding proteins (Palme, 1992), which in animals and yeasts are involved in signal transduction, membrane traffic, and cytoskeletal functions (Hall, 1990). A plant gene that encodes a member of a different class of small GTP-binding proteins (rho proteins) has been isolated from pea. The sequence of the product of the *Arabidopsis* gene, Arab6, is similar to those of mammalian rab6 and yeast ryh1 proteins, and Arab6 can suppress a budding yeast mutation (*Ypt6*) (see Ma, 1993).

4.3. The *myo*-Inositol Trisphosphate Receptor in Plants

myo-Inositol trisphosphate receptors ($InsP_3R$) from animal system exist in multiple isoforms and have been well characterized. The respective genes are cloned and sequenced. The functional domains are well delineated, and the cell-specific expression of different subtypes has been studied (see Mikoshiba, 1993; Yamamoto-Hino *et al.*, 1994; Ferris and Snyder, 1992; Chadwick *et al.*, 1990; Chapter 3). However, $InsP_3R$ from plant systems has not yet been isolated and characterized. The presence of only receptor from the microsomal/vacuolar membrane fraction of plant (mung bean) cells for Ins(1,4,5)P_3/Ins(2,4,5)P_3 has been reported (S. Biswas *et al.*, 1995). The K_d for Ins(1,4,5)P_3 is 1.5 nM, suggesting the presence of high-affinity sites. However, the affinity of $InsP_3$ for animal receptors varies from 1nM to 100 nM. The abundance of apparent

$Ins(1,4,5)P_3$-specific binding sites in mung bean (1.1 pmol/mg protein) is two orders of magnitude lower than that in cerebellum, suggesting that the mung bean microsomal fraction has a lower receptor density (Challis *et al.*, 1991). However, this value is also comparable to that from red beet storage root microsomes (Brosnan and Sanders, 1993).

The molecular mass of the subunit has been recorded as 110 kDa under denaturing conditions, which is similar to that of the receptor isolated from olfactory cilia (Kalinoski *et al.*, 1992), though the receptor from animal systems has a mass of around 260 kDa (Supattapone *et al.*, 1988; Maeda *et al.*, 1990; Ross *et al.*, 1989). Animal $InsP_3$ receptors appear to function as tetramers in the membrane (Chadwick *et al.*, 1990) which has also been found true in the case of the mung bean microsomal receptor. The mung bean $InsP_3$ receptor shows similarity with animal receptors with respect to $InsP_3$ binding affinity, selectivity for binding, and heparin binding. Whether same receptor is also present in the plasma membrane has not yet been ascertained. In the case of animals, the $InsP_3$ binding sites and channel domains were shown to be in the N-terminal and C-terminal regions, respectively (Maeda *et al.*, 1990). The remaining part of the receptor has phosphorylation sites recognized by protein kinase A. However, protein kinase C and Ca^{2+}/calmodulin-dependent protein kinase II can also phosphorylate the $InsP_3$ receptor *in vitro*. Alternative splicing has been considered to confer further functional complexity (Supattapone *et al.*, 1988).

These studies suggest that the $InsP_3$ receptor is, in general, a key molecule into which various kinds of signal pathways converge. Whether all these functional and structural complexities are retained by the plant $InsP_3$ receptor is yet to be worked out. What is now needed most is to establish the link among the microsomal receptor activity, Ca^{2+} release, and the signal transduction network in the plant cell, in spite of the evidence now accumulating that enhanced turnover of phosphatidylinositol-4,5-bisphosphate and an associated increase of $InsP_3$ level is a primary event in a variety of responses in plants (Drøbak, 1992). Receptors for other inositol phosphates, say $Ins(1,3,4,5)P_4$ and $InsP_6$, have been identified in other than plant systems (see Irvine, 1992; Theibert *et al.*, 1991; Raha *et al.*, 1995). The $InsP_6$-binding protein has been characterized from a plant system (Rajendran and Prakash, 1993). The $InsP_6$ receptor from animal was found to be identical to subunits of mouse AP-2 protein, which takes part in the assembly of clathrin into coated vesicles (Voglmaier *et al.*, 1992). However, the actual role of $InsP_6$ as an effector molecule is still unclear.

4.4. Cascades of Protein Kinases and Phosphatases in Plants

Roles of protein kinases and phosphatases in signal transduction in general have already been elucidated with animal systems (Hardie, 1990; Krebs and Beavo, 1979; Edelman *et al.*, 1987; Chapter 5). Furthermore, protein kinases and phosphatases are often themselves regulated by phosphorylation, either in direct response to cellular signals or via signal transduction pathways (Gilman, 1987). The Ca^{2+}-dependent, calmodulin-independent protein kinase (CDPK) has already been reported in plants (Harper *et al.*, 1991). The *Arabidopsis* membrane-associated CDPK (AK1) has been found to be stimulated by Ca^{2+} and lipids in a synergistic manner (see Ma, 1993). The role of CDPK in plant responses to light has also been suggested (Fallon *et al.*, 1993).

Receptor-like protein kinases such as tyrosine kinases and Ser/Thr kinases are involved in signal transduction networks in animal systems (Hunter and Cooper, 1985). A plant receptor-like gene, the S-locus receptor kinase of *Brassica oleracea,* encodes a functional Ser/Thr kinase (Stein and Nasralla, 1993). Chang and co-workers (1992) reported the presence of *Arabidopsis TMK1* gene, which encodes a Ser/Thr kinase resembling that of maize *ZmPK1* (Walker and Zhang, 1990), *Brassica oleracea SRK* (Stein *et al.*, 1991), and *Arabidopsis ARK1* Tobias *et al.*, 1992).

Furthermore, a component in the transduction pathway of ethylene has recently been isolated using *Arabidopsis* mutants. It has been shown to be a Raf-like protein kinase (Kieber *et al.*, 1993). Similarly, a blue-light-induced kinase is present in the membrane of pea seedlings (Short *et al.*, 1993). Cyclin-dependent protein kinases (CDKs) are homologues of $p34^{Cdc2}$ protein kinase involved in the regulation of the cell cycle, and plant CDKs have now been reported (Colasanti *et al.*, 1991). Four distinct CDKs were reported from pea. It appears that cdc2 is involved in the regulation of position and orientation of plant cell division. One of the *Arabidopsis* genes encoding cdc2a and cyclins revealed the interesting observation that cdc2a is produced not only in dividing cells but also in proliferation-competent cells (root pericycle), whereas the cyclin component (regulatory subunit) of the kinase is produced only in dividing cells. The phenotypes of transgenic plants that carry dominant mutations in the *Arabidopsis cdc2a* gene indicate that normal cdc2 function is involved in proper plant development (see Ma, 1993).

Protein phosphorylation does not depend only on the kinase but also on the phosphatase activities in the cell, and the turnover of phosphoproteins including receptors is determined by the kinase–phosphatase relay. The protein phosphatase inhibitor calyculinA mimics elicitor action in plant cells and induces rapid hyperphosphorylation of specific proteins (Felix *et al.*, 1994). There are multiple protein phosphatases (such as serine/threonine PP1, PP2A, and PP2C) reported from plant systems (R. Mackintosh *et al.*, 1990; C. Mackintosh *et al.*, 1991; Arino, 1993) analogous to those in animal systems (Cohen, 1989, 1992; Chapter 5). In *Arabidopsis,* eight PP1 genes have been identified (Smith and Walker, 1993). One of these genes (*TOPP2*) can suppress the mitotic defect of a yeast PP1 mutant (*glc7*) but not its failure to accumulate glycogen; other genes (*TOPP1* and *TOPP3*) can not suppress either defect, suggesting that the different *TOPP*s are functionally distinct. An abscisic-acid-resistant (*ABI1*) gene was cloned and shown to encode a protein with high homology with animal PP2C at the carboxyl end and a Ca^{2+}-binding domain at the amino terminus (Leung *et al.*, 1994; Meyer *et al.*, 1994). This phosphatase appears to be a Ca^{2+}-activated form of PP2C and possibly a candidate for mediating K^+ channel closure through Ca^{2+}-induced protein dephosphorylation. A protein tyrosine phosphatase (PTPase) activity in pea nuclei has recently been reported (see Ma, 1993). Though the roles of all these plant protein kinases and phosphatases are not known in detail, in the network of signal transduction a number of reports are now emerging from several laboratories identifying roles of these proteins in plant responses.

5. PHOSPHOINOSITIDE AND OTHER *myo*-INOSITOL PHOSPHATE SIGNALING SYSTEMS IN PLANTS

A body of evidence is now available to indicate that phosphoinositide pathways generating $InsP_3$ and DAG exist in plants. The phosphoinositide $PtdIns(4,5)P_2$ has been unequivocally identified in lower and higher plants. Kinases for the synthesis of phosphoinositides and the presence of phospholipase C for the production of $InsP_3$ and DAG from phosphoinositide along with the activation of PLC with G proteins have been reported (see Drøbak, 1992, 1993; Coté and Crain, 1993). Calcium has been implicated in the signal transduction cascades leading to different plant responses (Gilroy *et al.*, 1993). The *myo*-inositol trisphosphate receptor has also been reported and characterized (S. Biswas *et al.*, 1995). Because $InsP_6$ is present in high amounts in plant systems, generation of $InsP_3$ from $InsP_6$ might be a possibility. Does that $InsP_3$ generated mediate any physiological functions?

Initial evidence obtained in our laboratory indicates the existence of an interesting alternative pathway of generation of $InsP_3$ that can mediate Ca^{2+} mobilization from internal stores. One of the $InsP_3$s obtained by the action of phytase on $InsP_6$ has been identified as $Ins(2,4,5)P_3$ (Maitra *et al.*, 1988). The participation of $Ins(2,4,5)P_3$ in mobilization of Ca^{2+} was demonstrated when the $InsP_6$–phytase complex was added after a period of hydrolysis that coincided with the time of optimal production of $Ins(2,4,5)P_3$ bound to phytase. The $Ins(1,4,5)P_3$– or $Ins(2,4,5)P_3$–phytase complex formed *in vitro* has also been found effective in releasing Ca^{2+} from cellular stores; the release is 40–50% more than that by free $InsP_3$ under identical conditions (Samanta *et al.*, 1993). The question arises whether the $InsP_3$–phytase complex *per se* interacts with the receptor or whether $InsP_3$ generated by phytase is released and, because of its higher affinity for the receptor, interacts with the receptor. But the evidence obtained thus far favors the $InsP_3$–phytase complex interaction with the receptor (Dasgupta *et al.*, 1995). Spectrofluorometric studies of complex formation between different inositol phosphates and phytase revealed an important feature: there is a definite trend in the accessibility of tryptophan residues on the enzyme in presence of different inositol phosphates. There is a progressive decrease in the accessibility from $InsP_6$ to InsP with the exceptions of Ins(1,4,5)P_3 and Ins(2,4,5)P_3, indicating a detectable conformational change in the enzyme. Because K_d for $InsP_3$ with phytase is in the micromolar range, whereas that for the receptor is nanomolar range, the transfer of free $InsP_3$ to the receptor is likely unless it is shown that phytase has another high-affinity (noncatalytic) site for $InsP_3$. In fact, this has been shown by demonstrating the formation of a ternary complex *in vitro* among $InsP_3$, phytase, and the receptor (Dasgupta *et al.*, 1995). Thus, interaction of phytase with $InsP_3$ at the noncatalytic site could lead specifically to an altered conformation of the enzyme that is recognized by the receptor, implying a novel pathway for signaling Ca^{2+} homeostasis mediated by $InsP_6$–phytase activity in plant cell. How universal this signaling is has yet to be established.

6. SIGNALING IN PLANTS BY WOUNDING

A recent report by Wildon *et al.* (1989) demonstrated convincingly that in response to a wound, a proteinase inhibitor (*pin* gene product) spread from the site of injury to the unwounded regions of the plant within seconds; this has added further thought of a putative electrical signaling system. They proposed from their experimental results that an electrical signal is the mechanism underlying this systemic wound response. However, the relationship between the electrical signal and the second messengers such as Ca^{2+} or $InsP_3$ involved in *pin* induction is not known yet. Two possibilities for a physical signal might be considered: (1) a hydraulic signal with secondary electrical activity or (2) an electrical signal (action potential) *per se*. In fact, earlier work has shown that many plant species can transmit action potentials and variation potentials (see Pickard, 1973; Davies, 1987), but the biochemical events were not identified.

Wounding by pathogen has attracted considerable interest in the signaling mechanisms from recognitional events at the cell surface, such as elicitor binding to plasma membrane receptors, to the activation of inducible defenses. Calcium appears to be essential for the induction of phytoalexin biosynthetic genes in response to elicitors, and influx of Ca^{2+} leading to increased cytosolic levels has been observed in the early stages of response (Scheel *et al.*, 1991). H^+ influx and K^+ efflux precede the Ca^{2+} response, and experiments with polyene antibiotics suggest that these ion fluxes are likewise related causally to defense activation. Elicitors also cause rapid changes in the phosphorylation of specific plasma membrane and nuclear proteins, and protein kinase inhibitors block ethylene and PAL induction (Felix *et al.*, 1991; see Lamb *et al.*, 1992; Lamb, 1994). Wounding produces ethylene, and ethylene-dependent signal transduction involves both Ca^{2+} (Raj and Fluhr, 1992) and protein phosphorylation (Raj and Fluhr, 1993). In this connection, it is also noted that wounding elicits lipid-based signaling pathways (Farmer and Ryan, 1992). In fact, many cellular responses in eukaryotic cells appear to be under the control of lipid-based signaling pathways (Samuelson *et al.*, 1987). Jasmonic acid (JA) and methyl jasmonate (MJ) are present in most of the plant species (Meyer *et al.*, 1984; Staswick, 1992). Both JA and MJ induce proteinase inhibitors (Farmer and Ryan, 1992), soybean vegetative storage proteins (Mason *et al.*, 1992), lipoxygenases (Tranbarger *et al.*, 1991), anthocyanins, and expression of a number of wound-inducible genes (Hildmann *et al.*, 1992). The question now arises whether wounding of plants can generate signals to make the plants resistant to pests and pathogens. Recent observations suggest that MJ participates in the regulation of wound response programs, and the responses observed during a hypersensitivity reaction (HR) in potato are not simply an enhancement of wound responses but rather an induction of different cellular programs (Choi *et al.*, 1994). Methyl jasmonate and the fungal elicitor arachidonic acid induce different 3-hydroxy-3-methylglutarylcoenzyme A reductase genes and antimicrobial isoprenoids in *Solanum tuberosum*. It would be interesting to ascertain the downstream processes that follow the detection of electrical activity in the early events of wounding, Ca^{2+} and $InsP_3$ generation, and JA and MJ production, leading to the induction of the specific genes just mentioned.

7. SIGNALING IN LEAF MOVEMENT

Evidence is available that phosphoinositide turnover and Ca^{2+} release mediate K^+ channel closure and cell shrinkage in pulvini of *Mimosa* leaflets as well as in stomatal guard cells (see Assmann, 1993).

Although changes in cytosolic Ca^{2+} levels have not yet been demonstrated in pulvinar cells, there is evidence that when excised leaflets of *Albizzia lopantha* are floated in a medium containing Ca^{2+}, pulvini were closed. A23187, a Ca^{2+} ionophore, enhanced the closure of leaflets of *A. lopantha* (Moysset and Simon, 1989). It appears from the work of Morse *et al.* (1987) that white light increases the levels of inositol phosphates with a concomitant decrease of phosphoinositide level in the pulvini of *Samanea saman* leaves. This is suggestive that $Ins(1,4,5)P_3$ might regulate Ca^{2+} necessary for light-induced leaf movement. Kim *et al.* (1992, 1993) showed for the first time that extensor and flexor cell protoplasts isolated from *S. saman* retain their sensitivity to light signals similar to intact pulvini. If the protoplasts are kept in white light, a pulse of red light followed by darkness rapidly triggers closure of K^+ channels in extensor protoplasts and opening of the same in flexor protoplasts. The involvement of a G protein in photoreceptor activation of Ins(1,4,5)P_3 production by phospholipase C has been demonstrated indirectly by use of G protein activators (see Chapter 11).

Protein phosphorylation is also implicated in circadian rhythms (Carter *et al.*, 1991). Our studies indicate that an NTPase (mostly having ATPase activity) has a positive correlation with the sensitivity of *Mimosa* leaflets to external stimuli (S. Biswas and Bose, 1972; Mukherjee and Biswas, 1980). What has been revealed at present is that the NTPase is a complex of polypeptides, polysaccharides, and a cofactor similar to tetrahydrofolate. This complex formation is mediated by Ca^{2+}. The protein moiety *per se* is inactive (R. Ghosh, P. C. Sen, S. Biswas, unpublished data). This Ca^{2+}-modulated activity of NTPase has been found to be enhanced in the presence of microtubule (MT), implicating the involvement of cytoskeletal proteins in the downstream network of signal transduction. Further, the presence of a tetrahydrofolatelike moiety in NTPase might indicate its involvement in the blue light–mediated response of mimosa leaflets. Recently, putative blue light photoreceptors from *Arabidopsis thaliana* and *Sinapis alba* have been shown to have homology to DNA photolyase (Melhotra *et al.*, 1995). Despite the high degree of sequence identity to, and identical chromophore composition with photolyases, neither photo receptor has any photoreactivating activity.

8. CONCLUDING REMARKS

Evidence for the biosynthesis and degradation of *myo*-inositol phosphates, including the presence of a metabolic cycle involving glucose-6-P and *myo*-inositol phosphates during formation and germination of seeds, have been presented. However, the role of $InsP_7$ or $InsP_8$ (pyrophosphate derivatives of $InsP_6$) is yet to be established. Calcium mobilization and its role in plant responses have been established unequivocally. Although different Ca^{2+} channels in the plasma membrane in plants have been reported, the molecular components of Ca^{2+} channels are not known. *myo*-Inositol-trisphosphate-activated Ca^{2+} channels in the microsomal/vacuolar fraction, other specific Ca^{2+} channels, phosphoinositides, $InsP_3$ receptor, phospholipase C, G proteins, and different protein kinases and phosphatases are suggestive that an inositol-phosphate-mediated signal transduction pathway similar to that of animal systems is also present in plants. Besides phosphoinositide, $InsP_6$ has now been shown to be involved in generation of $InsP_3$, eliciting Ca^{2+} mobilization from intracellular stores of cells, in certain plant systems. It appears that the $InsP_3$–phytase complex is the elicitor in that case rather than free $InsP_3$. In case of acidic and alkaline phytase, $InsP_3$s other than Ins(1,4,5)P_3 and Ins(2,4,5)P_3 are produced. The role of those $InsP_3$s as second messengers has not yet been established. It is worth speculating that other $InsP_3$s, when complexed with phytase (at the high-affinity site), might modulate the Ins(1,4,5)P_3 receptor so as to elicit Ca^{2+} release from intracellular stores. However, in none of the plant systems thus far studied has a causual link between the generation of InsP_3 and the plant's response been conclusively demonstrated.

Two kinds of approaches for exploring the mechanisms of plant signal transduction are now being undertaken. First is the genetic approach, allowing the isolation of genes whose products are involved in a particular process of the signal transduction pathway. Second, analysis is being done of genes whose sequences resemble those of proteins of known functions in mammalian signal transduction. Environmental sensing is rather widely distributed throughout the plant body because plants have fewer specialized organs than animals. Similarly, a lack of extensive cellular specialization in plants may be compensated by the ability of plant cells to respond to many types of signals, thus exhibiting great complexity at the cellular level.

We believe that those approaches that are now being applied to some of the model plant systems will help in unraveling the mechanisms plant cells use to perceive signals as diverse as light, temperature, touch, phytohormones, pathogens, gravity, wind, water, minerals, gases, and soil structure. Plants are by far different systems than animals. Extrapolation of the results obtained from the animal systems may not always be appropriate in terms of the totality of the phenomenon occurring in the plant systems. It is likely that interesting signal perception and transduction in plants will be reported more and more in the near future, and the role of plant growth substances in signal transduction will add more excitement because both genetic and biochemical approaches can now be applied simultaneously to plant systems.

ACKNOWLEDGMENTS. The authors are grateful to the Department of Biotechnology and Council of Scientific and Industrial Research, Government of India, for financial support for the recent work originating in their laboratories, and to their colleagues for their help in preparation of this chapter.

9. REFERENCES

Alexandre, J., Lassalles, J. P., and Kado, R. T., 1990, Opening of Ca^{2+} channels in isolated red beet root vacuole membrane by inositol 1,4,5-trisphosphate, *Nature* **343**:567–570.

Anderson, J. M., 1989, Membrane derived fatty acids as precursors to second messengers, in *Second Messengers in Plant Growth and Development* (W. F. Boss and D. J. Moore, eds.), pp. 289–303, Alan R. Liss, New York.

Arino, J., 1993, Protein phosphatases in higher plants: Multiplicity of type 2A phosphatases in *Arabidopsis thaliana*, *Plant Mol. Biol.* **21**:475–485.

Assmann, S. M., 1993, Signal transduction in guard cells, *Annu. Rev. Cell Biol.* **9**:345–375.

Barrientos, L., Scott, J. J., and Murthy, P.P.N., 1994, Specificity of hydrolysis of phytic acid by alkaline phytase from lily pollen, *Plant Physiol.* **106:**1489–1495.

Berridge, M. J., and Irvine, R. F., 1989, Inositol phosphates and cell signalling, *Nature* **341:**197–205.

Birnbaumer, L., 1992, Receptor-to-effector signaling through G-proteins: Roles for βγ dimers as well as α-subunits, *Cell* **71:**1069–1072.

Biswas, B. B., Biswas, S., Chakraborti, S., and De, B. P., 1978, A novel metabolic cycle involving myoinositol phosphates during formation and germination of seeds, in *Cyclitols and Phosphoinositides* (W. W. Wells and Y. Eisenberg, Jr., eds.), pp. 57–58, Academic Press, New York.

Biswas, B. B., Ghosh, B., and Mazumder, A. L., 1984, Myoinositol polyphosphates and their role in cellular metabolism: A proposed cycle involving glucose-6-phosphate and *myo*-inositol phosphates, *Subcell. Biochem.* **10:**237–280.

Biswas, S., and Biswas, B. B., 1965, Enzymatic synthesis of guanosine triphosphate from phytin and guanosine diphosphate, *Biochim. Biophys. Acta* **108:**710–713.

Biswas, S., and Biswas, B. B., 1994, Myoinositol phosphates as implicated in metabolic and signal transduction processes in plants, *Ind. J. Chem.* **71:**495–499.

Biswas, S., and Bose, D. M., 1972, An ATPase from *Mimosa pudica* L. *Arch. Biochem. Biophys.* **148:**199–207.

Biswas, S., Burman, S., and Biswas, B. B., 1975, Inositol hexaphosphate guanosine diphosphate phosphotransferase from *Phaseolus aureus*, *Phytochemistry* **14:**373–375.

Biswas, S., Maity, I. B., Chakraborti, S., and Biswas, B. B., 1978, Purification and characterization of myoinositol hexaphosphate adenosine diphosphate phosphotransferase from *Phaseolus aureus*, *Arch. Biochem. Biophys.* **185:**557–566.

Biswas, S., Dalal, B., Sen, M., and Biswas, B. B., 1995, Receptor for myoinositol trisphosphate from microsomal fraction of *Vigna radiata*, *Biochem. J.* **306:**631–636.

Blackford, S., 1990, Voltage sensitivity of H^+/Ca^{2+} antiport in higher plant tonoplast suggests a role in vacuolar Ca^{2+} accumulation, *J. Biol. Chem.* **265:**1617–1620.

Blatt, M. R., Thiel, G., and Trentham, D. R., 1990, Reversible inactivation of K^+ channels of *Vicia* stomatal guard cells following the photolysis of caged inositol 1,4,5-trisphosphate, *Nature* **346:**766–768.

Blowers, D. P., and Trewavas, A. J., 1989, Second messengers: Their existence and relationship to protein kinases, in *Second Messengers in Plant Growth and Development* (W. F. Boss and D. J. Moore, eds.), pp. 298–303, Alan R. Liss, New York.

Boltman, O., Strother, S., and Hoffman-Ostenhof, O., 1980, The enzymes involved in the synthesis of phytic acid in *Lemna gibba*, *Mol. Cell Biol.* **30:**171–175.

Boss, W. F., and Massel, M. O., 1985, Polyphosphoinositides are present in plant tissue culture cells, *Biochem. Biophys. Res. Commun.* **132:**1018–1023.

Bowler, C., and Chua, N.-H., 1994, Emerging themes of plant signal transduction, *Plant Cell* **6:**1529–1541.

Brearly, C. A., and Honke, D. E., 1992, 3- and 4-phosphorylated phosphatidyl inositols in the aquatic plant *Spirodela polyrhiza L.*, *Biochem. J.* **283:**247–255.

Brosnan, J. M., and Sanders, D., 1990, Inositol trisphosphate mediated calcium release in beet microsomes is inhibited by heparin, *FEBS Lett.* **260:**70–72.

Brosnan, J. M., and Sanders, D., 1993, Identification and characterization of high affinity binding sites for inositol trisphosphate in red beet, *Plant Cell* **5:**931–940.

Bush, D. S., 1993, Regulation of cytosolic calcium in plants, *Plant Physiol.* **103:**7–13.

Bush, D. S., and Jones, R. L., 1990, Measuring intracellular calcium levels in plant cells using the fluorescent probes indo-1 and fura-2, *Plant Physiol.* **93:**841–845.

Canut, H., Carrasco, A., Graziana, A., Boudet, A. M., and Ranjeva, R., 1989, Inositol trisphosphate stimulated calcium release from *Acer* microsomal fractions involves the uptake of potassium, *FEBS Lett.* **253:**173–177.

Carter, P. J., Nimmo, H. G., Fewson, C. A., and Wilkins, M. B., 1991, Circadian rhythms in the activity of plant protein kinase, *EMBO J.* **10:**2063–2066.

Chadwick, C. C., Saito, A., and Fleischer, S., 1990, Isolation and characterization of the inositol trisphosphate receptor from smooth muscle, *Proc. Natl. Acad. Sci. USA* **87:**2132–2136.

Chakraborti, S., and Biswas, B. B., 1981b, Evidence for the existence of a phosphoinositol kinase in chicken erythrocytes, *Indian J. Biochem. Biophys.* 18:398–402.

Chakraborti, S., and Biswas, B. B., 1981, Two forms of phosphoinositol kinase from germinating mung bean seeds, *Phytochemistry* **20:**1815–1817.

Challis, J., Willcocks, A. L., Mulloy, B., Potter, B. V., and Nahorski, S. R., 1991, Characterization of inositol 1,4,5-trisphosphate and 1,3,4,5-tetrakisphosphate binding sites in rat cerebellum, *Biochem. J.* **274:**861–867.

Chang, C., Blecker, A. B., Kwok, S. F., and Meyerowitz, E. M., 1992, Molecular cloning approach for a putative receptor gene in *Arabidopsis*, *Biochem. Soc. Trans.* **20:**73–75.

Chen, I. W., and Charalampous, F. C., 1966, Biochemical studies on D-inositol 1-phosphate as intermediate in the biosynthesis of inositol from glucose 6-phosphate and characteristics of two reactions in this biosynthesis, *J. Biol. Chem.* **241:**2194–2199.

Choi, K.-Y., Satterberg, B., Lyons, D. M., and Elion, E. A., 1994, Ste 5 tethers multiple protein kinases in MAP kinase cascade required for mating in *S. cerevisiae*, *Cell* **78:**499–512.

Cohen, P., 1989, The structure and regulation of protein phosphatases, *Annu. Rev. Biochem.* **58:**453–508.

Cohen, P., 1992, Signal integration at the level of protein kinases, protein phosphatases and their substrates, *Trends Biochem. Sci.* **17:**408–413.

Colasanti, J., Tyers, M., and Sundaresan, V., 1991, Isolation and characterization of cDNA clones encoding a functional $p34^{cdc2}$ homologue from *Zea mays*, *Proc. Natl. Acad. Sci. USA* **88:**3377–3381.

Cosgrove, D. J., 1966, The chemistry and biochemistry of inositol polyphosphates, *Rev. Pure Appl. Chem.* **16:**209–224.

Cosgrove, D. J., 1980, *Studies in Organic Chemistry 4: Inositol Phosphates: Their Chemistry, Biochemistry and Physiology*, Elsevier, Amsterdam.

Cosgrove, D. J., and Hedrich, R., 1991, Stretch activated chloride, potassium and calcium channels coexisting in plasma membranes of guard cells of *Vicea faba*, *Planta* **186:**143–153.

Cote, G. G., and Crain, R. C., 1993, Biochemistry of phosphoinositides, *Annu. Rev. Plant Physiol. Plant Mol. Biol.* **44:**333–356.

Craxton, A., Erneux, C., and Shears, S. B., 1994, Inositol 1,4,5,6-tetrakisphosphate is phosphorylated in rat liver by a 3-kinase that is distinct from inositol 1,4,5-trisphosphate 3-kinase, *J. Biol. Chem.* **269:**4337–4342.

Dasgupta, S., Dasgupta, D., Sen, M., Biswas, S., and Biswas B. B., 1996, Myoinositol phosphates induced conformational change in phytase in interacting with the receptor for calcium mobilization from microsomal fraction, *Biochemistry* (in press).

Davies, E., 1987, Plant responses to wounding, in *The Biochemistry of Plants* (P. K. Stumf, ed.), pp. 143–164, Academic Press, New York.

De, B. P., and Biswas, B. B., 1979, Evidence for the existence of a novel dehydrogenase in *Phaseolus aureus*, *J. Biol. Chem.* **254:**8717–8719.

Dietzel, C., and Kurjan, J., 1987, The yeast SCG1 gene: A Ga-like protein implicated in the a and α factor response pathway, *Cell* **50:**1001–1010.

Drøbak, B. K., 1992, The plant phosphoinositide system, *Biochem. J.* **288:**679–712.

Drøbak, B. K., 1993, Plant phosphoinositides and intracellular signalling, *Plant Physiol.* **102:**705–709.

Drøbak, B. K., Allan, E. F., Comerford, J. G., Roberts, R., and Dawson, A. P., 1988, Presence of guanine nucleotide-binding proteins in a plant hypocotyl fraction, *Biochem. Biophys. Res. Commun.* **150:**899–903.

Edelman, A. M., Blumenthal, D. K., and Krebs, E. G., 1987, Protein serine/threonine kinases, *Annu. Rev. Biochem.* **56**:567–613.

Estevez, F., Paulford, D., Stark, M.J.R., Crater, A. N. and Downes, C. P., 1994, Inositoltrisphosphate metabolism in *Saccharomyces cerevisiae:* identification, purification and properties of inositol 1,4,5-tris phosphate 6-kinase, *Biochem. J.* **302**:709–716.

Fallon, K. M., Shacklock, P. S., and Trewavas, A. J., 1993, Detection *in vivo* of very rapid red light-induced calcium-sensitive protein phosphorylation in etiolated wheat (*Tritecum aestivum*) leaf protoplasts, *Plant Physiol.* **101**:1039–1945.

Farmer, E. E., and Ryan, C. A., 1992, Octadecanoid precursors of jasmonic acid activate the synthesis of wound-inducible proteinase inhibitors, *Plant Cell* **4**:129–134.

Felix, G., FrossKopf, G., Regengass, M., and Boller, T., 1991, Rapid changes of protein phosphorylation are involved in transduction of the elicitor in plant cells, *Proc. Natl. Acad. Sci. USA* **88**:8831–8834.

Felix, G., Regengass, M., Sapanu, P., and Boller, T., 1994. The protein phosphatase inhibitor calyculin A mimics elicitor action in plant cells and reduces rapid hyperphosphorylation of specific proteins as revealed by pulse labeling with [^{32}p] phosphate, *Proc. Natl. Acad. Sci. USA* **91**:952–956.

Ferris, C. D., and Snyder, S. H., 1992, Inositol 1,4,5-trisphosphate activated Ca^{2+} channels, *Annu. Rev. Physiol.* **54**:469–488.

Firtel, R. A., van Haastert, P.J.M., Kimmel, A. R., and Devreotes, P. N., 1989, G protein linked signal transduction pathways in development: *Dictyostelium* as an experimental system. *Cell* **58**:235–239.

Gehring, C. A., Irving, H. R., and Paris, R. W., 1990, Effects of auxin and abscisic acid on cytosolic calcium and pH in plant cells, *Proc. Natl. Acad. Sci. USA* **87**:9645–9649.

Ghosh, B., De, B. P., and Biswas, B. B., 1984, Purification and properties of myoinositol-1-phosphate dehydrogenase from germinating mung bean seeds, *Arch. Biochem. Biophys.* **228**:309–319.

Gilman, A. G., 1987, G proteins: Transducers of receptor generated signals, *Annu. Rev. Biochem.* **56**:615–649.

Gilroy, S., and Trewavas, T., 1994, A decade of plant signals, *BioEssays* **16**:677–682.

Gilroy, S., Read, N. D., and Trewavas, A. J., 1990, Elevation of cytoplasmic calcium by caged calcium or caged inositol trisphosphate initiates stomatal closure, *Nature* **346**:769–771.

Gilroy, S., Bethke, P. C., and Jones, R. L., 1993, Calcium homeostasis in plants, *J. Cell Sci.* **106**:453–462.

Gupta, S., Gallego, C., and Johnson, G., 1992, Mitogenic pathways regulated by G-protein oncogenes, *Mol. Biol. Cell* **3**:123–128.

Hadwiger, J. A., and Firtel, R. A., 1992, Analysis of Ga4, a G-protein subunit required for multicellular development in *Dictyostelium. Genes Dev.* **6**:38–49.

Hall, A., 1990, *ras* and *GAP*—Who's controlling whom, *Cell* **61**:921–923.

Hardie, D. G., 1990, Roles of protein kinases and phosphatases in signal transduction, *Symp. Soc. Exp. Biol.* **44**:241–255.

Harper, J. F., Sussman, M. R., Schaller, G. E., Putman-Evans, C., and Charboulav, H., 1991, A calcium dependent protein kinase with a regulatory domain similar to calmodulin, *Science* **252**:951–954.

Harwood, J. L., 1980, Lipids: Structure and function, in *The Biochemistry of Plants,* Vol. 4 (P. K. Stumf, ed.), pp. 1–55. Academic Press, New York.

Helen, M., and Goldsmith, M., 1993, Cellular signalling: New insights into the action of the plant growth hormone auxin, *Proc. Natl. Acad. Sci. USA* **90**:11442–11445.

Hepler, K. P., and Wayne, R. O., 1985, Calcium and plant development, *Annu. Rev. Plant Physiol.* **36**:397–439.

Hetherington, A. M., and Trewavas, A., 1984, Activation of pea membrane protein kinase by calcium ions, *Planta* **161:**409–417.

Hildmann, T., Ebneth, M., Pena-Cortes, H., Sanchez-Serrano, J. J., Willmitzer, L., and Prat, S., 1992, General roles of abscisic and jasmonic acids in gene activation as a result of mechanical wounding. *Plant Cell* **4:**1157–1170.

Hirayama, T., Ohto, C., Mizoguchi, T., and Shinozaki, K., 1995, A gene encoding a phosphatidylinositol-specific phospholipase C is induced by dehydration and salt stress in *Arabidopsis thaliana, Proc. Natl. Acad. Sci., USA* **92:**3903–3907.

Hunter, T., and Cooper, J. A. 1985, Protein-tyrosine kinases, *Annu. Rev. Biochem.* **54:**897–930.

Igaue, I., Shimizu, S., and Miyauchi, S., 1980, Formation of a series of myoinositol phosphates during growth of rice plant cells in suspension culture, *Plant Cell Physiol.* **21:**351–356.

Igaue, I., Miyauchi, S., and Saito, K., 1982, Formation of myoinositol phosphates in a rice cell suspension culture, in *Proceedings Fifth International Congress on Plant Tissue Cell Culture,* pp. 265–266, Tokyo.

Irvine, R. F., 1990, Quantal Ca^{2+} release and control of Ca^{2+} entry by inositol phosphates—a possible mechanism, *FEBS Lett.* **263:**5–9.

Irvine, R. F., 1992, Inositol phosphates and Ca^{2+} entry: Toward a proliferation or a simplification, *FASEB J.* **6:**3085–3091.

Irvine, R. F., Letcher, A. J., and Dowson, R.M.S., 1980, Phosphatidyl inositol phosphodiesterase in higher plants, *Biochem. J.* **192:**279–283.

Irvine, R. F., Letcher, A. J., Lander, D. J., Drobak, B. K., and Dowson, A. P., 1989, Phosphotidylinositol(4,5)bisphosphate and phosphatidylinositol(4)phosphate in plant tissues, *Plant Physiol.* **89:**888–892.

Isshiki, T., Mochizuki, N., Maeda, T., and Yamamoto, M., 1992, Characterization of a fission yeast gene *GAP2* that encodes a Ga subunit involved in the monitoring of nutrition, *Genes Dev.* **6:**2455–2462.

Johannes, E., Brosman, J. M., and Sanders, D., 1991, Calcium channels and signal transduction in plant cells, *BioEssays* **13:**331–336.

Johannes, E., Brosman, J. M., and Sanders, D., 1992, Parallel pathways for intracellular Ca^{2+} release from the vacuole of higher plants, *Plant J.* **2:**97–102.

Kalinoski, D. L., Aldinger, S. B., Boyle, A. G., Huque, T., Marecek, J. F., Prestwich, G. D., and Diego, R., 1992, Characterization of a novel inositol 1,4,5-trisphoshate receptor in isolated alfactory cilia, *Biochem. J.* **281:**449–456.

Kieber, J. J., Rothenberg, M., Roman, G., Feldmann, K. A., and Ecker, J. R., 1993, CTR1, a negative regulator of the ethylene response pathway in *Arabidopsis* encodes a member of the Raf family of protein kinases, *Cell* **72:**427–441.

Kim, N. Y., Cote, G. G., and Crain, R. C., 1992, Effects of light on the membrane potential of photoplasts from *Samanea saman* pulvini: Involvement of K^+ channel and H^+ ATPase, *Plant Physiol.* **99:**1532–1539.

Kim, H. Y., Cote, G. G., and Crain, R. E., 1993, Potassium channels in *Samanea saman* protoplasts by phytochrome and the biological clock, *Science* **260:**960–962.

Kinnard, R. L., Narsimhan, B., Pliskas Malyshak, G., and Murthy, P.P.N., 1995, Characterization of scylloinositol containing phosphotidyl inositol in plant Cells, *Biochem. Biophys. Res. Commun.,* **210:**549–553.

Knight, M. R., Campbell, A. K., Smith, S. M., and Trewavas, A. J., 1991, Transgenic plant acquarin reports the effects of touch and cold shock and elicitors on cytoplasmic calcium, *Nature* **352:**524–526.

Knight, M. R., Smith, S. M., and Trewavas, A. J., 1992, Wind induced plant motion immediately increases cytosolic calcium, *Proc. Natl. Acad. Sci. USA* **89:**4967–4971.

Krebs, E. G., and Beavo, J. A., 1979, Phosphorylation–dephosphorylation of enzymes, *Annu. Rev. Biochem.* **48:**923–959.

Lam, E., 1993, Heterotrimeric guanine nucleotide binding proteins and light responses in higher plants, in *Plant Signals in Interacting with Other Organisms* (J. K. Schultz and I. Raskin, eds.), pp. 7–13, American Society for Plant Physiology Baltimore, Maryland.

Lamb, C. J., 1994, Plant disease resistance genes in signal perception and transduction, *Cell* **76:**419–422.

Lamb, C. J., Ryals, J. A., Ward, E. R., and Dixon, R. A., 1992, Emerging strategies for enhancing crop resistance to microbial pathogens, *Biotechnology* **10:**1436–1445.

Lehle, I., 1990, Phosphatidyl inositol mechanism and its role in signal transduction in growing plants, *Plant Mol. Biol.* **15:**647–658.

Leung, J., Bouvier-Durand, M., Marris, P. C., Querrier, D., Chefdar, F., and Giraudat, J., 1994, *Arabidopsis* ABA response gene ABI 1: Features of a calcium-modulated protein phosphatases, *Science* **264:**1448–1452.

Li, H., Dauwalder, M., and Roux, S. J., 1991, Partial purification and characterization of a Ca^{2+} dependent protein kinase from pea nuclei, *Plant Physiol.* **96:**720–727.

Lijima, T., and Sibaoka, T., 1985, Membrane potentials in excitable cells of *Aldrovanda vesiculosa* trap lobes, *Plant Cell Physiol.* **26:**1–13.

Lim, P. E., and Tate, M. E., 1973, The phytases II: Properties of phytase fraction F_1 and F_2 from wheat bran and myoinositol phosphates produced by fraction F_2, *Biochim. Biophys. Acta* **302:**316–328.

Loewus, F. A., and Loewus, M. W., 1983, Myoinositol: Its biosynthesis and metabolism, *Annu. Rev. Plant Physiol.* **34:**137–161.

Loewus, F. A., Everard, J. D., and Young, K. A., 1990, Inositol metabolism: Precursor role and breakdown, in *Inositol Metabolism in Plants* (D. J. Moore, W. F. Boss, and F. A. Loewus, eds.), pp. 21–45, Wiley-Liss, New York.

Loewus, M. W., Sasaki, K., Leavitt, A. L., Munsell, L., Sherman, W. R., and Loewus, F. A., 1982, The enantiomeric form of myoinositol-1-phosphate produced by *myo*-inositol 1-phosphate synthase and myoinositol kinase in higher plants, *Plant Physiol.* **70:**1661–1663.

Ma, H., 1993, Protein phosphorylation in plants: Enzymes, substrates and regulators, *Trends Genet.* **9:**228–230.

Ma, H., Yanofski, M. F., and Meyerowitz, E. M., 1990, Molecular cloning and characterization of GPA 1, a G protein α-subunit gene from *Arabidopsis thaliana, Proc. Natl Acad. Sci. USA* **87:**3821–3825.

Ma, H., Yanofski, M. F., and Huang, H., 1991, Isolation and sequence analysis of TGA 1 cDNAs encoding a tomato G protein a subunit, *Gene* **107:**189–195.

Mackintosh, R. W., Haycox, G., Hardie, G., and Cohen, P.T.W., 1990, Identification by molecular cloning of two cDNA sequences from the plant *Brassica napus* which are very similar to mammalian protein phosphatases-1 and 2A, *FEBS Lett.* **276:**156–160.

Mackintosh, C., Coggins, J., and Cohen, P., 1991, Plant protein phosphatases. Subcellular distribution, detection of protein phosphatase 2C and identification of protein phosphatase 2A as the major quinate dehydrogenase phosphatase, *Biochem. J.* **273:**733–738.

Maeda, N., Niinobe, M., and Mikoshiba, K., 1990, A cerebellar purkinje cells marker P_{400} protein is an inositol 1,4,5-trisphosphate ($InsP_3$) receptor protein. Purification and characterization of $InsP_3$ receptor complex, *EMBO J.* **9:**61–67.

Maitra, R., Samanta, S., Mukherjee, M., Biswas, S., and Biswas, B. B., 1988, Isolation characterization and biological function of myoinositol trisphosphate generated by phytase action on myoinositol hexaphosphate, *Ind. J. Biochem. Biophys.* **25:**655–659.

Majerus, P. W., Connolly, T. M., Bansal, V. S., Inhorn, R. C., Ross, T. S., and Lips, D. L., 1988, Inositol phosphates: Synthesis and degradation, *J. Biol. Chem.* **263:**3051–3054.

Malhotra, K., Kim, Sang-Tae, Batschauer, A., Dawat, L., and Sancar, A., 1995, Putative blue light photoreceptors from *Arabidopsis thaliana* and *Sinapis alba* with a high degree of sequence homology to DNA photolyase contain the two photolyase cofactors but lack DNA repair activity, *Biochem.* **34:**6892–6899.

Mansfield, T. A., Hetherington, A. M., and Atkinson, C. J., 1990, Some current aspects of stomatal physiology, *Plant Mol. Biol.* **41:**55–75.

Mason, H. S., De Wald, D. B., Creelman, R. A., and Mullet, J. E., 1992, Coregulation of soyabean vegetative storage protein gene expression by methyl jasmonate and soluble sugars, *Plant Physiol.* **98:**859–867.

McAinsh, M. R., Brownlee, C., and Hetherington, A. M., 1990, Abscisic acid-induced elevation of guard cell cytosolic Ca^{2+} precedes stomatal closure, *Nature* **343:**186–188.

McAinsh, M. R., Brownlee, C., and Hetherington, A. M., 1992, Visualizing changes in cytosolic free Ca^{2+} during the response of stomatal guard cells to abscisic acid, *Plant Cell* **4:**113–122.

McMurray, W. C., and Irvine, R. F., 1985, Phosphatidyl 4,5-bisphosphate phosphodiesterase in higher plants, *Biochem. J.* **249:**877–881.

Melin, P. M., Sommarin, M., Sandelius, A. S., and Jergil, B., 1987, Identification of calcium stimulated polyphosphoinositide phospholipase C in isolated plant plasma membranes, *FEBS Lett.* **223:**87–91.

Meyer, A., Miersch, O., Buttner, C., Dathe, W., and Sembdner, G., 1984, Occurrence of the plant growth regulator jasmonic acid in plants, *J. Plant Growth Regul.* **3:**1–8.

Meyer, K., Leube, M. P., and Grill, E., 1994, A protein phosphatase 2C involved in ABA signal transduction in *Arabidopsis thaliana, Science* **264:**1452–1455.

Mikoshiba, K., 1993, Inositol 1,4,5-trisphosphate receptor, *Trends Pharmacol. Sci.* **14:**86–89.

Moore, T. S., 1982, Phospholipid biosynthesis, *Annu. Rev. Plant Physiol.* **33:**235–259.

Moore, T. S., 1990, Biosynthesis of phosphatidyl inositol, in *Inositol Metabolism in Plants* (D. J. Moore, W. F. Boss, and F. Loewus, eds.), pp. 107–112, Wiley-Liss, New York.

Morse, M. J., Crain, R. C., and Satter, R. L., 1987, Light stimulated inositol phospholipid turnover in *Samanea saman* leaf pulvini, *Proc. Natl. Acad. Sci. USA* **84:**7075–7078.

Moysset, L., and Simon, E., 1989, Role of calcium in phytochrome controlled nystinastic movement of *Albizzia laphantha* leaflets, *Plant Physiol.* **90:**1108–1114.

Mukherjee, J., and Biswas, S., 1980, Purification and characterization of a nucleoside trisphosphatase from *Mimosa pudica, Ind. J. Biochem. Biophys.* **17:**452–456.

Obara, T., Nakafuku, M., Yamamoto, M., and Kaziro, Y., 1991, Isolation and characterization of a gene encoding a G protein α-subunit from *Schizosaccharomyces pombe:* Involvement in mating and sporulation pathways, *Proc. Natl. Acad. Sci. USA* **88:**5877–5881.

Palme, K., 1992, Molecular analysis of plant signaling elements: Relevance of eukaryotic signal transduction models, *Int. Rev. Cytol.* **132:**223–283.

Parks, S., and Weischau, E., 1991, The *Drosophila* gastrulation gene concertina encodes a G α-like protein, *Cell* **64:**447–458.

Pentoja, O., Gilli, A., and Blumwald, E., 1992, Voltage dependent Ca^{2+} channels in plant vacuoles, *Science* **255:**1567–1569.

Pfaffmann, H., Hartman, E., Brightman, A. O., and Moore, D. J., 1987, Phosphatidyl inositol specific phospholipase C of plant stems, *Plant Physiol.* **85:**1151–1155.

Phillippy, B. Q., and Bland, J. M., 1988, Gradient ion chromotography of inositol phosphates, *Anal. Biochem.* **175:**162–166.

Phillippy, B. Q., Ulah, A.H.J., and Ehrlich, K. C., 1994, Purification and some properties of inositol 1,3,4,5,6-pentakisphosphate 2-kinase from immature soybean seeds, *J. Biol. Chem.* **269:**28393–28399.

Pickard, B., 1973, Action potentials in higher plants, *Bot. Rev.* **39:**172–201.

Polya, G. M., Nott, R., Klucis, E. M., and Chandra, S., 1990, Inhibition of plant Ca^{2+} dependent protein kinase by basic polypeptides, *Biochim. Biophys. Acta* **1037:**259–262.

Poovaiah, B. W., and Reddy, A.S.N., 1987, Calcium messenger system in plants, *CRC Crit. Rev. Plant Sci.* **6:**47–103.

Poovaiah, B. W., and Reddy, A.S.N., 1993, Calcium and signal transduction in plants, *CRC Crit. Rev. Plant Sci.* **12:**185–211.

Putman-Evanans, C. L., Harmon, A. C., and Cormier, M. J., 1990, Purification and characterization of a novel Ca^{2+} dependent protein kinase from soybean, *Biochem. J.* **29:**2488–2495.

Raboy, V., 1990, Biochemistry and genetics of phytic acid synthesis, in *Inositol Metabolism in Plants* (D. J. Moore, W. F. Boss, and F. A. Loewus, eds.), pp. 55–76, Wiley-Liss, New York.

Raha, S., Giri, B. B., Bhattacharya, B., and Biswas, B. B., 1995, Inositol (1,3,4,5) tetrakisphosphate has an important role in calcium mobilization from *Entamoeba histolytica, FEBS Lett.* **362:**316–318.

Raj, V., and Fluhr, R., 1992, Calcium requirement for ethylene dependent responses, *Plant Cell* **4:**1123–1130.

Raj, V., and Fluhr, R., 1993, Ethylene signal is transduced via protein phosphorylation events in plant cells, *Plant Cell* **5:**523–530.

Rajendran, S., and Prakash, V., 1993, Interaction of myoinositol hexaphosphate with β-globulin from *Sesamum indicum, Int. J. Peptide Res.* **42:**78–83.

Ranjeva, R., and Boudet, A. M., 1987, Phosphorylation of proteins in plants: Regulatory effects and potential involvement in stimulus/response coupling, *Annu. Rev. Plant Physiol.* **38:**73–93.

Ranjeva, R., Carrasco, A., and Boudet, A. M., 1988, Inositol trisphosphate stimulates the release of calcium from intact vacuoles isolated from *Acer* cells, *FEBS Lett.* **230:**137–141.

Reddy, N. R., Sathe, S. K., and Salunkhe, D. K., 1982, Phytates in legumes and cereals, *Adv. Food Res.* **28:**1–92.

Roberts, D. M., and Harmon, A. C., 1992, Calcium modulated proteins: Targets of intracellular calcium signals in higher plants, *Annu. Rev. Plant Physiol. Plant Mol. Biol.* **43:**375–414.

Roberts, D. M., Lukas, T. J., and Watterson, D. M., 1986, Structure function and mechanism of action of calmodulin, *CRC Crit. Rev. Plant Sci.* **4:**311–321.

Ross, C. A., Meldolesi, J., Milner, T. A., Satosh, T., Supattapone, S., and Snyder, S. H., 1989, Inositol 1,4,5-trisphosphate receptor localized to endoplasmic reticulum in cerebellar purkinje neurons, *Nature* **339:**468–470.

Samanta, S., Dalal, B., Biswas, S., and Biswas, B. B., 1993, Myoinositol trisphosphate phytase complex as an elicitor in calcium mobilization in plants, *Biochem. Biophys. Res. Commun.* **191:**427–434.

Samuelson, B., Dahlen, S. E., Lindgren, J. A., Rouzer, C. A., and Serhan, C. N., 1987, Leukotrienes and lipoxins: Structures, biosynthesis and biological effects, *Science* **237:**1171–1175.

Sandelius, A. S., and Sommarin, M., 1986, Phosphorylation of phosphatidyl inositol in isolated plant membranes, *FEBS Lett.* **201:**282–286.

Sandelius, A. S., and Sommarin, M., 1988, Phosphatidyl inositol and phosphatidyl inositol phosphate kinases in plant plasma membranes, *Biochim. Biophys. Acta* **958:**268–278.

Scheel, D., Colling, C., Hedrich, R., Kawalleck, P., Parker, J. E., Sacks, W. R., Somssich, I. E., and Hahlbrock, K., 1991, Signals in plant defence gene activation, in *Advances in Molecular Gene Plant–Microbial Interaction,* Vol. 1 (H. Hennecke and D.P.S. Verma, eds.), pp. 373–380, Luwen, Dordrecht.

Scheuerlein, R., Schmidt, K., Poenie, M., and Roux, S. J., 1991, Determination of cytoplasmic calcium concentration in *Dryopteris* spores, *Planta* **184:**166–171.

Schroeder, J. I., and Hagiwara, S., 1990, Repetitive increases in cytosolic Ca^{2+} of guard cells by abscisic acid activation of non selective Ca^{2+} permeable channels, *Proc. Natl. Acad. Sci. USA* **87:**9305–9309.

Schroeder, J. I., and Thuleau, P., 1991, Ca^{2+} channels in higher plants cells, *Plant Cell* **3:**555–559.

Schumaker, K. S., and Szi, H., 1987, Inositol 1,4,5-trisphosphate releases Ca^{2+} from vacuolar membrane vesicles of oat roots, *J. Biol. Chem.* **262:**3944–3946.

Scott, J. J., and Loewus, F. A., 1986, A calcium activated phytase from *Lilium longiflorum, Plant Physiol.* **82:**330–335.

Short, T. W., Reymond, P., and Briggs, W. R., 1993, A pea plasma membrane protein exhibiting blue light induced phosphorylation retains photosensitivity following Triton solubilization, *Plant Physiol.* **101:**647–655.

Shimazaki, K-I., Kinoshita, T., and Nishimura, M., 1992, Involvement of calmodulin and calmodulin-dependent myosin light chain kinase in blue light dependent H^+ pumping by guard cell protoplasts from *Vicia faba* L., *Plant Physiol.* **99:**1416–1421.

Simon, M. I., Strathmann, M. P., and Gautam, N., 1991, Diversity of G proteins in signal transduction, *Science* **252:**802–808.

Sistrunk, M. L., Antosiewicz, D. M., Purugganan, M. M., and Braan, J., 1994, Arabidopsis TCH_3 encodes a novel Ca^{2+} binding protein and shows environmentally induced and tissue specific regulation, *Plant Cell* **6:**1553–1565.

Smith, R. D., and Walker, J. C., 1993, Expression of multiple type 1 phosphatases in *Arabidopsis thaliana, Plant Mol. Biol.* **21:**307–316.

Sounders, M. J., and Helper, P. K., 1983, Calcium antagonists and calmodulin inhibitors block cytokinin induced bud formation in *Funaria, Dev. Biol.* **99:**41–49.

Staswick, P. E., 1992, Jasmonic acid signaling in plants, in *Plant Signals in Interacting with Other Organisms* (J. K. Schulz and I. Raskin, eds.), pp. 7–13, American Society for Plant Physiology, Baltimore, Maryland.

Stein, J. C., and Nasaralla, J. B., 1993, A plant receptor like gene, the S locus receptor kinase of *Brassica oleracea* L. encodes a functional serine–threonine kinase, *Plant Physiol.* **101:**1103–1106.

Stein, J. C., Howlett, B., Boyes, D. C., Nasarallah, M. E., and Nasarallah, J. B., 1991, Molecular cloning of a putative receptor protein kinase gene encoded at the self incompatibility locus of *Brassica oleracea, Proc. Natl. Acad. Sci. USA* **88:**8816–8820.

Stephens, L. R., and Irvine, R. F., 1990, Stepwise phosphorylation of myoinositol leading to myoinositol hexaphosphate in *Dictyostelium, Nature* **346:**580–583.

Stephens, L. R., Hawkins, P. T., Stanley, A. F., Moore, T., Poyner, D. R., Morris, P. J., Hanley, M. R., Kay, R. R., and Irvine, R. F., 1991, Myoinositol pentakisphosphates, structure, biological occurrence and phosphorylation to myoinositol hexakisphosphate, *Biochem. J.* **275:**485–499.

Stephens, L. R., Radenberg, T., Thiel, U., Vogel, G., Khoo, K-H., Dell, A., Jackson, T. R., Hawkins, P. T., and Mayr, G. W., 1993, The detection, purification, structural characterization and metabolism of diphosphoinositol pentakisphosphate(s) and bisdiphosphoinositol tetrakisphosphate(s), *J. Biol. Chem.* **268:**4009–4015.

Supattapone, S., Worley, P. F., Barban, J. M., and Snyder, S. H., 1988, Solubilization, purification and characterization of an inositol trisphosphate receptor, *J. Biol. Chem.* **263:**1530–1534.

Taylor, C. W., and Marshall, I.C.B., 1992, Calcium and inositol 1,4,5-trisphosphate receptor: A complex relationship, *Trends Biochem. Sci.* **17:**403–407.

Terryn, N., Van Montagu, M., and Inze, D., 1993, GTP binding proteins in plants, *Plant Mol. Biol.* **22:**143–152.

Theibert, A. B., Estevej, V. A., Ferris, C. D., Danoff, S. K., Barrow, R. K., Presturich, G. D., and Snyder, S. H., 1991, Inositol 1,3,4,5-tetrakisphosphate and inositol hexakisphosphate receptor proteins: Isolation and characterization from rat brain, *Proc. Natl. Acad. Sci. USA* **88:**3165–3169.

Thiel, G., Blatt, M. R., Fricker, M. D., White, I. R., and Miller, P., 1993, Modulation of K^+

channels in *Vicea* stomatal guard cells by peptide homologous to auxin binding proteins C terminus, *Proc. Natl. Acad. Sci. USA* **90:**11493–11497.

Tobias, C. M., Howlett, B., and Nasarallah, J. B., 1992, An *Arabidopsis thaliana* gene with sequence similarity to the S-locus receptor kinase of *Brassica oleracea, Plant Physiol.* **99:**284–290.

Tomlinson, R. V., and Ballou, C. E., 1962, *myo*-Inositol polyphosphate intermediates in the dephosphorylation of phytic acid by phytase, *Biochemistry* **1:**166–177.

Tranbarger, T. J., Franceschi, V. R., Hildebrand, D. F., and Grimes, H. D., 1991, The soybean 94-kilodalton vegetative storage protein is a lipoxygenase that is localized in paraveinal mesophyll cell vacuoles, *Plant Cell* **3:**973–987.

Trewavas, A. J., and Gilroy, S., 1991, Signal transduction in plant cells, *Trends Genet.* **7:**356–361.

Urao, T., Katagiri, T., Mizoguchi, T., Yamaguchi-Shinozaki, K., Hayashida, N., and Shinozaki, K., 1994, Two genes that encode Ca^{2+} dependent protein kinases are induced by draught and high salt stresses in *Arabidopsis thaliana, Mol. Gen. Genet.* **244:**331–346.

Veluthambi, K., and Poovaiah, B. W., 1984, Calcium promoted protein phosphorylation in plants, *Science* **223:**167–169.

Voglmaier, S. M., Keen, J. H., Murphy, J., Ferris, C. D., Prestwich, G. D., Snyder, S. H., and Theibert, A. B., 1992, Inositol hexakisphosphate receptor identified as the clathrin assembly protein AP-2, *Biochem. Biophys. Res. Commun.* **187:**158–163.

Walker, J. C., and Zhang, R., 1990, Relationship of a putative receptor protein kinase from maize to the S-locus glycoproteins of *Brassica, Nature* **345:**743–746.

Weiss, C. A., Huang, H., and Ma, H., 1993, Immunolocalization of the G protein a subunit encoded by the GPA1 gene in *Arabidopsis, Plant Cell* **5:**1513–1528.

Wells, W. W., and Eisenberg, F., Jr., eds., 1978, *Cyclitols and Phosphoinositides,* Academic Press, New York.

Wheeler, J. J., and Boss, W. F., 1987, Polyphosphoinositides are present in plasma membranes isolated from fusogenic carrot cells, *Plant Physiol.* **85:**389–392.

Wheeler, J. J., and Boss, W. F., 1990, Inositol lysophospholipids in inositol metabolism in plants, in *Inositol Metabolism in Plants* (D. J. Moore, W. F. Boss, and F. A. Loewus, eds.), pp. 163–172, Wiley-Liss, New York.

Wildon, D. C., Doherty, H. M., Eagles, G., Bowles, D. J., and Thain, J. F., 1989, Systemic responses arising from localized heat stimuli in tomato plants, *Ann. Bot.* **64:**691–695.

Xu, P., Lloyd, C. W., Staiger, C. J., and Drobak, B. K., 1992, Association of phosphatidyl inositol-4 kinase with plant cytoskelton, *Plant Cell* **4:**941–951.

Yamamoto-Hino, M., Sugiyama, T., Hikichi, K., Mattei, M. G., Hasegawa, K., Sekine, S., Sakurada, K., Miyawaki, A., Furuichi, T., Hasegawa, M., and Mikoshiba, K., 1994, Cloning and characterization of human type 2 and type 3 inositol 1,4,5-trisphosphate receptors, *Receptors Channels* **2:**9–22.

Chapter 11

Phosphoinositide Turnover and Its Role in Plant Signal Transduction

Gary G. Coté, Yir G. Yueh, and Richard C. Crain

1. INTRODUCTION

1.1. Plants' Perception of Their Environment

Plants often seem to be passive, background scenery of our world. Because they do not flee, attack, or vocalize, we often think of them as unresponsive. In reality, plants actively monitor their environment and respond to it. We often fail to recognize these responses because they are usually subtle: slow movements or developmental or biochemical changes. When a plant does respond rapidly and dramatically—a Venus's flytrap snares its prey, or a disturbed *Mimosa pudica* folds its leaves—it is a powerful reminder that plants are not scenery but active players.

Although plant responses are often slow, the reception and transduction of the regulatory signals may be rapid. For example, the massive redistribution of ions that opens stomata requires over an hour for completion, but the activation

Gary G. Coté, Yir G. Yueh, and Richard C. Crain Department of Molecular and Cell Biology, The University of Connecticut, Storrs, Connecticut 06269-3125. *Present address for G.G.C.:* Department of Biology, Millikin University, Decatur, Illinois 62522. *Present address for Y.G.Y.:* Department of Biochemistry and Molecular Biology, Mayo Clinic Scottsdale, Scottsdale, Arizona 85259.

Subcellular Biochemistry, Volume 26: myo-Inositol Phosphates, Phosphoinositides, and Signal Transduction, edited by B. B. Biswas and Susweta Biswas. Plenum Press, New York, 1996.

of this massive ion movement occurs in minutes. Similarly, because the genetic induction and protein synthesis that occurs in plant responses to pathogens requires a time scale of hours, the faster the plant initiates these responses, the greater its chances of thwarting the pathogen's attack.

Light is arguably the most important factor in a plant's environment because it provides energy for photosynthesis. Plants have photoreceptor pigments for red, far-red, blue, and ultraviolet light, with which they monitor the direction, brightness, and spectral quality of the light that reaches them. The light-absorbing protein phytochrome monitors red and far-red light and the balance between them. This pigment exists in two forms, which interconvert in reactions triggered by the absorption of red and far-red light. Blue and ultraviolet photoreceptors have yet to be identified.

There are multiple forms of phytochrome (Smith and Whitelam, 1990; Quail, 1994), and likely multiple forms of the blue and ultraviolet receptors (Deng, 1994). Each form of phytochrome or other photoreceptor may be specialized for monitoring specific qualities of the environment and initiating specific responses (Deng, 1994; Whitelam and Harberd, 1994; see references in Quail *et al.*, 1994). The signal transduction mechanisms by which light activation of phytochrome or other photoreceptors ultimately triggers appropriate responses remain unknown. Evidence suggests that phosphoinositide turnover may be one of several signal transduction mechanisms mediating plant responses to light, particularly in light regulation of turgor-mediated leaflet movement in legumes.

Specific tissues of the plant may regulate distant organs and tissues through hormonal messengers. Auxins, gibberellins, and cytokinins, for example, trigger growth of plant parts under various conditions, and ethylene triggers fruit ripening and senescence. Abscisic acid is sometimes considered a general stress hormone produced under unfavorable conditions such as water deficit. One of the actions of abscisic acid is to close stomata, the leaf pores through which carbon dioxide is absorbed and water is lost, so that stressed plants conserve water. It seems likely that phosphoinositide turnover mediates stomatal closure triggered by abscisic acid. Other abscisic acid signals, as well as other hormonal signals, might also be transduced by Ins(1,4,5)P_3 production.

Plant cells recognize and respond to pathogenic fungi, bacteria, and viruses. They initiate a hypersensitive response that includes necrosis and the rapid production of active oxygen species at the site of infection (Mehdy, 1994). At least one of the signal transduction pathways leading to this response appears to involve phosphoinositide turnover and the mobilization of Ca^{2+}. Long-term responses to pathogens include cell wall modification (Lamb *et al.*, 1989), production of antimicrobial agents (Darvill and Albersheim, 1984), and acquisition of broad-spectrum systemic resistance to subsequent infection (Ryals *et al.*, 1994). These responses require gene induction, which may be triggered by rapid phosphoinositide turnover and mobilization of Ca^{2+} (Smith, 1994).

Unicellular algae are particularly sensitive to environmental conditions. They are threatened, for example, by changes in osmotic strength or changes in pH. Charophyte algae and the green alga *Dunalliela salina* adjust their cytosolic osmolarity to counter sudden decreases in environmental osmolarity, a response that is likely mediated by phosphoinositide turnover. The green alga *Chlamydomonas reinhardtii* responds to acidification and other noxious stimuli by spontaneous loss of its flagella, a response almost certainly mediated by phosphoinositide turnover.

1.2. Evidence for a Phosphoinositide Signaling System in Plants

There is no doubt that signal transduction by phosphoinositide turnover is possible in plants. The key phospholipid precursor PtdIns(4,5)P_2* has been unequivocally identified in plant tissues and in green algal cells (reviewed in Coté and Crain, 1993). Kinases that catalyze the synthesis of this lipid from phosphatidylinositol have been demonstrated, and phosphoinositide-specific phospholipases C, which hydrolyze PtdIns(4,5)P_2 to yield Ins(1,4,5)P_3 and diacylglycerol, have been reported and partially purified (reviewed in Coté and Crain, 1993). In animal systems, phosphoinositide-specific phospholipases C in the β subfamily are activated by heterotrimeric G proteins. G proteins have been indirectly implicated in physiological processes in plants (see Terryn *et al.*, 1993), and a gene for a G protein α subunit has been cloned from *Arabidopsis thaliana* (Ma *et al.*, 1990). However, whether plant phospholipases C are regulated by G proteins remains an open question.

Inositol-1,4,5-trisphosphate and other inositol phosphates have been demonstrated in plant tissues, and the metabolism of these compounds is similar, although not identical, to the corresponding metabolism in animal cells (see Coté and Crain, 1993; Chapter 8). Furthermore, Ca^{2+} has been implicated in the signal transduction cascades leading to many plant responses (Bush, 1993; Gilroy *et al.*, 1993). This Ca^{2+} could be mobilized by an Ins(1,4,5)P_3-gated channel. Although a plant Ins(1,4,5)P_3 receptor has not yet been isolated and characterized, Ins(1,4,5)P_3 releases Ca^{2+} from isolated microsomal vesicles (Drøbak and Ferguson, 1985) and mobilizes Ca^{2+} in plant protoplasts (Gilroy *et al.*, 1990). However, the location of the Ca^{2+} pool mobilized by Ins(1,4,5)P_3 remains controversial. Ins(1,4,5)P_3-gated Ca^{2+} currents have been reported in vacuolar membranes (Alexandre *et al.*, 1990), and Ins(1,4,5)P_3-triggered Ca^{2+} release from vacuolar vesicles has been demonstrated (Schumaker and Sze, 1987; Canut *et al.*, 1993). In contrast, Ping *et al.* (1992) were unable to demon-

*The abbreviations used are: Ins(1,4,5)P_3, inositol-1,4,5-trisphosphate; PtdIns(4)P, phosphatidylinositol-4-phosphate; PtdIns(3)P, phosphatidylinositol-3-phosphate; PtdIns(4,5)P_2, phosphatidylinositol-4,5-bisphosphate.

strate an Ins(1,4,5)P_3-gated Ca^{2+} channel in vacuoles of either red beet or tobacco suspension culture cells, and Gelli and Blumwald (1993) showed that vacuoles of sugar beet suspension culture cells contain a highly specific Ca^{2+} channel that is insensitive to Ins(1,4,5)P_3.

Calcium-activated protein kinases have been characterized (Roberts and Harmon, 1992) and a role for them in plant responses to light has been suggested (Fallon *et al.*, 1993; Li *et al.*, 1991). A putative Ca^{2+}-dependent protein phosphatase similar to the animal protein phosphatase calcineurin has been implicated in the regulation of K^+ uptake in guard cells (Luan *et al.*, 1993). Ion channels regulated by Ca^{2+} have also been demonstrated (Schroeder and Hagiwara, 1989; Hedrich *et al.*, 1990). In contrast, although diacylglycerol has physiological effects in plants (reviewed by Coté and Crain, 1993), a classical diacylglycerol- and Ca^{2+}-activated protein kinase C has not been identified in plants by either biochemical or genetic criteria (reviewed by Roberts and Harmon, 1992).

2. STRESS RESPONSES IN UNICELLULAR GREEN ALGAE

2.1. Osmotic Stress

The volume (and, in plants, the resulting turgor pressure) of a cell is a function of the cell's permeability to water and solutes and of the forces driving these substances across the plasma membrane (McCarty and O'Neil, 1992). The plasma membrane is far more permeable to water than to solutes. Thus, in an environment of lower osmolarity than the cytosol (a hypotonic environment), cells absorb water and swell, whereas in an environment of higher osmolarity than the cytosol (a hypertonic environment), cells lose water and shrink. Swollen cells can recover by decreasing their osmolyte concentration and losing water, whereas shrunken cells can recover by increasing their osmolyte concentration and gaining water (McCarty and O'Neil, 1992). In this way, both plant and animal cells can, to an extent, maintain volume homeostasis despite exposure to media of varying osmolarity. Increased Ca^{2+} levels have been implicated in both algal (Okazaki and Tazawa, 1990) and animal (McCarty and O'Neil, 1992) cell responses to low osmolarity, but it is not clear what triggers the increase in Ca^{2+} levels. We present evidence that, at least in green algae, phosphoinositide turnover triggers subsequent Ca^{2+} mobilization, which mediates responses to hypotonic stress.

In no system is the need for volume regulation more apparent than in algae of brackish waters, which may be subjected to wide fluctuations in the osmotic strength of their environment. The charophyte *Lamprothamnium succinctum* of brackish coastal ponds responds to hypotonic stress with an efflux of Cl^- (Okazaki and Iwasaki, 1992), which depolarizes the plasma membrane and triggers K^+ efflux through depolarization-activated K^+ channels. Increased K^+ and Cl^- efflux decreases cellular osmolarity to balance the environmental osmotic strength. These changes in ion permeability may be triggered by Ca^{2+} mobilization because they are preceded by an increase in cytoplasmic Ca^{2+} levels (Okazaki and Iwasaki, 1992) and are inhibited by low extracellular Ca^{2+} levels (Okazaki and Iwasaki, 1992) or by Ca^{2+} chelators previously microinjected into the cell (Okazaki and Iwasaki, 1991). Whether Ins(1,4,5)P_3 is the second messenger mobilizing Ca^{2+} in this alga has not been directly tested; however, the related charophyte, *Chara corallina,* is depolarized by microinjection of either Ca^{2+} or Ins(1,4,5)P_3 (Thiel *et al.*, 1990). Similarly, the charophyte *Nitella translucens* exhibits depolarization at one end of its elongated cell when the other end is permeabilized in high-calcium buffer or in buffer containing Ins(1,4,5)P_3 (Thiel *et al.*, 1990). In both species, Ins(1,4,5)P_3-induced depolarization is abolished by simultaneous introduction of Ca^{2+} chelators (Thiel *et al.*, 1990).

Further evidence that phosphoinositides mediate responses to hypotonic stress comes from experiments with the green alga *Dunalliela salina,* a species that lacks a cell wall, yet survives in salinities ranging from 0.05 to 5 M. Over the long term it maintains osmotic equilibrium by adjusting glycerol levels. In the short term, rapid changes in ion fluxes and cell volume protect the cells following sudden changes in environmental osmolarity. Einspahr *et al.* (1988) showed that on hypotonic shock, *D. salina* rapidly reduced both the level of incorporated radioactive phosphate in PtdIns(4)P and PtdIns(4,5)P_2 and the absolute mass levels of these phosphoinositides, suggesting that hypotonic stress triggers PtdIns(4,5)P_2 hydrolysis. Concomitant with the decrease in phosphoinositide levels, diacylglycerol levels increased (Ha and Thompson, 1991).

In contrast, responses to hypertonic stress in algae are probably not mediated by phosphoinositide turnover. In *D. salina,* hypertonic shock increased rather than decreased the incorporation of radioactive phosphate into phosphoinositides. Similarly, although elevated Ca^{2+} in the medium improves survival of *C. corallina* in high salt, the lack of any antagonism by nonspecific Ca^{2+} channel blockers indicates that the action of Ca^{2+} is on the extracellular side of the membrane (Reid *et al.*, 1993).

Phosphoinositide metabolism may also mediate responses to osmotic stress in the tap roots of higher plants. Jacoby and colleagues have reported that hypertonic stress of red beet slices, either with mannitol or with salt water, elicited increased levels of Ins(1,4,5)P_3 within minutes (Srivastava *et al.*, 1989; Beno-Mualem *et al.*, 1994). In contrast, Boss and co-workers (Cho *et al.*, 1993) observed no significant change in Ins(1,4,5)P_3 levels in response to hypertonic stress of carrot (*Daucus carota*) root cells in culture, although they did observe, within minutes of stimulus, a 30% decrease in PtdIns(4)P levels, a corresponding 60% increase in phospholipase C activity against PtdIns(4)P, and a 50% decrease in phosphatidylinositol kinase activity. They suggested that these changes in polyphosphoinositide metabolism might mediate changes in the cytoskeleton involved in recovery of cell shape and volume (Cho *et al.*, 1993). The failure of Boss and co-workers to find an increase in Ins(1,4,5)P_3 levels in response to hypertonic stress may result from their method for Ins(1,4,5)P_3 detection. They utilized [^{3}H]inositol labeling and HPLC separation to quantitate inositol phosphates and observed no measurable Ins(1,4,5)P_3 in either control or osmotically stressed cells. Because incorporation of inositol into phosphoinositides is relatively poor in plant cells (reviewed in Coté *et al.*, 1990), it is possible that their methods were not sensitive enough, compared with the radioreceptor binding assay employed by Jacoby and co-workers (Srivastava *et al.*, 1989; Beno-Mualem *et al.*, 1994) to detect Ins(1,4,5)P_3, even when, as they observed, phospholipase C was stimulated (Cho *et al.*, 1993).

It is intriguing that the evidence, although preliminary, suggests a role for phosphoinositides in mediating responses to hypertonic stress in tap roots while phosphoinositides apparently mediate responses to hypotonic stress in algal and animal cells. This may reflect a diversity of response. Algal and animal cells need only to protect themselves from osmotic changes while roots, which provide water to the entire plant, would be expected to initiate other responses for the protection of the plant's water status.

2.2. Deflagellation of *Chlamydomonas reinhardtii*

C. reinhardtii, a unicellular, free-swimming green alga, casts off its two flagella in response to a variety of environmental signals, including acidification, mechanical shear, high temperature, and sublethal UV light. A few chemical substances, such as dibucaine and ethanol, also induce this flagellar excision or deflagellation. The loss of flagella could serve to decrease cell surface area and reduce exposure to the noxious stimulus. Sanders and Salisbury (1989) suggested that flagellar excision proceeds through contraction of the centrin-containing fibers at the transition zone where the flagella join the cell body, followed by shear-induced weakening of flagella microtubule attachment to the cell body.

The involvement of Ca^{2+} in deflagellation has been suggested (Huber *et al.*, 1986). Treatments that increase intracellular Ca^{2+} levels in the flagella induce excision, and reducing extracellular Ca^{2+} by repeated buffer washes prevents deflagellation of *C. reinhardtii* gametes in response to low pH (Goodenough *et al.*, 1993). Furthermore, intracellular Ca^{2+} appears to increase after signals that trigger flagellar excision, as indicated by changes in the length of filaments connecting the nucleus to the basal body; the length of these filaments has been shown to vary with Ca^{2+} concentration (Salisbury *et al.*, 1987).

These observations implicating Ca^{2+} in control of deflagellation led us to postulate that products of phosphoinositide hydrolysis might regulate cellular calcium levels during the deflagellation. Early experiments showed that PtdIns(4,5)P_2 was hydrolyzed during acid-induced deflagellation and that Ins(1,4,5)P_3 and phosphatidic acid (the phosphorylation product of diacylglycerol) accumulated. Both deflagellation and Ins(1,4,5)P_3 accumulation were inhibited by neomycin, an inhibitor of phospholipase-C-catalyzed hydrolysis of PtdIns(4,5)P_2 (Quarmby *et al.*, 1992; Quarmby, 1990). In addition, both flagellar excision and Ins(1,4,5)P_3 production were triggered by micromolar levels of mastoparan, a G-protein activator, suggesting that a G protein couples the putative acid receptor to phospholipase C (Quarmby *et al.*, 1992). However, in these studies, the earliest detectable increase in Ins(1,4,5)P_3 level was at 10 sec after acid treatment, whereas deflagellation was complete in less than 5 sec. We thus turned to continuous-flow, rapid-quench methods to examine the events in the deflagellation process during the first second following acid exposure. Deflagellation was very rapid, essentially complete within a second, and was preceded by an even more rapid, transient increase of Ins(1,4,5)P_3 levels (Yueh and Crain, 1993). This pulse of Ins(1,4,5)P_3 production might trigger deflagellation, whereas the slower Ins(1,4,5)P_3 accumulation that occurs subsequent to deflagellation might play a role in some other cellular response to acidification.

Acid treatment of *C. reinhardtii* induces rapid uptake of $^{45}Ca^{2+}$ from the medium (Quarmby and Hartzell, 1994). The nonspecific Ca^{2+} channel blocker La^{3+} inhibited both acid-induced deflagellation and rapid acid-induced uptake of $^{45}Ca^{2+}$ without blocking the rapid, transient increase in Ins(1,4,5)P_3 levels, suggesting that extracellular Ca^{2+} is required for Ins(1,4,5)P_3 to induce flagellar excision (Yueh and Crain, 1993; Quarmby and Hartzell, 1994). In addition, ruthenium red, which inhibits Ins(1,4,5)P_3-gated Ca^{2+} channels in olfactory ciliary membranes and muscle sarcoplasmic reticulum, completely blocked deflagellation induced by either acidification or mastoparan (Yueh and Crain, 1993). These results suggest that the Ins(1,4,5)P_3 receptor in *C. reinhardtii* may be at least partially localized in the plasma membrane, as has been suggested for olfactory neurons (Fadool and Ache, 1992), human T cells (Khan *et al.*, 1992), and rat hepatocytes (Striggow and Bohnensack, 1994). Lanthanum ions also completely inhibited the accumulation of Ins(1,4,5)P_3 observed after deflagellation (Quarmby and Hartzell, 1994; Yueh and Crain, 1993), suggesting that this late Ins(1,4,5)P_3 accumulation is dependent on uptake of extracellular Ca^{2+}.

We hypothesize that Ins(1,4,5)P_3 is the second messenger that mediates deflagellation, eliciting calcium entry from the extracellular medium. The rapid production of Ins(1,4,5)P_3 that precedes deflagellation supports this hypothesis, as does the sensitivity of deflagellation to ruthenium red and other Ca^{2+} channel blockers. That Ins(1,4,5)P_3 production is not simply a consequence of deflagellation is demonstrated by studies on two mutants that are unable to deflagellate. The *bald-2* mutant lacks flagella, and the *fa-1* mutant has flagella but does not lose them on acid treatment. Although neither of these mutants deflagellates, both generate substantial Ins(1,4,5)P_3 in response to acid treatment (Quarmby and Hartzell, 1994; Yueh and Crain, 1993). Another mutant, *adf-1,* is defective in the acid-activated signaling pathway: it does not produce Ins(1,4,5)P_3, accumulate $^{45}Ca^{2+}$, or deflagellate on acidification, but it does all three in response to mastoparan treatment (Quarmby and Hartzell, 1994; Yueh, 1994). This is consistent with the existence of a putative "pH receptor," lacking in the *adf-1* mutant, which is linked to a G-protein-activated phospholipase C.

3. PATHOGEN RESPONSES IN HIGHER PLANTS

The hypersensitive response of plants to pathogen invasion includes localized tissue necrosis, formation of hydrogen peroxide (Mehdy, 1994), and synthesis of phytoalexins, compounds with antimicrobial activity (Darvill and Albersheim, 1984; Smith, 1994). Hydrogen peroxide, which is rapidly produced in response to elicitors, may be the plant's first line of defense (Apostol *et al.*, 1989; Schwacke and Hager, 1992; Mehdy, 1994); it not only poisons invading pathogens but may act as a second messenger to stimulate subsequent plant defense pathways, including defensive modifications of the cell wall and induction of enzymes that synthesize phytoalexins. The hypersensitive response is triggered by elicitors, substances derived from the breakdown of the cell wall of an invading pathogen or of the cell walls of host plant tissue (Darvill and Albersheim, 1984), and this triggering appears to be mediated by increased intracellular Ca^{2+} levels. Knight *et al.* (1991) produced transgenic tobacco seedlings expressing the calcium-sensitive luminescent protein aequorin and demonstrated that elicitor treatment of the seedlings elicited transient increases in luminescence, reflecting transient increases in Ca^{2+} levels. Likewise, Bach *et al.* (1993) observed increased $^{45}Ca^{2+}$ influx into protoplasts of cultured carrot cells prior to accumulation of the phytoalexin 4-hydroxybenzoic acid in response to fungal elicitor. Furthermore, reducing extracellular Ca^{2+} concentrations inhibited both the elicitor-induced production of hydrogen peroxide (Schwacke and Hager, 1992) and the accumulation of 4-hydroxybenzoic acid (Bach *et al.*, 1993).

Inositol-1,4,5-trisphosphate may mediate at least some plant responses to elicitors (reviewed in Smith, 1994). Fungal elicitors stimulate turnover of PtdIns(4)P and PtdIns(4,5)P_2 in plasma membranes isolated from pea (*Pisum sativum*) leaves, while fungal suppressors of elicitor responses also suppress elicitor-induced phosphoinositide turnover in isolated plasma membranes (Toyoda *et al.*, 1992). An elicitor preparation from the fungus *Mycosphaerella pinodes* resulted in a neomycin-sensitive accumulation of Ins(1,4,5)P_3 in pea epicotyls within 10 min of application and a neomycin-sensitive accumulation of the phytoalexin pisatin within the subsequent 24 hr. Furthermore, neomycin treatment rendered pea leaves sensitive to infection by a fungal strain to which they would normally be resistant (Toyoda *et al.*, 1993).

Application of elicitor to cultured cells also triggers Ins(1,4,5)P_3 accumulation, but with more rapid kinetics than in intact plants. In soybean cell suspensions, the elicitor polygalacturonic acid produced a 2.6-fold increase in Ins(1,4,5)P_3 that occurred within 1 min, prior to formation of any detectable hydrogen peroxide (Legendre *et al.*, 1993). Similar results were observed in lucerne (*Medicago sativa*) suspension culture cells exposed to a glycoprotein elicitor from the fungus *Verticillium albo-atrum;* within 1 min of treatment Ins(1,4,5)P_3 increased approximately twofold while PtdIns(4,5)P_2 levels decreased (Walton *et al.*, 1993). Likewise, *Phytophthora nicotianae* elicitor stimulated phosphoinositide turnover within minutes in tobacco (*Nicotiana tabacum*) suspension culture cells (Kamada and Muto, 1994).

There is also evidence inconsistent with a role of phosphoinositides in elicitor-induced responses. Preisig and Moreau (1994) studied the production of sesquiterpene phytoalexin in tobacco suspension culture cells treated with cellulase, which releases cell wall fragments as elicitors. They concluded that Ca^{2+} and Mg^{2+} function at several steps in this pathway; however, neomycin or Li^+, potential inhibitors of Ins(1,4,5)P_3 production, had no effect on phytoalexin synthesis or even stimulated it. These results are inconclusive because the researchers did not show that these inhibitors were effective in tobacco cells at the concentrations used. Similarly, 100 μM neomycin inhibited hydrogen peroxide production in response to polygalacturonic acid treatment of soybean cell suspensions by only approximately 50% while completely abolishing Ins(1,4,5)P_3 accumulation (Legendre *et al.*, 1993). These negative results suggest that phosphoinositide turnover may be one of several pathways coupling elicitor stimulation to various cellular responses.

It recently has been suggested that the activation of phospholipase C by elicitor may be mediated by a heterotrimeric G protein (Legendre *et al.*, 1993). Mastoparan, a G protein activator, mimics elicitors in stimulating hydrogen peroxide production (Legendre *et al.*, 1992, 1993) and also triggers Ins(1,4,5)P_3 accumulation and PtdIns(4,5)P_2 turnover in soybean cell suspensions; these effects are all inhibited by neomycin (Legendre *et al.*, 1993). Furthermore, a fungal elicitor from *Phytophthora infestans* causes increased GTPγS binding to membranes from potato tuber tissue (Kawakita and Doke, 1994). In further support of the role of G proteins in elicitor responses, an antibody that recognizes both plant and animal G protein α subunits, enhances hydrogen peroxide formation in the presence of an elicitor (Legendre *et al.*, 1992), suggesting that at least two G proteins are involved in cell responses to elicitors and that one of these is an inhibitory G protein.

4. PLANT MOVEMENTS MEDIATED BY OSMOTIC CHANGES

4.1. Hydraulic Movement

Plants move by hydraulics. Water moves into or out of cells, respectively increasing or decreasing their internal pressure, or turgor, and leading to changes in cell shape and cell volume, which can move structures as large as entire leaves. Hydraulics open and close leaf stomata, the pores in the leaf surface through which gas exchange takes place. Each stomatal pore is flanked by a pair of guard cells. When these cells take up water, the increased turgor pressure is directed longitudinally by cellulose microfibrils (Aylor *et al.,* 1973) so that the guard cells, being attached to each other at their ends, bow away from each other and open the pore. Conversely, loss of water reduces turgor pressure in the guard cells, causing them to collapse together and close the pore.

A more dramatic example of plant hydraulics is the movement of leaves by members of the legume and sorrel families. These plants spread their leaves to the light by day and fold them together by night. In shade-loving species, the leaves also fold together when exposed to excess light. Particularly dramatic is the sudden collapse of the leaves of the sensitive plant *Mimosa pudica* triggered by mechanical disturbance. Other macroscopic plant movements, such as the closing of the traps of Venus's flytraps (*Dionaea muscipula*) and the movement of petals and other flower parts, are likely to be mechanistically similar to the movement of legume and sorrel leaves.

Leaf movement organs, called pulvini, are visible as discrete swellings at the base of leafstalks of legumes. In cross section, two groups of cortical cells, flexor cells and extensor cells, surround the central strand of vascular tissue running through the pulvinus. Extensor cells swell and flexor cells shrink to extend the leaves toward the light, whereas flexor cells swell and extensor cells shrink to fold the leaves together. Thus, these two groups of cells, visibly indistinguishable under the microscope, behave oppositely and in concert to produce leaf movement.

The biochemistry of hydraulic movement in stomatal guard cells and in leaf pulvini is similar, perhaps identical (reviewed in Assmann, 1993; Schroeder and Hedrich, 1989, for guard cells; Lee, 1990; Satter *et al.,* 1988, for pulvini). In both cases, water moves passively, by osmosis, following the transport of K^+ ions (Satter *et al.,* 1982; Zeiger, 1983). The main counterion to K^+ is Cl^-, although malate synthesized from starch breakdown also plays a role (Zeiger, 1983). During swelling the incoming ions are primarily sequestered into the vacuole, protecting the cytosolic contents from excess salt (Humble and Raschke, 1971). As much as 60% of the K^+ in a leaf pulvinus is transported from one side of the pulvinus to the other during leaf opening or closing (Satter *et al.,*

1982). The K^+ translocation in guard cells is also substantial; the vacuolar K^+ concentration increases from around 100 mM to between 400 and 800 mM during cell swelling (Taiz and Zeiger, 1991, p. 134). In addition to K^+, Cl^-, and malate, sucrose may be an important osmoticum in guard cells (Poffenroth *et al.*, 1992; Talbott and Zeiger, 1993; Tallman and Zeiger, 1988).

Potassium enters and leaves plant cells by different channels (reviewed in Blatt, 1991; Hedrich and Schroeder, 1989; Schroeder and Hedrich, 1989), which are differently regulated. The K^+ enters cells through hyperpolarization-activated channels, which have been demonstrated in *Samanea saman* pulvinar cells (Moran, 1990) and in guard cells of *Vicia faba* (Fairley-Grenot and Assmann, 1993; Schroeder and Fang, 1991; Schroeder *et al.*, 1987), *Commelina communis* (Assmann *et al.*, 1988), and *Zea mays* (Fairley-Grenot and Assmann, 1992, 1993). These channels open under hyperpolarizing conditions, and K^+ influx is driven by the negative membrane potential. Active extrusion of protons by the plasma membrane ATPase maintains the necessary hyperpolarization for long-term net K^+ influx (Assmann *et al.*, 1985; Schroeder, 1988).

Potassium leaves plant cells through depolarization-activated channels, which have been demonstrated in pulvinar cells of *S. saman* (Moran, 1990; Moran *et al.*, 1988) and *Mimosa pudica* (Stoeckel and Takeda, 1989), in trap lobe cells of Venus's flytrap (Iijima and Hagiwara, 1987), and in guard cells of *C. communis* (Assmann *et al.*, 1988), *V. faba* (Assmann *et al.*, 1988; Blatt, 1988; Hosoi *et al.*, 1988; Schauf and Wilson, 1987; Schroeder, 1989; Schroeder *et al.*, 1987), and *Z. mays* (Fairley-Grenot and Assmann, 1992), as well as in other plant cells (reviewed in Moran, 1990). These channels are regulated by both membrane voltage and external K^+ concentration so that they open only when the membrane is depolarized to values more positive than the K^+ equilibrium potential (Blatt, 1992), allowing K^+ to leave the cell. Opening of anion-specific channels could produce the necessary depolarization to trigger K^+ efflux, because anions, Cl^- in particular, are more concentrated inside plant cells. Anion-specific channels permeable to Cl^-, NO_3^-, and malate have been described in guard cells (Keller *et al.*, 1989; Linder and Raschke, 1992; Schroeder and Keller, 1992; Schroeder *et al.*, 1993).

4.2. Regulation of Stomatal Aperture

Stomata must open to allow intake of carbon dioxide during photosynthesis, but because water is lost through stomata, the stomatal aperture is carefully regulated to prevent desiccation. The stomatal aperture is sensitive to carbon dioxide levels, illumination, humidity, and the plant's water status (for review see Kearns and Assmann, 1993). The best-studied signals controlling stomata are blue light, which signals stomata to open, and the plant hormone abscisic acid, produced during drought stress, which signals stomata to close (reviewed in MacRobbie, 1992). Evidence strongly suggests that stomatal closure by abscisic acid is mediated by phosphoinositide turnover.

At the molecular level, abscisic acid inhibits inward K^+ currents in intact *V. faba* guard cells (Thiel *et al.*, 1992) and enhances outward K^+ currents (Blatt, 1990). Abscisic acid also enhances an inward, depolarizing "leak" current, which could be carried by anion channels (Blatt, 1991) or by a nonselective cation channel capable of carrying Ca^{2+} (Schroeder and Hagiwara, 1990a).

Increased Ca^{2+} levels mimic abscisic acid in stimulating stomatal closure. Gilroy and co-workers treated guard cells with high Ca^{2+} levels, low K^+ levels, or the Ca^{2+} ionophore Br-A23187 and demonstrated, in each case, both increased cytosolic Ca^{2+} levels and subsequent stomatal closure (Gilroy *et al.*, 1991). Stomata similarly close if the intracellular Ca^{2+} concentration is increased to greater than 600 nM (from a resting level of about 100 nM) by microinjecting guard cells with caged Ca^{2+} and releasing the Ca^{2+} by photolysis (Gilroy *et al.*, 1990).

Patch-clamp studies have shown that micromolar cytosolic Ca^{2+} closes the inward-directed K^+ channels of *V. faba* (Luan *et al.*, 1993; Schroeder and Hagiwara, 1989) and *Z. mays* (Fairley-Grenot and Assmann, 1992). Outward-directed K^+ channels, however, appear to be Ca^{2+}-insensitive in *V. faba* (Hosoi *et al.*, 1988; Schroeder and Hagiwara, 1989). Micromolar cytosolic Ca^{2+} enhances anion channel currents in plasma membranes of *V. faba* guard cells (Schroeder and Hagiwara, 1989; Hedrich *et al.*, 1990; Linder and Raschke, 1992; Schroeder and Keller, 1992). Opening of these channels would allow Cl^- and other anion efflux, depolarizing the cell, even to positive potentials (Schroeder and Hagiwara, 1990b), which would activate the outward-directed K^+ channels. Nonspecific ion channels in the vacuolar membrane are also Ca^{2+}-activated (Hedrich and Neher, 1987; Ward and Schroeder, 1994) and could mobilize vacuolar anions and K^+ for excretion through the plasma membrane. Elevated Ca^{2+} might also inhibit the hyperpolarizing activity of the plasma membrane ATPase that normally drives K^+ uptake because there is evidence implicating a Ca^{2+}-activated protein kinase in the regulation of the ATPase (Schaller and Sussman, 1988; Sussman *et al.*, 1990).

Inhibition of inward-directed K^+ channels by Ca^{2+} can be reversed in patch-clamp experiments by adding the immunosuppressant drug FK506 along with its specific binding protein, FK506 binding protein (FKBP), to the cytoplasmic side of the channels (Luan *et al.*, 1993). Another immunosuppressant drug, cyclosporin A, was also effective, although the addition of its specific binding protein (cyclophilin) was not required (Luan *et al.*, 1993). These immunosuppressant/binding-protein complexes are potent inhibitors of calcineurin (also called protein phosphatase 2B), a Ca^{2+}/calmodulin-activated protein phosphatase (Liu *et al.*, 1991; Clipstone and Crabtree, 1993). That the effect of FK506 was specific was demonstrated by the finding that its analogue 15-dimethyl-FK520, which binds FKBP but does not inhibit calcineurin, did not antagonize Ca^{2+}-induced closure of guard-cell K^+ channels when added along with FKBP (Luan *et al.*, 1993).

The effects of immunosuppressants on guard cells suggest that Ca^{2+} activates a calmodulin-dependent protein phosphatase, homologous to calcineurin, that dephosphorylates and inactivates the inward-directed K^+ channel. Guard cells contain a Ca^{2+}-stimulated protein phosphatase activity against a synthetic peptide substrate, and this activity is inhibited by FK506/FKBP and cyclosporin A/cyclophilin complexes (Luan *et al.*, 1993). An activated tryptic fragment of bovine brain calcineurin can substitute for Ca^{2+} in closing the K^+ channel in patch-clamp experiments (Luan *et al.*, 1993), demonstrating that the inward-directed K^+ channel is potentially a substrate for a calcineurin homologue.

The *Arabidopsis thaliana ABI1* locus encodes a product essential for several responses to abscisic acid, including abscisic acid regulation of stomata (Koornneef *et al.*, 1984). This gene has been cloned and shown to encode a protein with high homology to protein phosphatase 2C at the carboxyl terminus and a putative Ca^{2+}-binding domain at the amino terminus (Leung *et al.*, 1994; Meyer *et al.*, 1994). Unlike calcineurin, classical protein phosphatases 2C are Ca^{2+}-independent; however, the putative Ca^{2+}-binding domain of the *ABI1* product suggests that this protein may be a Ca^{2+}-activated version of protein phosphatase 2C and thus a good candidate for mediating K^+ channel closure through Ca^{2+}-induced protein dephosphorylation.

The preceding results clearly show that Ca^{2+} can induce turgor loss in guard cells and is thus a potential messenger in abscisic-acid-induced stomatal closure. Intracellular Ca^{2+} levels increase two- to tenfold following abscisic acid treatment of guard cells of *V. faba* (Schroeder and Hagiwara, 1990a), *C. communis* (Gilroy *et al.*, 1991; McAinsh *et al.*, 1990), and the orchid *Paphiopedilum tonsum* (Irving *et al.*, 1992), and this increase precedes abscisic-acid-induced stomatal closure. In fact, abscisic acid has been shown to increase Ca^{2+} levels in a variety of plant tissues, including parsley hypocotyls, corn roots, and corn coleoptiles (Gehring *et al.*, 1990), suggesting that Ca^{2+} regulation may be a common property of abscisic acid signaling.

In studies of abscisic acid effects on Ca^{2+} levels, only a fraction, ranging from 25% to 80%, of the guard cells typically respond to abscisic acid with an increase in Ca^{2+} levels, although essentially all respond with a loss of turgor, leading to the suggestion that Ca^{2+}-independent pathways may also trigger stomatal closure (Gilroy *et al.*, 1991). McAinsh and co-workers (1992), however, using a microinjection protocol designed to ensure the integrity and viability of tested cells, were able to demonstrate increased Ca^{2+} levels in over 80% of guard cells. The apparent lack of response of the remaining cells was attributed to methodological limitations preventing detection of transient localized changes. A similar variability of response is often seen in animal cells (discussed by McAinsh *et al.*, 1992).

The Ca^{2+} appearing in the cytoplasm during abscisic-acid-induced stomatal closure apparently comes from intracellular stores. Chelation of cytoplasmic Ca^{2+} with BAPTA prevented abscisic acid inactivation of the inward-directed K^+ channels in *V. faba* guard cells, but chelation of extracellular Ca^{2+} with BAPTA had no effect (Lemtiri-Chlieh and MacRobbie, 1994). It has been argued that at least some of the Ca^{2+} must come from extracellular sources because Ca^{2+} channel blockers, such as La^{3+}, verapamil, and nifedipine, added to the extracellular medium partially inhibit abscisic acid-induced stomatal closure in *C. communis* (De Silva *et al.*, 1985; McAinsh *et al.*, 1991). This conclusion must now be reconsidered in light of the finding that these Ca^{2+} channel blockers also directly inhibit outward-directed K^+ channels (Terry *et al.*, 1992; Thomine *et al.*, 1994). Inhibiting these channels would prevent stomatal closure regardless of any changes in Ca^{2+} levels.

Inositol-1,4,5-trisphosphate microinjected into guard cells as a caged derivative and then released into the cytosol by photolysis of the caging compound both elevated cytoplasmic Ca^{2+} levels, as demonstrated by the fluorescence of a Ca^{2+}-sensitive dye, and triggered subsequent stomatal closure (Gilroy *et al.*, 1990). On a molecular level, the inward-directed K^+ channels closed on Ins(1,4,5)P_3 release, and a depolarizing leak current, possibly carried by anion channels, was activated (Blatt *et al.*, 1990; Gilroy *et al.*, 1990). Extracellular La^{3+} did not block responses to photo-released Ins(1,4,5)P_3 (Gilroy *et al.*, 1990), indicating that the Ca^{2+} was released from internal stores. Guard cells thus contain the biochemical apparatus to respond to Ins(1,4,5)P_3 by releasing Ca^{2+} from intracellular stores.

Phosphatidylinositol-4-phosphate has been definitively demonstrated in guard cells of *C. communis* (Parmar and Brearley, 1993), and PtdIns(4,5)P_2 also appeared to be present, although it could not be rigorously identified because of the low levels present. Inositol trisphosphates were also shown to be present, although their stereochemistry was not determined (Parmar and Brearley, 1993). Abscisic acid treatment first triggered a rapid decrease in the amount of radioactive label in putative PtdIns(4,5)P_2 in *V. faba* guard cells prelabeled with [^{32}P]phosphate, followed by an increase in PtdIns(4,5)P_2 label (Choi *et al.*, 1992). The amount of label in PtdIns(4)P simultaneously declined (Choi *et al.*, 1992). This could be interpreted to indicate activation of PtdIns(4,5)P_2 hydrolysis and Ins(1,4,5)P_3 production by abscisic acid, followed by enhanced PtdIns(4,5)P_2 synthesis by phosphorylation of PtdIns(4)P. Similarly, if *C. communis* guard cell lipids were prelabeled with tritiated inositol, abscisic acid treatment led to a rapid increase of label in inositol phosphates and a rapid decrease of label in phosphoinositides relative to control cells (Parmar, 1991;

MacRobbie, 1992), again supporting an abscisic-acid-induced activation of PtdIns(4,5)P_2 hydrolysis.

In animal systems, activation of phosphoinositide hydrolysis is often mediated by G proteins; therefore, evidence for the involvement of G proteins in stomatal regulation has been sought using pharmacological agents that affect G proteins. The bacterial toxin pertussis toxin, which in animal cells irreversibly inactivates specific G proteins, induced opening of closed *C. communis* stomata (Lee *et al.*, 1993), indicating that G protein activity is necessary to maintain closed stomata. Surprisingly, cholera toxin, which in animal cells generally activates G proteins, also stimulated opening of stomata in epidermal peels of *V. faba* in the dark (Curvetto and Delmastro, 1990). To further complicate the picture, both toxins, when microinjected into guard cell protoplasts of *V. faba*, decreased the magnitude of the inward K^+ current (Fairley-Grenot and Assmann, 1991), which would be expected to facilitate closure of intact stomata. Guanosine 5′-(γ-thiotriphosphate) (GTPγS), a G protein activator, when microinjected into guard cells in epidermal peels of *C. communis*, induced opening, although guanosine 5′-(β-thiodiphosphate) (GDPβS), a G protein inhibitor, had no effect (Lee *et al.*, 1993). In contrast, GTPγS decreased the inward K^+ current when microinjected into *V. faba* guard cell protoplasts, whereas GDPβS had the opposite effect (Fairley-Grenot and Assmann, 1991). The effect of GTPγS on the K^+ current could be prevented by simultaneous injection of BAPTA, indicating that elevated, or at least physiological, Ca^{2+} levels are required for inhibition of the current by G proteins (Fairley-Grenot and Assmann, 1991). If none of these results are artifactual or represent nonspecific effects of the toxins and nucleotide analogues, the inconsistency between the results of the various groups might suggest the involvement of several different G proteins in stomatal regulation, each of which might be differentially important in different species or under different experimental conditions.

4.3. Regulation of Leaf Movement

Because turgor-mediated leaf movement appears to share the same biochemical mechanism as stomatal opening and closing, one might expect similar signal transduction pathways to control both types of movement. Evidence, in fact, indicates that phosphoinositide turnover and Ca^{2+} release mediate K^+ channel closure and cell shrinkage in pulvinar cells as well as in guard cells.

Pulvini in darkness, with their leaves folded, respond to blue light by extending their leaves; pulvini in the light, with their leaves extended, respond to a pulse of red light and subsequent darkness by folding their leaves (Satter *et al.*, 1981). The effect of red light, like many effects of red light in plants, is mediated by a phytochrome and can be prevented by far-red light (Satter *et al.*, 1981).

Pulvini also respond to internal regulation by a biological clock, folding and unfolding their leaves with a circadian or daily rhythm, even when kept under constant conditions, without cues to the external passage of time (reviewed by Satter and Morse, 1990).

Although changes in cytosolic Ca^{2+} levels have not yet been demonstrated in pulvinar cells, there is abundant circumstantial evidence that Ca^{2+} plays a role in regulating the movement of legume leaves. Pulvini closed, folding their leaves together, when excised leaflets of *Albizzia lophantha* were floated on a medium containing high levels of Ca^{2+} (Moysset and Simon, 1989). The Ca^{2+} ionophore A23187 promoted closure in *A. lophantha* (Moysset and Simon, 1989) and enhanced both dark-induced closure and light-induced opening in *Cassia fasciculata* (Roblin and Fleurat-Lessard, 1984). Similarly, the Ca^{2+} channel agonist BAY K684 promoted dark-induced closure (Roblin *et al.*, 1989), and calmodulin antagonists enhanced both light- and dark-induced movements (Roblin and Fleurat-Lessard, 1984) in *C. fasciculata*. On the other hand, the Ca^{2+} chelator EGTA inhibited dark-induced leaflet closure (Roblin and Fleurat-Lessard, 1984; Roblin *et al.*, 1989) and light-induced leaflet opening (Roblin and Fleurat-Lessard, 1984) in *C. fasciculata*.

Preliminary evidence suggests that white light slightly increases the levels of inositol phosphates and decreases the levels of phosphoinositides in whole intact pulvini of the tropical legume *Samanea saman* (Morse *et al.*, 1987), which supports the possibility that Ins(1,4,5)P_3 production regulates the Ca^{2+} needed for light-induced leaflet movement. Intact pulvini, however, are not a good system for studying turgor-mediated movement at the biochemical level. The same environmental signal must close inward-directed K^+ channels and trigger cell relaxation in one pulvinar cell type while opening channels and triggering cell swelling in the opposite cell type. This complexity could explain the ability of some pharmacological agents affecting Ca^{2+} levels to promote both leaflet opening and leaflet closure. It could also explain the relatively small changes in Ins(1,4,5)P_3 levels seen in intact pulvini, because an increase in Ins(1,4,5)P_3 levels in one cell type might be masked by no change or an opposite change in Ins(1,4,5)P_3 levels in the other cell type.

Fortunately, preparations of isolated *S. saman* extensor and flexor cell protoplasts can be made, and these protoplasts retain their sensitivity to environmental light signals (Kim *et al.*, 1992). If protoplasts are kept in darkness, a pulse of blue light triggers closure of inward-directed K^+ channels in flexor protoplasts and opening of similar channels in extensor protoplasts within 3 min of light exposure (Kim *et al.*, 1992). If protoplasts are kept in white light, a pulse of red light followed by darkness rapidly triggers closure of K^+ channels in extensor protoplasts and opening of similar channels in flexor protoplasts (Kim *et al.*, 1993). The effects of red light on protoplasts, like its effects on intact leaves, appear to be mediated by phytochrome because they can be prevented by far-red light (Kim *et al.*, 1993). Thus, inward K^+ channels respond to light signals oppositely in flexor and extensor protoplasts, and in both cell types the responses are consistent with the response of the intact pulvinus on the plant.

The Ins(1,4,5)P_3 levels, measured by a radioreceptor assay, increased in each kind of protoplast following those treatments that closed inward-directed K^+ channels, that is, in flexor protoplasts following a blue light pulse in darkness and in extensor protoplasts following a transfer from white light to darkness (Kim, 1994). Following identical light treatments, Ins(1,4,5)P_3 levels did not increase in the opposite cell types, in which the inward-directed K^+ channels were not induced to close. Neomycin, an inhibitor of PtdIns(4,5)P_2 hydrolysis, prevented both the increase in Ins(1,4,5)P_3 levels and the closure of inward-directed K^+ channels in both kinds of protoplasts (Kim, 1994). Thus, closure of inward-directed K^+ channels correlates with phosphoinositide turnover in both kinds of leaf-moving cells, as in stomatal guard cells.

Mastoparan, a G protein activator in animal cells, increased Ins(1,4,5)P_3 levels and closed inward-directed K^+ channels in both flexor cells in the light and extensor cells in the darkness. Neomycin prevented the mastoparan-induced changes in both Ins(1,4,5)P_3 levels and K^+ channels, just as it prevented similar light-induced changes (Kim, 1994). These results suggest that a G protein might couple photoreceptor activation to Ins(1,4,5)P_3 production by phospholipase C.

4.4. Plant Hydraulic Movement: A Common Signaling Mechanism?

The evidence reviewed above supports the following model for a common mechanism linking environmental signals to the loss of turgor pressure and consequent cell relaxation during plant hydraulic movements. Environmental signals, such as light and abscisic acid, trigger phosphoinositide turnover, possibly through G-protein-linked receptors. The resulting increase in Ins(1,4,5)P_3 levels triggers Ca^{2+} release from internal stores. The resulting increase in Ca^{2+} levels activates a Ca^{2+}-dependent protein phosphatase, possibly equivalent to the *ABI1* gene product of *A. thaliana,* which dephosphorylates and inactivates inward-directed K^+ channels. Calcium also opens anion channels, and anion efflux depolarizes the cell, activating the outward-directed K^+ channels. Potassium and Cl^- now exit the cell instead of entering, and water follows, relaxing the cell.

Although in guard cells there is some evidence for each step in this pathway, it remains to be demonstrated conclusively whether, in fact, this pathway does mediate stomatal closure by abscisic acid because only preliminary evidence links abscisic acid treatment to phosphoinositide turnover. In pulvinar cells, on the other hand, Ins(1,4,5)P_3 production triggered by environmental signals has been demonstrated and has been linked to closure of K^+ channels by the demonstration that neomycin prevents both responses. However, the intermediate steps between $Ins(1,4,5)P_3$ production and K^+ channel closure, including the role of Ca^{2+}, remain to be elucidated.

5. FUTURE PROSPECTS

5.1. Conclusive Demonstration of Phosphoinositide Signaling in Plants

We have reviewed the evidence implicating phosphoinositide turnover in the rapid transduction of information across the plasma membrane triggering cellular responses to environmental signals. In none of the systems studied has a causal link between phosphoinositide hydrolysis and the plant's response been conclusively demonstrated. We believe that it is only a matter of time before the detailed sequence of biochemical events from signal reception through phosphoinositide hydrolysis and Ca^{2+} accumulation to the ultimate cellular response is traced out in a plant system. In this effort, pharmacological inhibitors and activators of signal transduction pathways will continue to be important, especially when biochemical measurements can correlate inhibition of response with actual inhibition of the signaling pathway. The use of mutations, as exemplified by the *ABI1* locus of *A. thaliana,* will become increasingly important. In the end, a combination of physiological, biochemical, and genetic studies will be needed to elucidate the details of plant phosphoinositide signaling.

5.2. Other Possible Roles of Phosphoinositides

A role for $PtdIns(4,5)P_2$ as the substrate for phospholipase C-catalyzed production of the second messenger $Ins(1,4,5)P_3$ does not preclude other cellular roles for phosphoinositides. In animal cells, $PtdIns(4,5)P_2$ can potentially activate membrane-associated enzymes, including protein kinase C (Kochs *et al.*, 1993). Likewise, in carrot suspension cells and in sunflower hypocotyls, PtdIns(4)P and $PtdIns(4,5)P_2$ can potentially activate the plasma membrane proton ATPase (Memon and Boss, 1990), although it is not known whether such activation is physiologically relevant.

Phosphatidylinositol-3-phosphate, formed by the action of phosphatidylinositol 3-kinase on phosphatidylinositol, may mediate the action of growth factors in animal cells (Soltoff *et al.*, 1992). It has been identified in the green algae *Chlamydomonas reinhardtii* (Quarmby *et al.*, 1992) and *C. eugametos* (Munnik *et al.*, 1994), in the duckweed *Spirodela polyrhiza* (Brearley and Hanke, 1993), in stomatal guard cells of *Commelina communis* (Parmar and Brearley, 1993), and in barley (*Hordeum vulgare*) aleurone tissue (Brearley and Hanke, 1994). Although a signaling role for PtdIns(3)P or its polyphosphorylated derivatives has not yet been indicated for any of these systems, the rapid turnover of PtdIns(3)P relative to structural lipids led Munnik *et al.* (1994) to suggest such a signaling role in *C. eugametos*.

It has recently been suggested that phosphoinositides play a role in regulating vesicular traffic in mammalian cells and in yeast (Burgoyne, 1994). In the green alga *Dunaliella salina,* Thompson and co-workers (Einspahr *et al.*, 1988) have shown phosphoinositide turnover in response to hypotonic shock, a treatment that results in rapid fusion of vesicles to the plasma membrane and a corresponding increase in cell volume (Maeda and Thompson, 1986). Phosphoinositides and calcium may also be involved in regulating cytoskeletal dynamics in animal cells through association with proteins involved in actin assembly and disassembly, such as profilin and gelsolin, and through association with other actin-binding proteins (Janmey, 1994). Phosphatidylinositol kinase and phosphatidylinositol monophosphate kinase, the enzymes that synthesize PtdIns(4,5)P_2, associate with the plant cytoskeleton in carrot suspension culture cells (Tan and Boss, 1992). Furthermore, profilin occurs in plants (Valenta *et al.*, 1993; Staiger *et al.*, 1993) and inhibits plant phosphoinositide-specific phospholipase C (Drøbak *et al.*, 1994), just as it inhibits mammalian phospholipase C (Goldschmidt-Clermont *et al.*, 1990). Elucidating the role of actin-binding proteins in the cell and the significance of their association with phosphoinositides remains an important area for future exploration in both plant and animal systems.

The last decade has seen an explosion of information elucidating the mechanisms involved in plant signaling (Gilroy and Trewavas, 1994). The important roles of calcium and protein phosphorylation in plant signal transduction have become apparent. Phosphoinositide turnover has been implicated in the regulation of Ca^{2+} during signal transduction in many systems. We predict that in the next decade the important role of PtdIns(4,5)P_2 and its hydrolytic product Ins(1,4,5)P_3 will be confirmed, and the details of signal transduction in algae, stomata, pulvini, and pathogen-infected plant tissues will become increasingly clear. We also expect the elucidation of other roles of phosphoinositides in controlling enzyme activity and in the regulation of cytoskeletal dynamics and of vesicular traffic.

ACKNOWLEDGMENTS. Research support was provided by NSF and BARD to RCC and USDA to GGC.

6. REFERENCES

Alexandre, J., Lassalles, J. P., and Kado, R. T., 1990, Opening of Ca^{2+} channels in isolated red beet root vacuole membrane by inositol 1,4,5-trisphosphate, *Nature* **343:**567–570.

Apostol, I., Heinstein, P. F., and Low, P. S., 1989, Rapid stimulation of an oxidative burst during elicitation of cultured plant cells. Role in defense and signal transduction, *Plant Physiol.* **90:**109–116.

Assmann, S. M., 1993, Signal transduction in guard cells, *Annu. Rev. Cell Biol.* **9:**345–375.

Assmann, S. M., Simoncini, L., and Schroeder, J. I., 1985, Blue light activates electrogenic ion pumping in guard cell protoplasts of *Vicia faba, Nature* **318:**285–287.

Assmann, S. M., Moran, N., and Schwartz, A., 1988, Whole-cell K^+ currents in guard cell protoplasts (GCP) of *Vicia faba* and *Commelina communis, Plant Physiol.* **86**(Suppl.):83.

Aylor, D. E., Parlange, J.-Y., and Krikorian, A. D., 1973, Stomatal mechanics, *Am. J. Bot.* **60:**163–171.

Bach, M., Schnitzler, J.-P., and Seitz, H. U., 1993, Elicitor-induced changes in Ca^{2+} influx, K^+ efflux, and 4-hydroxybenzoic acid synthesis in protoplasts of *Daucus carota* L, *Plant Physiol.* **103:**407–412.

Beno-Mualem, D., Naveh, L., and Jacoby, B., 1994, Responses to hypertonic salt-shock: Inositol 1,4,5-trisphosphate, ATPase activation and protein phosphorylation, *Plant Physiol.* **105**(Suppl.):19.

Blatt, M. R., 1988, Potassium-dependent, bipolar gating of K^+ channels in guard cells, *J. Membr. Biol.* **102:**235–246.

Blatt, M. R., 1990, Potassium channel currents in intact stomatal guard cells: Rapid enhancement by abscisic acid, *Planta* **180:**445–455.

Blatt, M. R., 1991, Ion channel gating in plants: Physiological implications and integration for stomatal function, *J. Membr. Biol.* **124:**95–112.

Blatt, M. R., 1992, K^+ channels of stomatal guard cells, *J. Gen. Physiol.* **99:**615–644.

Blatt, M. R., Thiel, G., and Trentham, D. R., 1990, Reversible inactivation of K^+ channels of *Vicia* stomatal guard cells following the photolysis of caged inositol 1,4,5-trisphosphate, *Nature* **346:**766–769.

Brearley, C. A., and Hanke, D. E., 1993, Pathway of synthesis of 3,4- and 4,5-phosphorylated phosphatidylinositols in the duckweed *Spirodela polyrhiza* L., *Biochem. J.* **290:**145–150.

Brearley, C. A., and Hanke, D. E., 1994, Phosphoinositides in barley (*Hordeum vulgare* L.) aleurone tissue, *Plant Physiol.* **104:**1381–1384.

Burgoyne, R. D., 1994, Phosphoinositides in vesicular traffic, *Trends Biochem. Sci.* **19:**55–57.

Bush, D. S., 1993, Regulation of cytosolic calcium in plants, *Plant Physiol.* **103:**7–13.

Canut, H., Carrasco, A., Rossignol, M., and Ranjeva, R., 1993, Is vacuole the richest store of IP_3-mobilizable calcium in plant cells? *Plant Sci.* **90:**135–143.

Cho, M. H., Shears, S. B., and Boss, W. F., 1993, Changes in phosphatidylinositol metabolism in response to hyperosmotic stress in *Daucus carota* L. cells grown in suspension culture, *Plant Physiol.* **103:**637–647.

Choi, Y. B., Crain, R. C., Assmann, S. M., and Lee, Y., 1995, Abscisic acid-induced inositol-phospholipid turnover in guard cell protoplasts of *Vicia faba, Plant Physiol.*, in press.

Clipstone, N. A., and Crabtree, G. R., 1993, Calcineurin is a key signaling enzyme in T lymphocyte activation and the target of the immunosuppressive drugs cyclosporin A and FK506, *Ann. N.Y. Acad. Sci.* **696:**20–30.

Coté, G. G., and Crain, R. C., 1993, Biochemistry of phosphoinositides, *Annu. Rev. Plant Physiol. Plant Mol. Biol.* **44:**333–356.

Coté, G. G., Quarmby, L. M., Satter, R. L., Morse, M. J., and Crain, R. C., 1990, Extraction, separation, and characterization of the metabolites of the inositol phospholipid cycle, in *Inositol Metabolism in Plants,* (D. J. Morré, W. F. Boss, and F. A. Loewus, eds.), pp. 113–137, Wiley-Liss, New York.

Curvetto, N., and Delmastro, S., 1990, A biochemical and physiological proposal for stomatal movement: Possible involvement of adenosine 3′,5′-cyclic monophosphate, *Plant Physiol. Biochem.* **28:**367–378.

Darvill, A. G., and Albersheim, P., 1984, Phytoalexins and their elicitors—a defense against microbial infections in plants, *Annu. Rev. Plant Physiol.* **35:**243–275.

Deng, X.-W., 1994, Fresh view of light signal transduction in plants, *Cell* **76:**423–426.

De Silva, D.L.R., Cox, R. C., Hetherington, A. M., and Mansfield, T. A., 1985, Suggested involvement of calcium and calmodulin in the responses of stomata to abscisic acid, *New Phytol.* **101:**555–563.

Drøbak, B. K., and Ferguson, I. B., 1985, Release of Ca^{2+} from plant hypocotyl microsomes by inositol-1,4,5-trisphosphate, *Biochem. Biophys. Res. Commun.* **130:**1241–1246.

Drøbak, B. K., Watkins, P. A. C., Valenta, R., Dove, S. K., Lloyd, C. W., and Staiger, C. J., 1994, Inhibition of plant plasma membrane phosphoinositide phospholipase C by the actin-binding protein, profilin, *Plant J.* **6:**389–400.

Einspahr, K. J., Peeler, T. C., and Thompson, G. A. J., 1988, Rapid changes in polyphosphoinositide metabolism associated with the response of *Dunaliella salina* to hypoosmotic shock, *J. Biol. Chem.* **263:**5775–5779.

Fadool, D. A., and Ache, B. W., 1992, Plasma membrane inositol 1,4,5-trisphosphate-activated channels mediate signal transduction in lobster olfactory receptor neurons, *Neuron* **9:**907–918.

Fairley-Grenot, K., and Assmann, S. M., 1991, Evidence for G-protein regulation of inward K^+ channel current in guard cells of fava bean, *Plant Cell* **3:**1037–1044.

Fairley-Grenot, K. A., and Assmann, S. M., 1992, Whole-cell K^+ current across the plasma membrane of guard cells from a grass: *Zea mays*, *Planta* **186:**282–293.

Fairley-Grenot, K. A., and Assmann, S. M., 1993, Comparison of K^+-channel activation and deactivation in guard cells from a dicotyledon (*Vicia faba* L.) and a graminaceous monocotyledon (*Zea mays*), *Planta* **189:**410–419.

Fallon, K. M., Shacklock, P. S., and Trewavas, A. J., 1993, Detection *in vivo* of very rapid red light-induced calcium-sensitive protein phosphorylation in etiolated wheat (*Triticum aestivum*) leaf protoplasts, *Plant Physiol.* **101:**1039–1045.

Gehring, C. A., Irving, H. R., and Parish, R. W., 1990, Effects of auxin and abscisic acid on cytosolic calcium and pH in plant cells, *Proc. Natl. Acad. Sci. USA* **87:**9645–9649.

Gelli, A., and Blumwald, E., 1993, Calcium retrieval from vacuolar pools. Characterization of a vacuolar calcium channel, *Plant Physiol.* **102:**1139–1146.

Gilroy, S., and Trewavas, T., 1994, A decade of plant signals, *BioEssays* **16:**677–682.

Gilroy, S., Read, N. D., and Trewavas, A. J., 1990, Evaluation of cytoplasmic calcium by caged calcium or caged inositol trisphosphate initiates stomatal closure, *Nature* **346:**769–771.

Gilroy, S., Fricker, M. D., Read, N. D., and Trewavas, A. J., 1991, Role of calcium in signal transduction of *Commelina* guard cells, *Plant Cell* **3:**333–344.

Gilroy, S., Bethke, P. C., and Jones, R. L., 1993, Calcium homeostasis in plants, *J. Cell Sci.* **106:**453–462.

Goldschmidt-Clermont, P. J., Machesky, L. M., Baldassare, J. J., and Pollard, T. D., 1990, The actin-binding protein profilin binds to PIP_2 and inhibits its hydrolysis by phospholipase C, *Science* **247:**1575–1578.

Goodenough, U. W., Shames, B., Small, L., Saito, T., Crain, R. C., Sanders, M. A., and Salisbury, J. L., 1993, The role of calcium in the *Chlamydomonas reinhardtii* mating reaction, *J. Cell Biol.* **121:**365–374.

Ha, K. S., and Thompson, G. A., Jr., 1991, Diacylglycerol metabolism in the green alga *Dunaliella salina* under osmotic stress. Possible role of diacylglyerols in phospholipase C-mediated signal transduction, *Plant Physiol.* **97:**921–927.

Hedrich, R., and Neher, E., 1987, Cytoplasmic calcium regulates voltage-dependent ion channels in plant vacuoles, *Nature* **329:**833–836.

Hedrich, R., and Schroeder, J. I., 1989, The physiology of ion channels and electrogenic pumps in higher plants, *Annu. Rev. Plant Physiol.* **40:**539–569.

Hedrich, R., Busch, H., and Raschke, K., 1990, Ca^{2+} and nucleotide dependent regulation of voltage dependent anion channels in the plasma membrane of guard cells, *EMBO J.* **9:**3889–3892.

Hosoi, S., Iino, M., and Shimazaki, K.-I., 1988, Outward-rectifying K^+ channels in stomatal guard cell protoplasts, *Plant Cell Physiol.* **29:**907–911.

Huber, M. E., Wright, W. G., and Lewin, R. A., 1986, Divalent cations and flagellar autotomy in *Chlamydomonas reinhardtii* (Volvocales, Chlorophyta), *Phycologia* **25:**408–411.

Humble, G. D., and Raschke, K., 1971, Stomatal opening quantitatively related to potassium transport. Evidence from electron probe analysis, *Plant Physiol.* **48:**447–453.

Iijima, T., and Hagiwara, S., 1987, Voltage-dependent K channels in protoplasts of trap-lobe cells of *Dionaea muscipula, J. Membr. Biol.* **100:**73–81.

Irving, H. R., Gehring, C. A., and Parish, R. W., 1992, Changes in cytosolic pH and calcium of guard cells precede stomatal movements, *Proc. Natl. Acad. Sci. USA* **89:**1790–1794.

Janmey, P. A., 1994, Phosphoinositides and calcium as regulators of cellular actin assembly and disassembly, *Annu. Rev. Physiol.* **56:**169–191.

Kamada, Y., and Muto, S., 1994, Stimulation of inositol phospholipid turnover in tobacco suspension culture cells by fungal elicitor, *Plant Cell Physiol.* **35:**397–404.

Kawakita, K., and Doke, N., 1994, Involvement of a GTP-binding protein in signal transduction in potato tubers treated with the fungal elicitor from *Phytophthora infestans, Plant Sci.* **96:**81–86.

Kearns, E. V., and Assmann, S. M., 1993, The guard cell–environment connection, *Plant Physiol.* **102:**711–715.

Keller, B. U., Hedrich, R., and Raschke, K., 1989, Voltage-dependent anion channels in the plasma membrane of guard cells, *Nature* **341:**450–453.

Khan, A. A., Steiner, J. P., Klein, M. G., Schneider, M. F., and Snyder, S. H., 1992, IP_3 receptor: Localization to plasma membrane of T cells and cocapping with the T cell receptor, *Science* **257:**815–818.

Kim, H. Y., 1994, *Involvement of Inositol Phospholipid Metabolism in Regulation of Turgor in Samanea Pulvinar Cells,* Ph.D. Dissertation, University of Connecticut, Storrs.

Kim, H. Y., Coté, G. G., and Crain, R. C., 1992, Effects of light on the membrane potential of protoplasts from *Samanea saman* pulvini. Involvement of K^+ channels and the H^+-ATPase, *Plant Physiol.* **99:**1532–1539.

Kim, H. Y., Coté, G. G., and Crain, R. C., 1993, Potassium channels in *Samanea saman* protoplasts controlled by phytochrome and the biological clock, *Science* **260:**960–962.

Knight, M. R., Campbell, A. K., Smith, S. M., and Trewavas, A. J., 1991, Transgenic plant aequorin reports the effects of touch, cold-shock and elicitors on cytoplasmic calcium, *Nature* **352:**524–526.

Kochs, G., Hummel, R., Fiebich, B., Sarre, T. F., Marmé, D., and Hug, H., 1993, Activation of purified human protein kinase C α and βI isoenzymes *in vitro* by Ca^{2+}, phosphatidylinositol and phosphatidylinositol 4,5-bisphosphate, *Biochem. J.* **291:**627–633.

Koornneef, M., Reuling, G., and Karssen, C. M., 1984, The isolation and characterization of abscisic acid-insensitive mutants of *Arabidopsis thaliana, Physiol. Plant.* **61:**377–383.

Lamb, C. J., Lawton, M. A., Dron, M., and Dixon, R. A., 1989, Signals and transduction mechanisms for activation of plant defenses against microbial attack, *Cell* **56:**215–224.

Lee, H. J., Tucker, E. B., Crain, R. C., and Lee, Y., 1993, Stomatal opening is induced in epidermal peels of *Commelina communis* L. by GTP analogs or pertussis toxin, *Plant Physiol.* **102:**95–100.

Lee, Y., 1990, Ion movements that control pulvinar curvature in nyctinastic legumes, *in The Pulvinus: Motor Organ for Leaf Movement* (R. L. Satter, H. L. Gorton, and T. C. Vogelmann, eds.), pp. 130–141, American Society of Plant Physiologists, Rockville, MD.

Legendre, L., Heinstein, P. F., and Low, P. S., 1992, Evidence for participation of GTP-binding proteins in elicitation of the rapid oxidative burst in cultured soybean cells, *J. Biol. Chem.* **267**:20140–20147.

Legendre, L., Yueh, Y. G., Crain, R., Haddock, N., Heinstein, P. F., and Low, P. S., 1993, Phospholipase C activation during elicitation of the oxidative burst in cultured plant cells, *J. Biol. Chem.* **268**:24559–24563.

Lemtiri-Chlieh, F., and MacRobbie, E.A.C., 1994, Role of calcium in the modulation of *Vicia* guard cell potassium channels by abscisic acid: A patch-clamp study, *J. Membr. Biol.* **137**:99–107.

Leung, J., Bouvier-Durand, M., Morris, P.-C., Guerrier, D., Chefdor, F., and Giraudat, J., 1994, *Arabidopsis* ABA response gene *ABI1*: Features of a calcium-modulated protein phosphatase, *Science* **264**:1448–1452.

Li, H., Dauwalder, M., and Roux, S. J., 1991, Partial purification and characterization of a Ca^{2+}-dependent protein kinase from pea nuclei, *Plant Physiol.* **96**:720–727.

Linder, B., and Raschke, K., 1992, A slow anion channel in guard cells, activating at large hyperpolarization, may be principal for stomatal closing, *FEBS Lett.* **313**:27–30.

Liu, J., Farmer, J. D., Jr., Lane, W. S., Friedman, J., Weissman, I., and Schreiber, S. L., 1991, Calcineurin is a common target of cyclophilin-cyclosporin A and FKBP–FK506 complexes, *Cell* **66**:807–815.

Luan, S., Li, W., Rusnak, F., Assmann, S. M., and Schreiber, S. L., 1993, Immunosuppressants implicate protein phosphatase regulation of K^+ channels in guard cells, *Proc. Natl. Acad. Sci. USA* **90**:2202–2206.

Ma, H., Yanofsky, M. F., and Meyerowitz, E. M., 1990, Molecular cloning and characterization of *GPA1*, a G protein α subunit gene from *Arabidopsis thaliana*, *Proc. Natl. Acad. Sci. USA* **87**:3821–3825.

MacRobbie, E.A.C., 1992, Calcium and ABA-induced stomatal closure, *Phil. Trans. R. Soc. Lond.* **B338**:5–18.

Maeda, M., and Thompson, G. A., Jr., 1986, On the mechanism of rapid plasma membrane and chloroplast expansion in *Dunaliella salina* exposed to hypoosmotic shock, *J. Cell Biol.* **102**:289–297.

McAinsh, M. R., Brownlee, C., and Hetherington, A. M., 1990, Abscisic acid-induced elevation of guard cell cytosolic Ca^{2+} precedes stomatal closure, *Nature* **343**:186–188.

McAinsh, M. R., Brownlee, C., and Hetherington, A. M., 1991, Partial inhibition of ABA-induced stomatal closure by calcium-channel blockers, *Proc. R. Soc. Lond.* **B243**:195–201.

McAinsh, M. R., Brownlee, C., and Hetherington, A. M., 1992, Visualizing changes in cytosolic-free Ca^{2+} during the response of stomatal guard cells to abscisic acid, *Plant Cell* **4**:1113–1122.

McCarty, N. A., and O'Neil, R. G., 1992, Calcium signaling in cell volume regulation, *Physiol. Rev.* **72**:1037–1061.

Mehdy, M. C., 1994, Active oxygen species in plant defense against pathogens, *Plant Physiol.* **105**:467–472.

Memon, A. R., and Boss, W. F., 1990, Rapid light-induced changes in phosphoinositide kinases and H^+-ATPase in plasma membrane of sunflower hypocotyls, *J. Biol. Chem.* **265**:14817–14821.

Meyer, K., Leube, M. P., and Grill, E., 1994, A protein phosphatase 2C involved in ABA signal transduction in *Arabidopsis thaliana*, *Science* **264**:1452–1455.

Moran, N., 1990, The role of ion channels in osmotic volume changes in *Samanea* motor cells analysed by patch-clamp methods, *in The Pulvinus: Motor Organ for Leaf Movement* (R. L. Satter, H. L. Gorton, and T. C. Vogelmann, eds.), pp. 142–159, American Society of Plant Physiologists, Rockville, MD.

Moran, N., Ehrenstein, G., Iwasa, K., Mischke, C., Bare, C., and Satter, R. L., 1988, Potassium channels in motor cells of *Samanea saman*. A patch-clamp study, *Plant Physiol.* **88**:643–648.

Morse, M. J., Crain, R. C., and Satter, R. L., 1987, Light-stimulated inositolphospholipid turnover in *Samanea saman* leaf pulvini, *Proc. Natl. Acad. Sci. USA* **84**:7075–7078.

Moysset, L., and Simon, E., 1989, Role of calcium in phytochrome-controlled nyctinastic movements of *Albizzia lophantha* leaflets, *Plant Physiol.* **90:**1108–1114.

Munnik, T., Irvine, R. F., and Musgrave, A., 1994, Rapid turnover of phosphatidylinositol 3-phosphate in the green alga *Chlamydomonas eugametos:* Signs of a phosphatidylinositol 3-kinase signaling pathway in lower plants? *Biochem. J.* **298:**269–273.

Okazaki, Y., and Iwasaki, N., 1991, Injection of a Ca^{2+}-chelating agent into the cytoplasm retards the progress of turgor regulation upon hypotonic treatment in the alga, *Lamprothamnium, Plant Cell Physiol.* **32:**185–194.

Okazaki, Y., and Iwasaki, N., 1992, Net efflux of Cl^- during hypotonic turgor regulation in a brackish water alga *Lamprothamnium, Plant Cell Environ.* **15:**61–70.

Okazaki, Y., and Tazawa, M., 1990, Calcium ion and turgor regulation in plant cells, *J. Membr. Biol.* **114:**189–194.

Parmar, P., 1991, *Polyphosphoinositide Metabolism in Stomatal Guard Cells,* Ph.D. Dissertation, University of Cambridge, Cambridge.

Parmar, P. N., and Brearley, C. A., 1993, Identification of 3- and 4-phosphorylated phosphoinositides and inositol phosphates in stomatal guard cells, *Plant J.* **4:**255–263.

Ping, Z., Yabe, I., and Muto, S., 1992, Identification of K^+, Cl^-, and Ca^{2+} channels in the vacuolar membrane of tobacco cell suspension cultures, *Protoplasma* **171:**7–18.

Poffenroth, M., Green, D. B., and Tallman, G., 1992, Sugar concentrations in guard cells of *Vicia faba* illuminated with red or blue light. Analysis by high performance liquid chromatography, *Plant Physiol.* **98:**1460–1471.

Preisig, C. L., and Moreau, R. A., 1994, Effects of potential signal transduction antagonists on phytoalexin accumulation in tobacco, *Phytochemistry* **36:**857–863.

Quail, P. H., 1994, Phytochrome genes and their expression, *in Photomorphogenesis in Plants* (R. E. Kendrick and G.H.M. Kronenberg, eds.), pp. 71–104, Kluwer Academic Publishers, Dordrecht.

Quail, P. H., Briggs, W. R., Chory, J., Hangarter, R. P., Harberd, N. P., Kendrick, R. E., Koornneef, M., Parks, B., Sharrock, R. A., Schäfer, E., Thompson, W. F., and Whitelam, G. C., 1994, Spotlight on phytochrome nomenclature, *Plant Cell* **6:**468–471.

Quarmby, L. M., 1990, *Inositol Phospholipid Signalling in Chlamydomonas reinhardtii,* Ph.D. Dissertation, University of Connecticut, Storrs.

Quarmby, L. M., and Hartzell, H. C., 1994, Two distinct, calcium-mediated, signal transduction pathways can trigger deflagellation in *Chlamydomonas reinhardtii, J. Cell Biol.* **124:**807–815.

Quarmby, L. M., Yueh, Y. G., Cheshire, J. L., Keller, L. R., Snell, W. J., and Crain, R. C., 1992, Inositol phospholipid metabolism may trigger flagellar excision in *Chlamydomonas reinhardtii, J. Cell Biol.* **116:**737–744.

Reid, R. J., Tester, M., and Smith, F. A., 1993, Effects of salinity and turgor on calcium influx in *Chara, Plant Cell Environ.* **16:**547–554.

Roberts, D. M., and Harmon, A. C., 1992, Calcium-modulated proteins: Targets of intracellular calcium signals in higher plants, *Annu. Rev. Plant Physiol. Plant Mol. Biol.* **43:**375–414.

Roblin, G., and Fleurat-Lessard, P., 1984, A possible mode of calcium involvement in dark- and light-induced leaflet movements in *Cassia fasciculata* Michx., *Plant Cell Physiol.* **25:**1495–1499.

Roblin, G., Fleurat-Lessard, P., and Bonmort, J., 1989, Effects of compounds affecting calcium channels on phytochrome- and blue pigment-mediated pulvinar movements of *Cassia fasciculata, Plant Physiol.* **90:**697–701.

Ryals, J., Uknes, S., and Ward, E., 1994, Systemic acquired resistance, *Plant Physiol* **104:**1109–1112.

Salisbury, J. L., Sanders, M. A., and Harpst, L., 1987, Flagellar root contraction and nuclear movement during flagellar regeneration in *Chlamydomonas reinhardtii, J. Cell Biol.* **105:**1799–1805.

Sanders, M. A., and Salisbury, J. L., 1989, Centrin-mediated microtubule severing during flagellar excision in *Chlamydomonas reinhardtii, J. Cell Biol.* **108:**1751–1760.

Satter, R. L., and Morse, M. J., 1990, Light-modulated, circadian rhythmic leaf movements in nyctinastic legumes, *in The Pulvinus: Motor Organ for Leaf Movement* (R. L. Satter, H. L. Gorton, and T. C. Vogelmann, eds.), pp. 10–24, American Society of Plant Physiologists, Rockville, MD.

Satter, R. L., Guggino, S. E., Lonergan, T. A., and Galston, A. W., 1981, The effects of blue and far red light on rhythmic leaflet movements in *Samanea* and *Albizzia, Plant Physiol.* **67:**965–968.

Satter, R. L., Garber, R. C., Khairallah, L., and Cheng, Y.-S., 1982, Elemental analysis of freeze-dried thin sections of *Samanea* motor organs: Barriers to ion diffusion through the apoplast, *J. Cell Biol.* **95:**893–902.

Satter, R. L., Morse, M. J., Lee, Y., Crain, R. C., Coté, G. G., and Moran, N., 1988, Light- and clock-controlled leaflet movements in *Samanea saman:* A physiological, biophysical and biochemical analysis, *Bot. Acta* **101:**205–213.

Schaller, G. E., and Sussman, M. R., 1988, Phosphorylation of the plasma-membrane H^+-ATPase of oat roots by a calcium-stimulated protein kinase, *Planta* **173:**509–518.

Schauf, C. L., and Wilson, K. J., 1987, Effects of abscisic acid on K^+ channels in *Vicia faba* guard cell protoplasts, *Biochem. Biophys. Res. Commun.* **145:**284–290.

Schroeder, J. I., 1988, K^+ transport properties of K^+ channels in the plasma membrane of *Vicia faba* guard cells, *J. Gen. Physiol.* **92:**667–683.

Schroeder, J. I., 1989, Quantitative analysis of outward rectifying K^+ channel currents in guard cell protoplasts from *Vicia faba, J. Membr. Biol.* **107:**229–235.

Schroeder, J. I., and Fang, H. H., 1991, Inward-rectifying K^+ channels in guard cells provide a mechanism for low-affinity K^+ uptake, *Proc. Natl. Acad. Sci. USA* **88:**11583–11587.

Schroeder, J. I., and Hagiwara, S., 1989, Cytosolic calcium regulates ion channels in the plasma membrane of *Vicia faba* guard cells, *Nature* **338:**427–430.

Schroeder, J. I., and Hagiwara, S., 1990a, Repetitive increases in cytosolic Ca^{2+} of guard cells by abscisic acid activation of nonselective Ca^{2+} permeable channels, *Proc. Natl. Acad. Sci. USA* **87:**9305–9309.

Schroeder, J. I., and Hagiwara, S., 1990b, Voltage-dependent activation of Ca^{2+}-regulated anion channels and K^+ uptake channels in *Vicia faba* guard cells, *in Calcium in Plant Growth and Development* (R. T. Leonard and P. K. Hepler, eds.), pp. 144–150, American Society of Plant Physiologists, Rockville, MD.

Schroeder, J. I., and Hedrich, R., 1989, Involvement of ion channels and active transport in osmoregulation and signaling of higher plant cells, *Trends Biochem. Sci.* **14:**187–192.

Schroeder, J. I., and Keller, B. U., 1992, Two types of anion channel currents in guard cells with distinct voltage regulation, *Proc. Natl. Acad. Sci. USA* **89:**5025–5029.

Schroeder, J. I., Raschke, K., and Neher, E., 1987, Voltage dependence of K^+ channels in guard-cell protoplasts, *Proc. Natl. Acad. Sci. USA* **84:**4108–4112.

Schroeder, J. I., Schmidt, C., and Sheaffer, J., 1993, Identification of high-affinity slow anion channel blockers and evidence for stomatal regulation by slow anion channels in guard cells, *Plant Cell* **5:**1831–1841.

Schumaker, K. S., and Sze, H., 1987, Inositol 1,4,5-trisphosphate releases Ca^{2+} from vacuolar membrane vesicles of oat roots, *J. Biol. Chem.* **262:**3944–3946.

Schwacke, R., and Hager, A., 1992, Fungal elicitors induce a transient release of active oxygen species from cultured spruce cells that is dependent on Ca^{2+} and protein-kinase activity, *Planta* **187:**136–141.

Smith, C. J., 1994, Signal transduction in elicitation of phytoalexin synthesis, *Biochem. Soc. Trans.* **22:**414–420.

Smith, H., and Whitelam, G. C., 1990, Phytochrome, a family of photoreceptors with multiple physiological roles, *Plant Cell Environ.* **13:**695–707.

Soltoff, S. P., Carpenter, C. L., Auger, K. R., Kapeller, R., Schaffhausen, B., and Cantley, L. C., 1992, Phosphatidylinositol-3 kinase and growth regulation, *Cold Spring Harbor Symp. Quant. Biol.* **57:**75–80.

Srivastava, A., Pines, M., and Jacoby, B., 1989, Enhanced potassium uptake and phosphatidylinositol-phosphate turnover by hypertonic mannitol shock, *Physiol. Plant.* **77:**320–325.

Staiger, C. J., Goodbody, K. C., Hussey, P. J., Valenta, R., Drobak, B. K., and Lloyd, C. W., 1993, The profilin multigene family of maize. Differential expression of three isoforms, *Plant J.* **4:**631–641.

Stoeckel, H., and Takeda, K., 1989, Potassium currents in protoplasts from primary pulvinar motor cells of *Mimosa pudica:* A patch-clamp study, *in Plant Membrane Transport: The Current Position* (J. Dainty, M. I. De Michelis, E. Marrè, and F. Rasi-Caldogno, eds.), pp. 527–528, Elsevier, Amsterdam.

Striggow, F., and Bohnensack, R., 1994, Inositol 1,4,5-trisphosphate activates receptor-mediated calcium entry by two different pathways in hepatocytes, *Eur. J. Biochem.* **222:**229–234.

Sussman, M. R., Schaller, G. E., DeWitt, N., and Harper, J. F., 1990, Regulation of the plasma membrane proton pump, *in Plant Protein Phosphorylation, Protein Kinases, Calcium and Calmodulin* (D. D. Randall and D. G. Blevins, eds.), pp. 183–189, University of Missouri, Columbia, MO.

Taiz, L., and Zeiger, E., 1991, *Plant Physiology,* Benjamin Cummings, Redwood City, CA.

Talbott, L. D., and Zeiger, E., 1993, Sugar and organic acid accumulation in guard cells of *Vicia faba* in response to red and blue light, *Plant Physiol.* **102:**1163–1169.

Tallman, G., and Zeiger, E., 1988, Light quality and osmoregulation in *Vicia* guard cells. Evidence for involvement of three metabolic pathways, *Plant Physiol.* **88:**887–895.

Tan, Z., and Boss, W. F., 1992, Association of phosphatidylinositol kinase, phosphatidylinositol monophosphate kinase and diacylglycerol kinase with the cytoskeleton and F-actin fractions of carrot (*Daucus carota* L.) cells grown in suspension culture: Response to cell wall degrading enzymes, *Plant Physiol.* **100:**2116–2120.

Terry, B. R., Findlay, G. P., and Tyerman, S. D., 1992, Direct effects of Ca^{2+}-channel blockers on plasma membrane cation channels of *Amaranthus tricolor* protoplasts, *J. Exp. Bot.* **43:**1457–1473.

Terryn, N., Van Montagu, M., and Inze, D., 1993, GTP-binding proteins in plants, *Plant Mol. Biol.* **22:**143–152.

Thiel, G., MacRobbie, E.A.C., and Hanke, D. E., 1990, Raising the intracellular level of inositol 1,4,5-trisphosphate changes plasma membrane ion transport in characean algae, *EMBO J.* **9:**1737–1741.

Thiel, G., MacRobbie, E.A.C., and Blatt, M. R., 1992, Membrane transport in stomatal guard cells: The importance of voltage control, *J. Membr. Biol.* **126:**1–18.

Thomine, S., Zimmermann, S., Van Duijn, B., Barbier-Brygoo, H., and Guern, J., 1994, Calcium channel antagonists induce direct inhibition of the outward rectifying potassium channel in tobacco protoplasts, *FEBS Lett.* **340:**45–50.

Toyoda, K., Shiraishi, T., Yoshioka, H., Yamada, T., Ichinose, Y., and Oku, H., 1992, Regulation of polyphosphoinositide metabolism in pea plasma membranes by elicitor and suppressor from a pea pathogen, *Mycosphaerella pinodes, Plant Cell Physiol.* **33:**445–452.

Toyoda, K., Shiraishi, T., Yamada, T., Ichinose, Y., and Oku, H., 1993, Rapid changes in polyphosphoinositide metabolism in pea in response to fungal signals, *Plant Cell Physiol.* **34:**729–735.

Valenta, R., Ferreira, F., Grote, M., Swoboda, I., Vrtala, S., Duchêne, M., Deviller, P., Meagher, R. B., McKinney, E., Heberle-Bors, E., Kraft, D., and Scheiner, O., 1993, Identification of profilin as an actin-binding protein in higher plants, *J. Biol. Chem.* **268:**22777–22781.

Walton, T. J., Cooke, C. J., Newton, R. P., and Smith, C. J., 1993, Evidence that generation of inositol 1,4,5-trisphosphate and hydrolysis of phosphatidylinositol 4,5-bisphosphate are rapid responses following addition of fungal elicitor which induces phytoalexin synthesis in lucerne (*Medicago sativa*) suspension culture cells, *Cell. Sig.* **5**:345–356.

Ward, J. M., and Schroeder, J. I., 1994, Calcium-activated K^+ channels and calcium-induced calcium release by slow vacuolar ion channels in guard cell vacuoles implicated in the control of stomatal closure, *Plant Cell* **6**:669–683.

Whitelam, G. C., and Harberd, N. P., 1994, Action and function of phytochrome family members revealed through the study of mutant and transgenic plants, *Plant Cell Environ.* **17**:615–625.

Yueh, Y. G., 1994, *The Role of Phosphatidylinositol Metabolism and Calcium Mobilization in Deflagellation of Chlamydomonas reinhardtii*, Ph.D. Dissertation, University of Connecticut, Storrs.

Yueh, Y. G., and Crain, R. C., 1993, Deflagellation of *Chlamydomonas reinhardtii* follows a rapid, transitory accumulation of inositol 1,4,5-trisphosphate and requires Ca^{2+} entry, *J. Cell Biol.* **123**:869–875.

Zeiger, E., 1983, The biology of stomatal guard cells, *Annu. Rev. Plant Physiol.* **34**:441–475.

Chapter 12

Light-Induced Signal Transduction Pathway Involving Inositol Phosphates

Sudhir K. Sopory and Meena R. Chandok

1. INTRODUCTION

Signals such as light, hormones, and gravity control diverse physiological and developmental processes throughout the life cycle of plants. How a plant senses these signals and then responds in an appropriate manner has been a subject of great interest. In animal systems, the mechanism of transduction of extracellular signals into intracellular events has been studied in great detail. It involves two major signal pathways. The first pathway employs the messenger cyclic adenosine monophosphate (cyclic AMP), and the second pathway involves a combination of messengers that include inositol trisphosphate ($InsP_3$), diacylglycerol (DAG), and calcium ion (Berridge and Irvine, 1984; Nishizuka, 1986, 1988; Berridge, 1987).

Attempts have been made by plant physiologists to explore the presence of both signal pathways in plants. Although there are reports of the existence of cyclic AMP in higher plants, its physiological role as a second messenger remains a topic of debate (Brown and Newton, 1981; Newton and Brown, 1986).

Recent investigations from various laboratories indicate the existence of a

Sudhir K. Sopory and Meena R. Chandok School of Life Sciences, Jawaharlal Nehru University, New Delhi 110 067, India.

Subcellular Biochemistry, Volume 26: myo-Inositol Phosphates, Phosphoinositides, and Signal Transduction, edited by B. B. Biswas and Susweta Biswas. Plenum Press, New York, 1996.

calcium messenger system in plants (Hepler and Wayne, 1985; Roux *et al.*, 1986; Poovaiah and Reddy, 1987; Tretyn *et al.*, 1991) and show that a number of physiological responses elicited by such signals can be attained by manipulating cellular calcium levels. The discovery of various components of the calcium messenger system such as calmodulin and the regulatory role of protein phosphorylation by calcium-dependent protein kinases have led plant physiologists to propose calcium as a second messenger (Ranjeva and Boudet, 1989; Poovaiah and Reddy, 1990b; Roberts and Harmon, 1992).

In addition to calcium as a second messenger, phospholipids also play an important role in signal transduction in animals (Berridge and Irvine, 1984; Berridge *et al.*, 1983; Kurosaki *et al.*, 1987; Strasser *et al.*, 1986), including the breakdown products of phosphatidylinositol bisphosphate ($PtdInsP_2$), DAG and $InsP_3$ formed by phospholipase C. The primary effect of DAG is to activate protein kinase C (PKC), which in turn phosphorylates a range of cellular proteins.

In comparison to the literature on the involvement of calcium, that on the presence and involvement of inositol phospholipids in plants is smaller. Nevertheless, evidence is accumulating to suggest an important role for the inositol phospholipid cycle in signal transduction in plants.

The overall scheme of events operative in the animal system envisages that the exogenous stimulus would activate the cleavage of $PtdIns_2$ by a specific phosphodiesterase (phospholipase C) to release the second messengers $InsP_3$ and DAG. The α subunits of G proteins act as part of the transducing complexes involved in signal transduction from receptor to various effector systems such as adenylate cyclase and phospholipase C. When $InsP_3$ is released into the cytosol, it stimulates the mobilization of Ca^{2+} from endogenous stores such as endoplasmic reticulum. The increased level of Ca^{2+} results in the activation of enzymes (e.g., kinases) either directly or through Ca^{2+}-binding proteins (e.g., calmodulin), which initiate and modulate physiological responses. The other messenger, DAG, either activates a set of kinases in the presence of Ca^{2+} called "protein kinase C" to transduce the signal or is metabolized into other biochemical pathways.

Recent studies indicate that the animal model provides a kind of basic framework for understanding the role of phosphoinositides in signal transduction in plant systems. However, the interaction of components of signal transduction may be different, and each system will have to be examined within its own physiological milieu.

2. INOSITOL PHOSPHOLIPID TURNOVER AS SIGNAL TRANSDUCTION IN PLANTS

Changes in inositol phospholipid metabolism by environmental and/or internal stimuli, such as hormones, suggest that the components of the polyphosphoinositide (PPI) cycle have some regulatory function in cellular signaling in plants. Several reports indicate that signals such as light, hormones, and elicitors are capable of generating inositol phosphates and have a role in transmembrane signaling.

About 10 years ago, plant biologists began to investigate the possibility that accelerated PPI turnover is a transduction mechanism in plants. The evidence has accumulated in favor of this possibility as various components of the cycle have been identified (Drobak, 1992, 1993).

Polyphosphoinositides are present in all plant membranes and generally comprise 1–10% of the acyl lipids (Harwood, 1980). The presence of inositol phosphates and their role during seed germination were shown by Biswas *et al.* (1978), who proposed a metabolic cycle involving glucose-6-phosphate and *myo*-inositol phosphates. A number of other roles for inositol have also been implicated in plant systems. However, several investigators have recently suggested that inositol turnover may also be involved in the signal transduction chain in plants. Not only have most of the components of the PPI cycle and the enzyme been reported from plant sources but such exogenous factors as light and hormones have been shown to affect the PPI cycle.

By labeling cells with *myo*-[^{3}H]inositol, Boss and Mossel (1985) showed that wild carrot cells contained only 0.8% $PtdInsP_2$ compared to 1.7% PtdInsP and 93% PtdIns. In other systems as well, low levels of $PtdInsP_2$ have generally been reported; however, in some systems, such as brinjal leaves, a ratio of 1:5:5 has been shown (Wagh and Natarajan, 1990). The existence of these polyphosphoinositides in pulvini of *Samanea saman* (Morse *et al.*, 1987a) and plasma membrane of fusogenic carrot cells has also been reported (Wheeler and Boss, 1987). Irvine and colleagues (1989) deacylated *myo*-[^{3}H]inositol-labeled phosphatidylinositol and clearly showed that the resulting inositol phosphates were D-*myo*-inositol-1,4-bisphosphates. They concluded that plants contain phosphatidylinositol-4-phosphate (PtdInsP) and phosphatidylinositol-4,5-bisphosphate as the only polyphosphoinositides. In fact, phosphoinositides have since been isolated from different plant systems (see Boss and Moore, 1989; Irvine *et al.*, 1989; Hartmann and Pfaffmann, 1990).

Another approach for determining the localization of phosphoinositides was to study the enzymes involved in their synthesis and catabolism.

Sandelius and Sommarin (1986) isolated plasma membranes from wheat seedlings enriched in PtdIns and PtdInsP kinase activity. They have shown, using [γ-^{32}P]ATP and the endogenous substrates or exogenous PtdInsP, that phosphatidic acid (PA) was the primary radiolabeled product of the kinase reaction (Sandelius and Sommarin, 1986; Sommarin and Sandelius, 1988). Their data suggested that there was a very active plasma membrane phospholipase C that hydrolyzed PtdInsP to DAG, which was subsequently phosphorylated to PA. Sandelius and Sommarin went on to characterize the phospholipase C activity in wheat seedlings. Melin *et al.* (1987) also identified calcium-stimulated polyphosphoinositide phospholipase C in wheat seedlings and, like the kinase activity, the plasma membrane fractions were found to be enriched in PtdInsP and $PtdInsP_2$ phospholipase C activity. Although, phospholipase C activity was found in the cytosolic fractions, only the plasma membrane enzyme showed substrate specificity for PtdInsP and $PtdInsP_2$. At lower calcium concentrations, $PtdInsP_2$ was hydrolyzed more readily than PtdInsP. An increase in cytosolic calcium above 10μM preferentially decreased the hydrolysis of $PtdInsP_2$; however, other phosphoinositides were metabolized, producing $InsP_2$ and InsP (Tate *et al.*, 1989). As shown by Pfaffmann *et al.* (1987) PtdIns was hydrolyzed by plasma membrane phospholipase C, even at 1 mM calcium concentration.

The plasma-membrane-specific phospholipase C activity has also been characterized in bush bean and soybean (Pfaffmann *et al.*, 1987). It was found that the enzyme activity was stimulated by calcium with a broad optimum at 0.5 mM. Phospholipase activity, with PtdIns as a substrate, has also been found in the cytosolic fractions of celery, cauliflower, and daffodils (Irvine *et al.*, 1989) and in cytosolic and particulate fractions of lily pollen (Helsper *et al.*, 1986, 1987). Tate *et al.* (1989) have characterized the polyphosphoinositide phospholipase C from the plasma membrane of *Avena sativa*. They found that the enzyme was highly reactive toward inositol phospholipids but only minimally active toward phosphatidylethanolamine (PE) and phosphatidylcholine (PC). The activity of the enzyme was found to be highly dependent on calcium. At 10 μM calcium and 200 μM inositol phospholipid, the enzyme was specific for $PtdInsP_2$ and PtdInsP, which were hydrolyzed at rates ten and four times more, respectively, than the rate of PtdIns hydrolysis. The principal water-soluble products analyzed by them, using HPLC, were $InsP_3$ from $PtdInsP_2$, $InsP_2$ from PtdInsP, and InsP from PtdIns. Although the enzyme characterized by them shows more similarities with the animal phospholipase C (in its calcium and lipid requirement) than the plant phospholipase C (Melin *et al.*, 1987; Pfaffmann *et al.*, 1987), but considering all the evidence, their results suggest the presence of multiple forms of phospholipase C in plant tissues. A phospholipase C specific for polyphosphoinositides and requiring micromolar calcium may be localized in the plasma membrane while the phospholipase C specific for PtdIns and requiring millimolar calcium may be found predominantly in the cytosol. Phospholipase C has also been shown to be present in the purified plasma membrane of *Pharbitis nil* (Tretyn *et al.*, 1991) and *Dunaliella salina* (Einsphar *et al.*, 1989).

Changes in PtdIns metabolism by environmental and/or internal stimuli, such as hormones, suggest that the components of the PtdIns cycle have some regulatory function in cellular signaling in plants. Several reports indicate that such signals as light, hormones, and elicitors are capable of generating inositol phosphates.

The effect of light on the PPI cycle is discussed in the next section; however, in the following we briefly mention that other factors have also been shown to operate in some situations via inositol phosphates.

There is some evidence that hormones may also act by changing levels of PtdIns. Auxins have been implicated in the control of $PtdInsP_2$ degradation. Morre *et al.* (1984) showed that the degradation of PtdIns was enhanced in the presence of 2,4-dichlorophenoxyacetic acid (2,4-D); however, the growth-inactive analogue 2,3-D did not produce any effect. Ettlinger and Lehle (1988) demonstrated that auxin can generate transient changes in $InsP_2$ and $InsP_3$ within minutes of hormone treatment in *Catharanthus roseus* cells accompanied by a redistribution in the polyphosphoinositide fraction. Murthy *et al.* (1989) found stimulation of [^{3}H]inositol incorporation in PtdIns within 30 sec of treatment of aleurone layers with giberellic acid (GA_3). Zocchi (1990) showed the effect of indoleacetic acid (IAA) and fussiccocin (FC) on the breakdown of PtdIns in maize coleoptiles by using [^{3}H]*myo*-inositol. Prelabeled coleoptiles treated with IAA showed a loss of radioactivity incorporated in the phospholipid fraction. The treatment by FC did not modify the content of labeled PtdIns with respect to control, suggesting two different mechanisms of action for IAA and FC. Although FC is known to mimic some of the physiological effects of IAA, IAA seems to induce hydrolysis of PtdIns in plant tissues, although FC does not.

Kurosaki *et al.* (1987) provided evidence for the generation of $InsP_3$ by stimulation of carrot cells with an elicitor, 6-methoxymellein. Thus, it seems that a wide variety of stimuli can stimulate the generation of $InsP_3$ and other second messengers by affecting the phosphoinositides.

3. LIGHT-STIMULATED INOSITOL TURNOVER

Light is an important environmental signal that directs many morphogenetic and developmental processes in the life cycle of a plant in addition to its requirement for photosynthesis. For regulating photomorphogenesis there are two different photoreceptors with different light-absorbing pigments that perceive light signals from blue light and UV and from red and far-red (the phytochrome). Of these phytochrome has been studied in detail (Mohr, 1972). It exists in two forms, inactive (Pr) and active (Pfr). The Pr form of the phytochrome is converted to its Pfr form by absorbing red light (λ_{max} 660 nm), and Pfr is reconverted to Pr by absorbing far-red light (λ_{max} 730 nm). The molecular mechanism by which Pfr regulates various metabolic and morphogenetic processes is still elusive. One important question is how Pfr tranduces signals to affect development and the nature of the different components involved in the light-signaling processes. Similar to red light, blue light also affects many photomorphogenetic processes. The exact nature of the blue light receptor is not known; however, recently evidence in favor of photolyase attached to flavin was obtained (Ahmed and Cashmore, 1993). Attempts have also been made to find if the PPI cycle is involved in blue light responses.

Indications that the phosphoinositide cycle may be involved in light-

Table I
Reports on the Effect of Light on Phosphoinositide Turnover in Different Species

System	Observations	Reference
Samanea saman	WL 5–30 sec: decrease in $PtdInsP_2$, PtdInsP; increase in $InsP_3$, $InsP_2$	Morse *et al.*, 1987b
Maize	Roots, RL 3 min: increase in $InsP_3$	Reddy *et al.*, 1987
	Leaves, RL 15 sec: increase in $PtdInsP_2$ (involvement of Pfr shown)	Guron *et al.*, 1992
Wheat	PLC more in dark	Melin *et al.*, 1987
Sunflower	WL 10 sec: decrease in $PtdInsP_2$	Memon and Boss, 1990
Ceratodon purpureus	Involvement of Pfr shown; decrease in $PtdInsP_2$; increase in $InsP_3$, $InsP_2$	Hartmann and Pfaffmann, 1990
Brassica oleracea	WL 50 sec: decrease in $PtdInsP_2$, PtdInsP; increase in $InsP_3$, $InsP_2$	Acharya *et al.*, 1991
	RL 5 min: increase in all species	Pal *et al.*, 1993

mediated responses were given in 1985 by Das and Sopory. They used 5-hydroxytryptamine (5-HT, serotonin), an activator of the PPI cycle in animals, and showed that it replaced the light requirement for phytochrome regulation of calcium fluxes in maize protoplasts. A summary of the work on the role of inositol phospholipids in light-mediated responses is given in Table I.

3.1. Changes in Inositol Lipids and Inositol Phosphates

By using [^{3}H]inositol for labeling inositol phosphates, Morse *et al.* (1987b) found that brief irradiation of *Samanea saman* pulvini with light (15 sec) decreased the level of PtdInsP and $PtdInsP_2$ with a concomitant increase of $InsP_3$ and $InsP_2$. They suggested that inositol phospholipid turnover may act as a phototransduction mechanism in *Samanea* pulvini in a manner similar to that reported in animal systems.

On a parallel line, Acharya *et al.* (1991) showed similar kinds of changes in *Brassica* cultures and hypocotyls. White light stimulus of 50-sec duration led to a decrease in PtdInsP and $PtdInsP_2$ levels with increases in $InsP_2$ and $InsP_3$ levels. However, the experiments were done with ^{32}P labeling and not by inositol labeling, and the separation was not done by HPLC as done by Morse *et al.* (1987b). Reddy *et al.* (1987) also reported a light-mediated increase in $InsP_3$ in maize roots that show a gravitropic response but are also sensitive to red light. Memon and Boss (1990) showed that the $PtdInsP_2$ level decreased if sunflower hypocotyls were irradiated with white light.

The above reports indicated that phototransduction may be mediated via inositol turnover. However because of a lack of defined wavelength in the experiments, the photoreceptor involved in these processes could not be ascertained. Evidence that the active form of phytochrome (Pfr) affects inositol turnover in plants came from the studies of Guron *et al.* (1992) in *Zea mays*. Experiments were done with both *myo*-[^{3}H]inositol and ^{32}P labeling. The level of PtdInsP$_2$ increased by three- to sixfold when leaves were irradiated with red light for even 15 sec as compared to unirradiated leaves. The level of PtdInsP and PtdIns did not change significantly. The level of PtdInsP$_2$ was found to decrease with an increase in red light irradiation, but a significant increase in PtdInsP$_2$ was still noticed up to 5 min of irradiation as compared to unirradiated leaves. Red light irradiation followed by far-red irradiation decreased the level of PtdInsP$_2$. Irradiation with far-red light had no effect. Because the level of PtdInsP did not change significantly as compared to changes in the level of PtdInsP$_2$, the control appeared to be at the PtdInsP kinase and phospholipase level. Based on these studies, Guron *et al.* (1992) suggested that one of the initial early biochemical steps involved in Pfr-mediated signal transduction is to affect the turnover of the PPI cycle. Whether Pfr affects any of the components of the PPI cycle directly is not clear as yet. However, the initial site at which Pfr affects PPI turnover seems to be at the PtdInsP kinase/phospholipase level. In *Brassica* hypocotyls, it was found (Pal *et al.*, 1993) that red light not only decreased the PtdInsP$_2$ level but also led to a decrease in InsP$_2$ and InsP$_3$.

3.2. Changes in Kinases/Phosphatases and Phospholipase C

Sandelius and Sommarin's studies on isolated plasma membranes of wheat seedlings showed that the membranes were enriched in PtdIns and PtdInsP kinase activity and can use endogenous as well as exogenous substrates (Sandelius and Sommarin, 1986). A direct effect of light on kinase/phosphatase and phospholipase C activity to regulate PPI turnover has not been reported in the literature. However, the report by Memon and Boss (1990) indicated that irradiation of sunflower hypocotyls with white light led to a 51% decrease in plasma membrane PtdInsP kinase activity. As little as 10 sec of white light was sufficient to lower the PtdInsP$_2$ produced in the *in vitro* phosphorylation assay. This decrease was not caused by an increase in phospholipase C activity, because analysis of the water-soluble products did not show any increase in InsP$_2$ or InsP$_3$. Treatment of the plasma membrane with 200 μM vanadate prior to phosphorylation enhanced the PtdInsP kinase activity and overcame the light inhibition. In addition to decreasing the PtdInsP kinase activity, light irradiation resulted in a corresponding decrease in H$^+$-ATPase activity as compared to the dark control. The plasma membrane ATPase activity increased approximately twofold when PtdInsP or PtdInsP$_2$ was added to the isolated membranes. Thus, the effect of external stimuli on the levels of plasma membrane ATPase activity is direct, which provides an alternative mechanism for the control of cell growth. Their results do not exclude the phosphoinositide pathway; however, they suggested an alternative pathway for the inositol phosphates in regulating plant growth and development.

Although different reports show the presence of phospholipase C activity in various tissues and its regulation by calcium, no report on direct effect of light on the phospholipase C is yet available. Melin *et al.* (1987) found that plasma membrane fractions from wheat seedlings were enriched in PtdInsP and $PtdInsP_2$ phospholipase C activity. Also, similar to the kinase activity, the specific activity of polyphosphoinositide phospholipase C was greater in shoots than in roots (Melin *et al.*, 1987), and the specific activity of plasma membrane enzyme from dark-grown shoots was higher than that from the light-grown shoots.

Hartmann and Pfaffmann (1990) have shown that PLC in plasma membrane of *Ceratodon purpureus* protonemal tip cells is activated by phytochrome. In fact the protonemal cells do show phototropic responses in red light. This also indicates that phospholipase C (PLC) can be activated by light.

From the results available, it seems that the initial site of action of light may differ in different plant species (Figure 1). It could either affect PLC or PtdInsP kinase. Our results on isolated plasma membranes from dark and red-light-irradiated tissue (Figure 2) from measuring enzyme activity on exogenous and endogenous substrates confirms our previous observation (Guron *et al.*, 1992) that the initial affect of Pfr is to modulate PPI turnover by affecting PtdInsP kinase activity (M. R. Chandok and S. K. Sopory, unpublished data). However,

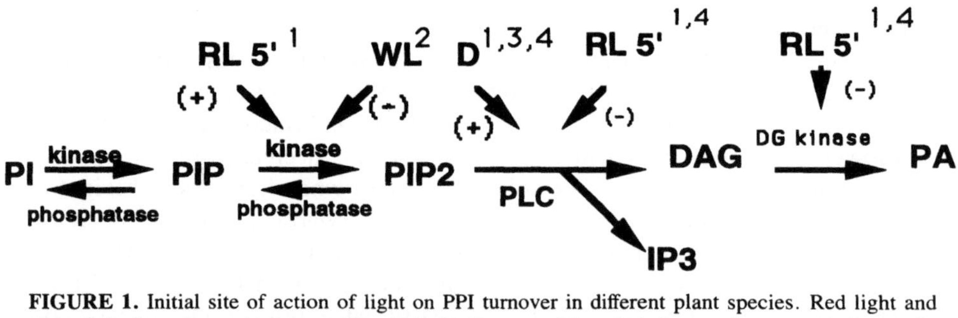

FIGURE 1. Initial site of action of light on PPI turnover in different plant species. Red light and white light have been shown to affect PtdInsP kinase activity: (+) denotes increase; (−) denotes decrease. In dark, PLC and DG kinase are more active. Red light (RL), besides increasing the activity of PtdInsP kinase, also leads to a decrease in PLC and DG kinase activity. Superscript denotes the reference cited. D, dark; RL, red light; WL, white light; PI, phosphatidylinositol; PIP, phosphatidylinositol phosphate; PIP_2, phosphatidylinositol bisphosphate; DAG, diacylglycerol; PA, phosphatidic acid; PLC, phospholipase C; IP_3, inositol trisphosphate. References: 1, Guron *et al.*, 1992; 2, Memon and Boss, 1990; 3, Melin *et al.*, 1987; 4, M. R. Chandok and S. K. Sopory, unpublished observations.

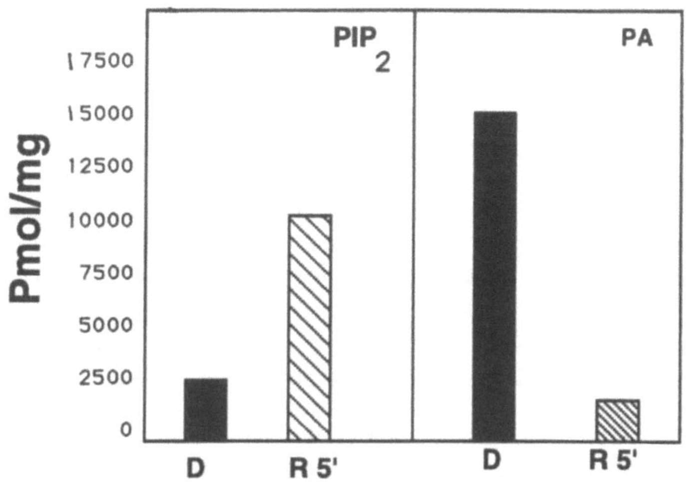

FIGURE 2. Effect of red light (5 min) on PtdInsP kinase and DAG kinase activity. Purified plasma membrane was taken as starting material. Assay was done using endogenous PtaInsP and DAG as substrates, and incorporation of labeled ^{32}P into products, $PtdInsP_2$ and PA, were monitored. Specific activity is expressed as picomole incorporated per milligram protein. A similar pattern was observed even with exogenous purified substrate from maize and rat brain (data not shown). D, dark; R 5′, red light 5 min; PIP_2, phosphatidylinositol bisphosphate; PA, phosphatidic acid.

it is not clear if the photoreceptor directly interacts with these enzymes or if some intermediates are involved.

3.3. Changes in $InsP_3$ and Calcium Level as a Consequence of Polyphosphoinositide Turnover

The two end products of the action of phospholipase C on $PtdInsP_2$ are DAG and $InsP_3$. Further, DAG affects activation of protein kinase C, and $InsP_3$ is involved in calcium mobilization from endogenous pools. As mentioned above, light has been shown to affect the levels of $InsP_3/InsP_2$ in a few plant systems, and it has been shown that an $InsP_3$-induced Ca^{2+} release mechanism exists in plants (Drøbak and Ferguson, 1985). Schumaker and Sze (1987) also provided evidence that $InsP_3$ but not $InsP_2$ or InsP caused a transient decrease of Ca^{2+} in tonoplast vesicles from oat roots. Similar findings have been reported in other systems (Cornelius and Nakashima, 1987; Ranjeva *et al.*, 1988; Rincon and Boss, 1987). Alexandre *et al.* (1990), using a patch-clamp technique, showed that a calcium current in the red beet root vacuole was induced by Ins(1,4,5)P_3. Neither Ins(1,4)P_2, Ins(1,3,4)P_3, Ins(2,4,5)P_3, nor $InsP_4$ was able to mimic the effect. Although reports of $InsP_3$-mediated Ca^{2+} release and light effects on $InsP_3$ levels exist, there is no strong evidence for a role of $InsP_3$ in signal transduction in light-mediated responses. However, in guard cells, which regulate opening and closing of stomata in a light-dependent (among other factors) manner, evidence for a role of $InsP_3$ in signaling in plants was provided by Blatt *et al.* (1990). They showed that guard cells are competent to use $InsP_3$ in signal cascades by controlling ion flux through K^+ channels. Later, Gilroy *et al.* (1990), by using a fluorescent calcium indicator, showed that calcium and $InsP_3$ may act as second messenger leading to guard cell closure in *Commelina communis*. The precise mechanism by which $InsP_3$ releases calcium from intracellular stores is not completely understood, and the presence of an $InsP_3$ receptor is not known.

3.4. Changes in DAG Level and Role of Protein Kinases

Another second messenger released through the breakdown of $PtdInsP_2$ by phospholipase C is DAG, which is the physiological activator of protein kinase C. Proteins with similar characteristics to PKC have been identified in *Neurospora crassa* (Favre and Turian, 1987) and some higher plants. The presence of a calcium- and lipid-dependent protein kinase was first reported by Schafer *et al.* (1985) in zucchini. Preliminary results suggesting that a protein immunologically related to PKC is present in plants has come from the analysis of partially purified extracts from *Amaranthus tricolor* (Elliot and Kokke, 1987; Elliot *et al.*, 1988). Western blot analysis using antibodies to bovine brain PKC gave bands suspected to be the intact PKC (84.5 kDa), a proteolytic product (65 kDa), and a regulatory fragment (40 kDa) (Elliot and Kokke, 1987). Based on these and other properties it was suggested that a PKC-type enzyme may be present in plants. Recently, a cDNA clone having homology to protein kinase C has been isolated from rice (Sasaki *et al.*, 1994). A kinase activity that was dependent on phorbol myristate acetate (PMA) has also been reported from a wheat cell homogenate (Olah and Kiss, 1986). A partially purified PMA-stimulated kinase activity was also reported in maize (Chandok and Sopory, 1992). This property further supports the presence of a PKC-type enzyme in plants.

In fact, in many systems PMA has been shown to affect physiological responses. The effect of phorbol esters on the coupling of the rhythm of oxygen emission to other biological clocks suggested that $InsP_3$ and DAG signaling systems are required for the expression of oxygen generation (Lonergan, 1990). In wheat, phytochrome regulates the cell volume in isolated protoplasts. In the same system, PMA was found to stimulate swelling in darkness or after red light irradiation. The red-light- or PMA-induced increase in protoplast volume was also inhibited by PKC inhibition (Bossen *et al.*, 1990). In maize, PMA stimulated nitrate-induced nitrate reductase (NR) activity in dark-grown leaves in a manner comparable to that obtained with red light (Chandok and Sopory, 1992; Sharma *et al.*, 1994). Although a PMA effect on some physiological processes has been shown, so far no conclusive evidence for PMA-stimulated protein kinase has been reported. In maize we have obtained a kinase activity that showed binding with labeled PMA (Chandok, 1993). Currently there are not enough reports to establish the presence of a PKC-type enzyme in plants as in animal systems. Nevertheless, it is possible that its properties may not be totally identical to those in an animal system.

Although more work needs to be done to show the involvement of DAG as a second messenger, changes in the levels of DAG have been observed in *Samanea saman* pulvini in response to white light irradiation (Morse *et al.*, 1989). Exposure to 30 sec of white light increased the DAG level to 19%, which declined to 90% of dark control values after 10 min of exposure to light. It was also found that the transient rise in DAG equaled the concurrent increase in $InsP_2$ and $InsP_3$, indicating that the changes in DAG level could be caused by changes in PPI turnover. However, it was not unequivocally shown whether changes in DAG in response to light were the result of PPI turnover or occurred through other phospholipids (PC, PE). In maize, we separated PtdIns from other phospholipids by TLC, and after formation of the methyl ester (with 5% H_2SO_4 and 95% methanol at 80°C for 3–4 hr), the fatty acids were analyzed by GLC on a diethylene glycol succinate column. This was done from leaves grown in darkness and from those exposed to red light for a short period (5 min). The fatty acid profile obtained in the two cases was different (M. R. Chandok and S. K. Sopory, 1995, unpublished results). Though mainly quantitative changes were observed, for a few fatty acids there were qualitative differences. For example, no oleic acid or myristic acid peak was observed in red-light-irradiated tissue. An increase in palmitic and linoleic acids was observed, whereas there was a decrease in linolenic acid (Table II). We have also gotten indications that different fatty acids can activate different kinases (M. R. Chandok and S. K. Sopory, unpublished data). These results point out that quantitative and qualitative

Table II
Effect of Red Light Irradiation on Fatty Acid Content of Phospholipids[a]

	Fatty acid (%)					
	12:0	14:0	16:0	18:1	18:2	18:3
D	15.6	0.5	8.5	5.2	27.5	42.4
R	15.9	0.0	12.0	0.0	36.3	35.7

[a]Red-light irradiation for 5 min was given to 8–9-day-old etiolated leaves, and tissues were frozen immediately. Purification of phosphatidylinositol was done. For fatty acid analysis, purified lipids were converted to methyl esters, and analysis was done on GLC. D, dark; R, red light.

changes induced by light could initiate different signaling pathways and may explain how single-light irradiation can lead to multiple responses in plants.

3.5. Is Polyphosphoinositide Turnover in Plants Coupled to G Protein?

In animal systems there are reports that the turnover of the PPI cycle is coupled to the activation of G proteins. The G_q type of G proteins activate phospholipase C and thereby affect the breakdown of $PtdInsP_2$ to release second messenger. In plant systems, the existence of G proteins has been shown by use of [^{35}S]GTPγS binding studies (Drøbak *et al.*, 1988; Hasunuma *et al.*, 1987a,b), ribosylation (Hasunuma and Funadera, 1987), antibodies (Blum *et al.*, 1988; Jacobs *et al.*, 1988), and purification of the protein (Bilush *et al.*, 1991).

There are also reports that show that the binding of G proteins with [^{35}S]GTPγS is affected by light. Blue light activated binding in pea (Warpeha *et al.*, 1991), and both red and far-red light inhibited binding in *Lemna* (Hasunuma *et al.*, 1987a). Hasunuma *et al.* (1987a,b) proposed that phytochrome may function in the same way as photoexcited rhodopsin in animals. Bossen *et al.* (1990) have shown that red-light-induced calcium-dependent swelling of etiolated wheat mesophyll protoplasts was nullified by an inhibitor of G proteins, GDPβS, whereas an activator of G proteins, GTPγS, induced swelling of the protoplasts to the same extent as obtained after red-light irradiation. Evidence for the possible involvement of G proteins in the phytochrome signal transduction chain has also come from the work of Romero *et al.* (1991a,b) and Bowler *et al.* (1994).

Clark *et al.* (1993) showed that monomeric G protein may be involved in phytochrome-mediated signal transduction chain in nuclei. Three GTP-binding proteins of molecular mass 27, 28, and 30 kDa were detected in nuclei of etiolated pea plumules. The level of [^{32}P]GTP binding of all the three protein bands increased in plumules irradiated by red light. Far-red light was able to reverse this effect, indicating that GTP-binding activity associated with the nuclear envelope is regulated by phytochrome.

Although the above reports indicate the presence and possible role of G protein in light-mediated signal transduction in plants, there is yet no direct report regarding the involvement of G proteins in the modulation of the PPI cycle.

Studies of Einspahr *et al.* (1989) on the activation of polyphosphoinositide-specific PLC from plasma membrane of *Dunaliella salina* by 100 μM GTPγS indicate a possible involvement of G protein in PPI turnover. Similarly, Dillenschneider *et al.* (1986) showed that guanine nucleotides could stimulate the release of inositol phosphates from [^{3}H]inositol-labeled membrane of *Acer pseudoplatanus*. However, in contrast to these results, McMurray and Irvine (1988) and Biffen and Hanke (1990) were unable to stimulate celery and soybean PLC by GTP or GTP analogues. Similar observations were reported by Melin *et al.* (1987), Tate *et al.* (1989), and Pical *et al.* (1992). Surely more work is required to find out if G proteins are an intermediate between photoreceptor and PPI cycle or if the two processes get triggered independently. As suggested by Coté and Crain (1993), reconstitution experiments with purified PLC and G protein will probably help in making the picture more clear.

4. RELATIONSHIP BETWEEN INOSITOL PHOSPHOLIPID TURNOVER AND PHYSIOLOGICAL RESPONSES

In the previous section, we learned that some data show that light can trigger changes in the components of the PPI cycle, although the exact site and mechanism of action are not clear. What is also not very obvious currently is whether signals generated by the PPI cycle play an important role in stimulus–response coupling.

The earlier studies of Fein and his co-workers with horseshoe crab (*Limulus,* an invertebrate) showed that phototransduction could be mediated by the turnover of inositol phospholipids (Fein *et al.*, 1984). Later, Corson and Fein (1987) showed that $InsP_3$ produced as a consequence of breakdown of $PtdInsP_2$ led to an increase in the intracellular free calcium concentration in *Limulus*. Therefore, the phototransduction process can be mediated via the PPI cycle in different systems. Does such a situation exist in plants too?

In order to understand the relationship between inositol phospholipid turnover and physiological or molecular responses, different workers have carried out studies in different ways. Generally, in systems that are sensitive to light, changes in phosphatidylinositol and inositol lipids have been monitored and then correlated with a specific response. In other studies, attempts were made to replace light by activators of G protein or of components of the PPI cycle or to inhibit the PPI cycle with compounds like lithium to establish a relationship between the signal and the response.

In the following we describe a few systems where such attempts have been made.

4.1. *Samanea* Leaf

The opening and closing of leaflets of *S. saman* are regulated by light (Satter *et al.*, 1974). It was shown that a rhythm phase shift in leaflet movement can be brought about by red or blue light (Simon *et al.*, 1976; Satter *et al.*, 1981). The mechanism of leaflet movement has been studied in detail, and it has been shown to be related to changes in membrane potential and to K^+ and H^+ fluxes in the leaf pulvinus, which includes flexor and extensor cells (H. Kim *et al.*, 1992). Using single cells prepared from both flexor and extensor tissue, H. Kim *et al.* (1992, 1993) showed that when K^+ channels are open in flexor cells, the channels in extensor cells are closed, and vice versa. Blue light opens an inward channel in extensor cells and closes a channel in the flexor cells.

Based on their work on PPI metabolism, mentioned in previous sections, Morse *et al.* (1990) proposed a phototransduction scheme involving inositol phospholipid turnover for leaflet movement. They had found that the turnover of the PPI cycle was faster: a 30-sec treatment led to an increase in $InsP_3/InsP_2$ accompanied by a decrease in $PtdInsP_2$. Whether these changes have a direct relationship with leaflet opening is not yet clear. However, it has been shown earlier and cited by Coté and Crain (1994) that a pulse of blue light led to the production of $InsP_3$ in flexor protoplasts while there was an inhibition in the release of $InsP_3$ in the extensor protoplasts. A negative relationship between $InsP_3$ levels and opening of K^+ channel has been reported in another system (Blatt *et al.*, 1990). On the basis of these studies, it can be argued that the production of IP_3 may be related to K^+ uptake, which in turn would affect the leaf movement in *S. saman*.

4.2. *Ceratodon purpureus*

The moss *Ceratodon purpureus* is another interesting system in which the possibility of inositol phospholipids as a phototransduction mechanism was explored. Hartmann and Jenkins (1984) have shown that phytochrome acts as photoreceptor for positive phototropism in the moss. When *C. purpureus* protonemal tips are unilaterally irradiated with red light, the tips grow toward the light source (show positive phototropism). This effect is blocked by far-red light (Hartmann and Weber, 1988).

Based on their studies with CTC fluorescence and an ionophore (monensin), Hartmann and Webber (1988) suggested that a steep calcium gradient at the tip of apical cell can be correlated with the processing of polar growth but can not carry information for the directional morphogenetic response. Based on the kinetics of normal and memorized phototropic response, they suggested that the photoreceptor itself does not store signal but that specific metabolites synthesized by Pfr may be important for the transduction of the signal.

To further characterize the nature of the components involved in the Pfr-mediated signal transduction chain, Hartmann and Pfaffmann (1990) explored the possibility of the involvement of inositol phospholipids. As mentioned earlier, they found that PtdIns-specific phospholipase C activity in this moss is controlled by phytochrome. A decrease in $PtdInsP_2$ and an increase in $InsP_2$ and $InsP_3$ levels indicated that inositol phospholipid turnover may be involved in signal transduction in the moss cells. It was further observed, cited in Hartmann and Pfaffmann, 1990) that monensin-coupled signal storage in moss protonemata was abolished by lithium. The lithium effect was more pronounced if it was present during and after the irradiation. When tested for its effect on inositol metabolism, the level of InsP increased after lithium treatment, indicating that the inhibition of inositol monophosphatase may be interrupting the regeneration of $PtdInsP_2$. Therefore, it is possible that phototransduction in *C. purpureus* may be utilizing the inositol phospholipids as one component of its signal chain.

A recent study by Algara *et al.* (1993) on sequence analysis of the phytochrome gene of *C. purpureus* revealed that the carboxyl domain of 300 amino acids of encoded protein has a homology to the catalytic domain of protein kinase. This protein seems to have an autophosphorylation activity that is modulated by light, indicating that Pfr may be directly linked to a phosphorylation event. It would be interesting to check if there is a direct link, if any, among Pfr, its autophosphorylating capacity, and inositol phospholipid turnover and if this interaction leads to phototropic response in *C. purpureus*.

4.3. Maize (*Zea mays*)

In 1985, Das and Sopory showed that in maize protoplasts 5-HT could stimulate calcium uptake and suggested the involvement of the PPI cycle in Pfr-mediated responses. Reddy *et al.* (1987) later suggested that gravitropic sensitivity in corn roots required turnover of inositol phospholipids. It had been shown earlier that red light is required for corn roots to become gravitropically sensitive (Roux and Slocum, 1982; Roux and Serlin, 1987; Perdue *et al.*, 1988), and earlier, an increase in calcium concentration was suggested as an initial event in this process (Poovaiah *et al.*, 1987; Roux and Serlin, 1987). However, the mechanism by which changes in calcium concentration occurred was not clear.

Reddy *et al.* (1987) tested the effect of light on the hydrolysis of phosphoinositides. Apical roots segments were preincubated in [^{3}H]inositol for 6 hr, and the level of InsP, $InsP_2$, and $InsP_3$ were measured after 3 min of exposure to light. The level of $InsP_3$ was higher in light-treated roots than in dark controls, suggesting that light could promote hydrolysis of $PtdInsP_2$, thereby producing $InsP_3$ and DAG (Poovaiah and Reddy, 1990a). In turn, $InsP_3$ could raise cytosolic calcium, thus initiating a cascade of biochemical events. Depletion of calcium from the medium by EGTA led to the loss of sensitivity to gravity, and restoration was achieved by calcium replenishment.

Support for the involvement of PPI cycle was also obtained from the use of 5-HT, which has been shown to promote $PtdInsP_2$ hydrolysis in blowfly (Litosh *et al.*, 1985), and 5-HT-treated roots also showed an increase in $InsP_3$ levels and gravitropic sensitivity (Reddy *et al.*, 1987; Poovaiah and Reddy, 1990a,b). Although these results do indicate a role of inositol phospholipid turnover in the gravitropism of corn roots, the possibility of a direct increase in calcium concen-

Table III
Reports on the Use of 5-Hydroxytryptamine to Replace Light Effect

System	Observations	References
Maize	5-HT replaces Pfr effect on Ca^{2+} uptake	Das and Sopory, 1985
	5-HT increases $PtdInsP_2$ level initially; longer treatment decreases; correspondingly, $InsP_3$ level increases	Chandok and Sopory, 1994
	5-HT replaces light requirement for nitrate reductase activity	
	In roots, light-mediated gravitropic response brought about by 5-HT, which also increases $InsP_3$ level	Reddy *et al.*, 1987 Poovaiah and Reddy, 1990a
Moss	5-HT affects bud formation	Bhatla, 1994

tration by red light cannot be ruled out. 5-Hydroxytryptamine has been shown to replace the light requirement in different processes (Table III).

In maize shoots, Guron *et al.* (1992) have shown, as mentioned earlier, changes in the inositol phospholipids by red light (5 min). In this system, 5-HT also affected PPI turnover in dark-grown etiolated maize leaves, and the overall changes monitored were similar to those obtained after red light irradiation (Chandok and Sopory, 1994a; Figure 3). The level of $PtdInsP_2$ increased in both red-light-irradiated and 5-HT-treated tissues. Therefore, in both cases the initial site of action could be at the PtdInsP kinase level. As noted by Poovaiah and Reddy (1990a), we also found an increase in $InsP_3$ in response to 5-HT treatment (Chandok and Sopory, 1994b).

Because 5-HT could affect the turnover of the PPI cycle in a manner comparable to red light irradiation, Chandok and Sopory (1994a) tested for a correlation between changes in PPI levels and a biochemical response [nitrate reductase (NR) gene expression]. In this system, NR is stimulated by light in the presence of nitrate. In the absence of light, NR is not produced. When tested, 5-HT was also found to mimic the effect of red-light irradiation to stimulate nitrate reductase activity (NRA) (Chandok and Sopory, 1994, 1995) as well as NR transcript (Raghuram and Sopory, 1995, and M. R. Chandok and S. K. Sopory, unpublished data; see Figure 3) in dark-grown leaves. This showed that 5-HT, which affected the PPI cycle, could replace the requirement for light for the stimulation of NRA.

Indirect support for the involvement of the PPI cycle in this light-mediated response was also obtained from the use of lithium. Lithium was found to block the increase in $PtdInsP_2$ in response to light, and it also blocked light-mediated NR gene expression. Further, as noted by Hartmann and Pfaffman (1990), we also found that the lithium effect was more pronounced if it was present immediately after red-light irradiation. If added after 1 hr of irradiation, its effect was much less (M. R. Chandok and S. K. Sopory, unpublished data). From the data

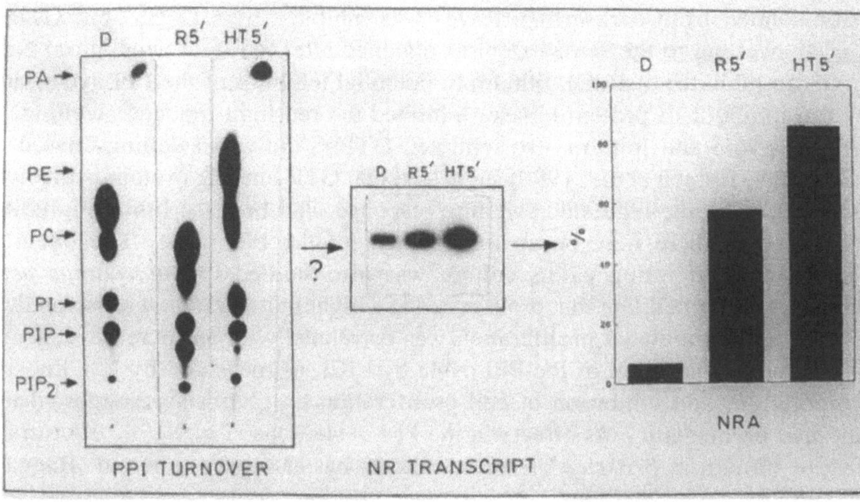

FIGURE 3. Replacement of light effect by 5-HT in maize leaves. 5-Hydroxytryptamine (5 min) could bring out similar changes in PPI turnover as were obtained with red light (5 min). A similar effect was observed on NR transcript and NRA level. The NR transcript was analyzed after 4 hr of nitrate treatment, and NRA assay was done after 6 hr of nitrate treatment. D, dark, R 5′, red light 5 min; 5-HT, 5-hydroxytryptamine; NR, nitrate reductase; NRA, nitrate reductase activity.

of Reddy *et al.* (1987) and our own, it seems that some of the Pfr-mediated responses in maize may involve the PPI cycle as one of the initial components of the signal transduction pathway. A further role of the second messengers $InsP_3$ and DAG is not clear; however, our results, as mentioned earlier, where the light effect could also be replaced by PMA (Chandok and Sopory, 1992), indicate that a DAG-mediated PKC pathway may be involved in this response.

Using [^{14}C]inositol or [^{14}C]acetate, Chicha *et al.* (1994) also showed the presence of PtdIns, PtdInsP, and $PtdInsP_2$ in coleoptiles and leaves of maize. The molecular species analysis of PtdIns and PtdInsP by radio-HPLC combined with gas chromatography of fatty acids showed that 16:0/18:2 fatty acids were more predominant in PtdIns and PtdInsP. In addition to these, other less abundant species were also identified in PtdIns and PtdInsP inositol lipids. Based on this observation, they suggested that PtdIns kinase has no clear selectivity with respect to various molecular species of PtdIns for the inositol lipid signaling pathway.

4.4. Tissue and Protoplast Culture

The role of G protein in Pfr-mediated protoplast swelling was investigated by Bossen *et al.* (1990). Red-light-induced swelling of wheat mesophyll protoplasts isolated from dark-grown leaves was inhibited by GDPβS, but GTPγS induced swelling to the same extent as obtained after red-light irradiation. Neomycin (an inhibitor of PLC), lithium (a potential inhibitor of the PPI cycle), and H-7 (an inhibitor of protein kinase) inhibited the red-light-induced swelling. In fact, neomycin and lithium also inhibited GTPγS-induced swelling. Based on their results, Bossen *et al.* (1990) suggested that GTP-binding protein is involved in the phytochrome-regulated swelling response, and because lithium inhibited this response, there may be an involvement of the PPI cycle. The effect of lithium on proliferating callus culture was also studied in *Amaranthus paniculatus*. A callus cell line that proliferated at a higher rate in light was isolated. In this, the light-stimulated proliferation was correlated with an increase in glyoxalase activity. Inhibition of the PPI cycle by LiCl, as measured by [^{14}C]inositol incorporation, and inhibition of cell proliferation by glyoxalase suggested that light may be mediating its effect via the PPI cycle (Das *et al.*, 1987). A similar effect of lithium in *Brassica oleracea* cultures has also been reported (Bagga *et al.*, 1987).

4.5. Bud Formation in *Funaria*

The effect of 5-HT and neomycin on bud formation in *Funaria* was also studied (see Bhatla, 1994). The 5-HT led to an increase in the number of caulonema cells, but buds that are formed did not develop normally, and reversion to filamentous growth was often observed. When PMA and A23187 were given together, a dramatic increase in the number and development of buds was observed in comparison to the situation where PMA and A23187 were given separately (cited in Bhatla, 1994).

5. ALTERNATE ROLES FOR INOSITOL PHOSPHOLIPIDS IN SIGNAL TRANSDUCTION

Recent studies indicate that besides acting as signal transducers, inositol phospholipids may be involved in other processes.

The membrane-labeling studies in carrot cells grown in suspension cultures showed that the plasma membrane is not only enriched in PtdInsP and PtdInsP$_2$ but also in lyso derivatives such as lyso-PtdInsP (LPIP) [but not lyso-PtdIns (LPI)] (Wheeler and Boss, 1987). The presence of LPIP suggests that there is a phospholipase A$_2$ activity associated with the plasma membrane (Scherer, 1990). The synthesis of LPIP is either by the hydrolysis of PtdInsP or by phosphorylation of LPI. The LPIP can play an important role in signal transduction by directly enhancing cation transport. Hayashi *et al.* (1978) have shown that lysophosphoinositides enhance K^+ transport in lipid bilayers and suggested that

$LPIP_2$ facilitated monovalent cation transport *in vivo* by forming an ion-conducting channel. Wheeler and Boss (1990) have found that LPIP can be positively correlated with the fusion-permissive state of plasma membrane. Blowers and Trewavas (1988) have proposed that LPIP can act as an anchor for protein kinase. Biosynthesis and subcellular localization of inositol lysophospholipids have been reviewed in detail by Wheeler and Boss (1990).

Polyphosphoinositides can also act as modulators of enzyme activity (Choquette *et al.*, 1984; Schafer *et al.*, 1987). Memon and Boss (1990) found that 10 sec of white light in sunflower led to decreased PtdIns and PtdInsP kinase activity. The decrease in phosphoinositide kinase activity was accompanied by a 50% decrease in H-ATPase activity. Exogenous addition of PtdInsP and $PtdInsP_2$ was able to restore the activity, indicating that phosphoinositides may directly affect the enzyme activity. Their results do not exclude the possibility of PPI turnover in this process. These data point out that besides being a source for the release of such messengers as $InsP_3$ and DAG, phosphoinositides or their lyso derivatives can directly affect other enzymes in the membranes or alter membrane permeability by affecting various ion channels.

6. CONCLUSIONS

In recent years it has become apparent that in order to control a large number of processes in plants, light probably triggers different signal transduction pathways. Involvement of calcium- and cyclic-GMP-mediated signaling pathways has been suggested. In this review we have tried to report and analyze the data so far available regarding the involvement of phosphoinositol phospholipids in light-mediated pathways.

From a number of papers published it is clear that the basic components of polyphosphoinositide cycle are present in plants, and in response to light irradiation, changes do occur in the levels of PtdIns, PtdInsP, and $PtdInsP_2$ and in the production of the second messengers $InsP_3$ and DAG. The mechanism by which photoreceptors bring about these changes and the site of primary action are not clear. In some reports, PtdInsP kinase has been shown to be the target enzyme, whereas in others it is PLC. It is also possible that in different systems the response may be differently mediated. Besides this, it is also not yet clear whether the effect of photoreceptor on the turnover of the PPI cycle is direct or indirect. There is no clear-cut evidence of mediation of G proteins in this activation, although such proteins are present. Recently Drøbak *et al.* (1994) have shown the possibility of interaction between PtdIns and PLC by the actin-binding protein profilin. This could be an intermediate step in involvement of the PPI cycle in cytoskeletal organization.

Another area where more work is needed is to show how the signal transduction occurs following $InsP_3$ and DAG release. There is no evidence of $InsP_3$ receptors in plants; however, the results of Biswas *et al.* reported in Chapter 10 may give an indication of the existence of these receptors. Also, there is no strong evidence in favor of DAG-dependent kinases (PKC types). Nevertheless, a number of reports and our own data (unpublished) do favor the existence of such a kinase, which may not be identical to animal PKC.

At present there are only a few systems where attempts have been made to correlate the initial changes in the PPI cycle induced by light with physiological or cellular responses. It has been shown that chemical activators of the PPI cycle such as 5-HT and activators of PKC such as PMA can bring about similar responses as induced by light. Further, neomycin and lithium, shown to inhibit PPI levels, were also shown to inhibit light-mediated responses. Although these data do indicate some role of inositol phospholipids, there is a lot more work needed to obtain direct proof.

It is our hope that in the future more workers interested in understanding light-mediated signal transduction will look for the possibility of involvement of PPI cycle in their systems. It would be of interest to purify the components of this pathway and make use of antisense technology to understand their role. Also, the analysis of photomorphogenetic mutants for defects in the PPI cycle may be rewarding.

ACKNOWLEDGMENTS. The research work of the authors cited in this chapter was supported by research grants from the Department of Biotechnology and Department of Science and Technology, Government of India, New Delhi.

7. REFERENCES

Acharya, M. K., Dureja-Munjal, I., and Mukherjee, S., 1991, Light-induced rapid changes in inositol phospholipids and phosphatidyl choline in *Brassica* seedlings, *Phytochemistry* **30:**2895–2897.

Ahmad, M., and Cashmore, A. R., 1993, HY4 gene of *A. thaliana* encodes a protein with characteristics of a blue-light photoreceptor, *Nature* **366:**162–164.

Alexandre, J., Lassalles, J. P., and Kado, R. T., 1990, Opening of Ca^{2+} channels in isolated red beet root vacuole membrane by inositol 1,4,5-triphosphate, *Nature* **343:**567–570.

Algarra, P., Linder, S., and Thummler, F., 1993, Biochemical evidence that phytochrome of the moss *Ceratodon purpureus* is a light regulated protein kinase, *FEBS Lett.* **315:**69–73.

Bagga, S., Das, R., and Sopory, S. K., 1987, Inhibition of cell proliferation and glyoxalase-I activity by calmodulin inhibitors and lithium in *Brassica oleracea, J. Plant Physiol.* **129:**149–153.

Berridge, M. J., 1987, Inositol trisphosphate and diacylglycerol: Two integrating second messengers, *Annu. Rev. Biochem.* **56:**159–193.

Berridge, M. J., and Irvine, R. F., 1984, Inositol triphosphate, a novel second messenger in cellular signal transduction, *Nature* **312:**315–321.

Berridge, M. J., Dawson, R.M.C., Downes, C. P., Heslop, J. P., and Irvine, R. F., 1983, Changes in the levels of inositol phosphates after agonist-dependent hydrolysis of membrane phosphoinositides, *Biochem. J.* **212:**473–482.

Bhatla, S. C., 1994, *Moss Protonema Differentiation,* Research Studies Press, Taunton, England.

Biffen, M., and Hanke, D., 1990, Polyphosphoinositide activity in soybean membrane is Ca^{2+} dependent and shows no requirement for guanine nucleotides, *Plant Sci.* **69:**147–155.

Biswas, B. B., Biswas, S., Chakrabarty, S., and De, B. P., 1978, A novel metabolic cycle involving *myo*-inositol phosphates during formation and germination of seeds, in *Cyclitols and Phosphosphoinositides* (W.W.F. Eisenberg, Jr., ed.), pp. 57–68, Academic Press, New York.

Bilush, S. V., Shebunin, A. G., and Babakov, A. V., 1991, Purification and subunit composition of a GTP-binding protein from maize root plasma membrane, *FEBS Lett.* **291:**219–221.

Blatt, M. R., Thiel, G., and Trentham, D. R., 1990, Reversible inactivation of K^+ channels of *Vicia* stomatal guard cells following the photolysis of caged inositol 1,4,5-trisphosphate, *Nature* **346:**766–768.

Blowers, D. P., and Trewavas, A. J., 1988, Phosphatidylinositol kinase activity of a plasma membrane-associated calcium-activated protein kinase from pea, *FEBS Lett.* **238:**87–89.

Blum, W., Hinsch, K. D., Schultz, G., and Weiler, E. W., 1988, Identification of GTP binding proteins in the plasma membrane of higher plants, *Biochem. Biophys. Res. Commun.* **156:**954–959.

Boss, W. F., and Massel, M. O., 1985, Polyphosphoinositides are present in plant tissue culture cells, *Biochem. Biophys. Res. Commun.* **132:**1018–1023.

Boss, W. F., and Moore, J. D., 1989, *Second Messengers in Plant Growth and Development,* Alan R. Liss, New York.

Bossen, M. G., Kendrick, R. E., and Vredenberg, W. J., 1990, The involvement of a G-protein in phytochrome-regulated Ca^{2+}-dependent swelling of etiolated wheat protoplasts, *Physiol. Plant.* **80:**55–62.

Bowler, C., Neuhaus, G., Yamagata, H., and Chua, N. H., 1994, Cyclic GMP and calcium mediated phytochrome phototransduction, *Cell* **77:**73–81.

Brown, E. F., and Newton, R. P., 1981, Cyclic AMP in higher plants, *Phytochemistry* **20:**2453–2463.

Chandok, M. R., 1993, *Molecular Nature of Signal Transduction in Phytochrome Mediated Stimulation of Nitrate Reductase in Maize: Involvement of Protein Kinase C and Phosphoinositides,* Ph.D. thesis, Jawaharlal Nehru University, New Delhi.

Chandok, M. R., and Sopory, S. K., 1992, Phorbol myristate acetate replaces phytochrome-mediated stimulation of nitrate reductase in maize, *Phytochemistry* **31:**2255–2258.

Chandok, M. R., and Sopory, S. K., 1994, 5-Hydroxytryptamine affects turnover of phosphoinositides and stimulates nitrate reductase in absence of light, *FEBS Lett.* 356:39–42.

Chandok, M. R., and Sopory, S. K., 1995, Signal response coupling in light mediated stimulation of enzymes, *Trans. Bose Res. Inst. Calcutta* (in press).

Chicha, A., Demandre, C., Justin, A., and Mazliak, P., 1994, The molecular species of phosphatidylinositol and phosphatidylinositol phosphate evidenced in the coleoptiles and first leaves of maize (*Zea mays* L.), *C. R. Acad. Sci. Paris* **317:**419–423.

Choquette, D., Hakim, G., Foloteo, A. G., Plisher, G. A., Bostwick, J. R., and Penniston, J. T., 1984, Regulation of plasma membrane Ca^{2+} ATPases by phosphatidylinositol cycle, *Biochem. Biophys. Res. Commun.* **125:**908–915.

Clark, G. B., Memon, A. R., Tong, C. G., Thompson, G. A., Jr., and Roux, S. J., 1993, Phytochrome regulates GTP-binding protein activity in the envelope of pea nuclei, *Plant J.* **4:**399–402.

Cornelius, G., and Nakashima, H., 1987, Vacuoles play a decisive role in calcium homeostasis in *Neurospora crassa, J. Gen. Microbiol.* **133:**2341–2347.

Corson, D. W., and Fein, A., 1987, Inositol 1,4,5-trisphosphate induces bursts of calcium release inside *Limulus* ventral photoreceptors, *Brain Res.* **423:**343–346.

Coté, G. G., and Crain, R. C., 1993, Biochemistry of phosphoinositides, *Annu. Rev. Plant Physiol. Plant Mol. Biol.* **44:**333–356.

Coté, G. G., and Crain, R. C., 1994, Why do plants have phosphoinositides? *Bioessays* **16:**39–46.

Das, R., and Sopory, S. K., 1985, Evidence of regulation of calcium uptake by phytochrome in maize protoplasts, *Biochem. Biophys. Res. Commun.* **128:**1455–1460.

Das, R., Bagga, S., and Sopory, S. K., 1987, Involvement of polyphosphoinositides, calmodulin and glyoxalase I in cell proliferation in callus cultures of *Amaranthus paniculatus, Plant Sci.* **53:**45–57.

Dillenschneider, A., Hetherington, A., Graziana, A., Alibert, G., Berta, P., Haiech, J., and Ranjeva, R., 1986, The formation of inositol phosphate derivatives by isolated membranes from *Acer pseudoplatanus* is stimulated by guanine nucleotides, *FEBS Lett.* **208:**413–417.

Drøbak, B. K., 1992, The plant phosphoinositide system, *Biochem. J.* **288:**697–712.

Drøbak, B. K., 1993, Plant phosphoinositides and intracellular signalling, *Plant Physiol.* **102:**705–709.

Drøbak, B. K., and Ferguson, I. B. 1985, Release of calcium from plant hypocotyl microsomes by inositol 1,4,5-trisphosphate, *Biochem. Biophys. Res. Commun.* **130:**1241–1246.

Drøbak, B. K., Allan, E. F., Comerford, J. G., Roberts, K., and Dawson, A. P., 1988, Presence of guanine nucleotide-binding proteins in plant hypocotyl microsomal fraction, *Biochem. Biophys. Res. Commun.* **150:**899–903.

Drøbak, B. K., Watkins, P.A.C., Valenta, R., Dove, S. K., Lloyd, C. W., and Straiger, C. J., 1994, Inhibition of plant plasma membrane phosphoinositide phospholipase C by the actin-binding, profilin, *Plant J.* **6:**389–400.

Einspahr, K. J., Peeler, T. C., and Thompson, G. A., Jr., 1989, Phosphatidylinositol 4,5-bisphosphate, phospholipase C, and phosphomonoesterase in *Dunaliella salina* membranes, *Plant Physiol.* **90:**1115–1120.

Elliot, D. C., and Kokke, Y. S., 1987, Partial purification and properties of a protein kinase C type enzyme from plants, *Phytochemistry* **26:**2929–2935.

Elliot, D. C., Fournier, A., and Kokke, Y. S., 1988, Phosphatidylserine activation of plant protein kinase C, *Phytochemistry* **27:**3725–3730.

Ettlinger, C., and Lehle, L., 1988, Auxin induces rapid changes in phosphatidylinositol metabolites, *Nature* **331:**176–178.

Favre, B., and Turian, G., 1987, Identification of a calcium and phospholipid dependent protein kinase (protein kinase C) in *Neurospora crassa, Plant Sci.* **49:**15–21.

Fein, A., Payne, R., Corson, D. W., Berridge, M. J., and Irvine, R. F., 1984, Photoreceptor excitation, adaptation by inositol 1,4,5-trisphosphate, *Nature* **311:**157–160.

Gilroy, S., Read, N. D., and Trewavas, A. J., 1990, Elevation of cytoplasmic calcium by caged calcium or caged inositol trisphosphate initiates stomatal closure, *Nature* **346:**768–770.

Guron, K., Chandok, M. R., and Sopory, S. K., 1992, Phytochrome-mediated rapid changes in the levels of phosphoinositides in etiolated leaves of *Zea mays, Photochem. Photobiol.* **56:**691–695.

Hartmann, E., and Jenkins, G., 1984, Photomorphogenesis of mosses and liverworts, in *The Experimental Biology of Bryophytes* (A. F. Dyer, ed.), pp. 203–228, Academic Press, London.

Hartmann, E., and Pfaffmann, H., 1990, Phosphatidylinositol and phytochrome-mediated phototropism of moss protonema tip cells, in *Inositol Metabolism in Plants* (D. J. Moore, W. F. Boss, and F. A. Loewus, eds.), pp. 259–275, Wiley–Liss, New York.

Hartmann, E., and Weber, M., 1988, Storage of phytochrome-mediated phototropic response of moss protonemal cells, *Planta* **175:**39–49.

Harwood, J. L., 1980, Lipids: Structure and function, in *The Biochemistry of plants,* Vol. 4 (P. K. Stumpf, ed.), pp. 1–55, Academic Press, New York.

Hasunuma, K., and Funadera, K., 1987, GTP-binding protein(s) in green plant, *Lemna paucicostata, Biochem. Biophys. Res. Commun.* **143:**908–912.

Hasunuma, K., Furaukawa, K., Funadera, K., Kubota, M., and Watanabe, M., 1987a, Partial characterization and light-induced regulation of GTP-binding proteins in *Lemna paucicostata, Photochem. Photobiol.* **46:**531–535.

Hasunuma, K., Furaukawa, K., Tomita, K., Mukai, C., and Nakamura, T., 1987b, GTP-binding proteins in etiolated epicotyls of *Pisum sativum* (Alaska) seedlings, *Biochem. Biophys. Res. Commun.* **148:**133–139.

Hayashi, F., Sokabe, M., Takagi, M., Hayashi, K., and Kishimoto, U., 1978, Calcium-sensitive univalent cation channel formed by lysotriphosphoinositide in bilayer lipid membranes, *Biochim. Biophys. Acta* **510:**305–315.

Helsper, J.P.F.G., DeGroot, P.F.M., Linskens, H. F., and Jackson, J. F., 1986, Phosphatidylinositol phospholipase C activity in pollen in *Lilium longiflorum, Biochem. J.* **25:**2053–2055.

Hepler, P. K., and Wayne, R. O., 1985, Calcium and plant development, *Annu. Rev. Plant Physiol.* **35:**397–439.

Irvine, R. F., Letcher, A. J., Lander, D. J., Drobak, B. K., Dawson, A. P., and Musgraue, A., 1989, Phosphatidylinositol(4,5)biophosphate and phosphatidylinositol(4)phosphate in plant tissues, *Plant Physiol.* **89:**888–892.

Jacobs, M., Thelen, M. P., Farndale, R. W., Astle, M. C., and Rubey, P. H., 1988, Specific guanine nucleotide binding by membranes from *Curcurbita pepo* seedlings, *Biochem. Biophys. Res. Commun.* **155:**1478–1484.

Kim, H. Y., Coté, G. G., and Crain, R. C., 1992, Effect of light on the membrane potential of protoplasts from *Samanea saman* pulvini: Involvement of K^+ and H^+-ATPase, *Plant Physiol.* **99:**1532–1539.

Kim, H. Y., Coté, G. G., and Crain, R. C., 1993, Potassium channels in *Samanea saman* protoplasts controlled by phytochrome and biological clock, *Science* **260:**960–962.

Kurosaki, F., Tsurusawa, Y., and Nishi, A., 1987, Breakdown of phosphatidyl inositol during elicitation of phytoalexin production in cultured cells, *Plant Physiol.* **85:**601–604.

Litosh, I., Wallis, C., and Fain, J. N., 1985, 5-Hydroxytryptamine stimulates inositol phosphate production in a cell-free system from blowfly salivary glands: Evidence for a role of GTP in coupling receptor activation to phosphoinositide breakdown, *J. Biol. Chem.* **260:**5464–5471.

Longeran, T. A., 1990, Steps linking the photosynthetic light reaction to the biological clock require calcium, *Plant Physiol.* **93:**110–115.

McMurray, W. C., and Irvine, R. F., 1988, Phosphatidyl-4,5-biphosphate phosphodiesterase in higher plants, *Biochem. J.* **249:**877–881.

Melin, P. M., Sommarin, M., Sandelius, A. S., and Jergil, B., 1987, Identification of Ca^{2+} stimulated polyphosphoinositide phospholipase C in isolated plant plasma membranes, *FEBS Lett.* **223:**87–91.

Memon, A. R., and Boss, W. F., 1990, Rapid light-induced changes in phosphoinositide kinases and H^+-ATPase in plasma membrane of sunflower hypocotyls, *J. Biol. Chem.* **265:**14817–14821.

Mohr, H., 1972, *Lectures on Photomorphogenesis,* Springer-Verlag, Berlin.

Morre, D. J., Gripshover, B., Monroe, A., and Morre, J. T., 1984, Phosphatidyl turnover in isolated soybean membranes stimulated by synthetic growth hormone, 2,4-dichlorophenoxyacetic acid, *J. Biol. Chem.* **259:**15364–15368.

Morse, M. J., Crain, R. C., and Satter, R. L., 1987a, Phosphatidylinositol cycle metabolites in *Samanea saman* pulvinus, *Plant Physiol.* **83:**640–644.

Morse, M. J., Crain, R. C., and Satter, R. L., 1987b, Light-stimulated inositol phospholipid turnover in *Samanea saman* leaf pulvini, *Proc. Natl. Acad. Sci. USA* **84:**7075–7078.

Morse, M. J., Satter, R. L., Crain, R. C., and Coté, G. G., 1989, Signal transduction and phosphatidylinositol turnover in plants, *Physiol. Plant.* **76:**118–121.

Morse, M. J., Crain, R. C., Coté, G. G., and Satter, R. L., 1990, Light-signal transduction via accelerated inositol phospholipid turnover in *Samanea pulvini,* in *Inositol Metabolism in Plants* (J. D. Morre, W. F. Boss, and F. A. Loewus, eds.), pp. 201–205, Wiley-Liss, New York.

Murthy, P.P.N., Renders, J. M., and Keranen, L. M., 1989, Phosphoinositides in barley aleurone layers and gibberellic acid induced changes in metabolism, *Plant Physiol.* **91:**1266–1269.

Newton, R. P., and Brown, E. G., 1986, The biochemistry and physiology of cyclic AMP in higher plants, in *Hormones, Receptors and Cellular Interactions* (C. M. Chadwick and D. R. Garrod, eds.), pp. 115–153. Cambridge University Press, Cambridge, England.

Nishizuka, Y., 1986, Studies and perspectives of protein kinase C, *Science* **233:**305–312.

Nishizuka, Y., 1988, The molecular heterogeneity of protein kinase C and its implications for cellular regulation, *Nature* **334:**661–665.

Olah, Z., and Kiss, Z., 1986, Occurrence of lipid and phorbol ester activated protein kinase in wheat cell, *FEBS Lett.* **195:**33–37.

Pal, S., Acharaya, M. K., and Mukherjee, S. G., 1993, Red light-mediated changes in inositol phospholipids and phosphatidylcholine in *Brassica* hypocotyls, *Phytochemistry* **32:**832–825.

Perdue, D. O., LaFaure, A. K., and Leopold, A. C., 1988, Calcium in the regulation of gravitropism by light, *Plant Physiol.* **86:**1276–1280.

Pfaffmann, H., Hartmann, E., Brightman, A. O., and Morre, D. J., 1987, Phosphatidylinositol specific phospholipase C of plant stems: Membrane associated activity concentrated in plasma membranes, *Plant Physiol.* **85:**1151–1155.

Pical, C., Sandelius, A. S., Melin, P. M., and Sommarin, M., 1992, Polyphosphoinositide phospholipase C in plasma membranes of wheat, *Plant Physiol.* **100:**1296–1303.

Poovaiah, B. W., McFadden, J. J., and Reddy, A.S.N., 1987, The role of calcium ions in gravity signal perception and transduction, *Physiol. Plant,* **71:**401–407.

Poovaiah, B. W., and Reddy, A.S.N., 1987, Calcium messenger system in plants, *CRC Crit. Rev. Plant Sci.* **6:**47–103.

Poovaiah, B. W., and Reddy, A.S.N., 1989, Calcium and root development: Importance of calcium in signal transduction, in *Plant Roots* (Y. Woisel, A. Eshel, and V. Kaffafi, eds.), pp. Marcel Dekker, New York.

Poovaiah, B. W., and Reddy, A.S.N., 1990a, Turnover of inositol phospholipids and calcium dependent protein phosphorylation in signal transduction, in *Inositol Metabolism in Plants* (D. J. Morre, W. F. Boss, and F. A. Loewus, eds.), pp. 335–350, Wiley–Liss, New York.

Poovaiah, B. W., and Reddy, A.S.N., 1990b, The role of calcium in signal transduction, in *Proceedings of the International Congress on Plant Physiology* (S. K. Sinha, P. V. Sane, S. C. Bhargava, and P. K. Agarwal, eds.), pp. 735–749, Society for Plant Physiology and Biochemistry, New Delhi.

Raghuram, N., and Sopory, S. K., 1995, Evidence for some common signal transduction events for opposite regulation of nitrate reductase and phytochrome-I gene expression by light, *Plant Mol. Biol.* **29:**25–35.

Ranjeva, R., and Boudet, A. M., 1989, *Signal Perceptions and Transduction in Higher Plants,* Springer-Verlag, New York.

Ranjeva, R., Carrasco, A., and Boudet, A. M., 1988, Inositol triphosphate stimulates the release of calcium from intact vacuoles isolated from *Acer* cells, *FEBS Lett.* **230:**137–141.

Reddy, A.S.N., McFadden, J. J., Friedmann, M., and Poovaiah, B. W., 1987, Signal transduction in plants: Evidence for the involvement of calcium and turnover of inositol phospholipids, *Biochem. Biophys. Res. Commun.* **149:**334–339.

Rincon, M., and Boss, W. F., 1987, *myo*-Inositol triphosphate mobilizes calcium from fusogenic carrot (*Daucus carota* L.) protoplasts, *Plant Physiol.* **83:**395–398.

Roberts, D. M., and Harmon, A. C., 1992, Calcium-modulated proteins: Targets of intracellular calcium signals in higher plants, *Annu. Rev. Plant Physiol. Plant Mol. Biol.* **43:**375–414.

Romero, L. C., Biswal, B., and Song, P. S., 1991a, Protein phosphorylation in isolated nuclei from etiolated *Avena* seedlings. Effects of red/far-red light and cholera toxin, *FEBS Lett.* **282:**347–380.

Romero, L. C., Sommer, D., Gotor, C., and Song, P. S., 1991b, G-protein in etiolated *Avena* seedlings. Possible phytochrome regulation, *FEBS Lett.* **282:**341–346.

Roux, S. J., and Serlin, B. S., 1987, Cellular mechanisms controlling light stimulated gravitropism: Role of calcium. *CRC Crit. Rev. Plant Sci.* **5:**205–236.

Roux, S. J., and Slocum, R. D., 1982, Role of calcium in mediating cellular functions important for growth and development in higher plants, in *Calcium and Cell Function,* Vol. III (W. Y. Cheung, ed.), pp. 409–453, Academic Press, New York.

Roux, S. J., Wayne, R. C., and Datta, N., 1986, Role of calcium ions in phytochrome responses: An update, *Physiol. Plant.* **66:**344–348.

Sandelius, A. S., and Sommarin, M., 1986, Phosphorylation of phosphatidylinositols in isolated plant membranes, *FEBS Lett.* **201:**282–286.

Sasaki, T., Song, J., Koga-Ban, Y., Matsui, E., Fang Fang, Higo, H., Nagasaki, H., Hori, M., Miya, M., Murayamakayano, E., Takiguchi, T., Takasuga, A., Niki, T., Ishimaru, K., Ikeda, H., Yamamoto, Y., Mukai, Y., Ohta, Isamu, Miyadera, N., Havukkala, I. and Minobe, Y., 1994, Toward cataloguing all rice genes: Large scale sequencing of randomly chosen rice cDNAs from a callus cDNA library, *Plant J.* **6:**615–624.

Satter, R. L., Geballe, G. T., Applewhite, P. B., and Glaston, A. W., 1974, Potassium flux and leaf movement in *Samanea saman.* I. Rhythmic movement, *J. Gen. Physiol.* **64:**413–430.

Satter, R. L., Guggino, S. E., Lonergan, T. A., and Galston, A. W., 1981, The effect of blue and far-red light on rhythmic leaflet movements in *Samanea* and *Albizzia, Plant Physiol.* **64:**965–968.

Schafer, A., Bygrave, F., Matzenauer, S., and Marme, D., 1985, Identification of a calcium and phospholipid-dependent protein kinase in plant tissue, *FEBS Lett.* **187:**25–28.

Schafer, M., Behele, G., Varsannyi, M., and Heilmeyer, L.M.G., Jr., 1987, Ca^{+2}-regulation of 1-(*sn*-phosphatidyl)-1D-*myo*-inositol-4-phosphate formation and hydrolysis on sarcoplasmic reticulum Ca^{+2}-transport ATPase: A new principle of phospholipid turnover regulation, *Biochem. J.* **249:**579–587.

Scherer, G., 1990, Phospholipid-activated protein kinase in plants: Coupled to phospholipase A_2? in *Signal Perception and Transduction in Higher Plants* (R. Ranjeva and A. M., Boudet, eds.) pp. 69–82, Springer-Verlag, New York.

Schumaker, K. S., and Sze, H., 1987, Inositol 1,4,5-triphosphate releases calcium from vacuolar membrane vesicles of oat roots, *J. Biol. Chem.* **262:**3944–3946.

Sharma, A. K., Raghuram, N., Chandok, M. R., Das, R., and Sopory, S. K., 1994, Characterization of phytochrome induced transmitter for the stimulation of nitrate reductase in etiolated leaves of maize, *J. Exp. Bot.* **45:**485–490.

Simon, E., Satter, R. L., and Galston, A. W., 1976, Circadian rhythmicity in excised *Samanea* pulvini II. Resetting the clock of phytochrome conversion, *Plant Physiol.* **58:**421–425.

Sommarin, M., and Sandelius, A. S., 1988, Phosphatidylinositol and phosphatidylinositol phosphate kinases in plant plasma membranes, *Biochim. Biophys. Acta* **958:**268–278.

Strasser, H., Hoffman, C., Grisebach, H., and Matern, U., 1986, Are polyphosphoinositides induced in signal transduction of elicitor-induced phytoalexin synthesis in cultured plant cells? *Z. Naturforsch.* **41:**717–724.

Tate, B. F., Schaller, G. E., Sussman, M. R., and Crain, R. C., 1989, Characterization of polyphosphoinositide phospholipase C from plasma membrane of *Avena sativa, Plant Physiol.* **91:**1275–1279.

Tretyn, A., Kendrick, R. E., and Wagner, G., 1991, The role(s) of calcium ions in phytochrome action, *Photochem. Photobiol.* **53:**1135–1156.

Wagh, S. S., and Natarajan, V., 1990, Phosphoinositide metabolism and calcium mobilization in plants, in *Proceedings of the International Congress on Plant Physiology* (S. K. Sinha, P. V. Sane, S. C. Bhargava, and P. K. Agarwal, eds.), pp. 680–687, Society for Plant Physiology and Biochemistry, New Delhi.

Warpeha, K.M.F., Hamm, H. E., Rasenick, M. M., and Kaufman, L. S., 1991, A blue-light-activated GTP-binding protein in the plasma membranes of etiolated peas. *Proc. Natl. Acad. Sci. USA* **88:**8925–8929.

Wheeler, J. J., and Boss, W. F., 1987, Polyphosphoinositides are present in plasma membranes isolated from fusogenic carrot cells, *Plant Physiol.* **85:**389–392.

Wheeler, J. J., and Boss, W. F., 1990, Inositol lysophospholipids in inositide metabolism in plants, in *Inositol Metabolism in Plants* (D. J. Morre, W. F. Boss, and F. A. Loewus, eds.), pp. 163–172, Wiley–Liss, New York.

Zocchi, G., 1990, Comparison of the effect of indoleacetic acid and fusicoccin on the breakdown of phosphatidylinositol in maize coleoptiles, *Plant Physiol.* **94:**1009–1011.

Chapter 13

Synthesis, Separation, and Identification of Different Inositol Phosphates

Carsten Schultz, Anne Burmester, and Christoph Stadler

1. INTRODUCTION

This review discusses recent progress in the synthesis of inositol phosphates and their analogues. Most of the research published prior to 1992 has been extensively reviewed (Potter and Nahorski, 1992; Potter, 1990; Billington, 1989), probably most completely in the book by Billington (1993). Therefore, mainly findings published since then and some of the most important earlier results are covered by this chapter. A brief discussion of the biochemistry of intracellular inositol phosphates in respect to cell signaling is presented, followed by some of the most crucial stereochemical aspects of inositol phosphates. The synthesis of naturally occurring *myo*-inositol phosphates and their synthetic derivatives is covered. In particular, the efforts to find agonists and antagonists of the most prominent inositol phosphate, *myo*-inositol-1,4,5-trisphosphate [Ins(1,4,5)P_3], are summarized, including some of their biochemical properties. Finally, some of the most widely used techniques for separation and identification of inositol phosphates are presented.

Carsten Schultz, Anne Burmester, and Christoph Stadler Institute for Organic Chemistry, Laboratory of Bioorganic Chemistry, University of Bremen, 28359 Bremen, Germany.
Subcellular Biochemistry, Volume 26: myo-Inositol Phosphates, Phosphoinositides, and Signal Transduction, edited by B. B. Biswas and Susweta Biswas. Plenum Press, New York, 1996.

1.1. Inositol Phosphate Turnover

The metabolism of inositol phosphates (Figure 1) gave rise to considerable research activity during the last decade (Shears, 1992). Attention was mainly attracted by Ins(1,4,5)P_3 because it is known to release calcium from internal stores (Berridge, 1993) via a specific Ins(1,4,5)P_3 receptor located in the endoplasmatic reticulum (Mikoshiba, 1993; Sharp *et al.*, 1992). Thus, Ins(1,4,5)P_3 was considered a second messenger (Berridge, 1983). Figure 1 illustrates the

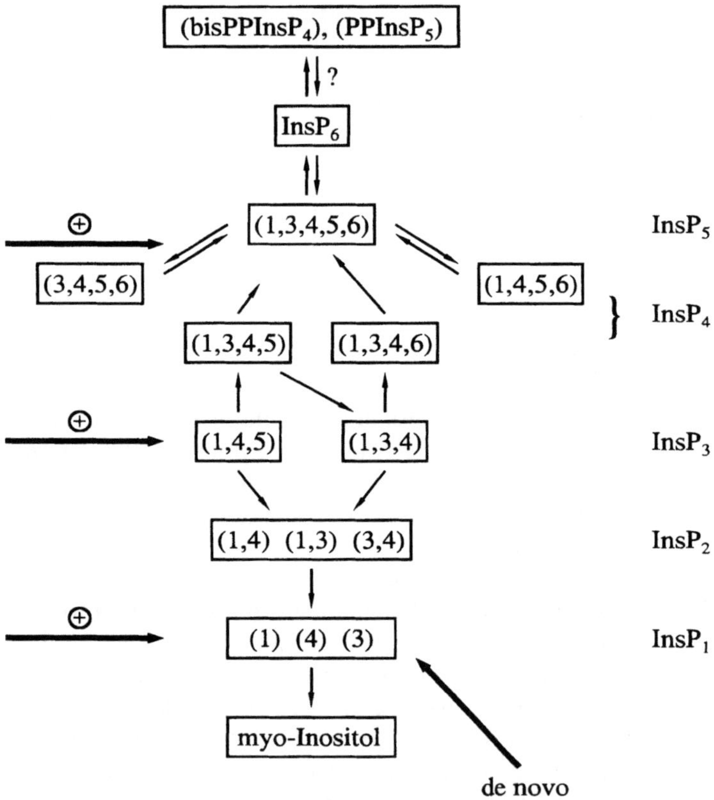

FIGURE 1. The metabolism of inositol phosphates. It appears that, apart from the *de novo* synthesis of inositol-3-phosphate [Ins(3)P] from D-glucose-6-phosphate, three agonist-triggered signaling pathways give rise to intracellular levels of inositol phosphates. Phospholipase-C-mediated hydrolysis of phosphatidylinositol (PI) and phosphatidylinositol-4,5-bisphosphate [PtdIns(4,5)P_2] elevates levels of *myo*-inositol-1-phosphate [Ins(1)P] and *myo*-inositol-1,4,5-trisphosphate [Ins(1,4,5)P_3], respectively. Additionally, the agonist-stimulated hydrolysis of *myo*-inositol-1,3,4,5,6-pentakisphosphate [Ins(1,3,4,5,6)P_5] leads to *myo*-inositol-3,4,5,6-tetrakisphosphate [Ins(3,4,5,6)P_4]. Numerous kinases and phosphatases are involved in the subsequent conversion of these inositol phosphates to give the complex mixture depicted.

generation of Ins(1,4,5)P_3 that derives from receptor-mediated breakdown of phosphatidylinositol-4,5-bisphosphate and its subsequent turnover. In a similar manner, the stimulation of cells by extracellular agonists results in the hydrolysis of phosphatidylinositol to give *myo*-inositol-1-phosphate [Ins(1)P] (Berridge, 1981). Additionally, by an independent but unknown mechanism, the hydrolysis of Ins(1,3,4,5,6)P_5 to Ins(3,4,5,6)P_4 is triggered (Menniti *et al.*, 1993a). The rise of intracellular levels of Ins(1,4,5)P_3 as well as the immediately subsequent elevation of the intracellular free calcium concentration ($[Ca^{2+}]_i$) is directly dependent on the rapid turnover of Ins(1,4,5)P_3, which terminates the signal. The metabolism of Ins(1,4,5)P_3 and Ins(1,3,4,5,6)P_5 creates the large variety of different inositol phosphates in living cells. Unfortunately, the functions of all but a few inositol phosphates still remain unclear.

The Ins(1,4,5)P_3 and all other inositol trisphosphates are eventually dephosphorylated to bis- and monophosphates and further to free *myo*-inositol. On a second pathway, Ins(1,4,5)P_3 is phosphorylated by means of a specific 3-kinase (Irvine *et al.*, 1986), yielding Ins(1,3,4,5)P_4 (Batty *et al.*, 1985). The Ins(1,3,4,5)P_4 has the reputation of being involved in regulating calcium signaling (Irvine, 1992; Tsien and Tsien, 1990). It is dephosphorylated by a 5-phosphatase (Erneux and Takazawa, 1991) to Ins(1,3,4)P_3, which is the precursor for Ins(1,3,4,6)P_4 (Balla *et al.*, 1987). The latter is the substrate for a 5-kinase to generate Ins(1,3,4,5,6)P_5 (Menniti *et al.*, 1993a), which is probably the major precursor for Ins(3,4,5,6)P_4 (Menniti *et al.*, 1990) and Ins(1,4,5,6)P_4 (Oliver *et al.*, 1992). The dephosphorylation to Ins(3,4,5,6)P_4 appears to be a result of agonist-stimulated modulation of the activity of the involved phosphatase (Menniti *et al.*, 1993a). Very little is known about even more highly phosphorylated inositols. Inositol hexakisphosphate is present in respectable amounts in all mammalian cells (Menniti *et al.*, 1993a), but there are different opinions with regard to its precursor (Menniti *et al.*, 1993a; Shears, 1992; Downes and MacPhee, 1990). For some time it has been believed that cells synthesize inositol pyrophosphates from their designated precursors, Ins(1,3,4,5,6)P_5 and InsP_6 (Menitti *et al.*, 1993b).

However, apart from Ins(1,4,5)P_3 with its known role in calcium homeostasis (Berridge, 1993; Gill *et al.*, 1989), functions for the higher phosphorylated inositols are widely discussed with regard to the various inositol tetrakisphosphates (Irvine, 1992) and completely unknown for InsP_5, InsP_6, and inositol pyrophosphates (Menniti *et al.*, 1993a). Because there is evidence of multiple molecular forms of the Ins(1,4,5)P_3 receptor (Maranto, 1994; Mikoshiba, 1993; Ferris and Snyder, 1992), it becomes comprehensible that the influence of Ins(1,3,4,5)P_4 on the calcium signal might differ strongly with the cellular system investigated. It has been discussed that to some degree Ins(1,3,4,5)P_4 is involved in mobilizing calcium from internal stores in conjunction with Ins(1,4,5)P_3 (Cullen *et al.*, 1990; Changya *et al.*, 1989; Irvine and Moor, 1986).

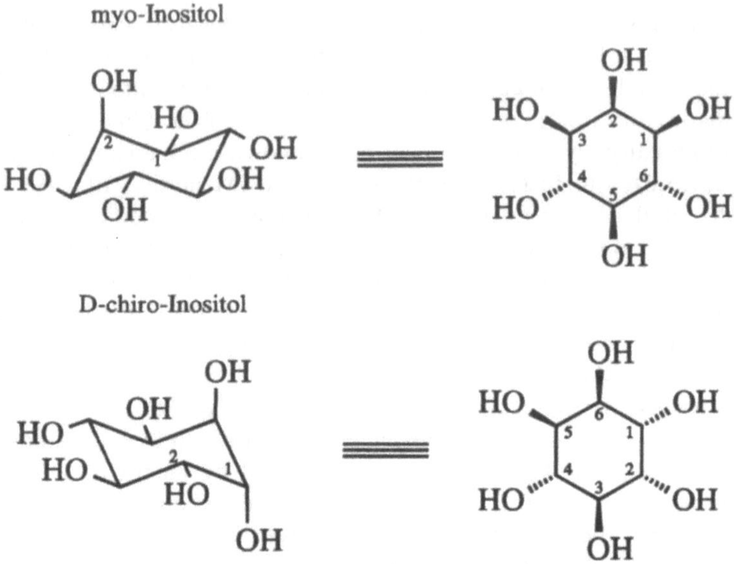

FIGURE 2. The structures of *myo-*, D- and L-*chiro-*, and *scyllo*-inositol in their pseudo-three-dimensional chair conformations and the corresponding two-dimensional projections used throughout the chapter.

Controversially, Balla *et al.* (1991) showed that in fibroblasts elevated levels of Ins(1,3,4,5)P_4 significantly reduced calcium signaling. Furthermore, it was shown that in a few cases Ins(1,3,4,5)P_4 is able to exhibit a distinct second messenger function independently (Ely *et al.*, 1990) or via the Ins(1,4,5)P_3 receptor (Wilcox *et al.*, 1993). Very recent studies revealed that other InsP_4 isomers, such as Ins(1,4,5,6)P_4 and especially Ins(3,4,5,6)P_4, seem to be involved in the modulation of cellular responses after receptor stimulation (Hashii *et al.*, 1994; Vajanaphanich *et al.*, 1994; Kachintorn *et al.*, 1993).

1.2. The Structure of *myo*-Inositol

The structures depicted in this review are presented in a two-dimensional projection rather than in the more complicated pseudo-three-dimensional form (Figure 2). The two-dimensional projection is a from-top view onto the molecule in which all hydroxyl groups, except the one in the 2-position, are roughly positioned in the plane of the paper. Actually, the hydroxyl groups in positions 1, 3, and 5 are somewhat directed toward, and the hydroxyl groups in positions 4 and 6 away from, the observer, each at an angle of 30° from the plane of the paper. The single axial hydroxyl group, by convention in position 2, is positioned vertically toward the observer. A plane of symmetry runs through C-2 and C-5 of

L-chiro-Inositol

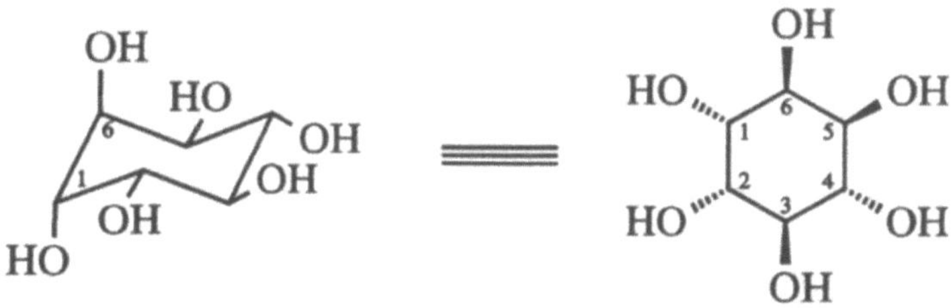

scyllo-Inositol

FIGURE 2. *Continued*

myo-inositol. If there is no element of symmetry, each of the enantiomers is understood as D-configured (Figure 2), according to IUPAC recommendations (IUPAC, 1989). This is indicated by means of a counterclockwise numbering of the ring carbons. Therefore, throughout this chapter, the prefix D is left out for *myo*-inositol derivatives, though D and L are used for the not so common *chiro*-inositols (Figure 2). Racemates are marked by the prefix DL.

Of the nine possible stereoisomeric inositols, the *myo* form is the most common. This review deals almost exclusively with its derivatives, though two slightly differing structures are of interest for the inositol phosphate analogue section: *chiro*-inositol and *scyllo*-inositol (Figure 2). The only optically active isomer, *chiro*-inositol, has two axial hydroxyl groups in positions 1 and 2 (D), respectively, positions 1 and 6 (L), whereas *scyllo*-inositol carries exclusively equatorial hydroxyl groups. More details on this topic are covered in the review by Parthasarathy and Eisenberg (1986).

2. SYNTHETIC INOSITOL PHOSPHATES

Since the characterization of Ins(1,4,5)P_3 as a second messenger in 1983 (Streb *et al.*), the synthesis of inositols and inositol phosphates has experienced an enormous renaissance. The interest of chemists was further fueled by the rapid identification of the metabolism of Ins(1,4,5)P_3 (Shears, 1992; Erneux and Takazawa, 1991) and more recently Ins(1,3,4,5,6)P_5 (Menniti *et al.*, 1993a), which was shown to lead to the great number of different inositol phosphates now known (for an example, see Radenberg *et al.*, 1989). Their degree of phosphorylation varies from monophosphates to InsP_6, and was most recently extended to even higher phosphorylated inositol pyrophosphates such as 1-diphosphoinositol pentakisphosphate (PPInsP_5) and bis-1,4-diphosphoinositol tetrakisphosphate (bisPPInsP_4) (Stephens *et al.*, 1993).

Inositol phosphates can be isolated from natural sources (Irvine *et al.*, 1985) as well as prepared by semisynthetic procedures from the corresponding phospholipids (Bird *et al.*, 1989) or enzymatic hydrolysis of phytic acid (Grado and Ballou, 1961). Nevertheless, in recent years fully synthetic procedures became very common.

A synthetic approach toward inositol phosphates usually follows a strategic sequence as depicted in Figure 3. It should be noted that general synthetic considerations are presented in great detail in Billington's *The Inositol Phosphates* (1993). In brief, the design of a synthetic pathway is determined by the choice of the starting material. The majority of procedures have started from the symmetric *myo*-inositol, but numerous routes have been described proceeding from chiral starting materials such as galactinol (Ballou and Pizer, 1960), L-quebrachitol, D-pinitol (Tegge and Ballou, 1989), (−)-quinic acid (Falck and Abdali, 1993; Falck and Yadagiri, 1989), or D-glucose (Estevez and Prestwich, 1991a), just to name a few. When *myo*-inositol is chosen as a starting material, the cyclitol has to be selectively protected, meaning that hydroxyl groups have to be spared from protection where the introduction of the phosphates is planned. This procedure usually involves several protecting/deprotecting steps. In its course, a preferentially single hydroxyl group opens the possibility of separating the chiral antipodes as diastereomeric esters. Most frequently, the chiral groups are camphanic acid esters (Aneja and Parra, 1994; Baker *et al.*, 1990), menthoxyacetates (Aguilo *et al.*, 1992), orthoesters of D-mannose (Stepanov and Shvets, 1979), or tartrates (Watanabe *et al.*, 1991), of which the first approach appears to be most popular (Billington, 1993). More recently, such enzymes as cholesterol esterase and several lipases have been employed for the regio- and stereoselective hydrolysis or formation of acetates (Ozaki *et al.*, 1994; Andersch and Schneider, 1993; Ling *et al.*, 1992; Liu and Chen, 1989). Because ideally only one of the two enantiomers is a substrate for the enzyme, chromatographic separation of the products is usually more facile than for diastereomers.

Beginning the reaction sequence with a chiral starting material avoids the resolution of enantiomers. Somewhere along the synthetic pathway, the derivative has to be modified to the *myo*-inositol configuration. For example, derivatives of quebrachitol or pinitol possess the *chiro*-inositol configuration, which can be converted to the *myo*-inositol structure by the selective inversion of a particular hydroxyl group (Tegge and Ballou, 1989).

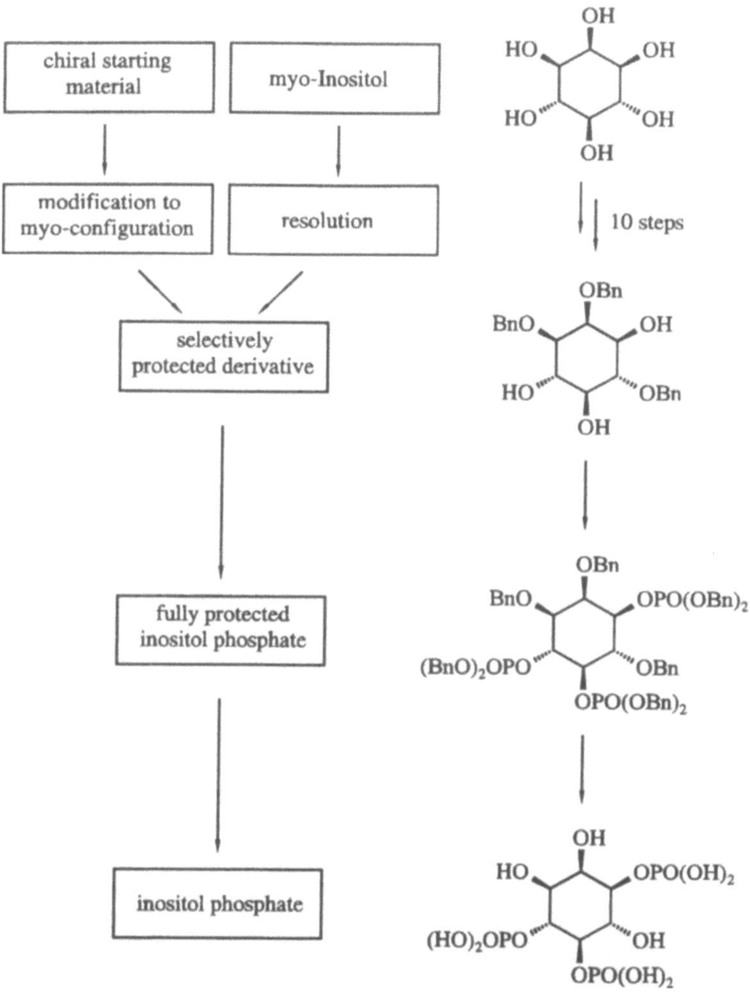

FIGURE 3. Schematic presentation of a general synthetic approach to inositol phosphates. Next to the scheme is one of the classic synthetic pathways from *myo*-inositol to *myo*-inositol trisphosphate. Though the step-by-step procedures varied, several groups described a similar reaction sequence, including some of the intermediates shown. For the many references on the synthesis of Ins(1,4,5)P_3, see text.

Both procedures, from *myo*-inositol or chiral starting material, eventually result in the preparation of the desired selectively protected *myo*-inositol derivative (Figure 3). The protection group left on the molecule will in most cases be a benzyl ether or a ketal. These are preferred over others for their stability under phosphorylating conditions and are quantitatively removable at the end of the synthetic pathway. Quite frequently, though, the use of other protecting groups including esters was reported (Billington, 1993).

The next problem to tackle is the introduction of the phosphate group(s). The choice of the phosphorylating reagent is mainly determined by the type of protecting groups abundant in the inositol derivative and the question of whether vicinal hydroxyl groups are intended to be phosphorylated. Vicinal hydroxyl groups do not permit the use of classical phosphorus (V) reagents such as diphenylchlorophosphate because they tend to form cyclic phosphates as a result of the required activation by strong nucleophiles such as pyridine or imidazole (Eckstein and Gindl, 1968). The frequently used and now commercially available tetrabenzyl pyrophosphate must be considered an exception. No formation of cyclic phosphates has been detected, probably because such strong bases as butyl lithium, sodium hydride, or potassium hydride are required to produce the alkoxide anions. The alternative to phosphorus (V) reagents are the more reactive phosphorus amidites [P(III)], originally developed for DNA synthesis (Sinha *et al.*, 1984). They usually contain chloro- or dialkylamino leaving groups (or both) and give excellent yields without cyclic phosphate formation. The intermediate inositol phosphite triesters are easily oxidized by peroxides to yield the fully protected inositol phosphate (Figure 3). This methodology also allows the synthesis of phosphorothioate analogues if the oxidation of the phosphites is carried out with sulfur (Cooke *et al.*, 1987b) or bis(phenylacetyl)thioate (Hansbro *et al.*, 1994) instead of peroxide.

The subsequent cleavage of phosphate- and hydroxyl protecting groups yields the desired inositol phosphate (Figure 3). As mentioned earlier, the deprotection might be a single step if the protection relies exclusively on benzyl groups. These can be quantitatively removed by catalytic hydrogenation (palladium on charcoal) without migration of the phosphates. For phosphorothioates, the use of a metal catalyst is prohibited. In this case, 2-cyanoethyl groups on the phosphorothioate allow easy deprotection under basic conditions, preferentially with sodium in liquid ammonia at $-78°C$ (Cooke *et al.*, 1987b).

2.1. Synthesis of Natural and Unnatural Inositol Phosphates

The theoretical number of different D-*myo*-inositol phosphates is 63, not counting cyclic phosphates and pyrophosphates. About one-third of them have been detected in tissue (Radenberg *et al.*, 1989), and most of the inositol phosphates found are listed in Figure 1. The amount of each inositol phosphate is small; concentrations are mostly in the submicromolar range, except for Ins(1,3,4,5,6)P_5 and InsP_6. Therefore, efficient methods for the synthetic preparation of inositol phosphates were in high demand when the biological relevance of Ins(1,4,5)P_3 was discovered in 1983. To this date, only inositol monophosphates (Kiely *et al.*, 1974; Géro *et al.*, 1972; Gigg and Warren, 1969; Mercier *et al.*, 1969; Mercier and Géro, 1968; Kurihara *et al.*, 1967; Géro, 1966) and bisphosphates (Shvets *et al.*, 1973; Angyal and Tate, 1961) had been prepared, but soon numerous procedures were reported for the synthesis of Ins(1,4,5)P_3 (Aguilo *et al.*, 1992; Salamonczyk and Pietrusiewicz, 1991; Ley *et al.*, 1990; Liu and Chen, 1989; Tegge and Ballou, 1989; Yu and Fraser-Reid, 1988; Cooke *et al.*, 1987a; Reese and Ward, 1987; Ozaki *et al.*, 1986), and lately even novel synthetic pathways have been added (Ling and Ozaki, 1994). In the meantime, all of the known natural inositol phosphates believed to be part of the inositol phosphate metabolism, including Ins(1,3,4,5)P_4 (Watanabe *et al.*, 1990; Baudin *et al.*, 1988) and Ins(1,3,4,6)P_4 (Watanabe *et al.*, 1989), were synthesized, and their preparation, if published in 1992 or earlier, has been extensively reviewed (Billington, 1993). More recently, the syntheses of Ins(3,4,5,6)P_4 and Ins(1,4,5,6)P_4 by conventional methodology (Pietrusiewicz *et al.*, 1992) and chemoenzymatic procedures (Ling and Ozaki, 1994) were reported.

Because of the large number of different inositol phosphates needed for biochemical evaluations, e.g., of the inositol phosphate metabolism, synthetic approaches are desirable that allow the preparation of not only one but a number of different inositol phosphates. For this purpose, a multifunctionally modified precursor is required. The inositol derivative should ideally carry a distinct protecting group on each of the six hydroxyl groups. The protecting groups should be chosen in a way that each of them can be cleaved separately. An approach close to this, offering considerable versatility, was reported by Falck and Abdali (1993). The synthetic sequence, starting from (−)-quinic acid, resulted in a *myo*-inositol derivative with four different protecting groups, and one free hydroxyl group, which was successfully converted to Ins(1,3,4,5)P_4 and Ins(3,4,5)P_3.

Apart from the naturally occurring inositol phosphates, several unnatural isomers have been prepared by total synthesis (see below) or purified from a partial hydrolysate of phytic acid (for a recent example, see DeLisle *et al.*, 1994). The term "unnatural inositol phosphate" refers to an inositol phosphate isomer that has not (yet) been found in cells or whose appearance in a particular cell type has to be considered unusual at the moment, including L-enantiomers of inositol phosphates usually found in D-configuration [like Ins(3,5,6)P_3, the L-enantiomer of D-Ins(1,4,5)P_3]. Both enantiomers have often been prepared together because the resolution of the respective precursor normally gives way for both antipodes. Ins(3,5,6)P_3 was shown to be practically inactive in releasing Ca^{2+} from internal stores (see Table I). Ins(2,4,5)P_3, a moderately potent agonist for elevating intracellular Ca^{2+} levels, was synthesized by Tegge *et al.* (1991), together with its enantiomer Ins(2,5,6)P_3 and the enantiomers of *chiro*-Ins(1,3,4)P_3. A derivative with very interesting features was introduced recently: the synthetic Ins(1,4,6)P_3 (Watanabe *et al.*, 1994). As might be deduced from Table 1, Ins(1,4,6)P_3 was shown to be a good inhibitor of the 3-kinase from rat

Table I
Binding Data of Inositol Polyphosphates and Their Analogues to the Ins(1,4,5)P_3 Receptor, the 5-Phosphatase, and the 3-Kinase, as Well as Their Potency to Release Ca^{2+} from Intracellular Stores

Ins(1,4,5)P_3 Analogue	IP_3-Receptor binding [$IC_{50}(K_i)$] [nM], tissue[a]	Ca^{2+}-Release EC_{50} [nM], cell type[a]	5-Phosphatase binding (K_i) [μM], tissue[a]	3-Kinase binding [$IC_{50}(K_i)$], [μM] tissue[a]	Reference[b]
Ins(1,4,5)P_3	61 pcm (pH 7.6)	110 SH-SY5Y			12
	6.3 (K_d) bacm	100 132IN1	32 (K_m) heg	1.5 (K_m) rbc	24
		52 SH-SY5Y			13
	6.5 (K_i) bacm	180 SH-SY5Y			21
	56 rcm				1
		300 basm	12 (K_m) bac	2.6 bac	15
	15 rcm	200 macro	14 heg	1.0 (K_i) rbc	8
	1.4 bacm		57 bc		8
	10 (K_i) rcm	50 reconst. rc receptors			16
	23 rcm	170 rcm			22
		1,700 gppa			7
	40 rcm	120 SH-SY5Y			5
	6.2 rcm	~500 RBL[c]	16 heg		9
		170 RBL			23
	55 rcm	210 Swiss 3T3			28
Ins(1:2cyclo 4,5)P_3	150 rcm	2,600 macro	38 heg (s)[d]		8
	240 rcm	3,600 Swiss 3T3			28
D-Ins(3,5,6)P_3/L-Ins(1,4,5)P_3		22,000 basm	39 bac	292 bac	15
			124 heg (i)[d]		3
	100,000 rcm				27
		>2,000,000 RBL			23
Gro*P*InsP_2		5,500 gppa			7
	1,800 rcm				1
	3,000 rcm	2,200 Swiss 3T3			28

DL-Ins(2,4,5)P_3		8,700 basm	130 bac (s)	1,868 bac	15
	500 rcm	9,800 macro	23 heg (s)		8
	45 bacm		66 bc (s)		8
	86 rcm	2,800 Swiss 3T3			28
Ins(2,4,5)P_3		4,300 RBL			23
Ins(2,5,6)P_3		110,000 RBL			23
DL-Ins(1,3,4)P_3		120,000 basm	32 bac (s)	1,847 bac	15
Ins(1,4,6)P_3	98 rcm	~63,000 RBL[c]	9.2 heg (s)	3.0[c,e] (i) rbc	9
Ins(1,3,6)P_3	550 rcm	>300,000 RBL	27.4 heg (s)	>30 rbc	9
Ins(1,3,5)P_3		630,000 basm	45 bac	1,687 bac	15
DL-Ins(1,3,4,5)P_4	1,100 rcm	2,800 SH-SY5Y			5
		65,000 basm	1.4 bac (s)	90 bac	15
Ins(1,3,4,5)P_4	21,000 rcm				1
			3.9 heg (s)		13
		880 HS-SY5Y			26
Ins(1,3,4,6)P_4	410 rcm				5
	3,400 rcm	2000 SH-SY5Y			1
					25
DL-Ins(1,4,5,6)P_4	8,500 rcm				5
	35,000 rcm				1
DL-Ins(1,2,4,5)P_4	26 (K_i) bacm	170 SH-SY5Y	2.9 heg (i)	100 (K_i) rbc	17
DL-3,6-Bz_2-Ins(1,2,4,5)P_4	>3,000 (K_i) bacm	>100,000 SH-SY5Y	16 heg (i)	>300 (K_i) rbc	13
scyllo-Ins(1,2,4,5)P_4	14 bac	77 SH-SY5Y			25
Ins(1,3,4,5,6)P_5	>10,000 rcm				1
InsP_6	36,000 rcm				1
DL-Ins(1,4,5)P_3S_3	2,000 rcm				27
		2,500 SH-SY5Y	1.7 heg (i)	230 (K_i) rbc (i)	17
			6.0 heg (i)		3
Ins(1,4,5)P_3S_3	130 rcm				1
6-Deoxy-Ins(1,4,5)P_3S_3	5,300 (K_i) bacm	16,000 SH-SY5Y			21

(*continued*)

Table I (*Continued*)

$Ins(1,4,5)P_3$ Analogue	IP_3-Receptor binding [$IC_{50}(K_i)$] [nM], tissue[a]	Ca^{2+}-Release EC_{50} [nM], cell type[a]	5-Phosphatase binding (K_i) [μM], tissue[a]	3-Kinase binding [$IC_{50}(K_i)$], [μM] tissue[a]	Reference[b]
DL-Ins(1)P(4,5)P_2S_2			1.3		14
DL-Ins(1,4)P_2(5)PS		2,500 SH-SY5Y	6.8 heg (i)	5.0 (K_i) rbc (s)	17
Ins(1,4)P_2 5-methylphosphonate		Proposed antagonist (hp)			4
DL-1-(3-Aminopropyl-1-phospho)-Ins(4,5)P_2	300 (K_i) rcm	420 reconst. rc receptor	No substrate	No substrate	16
D-*chiro*-Ins(1,3,4)P_3		4,200 RBL			23
L-*chiro*-Ins(1,4,6)P_3			6.4 psm (i)	Little effect	6
	>1,000 (K_i) rcm, bacm	> 30,000 132 IN1			24
L-*chiro*-Ins(1,4,6)P_3S_3		>30,000 hn	0.67 psm (i)	No effect	6
L-*chiro*-Ins(2,3,5)P_3S_3	500 (K_i) bacm	5,300 SH-SY5Y			21
L-*chiro*-Ins(2,3,5)P_3	160 (K_i) bacm	1,000			24
	160 (K_i) rcm	1321N1			
	60 (K_i) bacm	1,400 SH-SY5Y	7.1 (aff)[f] heg (i)	7.7 (aff)[f] heg (i)	19
3-Deoxy-Ins(1,4,5)P_3	23 bacm	160 SH-SY5Y			11
3-Deoxy-3-fluoro-Ins(1,4,5)P_3	13 (K_i) bacm	280 1321N1			24
	21 (K_i) rcm				
	8 bacm	370 SH-SY5Y	8.6 (aff)[f] heg (s)	3.9 (aff)[f] heg	19
	13 (K_i) bacm	120 SH-SY5Y			11
3-Azido-3-deoxy-Ins(2,4,5)P_3	19,000 (K_i) bag	Weak agonist			10
3-Amino-3-deoxy-Ins(1,4,5)P_3	5,700 pcm	1,500 SH-SY5Y			12
DL-2,2-Difluoro-Ins(1,4,5)P_3	38 (K_1) bacm	350 1321 N1			24
	58 (K_i) rcm				
2,2-Difluoro-Ins(1,4,5)P_3		210 SH-SY5Y	60 (K_1) heg (s)	6.6 (K_i) (s)	20
2,2-Difluoro-Ins(3,5,6)P_3		53,000 SH-SY5Y	19 (K_i) heg (i)	12 (K_i)	20
DL-2-Fluoro-*scyllo*-Ins(1,4,5)P_3	100 (K_i) bacm	1,200 1321N1			24
	58 (K_i) rcm				

DL-2-Deoxy-2-fluoro-Ins(1,4,5)P_3	26 bacm	100 SH-SY5Y			25
DL-*scyllo*-Ins(1,2,4)P_3	52 bacm	170 SH-SY5Y			25
DL-2-Deoxy-Ins(1,4,5)P_3	55 rcm	500 macro	13 heg (s)	1.6 (K_i) rbc	8
2,3-Dideoxy-Ins(1,4,5)P_3	40 (K_i) bacm	190 SH-SY5Y			11
DL-2-*p*-Azidobenzoyl-Ins(1,4,5)P_3	95 rcm	900 macro	2.1 heg (s)		8
	5 bacm		16 bc		
DL-2-(4-Aminocyclohexanecarboxylic acid)-Ins(1,4,5)P_3	44 rcm 3 bacm	1,200 macro	16 heg, 53 bc	2.4 (K_i) rbc	8
DL-6-Methyl-Ins(1,4,5)P_3		65,000 basm	52 bac	319 bac	15
6-Deoxy-Ins(1,4,5)P_3	3,800,000 (K_i) rcm	6,400 SH-SY5Y			24
			76 heg (i)	5.7 (K_i) rbc (s)	18
6-Deoxy-Ins(1,4,5)P_3S_3	5,300 (K_i) bacm	16,000 SH-SY5Y			21
2,3,6-Trideoxy-InsP_3		40,000 basm	137 bac	327 bac	15
	3,900 (K_i) bacm	>10,000 SH-SY5Y			11
β-D-Mannopyranosyl-(1)-methylphosphonate(4,6)P_2				26 (K_i) clone from rb	2
Adenophostin A + B	0.18 (K_i) rcm	1.4 rcmi	Resistant	Resistant	22
	1.3 rcm	55 NG108-15			
	0.95 jpm[g]				
Heparin	17 mg/ml rcm				1

[a]Cell types or tissue used as source of enzymes are abbreviated as follows: pcm, porcine cerebellar membranes; SH-SY5Y, neuroblastoma cell line; bacm, bovine adrenal cortical membranes; 1321N1, human astrocytoma cell line; rb, rat brain homogenate; heg, human erythrocyte ghosts; rbc, rat brain cytosol; rcm, rat cerebellar membranes; basm, bovine aortic smooth muscle cells; bac, bovine aorta cytosol; macro, guinea pig peritoneal macrophages; bc, bovine cortex; reconst. rc, reconstituted rat cerebellum receptors; gppa, guinea pig parotid acinar cells; RBL, rat basophilic leukemia cells; Swiss 3T3, cell line; bag, bovine adrenal gland P_2 membranes; hp, human platelets; psm, porcine skeletal muscle; jpm, jurkhat plasma membranes; rcmi, rat cortex microsomes.

[b]References: 1. Challiss *et al.*, 1991; 2. Chung and Moon, 1994; 3. Cooke *et al.*, 1989; 4. Dreef *et al.*, 1991b; 5. Gawler *et al.*, 1991; 6. Hansbro *et al.*, 1994; 7. Henne *et al.*, 1988; 8. Hirata *et al.*, 1989; 9. Hirata *et al.*, 1993; 10. Kozikowski *et al.*, 1992; 11. Kozikowski *et al.*, 1993; 12. Kozikowski *et al.*, 1994a; 13. Mills *et al.*, 1993; 14. Noble *et al.*, 1992b; 15. Polokoff *et al.*, 1988; 16. Preswich *et al.*, 1988; 17. Safrany *et al.*, 1991a; 18. Safrany *et al.*, 1991b; 19. Safrany *et al.*, 1992a; 20. Safrany *et al.*, 1992b; 21. Safrany *et al.*, 1993; 22. Takahashi *et al.*, 1994; 23. Tegge *et al.*, 1991; 24. Wilcox *et al.*, 1992; 25. Wilcox *et al.*, 1993; 26. Wilcox *et al.*, 1994; 27. Willcocks *et al.*, 1988; 28. Willcocks *et al.*, 1989.

[c]Values are estimated from figures in ref. 9.

[d](s) Refers to substrate, (i) refers to inhibitor of the enzyme.

[e]At a calcium concentration <0.01 M.

[f]Binding affinity.

[g]Receptor protein located in the plasma membrane.

brain cytosol and also inhibits the 5-phosphatase, though Ins(1,4,6)P_3 was only slowly dephosphorylated by the latter. The ability to release Ca^{2+} from internal stores of saponin-permeabilized RBL cells was 1/100 that of Ins(1,4,5)P_3 (Hirata *et al.*, 1993). Therefore, Ins(1,4,6)P_3 appears to be one of the better candidates to dissect the Ins(1,4,5)P_3 response (Ca^{2+} release) from the potential effects of other inositol phosphates. The synthesis of racemic Ins(1,2,4,5)P_4 (Mills *et al.*, 1993; Carless and Busia, 1990) yielded a potent Ca^{2+}-mobilizing agonist that represented a good inhibitor of the 5-phosphatase but only poorly affected the 3-kinase (Wilcox *et al.*, 1994). DeLisle *et al.* (1994) investigated 47 highly purified positional isomers from a phytic acid lysate and compared their potency in releasing intracellular Ca^{2+} after microinjection into *Xenopus* oocytes.

The synthetic derivatives and isomers just mentioned might all be considered analogues of Ins(1,4,5)P_3 or Ins(1,3,4,5)P_4. The basic chemical tools to prepare a wider variety of these biochemically interesting compounds as well as the results of some of their biological evaluations are covered in the following section.

2.2. Synthesis of Inositol Phosphate Analogues

Because most of the known natural and many unnatural inositol phosphates have been synthesized, chemists concentrate more and more on the preparation of inositol phosphate analogues. These analogues can be divided into two groups, mostly serving two different intentions: (1) analogues with predominantly functional modifications to provide tools for biomedical research and (2) analogues to map the respective binding proteins, e.g., the Ins(1,4,5)P_3 receptor or those proteins involved in inositol phosphate turnover.

The functional analogues include caged Ins(1,4,5)P_3 (Walker *et al.*, 1989), tethered Ins(1,4,5)P_3 and Ins(1,3,4,5)P_4 derivatives to prepare affinity columns (Ozaki *et al.*, 1992; Tegge and Ballou, 1992; Estevez and Prestwich, 1991a), and derivatives equipped with photolabels (Tegge and Ballou, 1992; Estevez and Prestwich, 1991b; Schäfer *et al.*, 1990). Furthermore, the search continues for inhibitors of the various kinases and phosphatases involved in inositol phosphate metabolism (see below) and for potent agonists and antagonists for those inositol phosphates directly regulating intracellular Ca^{2+} homeostasis like the inositol tris- and tetrakisphosphates. Inhibitors, agonists, and antagonists are of potential interest for future drug development and could in principle derive from screening experiments, though the great majority of derivatives prepared so far relied on the inositol structure.

Analogues for mapping inositol phosphate binding proteins require the systematic modification of one of the functional groups, namely, the hydroxyl and phosphate moieties, to alter particular sites of interaction with the binding protein(s) of interest. As described below, large sets of analogues have been prepared for mapping the inositol monophosphatase (IMPase) and the major Ins(1,4,5)P_3 binding proteins, the Ins(1,4,5)P_3 receptor, the 5-phosphatase, and the 3-kinase. Other proteins involved in inositol phosphate turnover still await closer investigation.

A description of the enormous number of synthetic pathways leading to inositol phosphate analogues (Billington, 1993) would by far exceed the scope of this chapter. Nevertheless, the following will sum up some of the principal synthetic approaches to introduce modifications into inositol phosphates: (1) modification of hydroxyl groups, (2) modification of the phosphate moieties, and (3) alterations of the basic cyclohexane skeleton, including exocyclic additions. All of the possibilities listed have been used to synthesize inositol phosphate analogues, quite often in combination.

2.2.1. Replacing Hydroxyl Groups

At first glance, each of the hydroxyl groups in a given inositol phosphate has the same potency to interact with a target protein. The simplest approach to find out which of the hydroxyl groups is actually involved in binding or biochemical activity appears to be a systematic replacement of each of them by a hydrogen. In the case of Ins(1,4,5)P_3, this leads to a set of the three deoxy analogues, 2-, 3-, and 6-deoxy-Ins(1,4,5)P_3. All of them have been synthesized (Kozikowski *et al.*, 1994b; Ley *et al.*, 1990; Seewald *et al.*, 1990; Hirata *et al.*, 1989; for their biochemical properties, see below and Table I). In general, the preparation from inositol required a single unprotected hydroxyl group, which was replaced by hydrogen in a Barton deoxygenation (Seewald *et al.*, 1990) or by reductive dehalogenation with tributyltin hydride (Baker *et al.*, 1989). Nucleophilic substitution of a hydroxyl group either with iodide or by replacement of its trifluoromethyl sulfonic acid ester opens the way to a large number of different functional groups. With these procedures, deoxyazido (Kozikowski *et al.*, 1992) and then deoxyamino derivatives of inositol (Guédat *et al.*, 1994; Kozikowski *et al.*, 1994b) have been synthesized. Furthermore, inositols carrying fluoro (Marecek and Prestwich, 1989), chloro (Lampe and Potter, 1993), and mercapto modifications (Guidot and Le Gall, 1993a,b) were prepared. It should be noted that the nucleophilic substitution of a secondary hydroxyl group, at least under the conditions mentioned, results in an inversion from an axial to an equatorial group or vice versa. From all but the deoxymercaptoinositols, different phosphates have been prepared, resulting in a large toolbox for mapping the respective binding proteins.

Another possibility to introduce modifications of hydroxyl groups is the selective opening of epoxides. By this method, deoxy, fluoro, deoxymethyl, and methyl analogues of Ins(1,4,5)P_3 have been synthesized (Ley *et al.*, 1990). Finally, simple alkylation of the respective hydroxyl group yields analogues of increased lipophilicity. This reaction does not alter the stereochemistry, and the acceptor potential for hydrogen bonds of the former hydroxyl group remains intact.

2.2.2. Modification of the Phosphates

Most modifications of the phosphate group are aimed at reducing the electrophilicity of the phosphorus and in that way decreasing the rate of (enzymatic) hydrolysis. The most often used replacements were alkyl groups to give phosphonates instead of the acidic hydroxyl group or the replacement of an oxygen by sulfur to give phosphorothioates. Other modifications included the formation of phosphate diesters and the chemically more demanding replacement of the bridging oxygen to the inositol ring by a CH_2 group. The replacement of all three phosphates of Ins(1,4,5)P_3 by sulfates, sulfamates, methylphosphonates, or carboxymethyl groups yielded completely inactive analogues of Ins(1,4,5)P_3 (Westerduin *et al.*, 1992).

2.2.3. Inositol Phosphate Analogues with Altered Carbon Backbone

Two approaches in the synthesis of inositol phosphate analogues involve the alteration of the six-membered carbon ring structure. The first represents an exocyclic extention by a CH_2-group (for an early example see Paulsen and Röben, 1983), mainly for the synthesis of C-phosphonates as analogues with increased stability against hydrolysis (Kozikowski *et al.*, 1993; Falck *et al.*, 1990; Kulagowski, 1989). Most recently, the synthesis of 3-C-trifluoromethyl-Ins(1,4,5)P_3 was reported, in which the CF_3-group is placed vertically to the molecule plane. The compound was only slightly less active in releasing Ca^{2+} than Ins(1,4,5)P_3 itself (Kozikowski *et al.*, 1994b). The second approach is based on the use of pyranoses instead of *myo*-inositol as a starting material. Because *myo*-inositol is biosynthesized *de novo* from D-glucose, it seems obvious that several of the chiral centers possess the same configuration as in *myo*-inositol. Depending on whether the axial 2-hydroxyl group of a given inositol phosphate is intended to be mimicked, other hexoses such as mannose or galactose might serve as starting materials as well.

As a first example, Chung and Moon (1994) prepared an Ins(3,5,6)P_3 analogue based on D-mannose. Biochemical evaluations showed no potency to release Ca^{2+}, but the analogue was a potent inhibitor of the 3-kinase. The recently discovered adenophostins from *Penicillium breviocompactum,* the most potent agonists for intracellular Ca^{2+} release known to date, also rely on a glucose-based structure (M. Takahashi *et al.*, 1994; S. Takahashi *et al.*, 1994; see below and Figure 6).

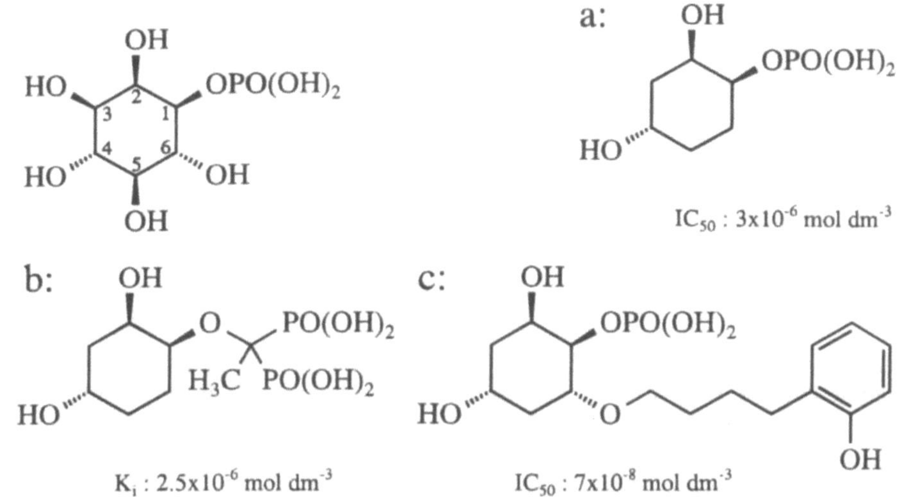

FIGURE 4. Structures of *myo*-inositol-1-phosphate [Ins(1)P] and the most potent inhibitors of the inositol monophosphatase (IMPase) known to date, including the apparent K_i or IC_{50} values, respectively.

2.3. Inositol Monophosphate Analogues as Inhibitors of the Inositol Monophosphatase

A target of particular pharmacological interest is the *myo*-inositol monophosphatase (IMPase). This enzyme, responsible for the dephosphorylation of inositol monophosphates, is believed to be a target for lithium in the treatment of manic depression (Nahorski *et al.*, 1991; Drummond, 1987). Lithium inhibits a variety of inositol mono- and polyphosphatases in an uncompetitive manner; analogues of Ins(1)P are aimed to block the enzyme(s) competitively. For potent inhibition, the analogue should be optimized for binding properties and stability against hydrolysis.

The synthesis of a first group of deoxy analogues of Ins(1)P bearing deletions of one or more hydroxyl functions revealed that the hydroxyl groups in position 2 and, to a lesser extent, in position 4 are essential for binding to the enzyme, whereas the 6-hydroxyl group is required for enzymatic hydrolysis (Russell *et al.*, 1992; Baker *et al.*, 1989). Hydroxyl groups 3 and 5 did not seem to be strongly involved in interactions with the enzyme. These results led eventually to the synthesis of 3,5,6-trideoxy-Ins(1)P (Figure 4a), a potent inhibitor of the IMPase with an IC_{50} of 3 μM (Baker *et al.*, 1990). The structural requirements listed above are also met in part by 2′-AMP, a substrate for the IMPase (Leeson *et al.*, 1993; Attwood *et al.*, 1988), implying that bulky groups could be

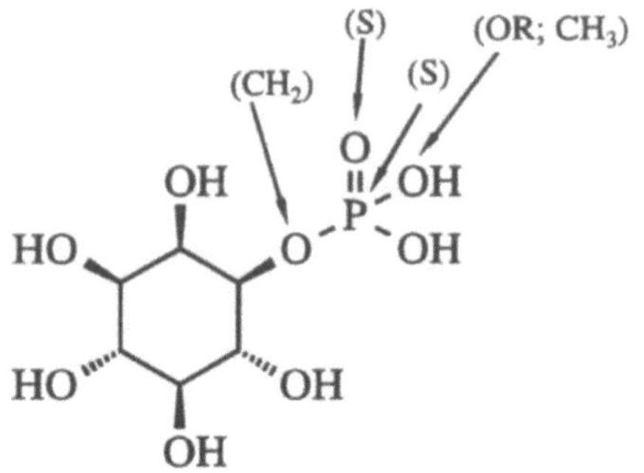

FIGURE 5. Modifications of the phosphate moiety of *myo*-inositol-1-phosphate.

tolerated in the 6-position. Consequently, the synthesis of several 6-alkyl and 6-alkoxy derivatives of 3,5,6-trideoxy-Ins(1)P yielded very potent inhibitors, of which the D/L-4-(2-hydroxyphenyl)butoxy analogue (Figure 4c) had an IC_{50} of 70 nM (Baker *et al.*, 1991).

A rational approach toward the modification of the phosphate faces five atoms/functional groups as shown in Figure 5. Replacement of the oxygen bridging to the inositol ring by a CH_2-group (Johnson *et al.*, 1993), which should ensure stability against hydrolysis, led to an analogue without inhibitory effect (Kulagowski *et al.*, 1991). The phosphorothioate analogue showed little inhibition and was a substrate for the enzyme, although the 1-sulfate showed no binding to the IMPase (Metschies, 1990). Replacement of one of the hydroxyl groups by a methyl (Dreef *et al.*, 1991b; Falck *et al.*, 1990) or alkoxyl group (Schultz *et al.*, 1990) was reported. Tests with the methylphosphonate were not published, but the racemic phosphate diesters were shown to be very weak inhibitors with a K_i of 3.6 mM for the methyl derivative and somewhat higher values for the ethyl and isopropyl compounds (Metschies, 1990).

Screening experiments in combination with the structural requirements extracted from compounds like 3,5,6-trideoxy-Ins(1)P led to the synthesis of a series of bisphosphonate analogues of which the most potent is depicted in Figure 4c (Kulagowski *et al.*, 1991).

2.4. Ins(1,4,5)P_3 Analogues and Some of Their Biological Properties

Because of its pivotal role in Ca^{2+} homeostasis and the fact that it represents the only inositol phosphate with an unambiguously defined biological function, Ins(1,4,5)P_3 was considered a major target for synthetic efforts. The most obvious goal was and is the synthesis of an Ins(1,4,5)P_3 antagonist directed against the Ins(1,4,5)P_3 receptor. Furthermore, there is serious interest in small molecule inhibitors of the other well-characterized Ins(1,4,5)P_3 binding proteins: the 5-phosphatase and the 3-kinase (Nahorski and Potter, 1989). All three proteins exhibited different specificities toward Ins(1,4,5)P_3 analogues as well as different inositol tris- and tetrakisphosphate isomers. Most of the available binding and activity data are summarized in Table I. Included are values from binding studies to the Ins(1,4,5)P_3 receptor, Ca^{2+} release experiments, and binding studies with the 5-phosphatase and the 3-kinase. Inhibitory effects toward the latter enzymes are marked. Data from different tissues and different animals are compared in Table I. Though there is a good chance that different cell types contain different isoforms of proteins, a comparison of the values for binding and activity of Ins(1,4,5)P_3 in different cell types shows very similar results (Table I). Furthermore, the consistency of cells and cell lines used is fortunately high; e.g., the great majority of studies with the Ins(1,4,5)P_3 receptor were performed with rat cerebellar membranes or bovine adrenal cortex membranes. Nevertheless, a comparison of the data for the analogues and the values determined for Ins(1,4,5)P_3 with the particular cell type is advised.

No analogue except the 5-phosphonate of Ins(1,4,5)P_3 (Figure 6) was claimed to be an antagonist for intracellular Ca^{2+} release (Dreef *et al.*, 1991b), nor were any of the inositol phosphate isomers or analogues more potent agonists than Ins(1,4,5)P_3 itself, except the very recently discovered adenophostins (Figure 6) (M. Takahashi *et al.*, 1994; S. Takahashi *et al.*, 1994). More successful was the search for specific inhibitors of the 5-phosphatase and the 3-kinase. The first synthetic analogue of Ins(1,4,5)P_3 was its trisphosphotothioate DL-Ins(1,4,5)P_3S_3 (Cooke *et al.*, 1987b). The compound was shown to be a moderate agonist in respect to mobilizing $[Ca^{2+}]_i$ but was no substrate for either the 5-phosphatase or the 3-kinase (Safrany *et al.*, 1991a). Hence, its application to permeabilized cells resulted in a sustained elevation of $[Ca^{2+}]_i$ rather than the Ca^{2+} spike typical for Ins(1,4,5)P_3. The sustained elevation of $[Ca^{2+}]_i$ was reduced for DL-Ins(1,4)P_2(5)PS (Safrany *et al.*, 1991a), a derivative in which only the phosphate in position 5 had been altered to a phosphorothioate (Noble *et al.*, 1992a). DL-Ins(1,4)P_2(5)PS, though as potent as the trisphosphorothioate in releasing Ca^{2+}, was shown to be a substrate for the 3-kinase and was an inhibitor of the 5-phosphatase (Safrany *et al.*, 1991a). In a similar way, DL-Ins(1,2,4,5)P_4 (Figure 7) represented a potent agonist of Ins(1,4,5)P_3 in respect to Ca^{2+} release and was also a strong inhibitor of the 5-phosphatase, although the 3-kinase was only moderately affected (Mills *et al.*, 1993). More recently, two very potent and highly selective inhibitors of the 5-phosphatase were presented: L-*chiro*-Ins(1,4,6)P_3 and its trisphosphorothioate L-*chiro*-Ins(1,4,6)P_3S_3 (Hansbro *et al.*, 1994; Figure 7). Both neither directly raised $[Ca^{2+}]_i$ nor affected the 3-kinase. Derivatives of Ins(1,4,5)P_3 specific for inhibiting the 3-kinase were shown to be Ins(1,4,6)P_3 (Hirata *et al.*, 1993) and 3-deoxy-3-fluoro-Ins(1,4,5)P_3 (Safrany *et al.*, 1992a, Figure 8). Ins(1,4,6)P_3 (Figure 8), only rarely reported to be abundant in tissue (Stephens and Downes, 1990; Radenberg *et al.*, 1989), appeared to be a weak but full agonist for Ca^{2+} release. The isomer was more slowly

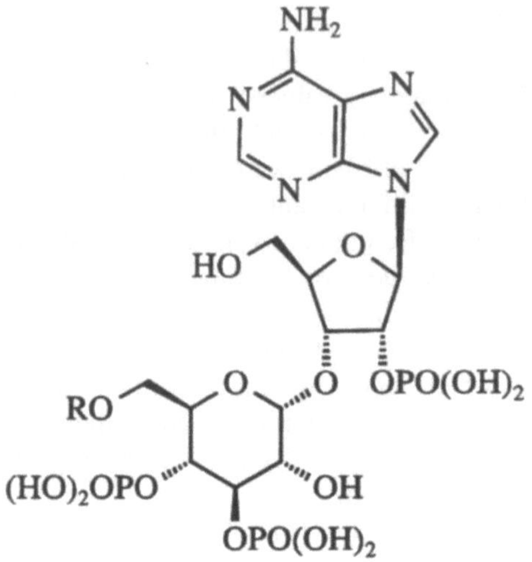

adenophostin A: R = H

adenophostin B: R = $COCH_3$

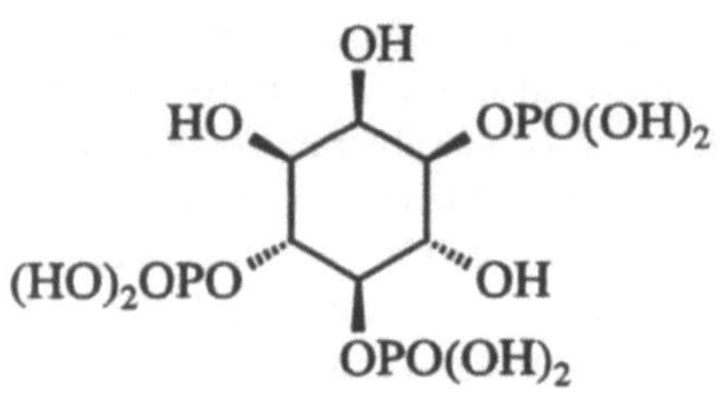

Ins(1,4,5)P_3

FIGURE 6. Structures of Ins(1,4,5)P_3 and of adenophostin A and B (M. Takahashi *et al.*, 1994), the most potent agonists for elevating intracellular Ca^{2+} levels known so far. The adenophostins were about 100 times more potent in releasing Ca^{2+} from cerebellar microsomes than Ins(1,4,5)P_3 and showed a sustained Ca^{2+} elevation because of their stability to metabolic enzymes (S. Takahashi *et al.*, 1994).

hydrolyzed by the 5-phosphatase than Ins(1,4,5)P_3 but was a competitive inhibitor of the 3-kinase. 3-Deoxy-3-fluoro-Ins(1,4,5)P_3 showed similar specificity but exhibited more potency in releasing Ca^{2+}. Both compounds are promising tools for biological evaluations, as are the *chiro*-inositol trisphosphates and the now commercially available phosphorothioates.

Apart from their functional potency, the Ins(1,4,5)P_3 analogues mentioned as well as those listed in Table I provide valuable insight into structure–activity relationships of the three Ins(1,4,5)P_3 binding proteins: the Ins(1,4,5)P_3 receptor, the 5-phosphatase, and the 3-kinase. From the vast amount of binding and activity data, it becomes clear that certain functional groups in the Ins(1,4,5)P_3 molecule are essential for binding, whereas others might be modified with no

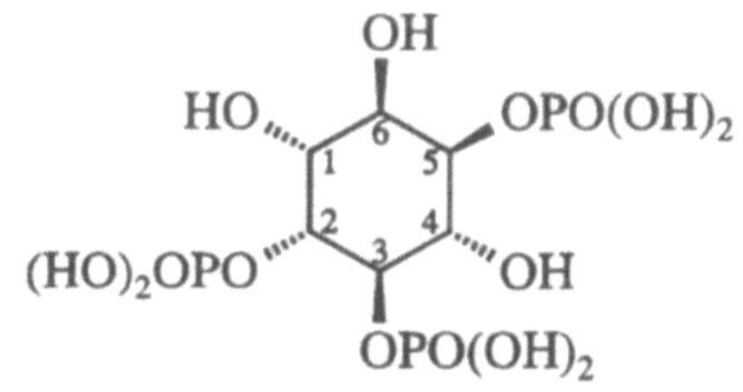

L-chiro-Ins(2,3,5)P_3

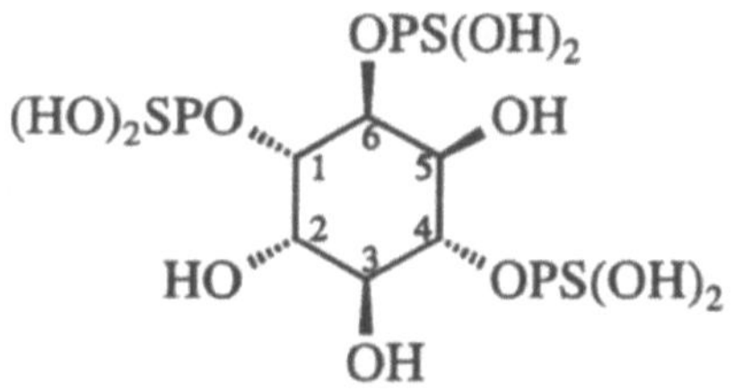

L-chiro-Ins(1,4,6)P_3S_3

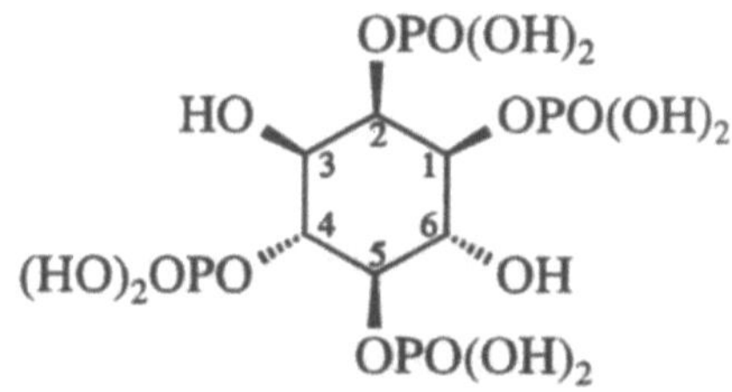

Ins(1,2,4,5)P_4

FIGURE 7. Structures of inhibitors of the 5-phosphatase with different selectivities. L-*chiro*-Inositol-2,3,5-trisphosphate [L-*chiro*-Ins(2,3,5)P_3] represents a potent agonist for Ca^{2+} release and inhibits the 5-phosphatase and the 3-kinase. DL-*myo*-Inositol-1,2,4,5-tetrakisphosphate [DL-Ins(1,2,4,5)P_4, D-enantiomer shown] has similar properties, except it hardly affects the 3-kinase. L-*chiro*-Inositol-1,4,6-phosphorothioate [L-*chiro*-Ins(1,4,6)P_3S_3] was shown to be a specific inhibitor of the 5-phosphatase without binding to the $InsP_3$ receptor or the 3-kinase (see text for references).

substantial effect on the binding/activity properties. The most important findings are summarized for each functional group in the following paragraphs (see Table I for detailed data and references).

2.4.1. 6-Hydroxyl Group

6-Deoxy and 6-O-methyl analogues, though full agonists, are both reduced in their ability to release Ca^{2+} by several orders of magnitude. The same analogues exhibited a slightly reduced affinity for the 5-phosphatase but were not hydrolyzed. Ins(1,3,6)P_3, a derivative with an axial hydroxyl group mimicking the 6-position of Ins(1,4,5)P_3, was a good substrate for the 5-phosphatase with a binding affinity comparable to that of Ins(1,4,5)P_3 (Hirata *et al.*, 1993). Therefore, the 6-hydroxyl group must be considered essential for the hydrolytic activity of the 5-phosphatase. 6-Deoxy- and 6-O-methyl-Ins(1,4,5)P_3 were both accepted by the 3-kinase, though only the 6-deoxy analogue exhibited an affinity comparable to Ins(1,4,5)P_3 (Table I).

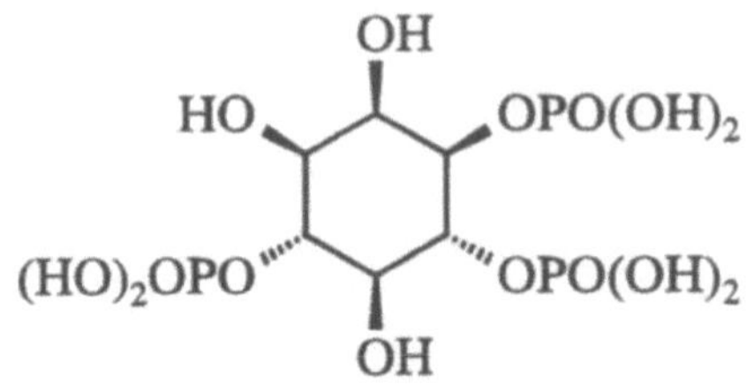

Ins(1,4,6)P_3

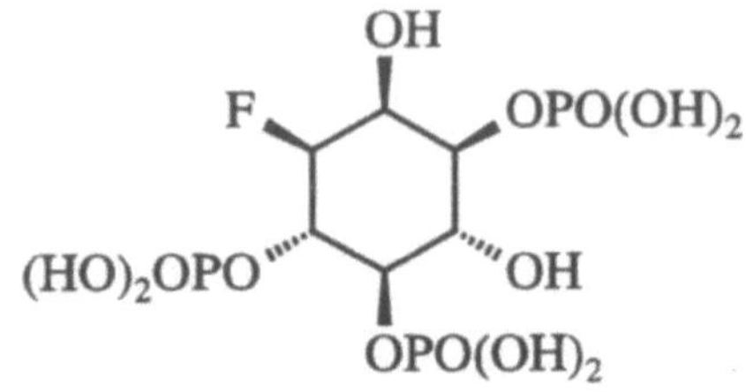

3-deoxy 3-fluoro Ins(1,4,5)P_3

FIGURE 8. Structures of inhibitors of the 3-kinase. Both compounds, *myo*-inositol-1,4,6-trisphosphate [Ins(1,4,6)P_3] and 3-deoxy-3-fluoro-*myo*-inositol-1,4,5-trisphosphate [3-deoxy-3-fluoro-Ins(1,4,5)P_3] are substrates for the 5-phosphatase, but only the latter is a potent agonist for Ca^{2+} release. For details and references see text and Table I.

2.4.2. 3-Hydroxyl Group

This group appears to contribute to only a minor extent to the interaction of Ins(1,4,5)P_3 with its receptor. Substitutions of the hydroxyl group by fluorine or hydrogen led to potent agonists (Table I), and 3-amino-3-deoxy-Ins(1,4,5)P_3 was a pH-dependent partial agonist. L-*chiro*-Ins(2,3,5)P_3, formally identical to Ins(1,4,5)P_3 except that the 3-hydroxyl group is inverted from the equatorial to the axial position, was also demonstrated to be a potent agonist for Ca^{2+}-release (Figure 7). This effect was reduced when the axial 3-hydroxyl group was phosphorylated (Kozikowski *et al.*, 1994a). At the same time, L-*chiro*-Ins(2,3,5)P_3 was a good inhibitor for the 5-phosphatase and the 3-kinase. In contrast, 3-deoxy-3-fluoro-Ins(1,4,5)P_3 was a substrate for the 5-phosphatase but a potent inhibitor of the 3-kinase. It appears that the 5-phosphatase, often considered nonspecific, was sensitive to modifications in the 3-position with respect to its hydrolytic activity. The lack of the substrate hydroxyl group was not preventing binding to the 3-kinase.

2.4.3. 2-Hydroxyl Group

Modifications of the 2-hydroxyl group altered the binding properties to the Ins(1,4,5)P_3 receptor to only a very small degree, even if substitutions were bulky or if a phosphate group was added as in DL-Ins(1,2,4,5)P_4. However, this compound was a potent inhibitor of the 5-phosphatase. The binding constant of the 3-kinase was higher by a factor of 100 than for Ins(1,4,5)P_3. In contrast, bulky carboxylic acid esters hardly affected the binding properties to the 3-kinase and the 5-phosphatase, though hydrolysis by the latter enzyme was reduced. Therefore, this position was one of the obvious choices to anchor Ins(1,4,5)P_3 to resins for affinity chromatography or for the introduction of photolabels (Ozaki *et al.*, 1992).

Interesting properties could be assigned to the enantiomeric 2,2-difluoro-Ins(1,4,5)P_3: While the D-enantiomer behaved similarly to Ins(1,4,5)P_3, the L-enantiomer [2,2-difluoro-Ins(3,5,6)P_3] was potently inhibiting the 3-kinase and the 5-phosphatase without mobilizing intracellular Ca^{2+}.

The relevance of all three hydroxyl groups of Ins(1,4,5)P_3 has recently been discussed in great detail (Hirata *et al.*, 1993; Kozikowski *et al.*, 1993; Wilcox *et al.*, 1992; Nahorski and Potter, 1989).

2.4.4. 5-Phosphate Group

Several analogues with specific modifications of the 5-phosphate have been introduced: the monophosphorothioate [Ins(1,4)P_2(5)PS], the Ins(1,4)P_2 5-deoxy-5-methylenephosphonate [Ins(1,4)P_2(5)CH_2P], and the Ins(1,4)P_2 methyl and difluoromethyl phosphonates [Ins(1,4)P_2(5)PCH_3 and Ins(1,4)P_2(5)$PCHF_2$]. Although the 5-phosphorothioate was remarkably reduced in its ability to release Ca^{2+}, it was a potent inhibitor of the 5-phosphatase. The 3-kinase accepted the analogue as a substrate with only slightly reduced affinity. The 5-methylene-phosphonate elevated $[Ca^{2+}]_i$ in a long-lasting manner (Falck *et al.*, 1990). The 5-methylphosphonate was reported to be an antagonist of Ins(1,4,5)P_3, whereas the corresponding difluoromethyl phosphonate showed no activity (Dreef *et al.*, 1991a). More detailed evaluations of the antagonistic properties have not been reported yet.

The 5-phosphate appears to be an essential feature of Ins(1,4,5)P_3. It should be mentioned that Ins(1,4)P_2 has no reasonable affinity for any of the three proteins (Shears, 1992; Berridge and Irvine, 1989). Therefore, future investigations with the analogues named above as well as novel derivatives carrying modifications of the 5-phosphate might be of particular interest.

2.4.5. 4-Phosphate Group

The vicinal bisphosphate motif was stated to be the minimum structural feature required for biological activity of Ins(1,4,5)P_3 analogues. Consequently, the synthetic Ins(1,5)P_2 was as inactive (Westerduin *et al.*, 1992) as was Ins(1,3,5)P_3. The latter was a moderate inhibitor of the 5-phosphatase. To our knowledge no other analogues with a specific modification of the 4-phosphate were reported.

2.4.6. 1-Phosphate Group

The 1-phosphate group must be considered somewhat open to modifications because the glycerol ester of Ins(1,4,5)P_3, Gro(1)PIns(4,5)P_2, was shown to be a moderate activator of Ca^{2+} release. In the same way, several ethylamino esters and the 3-aminopropyl ester were accepted. Because the latter compound was a substrate for neither the 5-phosphatase nor the 3-kinase, 1-phosphate esters were premier candidates for the preparation of affinity columns and photoaffinity-labeled Ins(1,4,5)P_3 analogues (Prestwich *et al.*, 1991; Schäfer *et al.*, 1990; Henne *et al.*, 1988). Lampe and Potter (1990) synthesized DL-Ins(4,5)P_2(1)PS and subsequently introduced a fluorescent label by a sulfur linkage, because the phosphate diesters might be sensitive to hydrolysis by phosphodiesterases.

Summing up the preceeding, it seems that any modification of Ins(1,4,5)P_3 will interfere, if not with all three, with at least one of the Ins(1,4,5)P_3 binding proteins. The 5-phosphatase, though rather unspecific in terms of binding Ins(1,4,5)P_3 analogues, appears to be very sensitive to some modifications with respect to its hydrolytic activity. The 3-kinase tolerates alterations of the 5-phosphate and the hydroxyl groups, especially deletions of OH. The Ins(1,4,5)P_3 receptor requires mainly the motif of the vicinal 4,5-bisphosphates plus the free 6-hydroxyl group and probably one of the negative charges of the 1-phosphate. Interestingly, all these features are offered by the adenophostins (Figure 6). Their high potency in releasing intracellular Ca^{2+} in combination with their resistance to the metabolic enzymes makes them a very promising target for future investigations.

2.5. Membrane-Permeant Derivatives of Inositol Phosphates

It is probably not too far-fetched to state that *myo*-inositol polyphosphates or their derivatives might become valuable tools for medical treatment in the future. Before these (future) compounds will be ready to leave the bench for application to living animals or even men, a number of pharmacological prerequisites for a drug will have to be fulfilled, including tissue specificity and membrane permeability. The latter seems to be a particular challenge for inositol phosphates because of their more or less large number of negative charges, which leaves them impermeable to lipid bilayers.

The most promising solution to this problem up to now seems to be the use of intracellularly bioactivatable masking groups such as acetoxymethyl esters (AM-esters, Figure 9). The AM-esters are stable outside cells but are readily cleaved inside cells by intracellular esterases. They have frequently been employed to mask carboxylic acids (Tsien, 1981; Jansen and Russell, 1965). Several applications of AM-esters to phosphates were successfully performed by Farquhar and co-workers (Sastry *et al.*, 1992; Freed *et al.*, 1989; Srivastva and Farquhar, 1984). More recently, AM-esters of the cyclic nucleotides bt_2cyclic AMP (Schultz *et al.*, 1993), cyclic AMP (Schultz *et al.*, 1994), and cyclic GMP (Zhuo *et al.*, 1994) showed high biological potency when applied extracellularly.

The synthesis of membrane-permeant derivatives of inositol polyphosphates

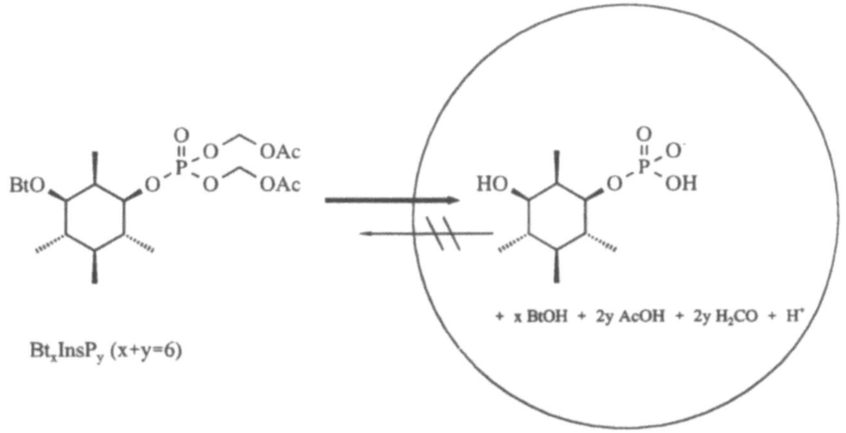

FIGURE 9. Principle of acetoxymethyl ester loading into living cells. The fully protected inositol phosphate is applied extracellularly and passes the cell membrane. The acetoxymethyl esters and butyrates are hydrolyzed by intracellular esterases to give the corresponding inositol phosphate, which is trapped inside the cell.

was expected to be a more challenging problem. Bundgaard (1993) masked one of the negative charges of each of the phosphates of Ins(1,2,6)P_3 with 4-acetoxy-butyryloxymethyl groups, resulting in a moderately convertible derivative. First attempts to use six AM-esters on Ins(1,4,5)P_3 needed the help of butyrates covering the hydroxyl groups for increased lipophilicity as well as for synthetic purposes. When applied to Ref-52 fibroblasts, only high extracellular doses (400 μM) of 2,3,6-tri-O-butyryl-Ins(1,4,5)P_3/AM were able to induce intracellular Ca^{2+} release (Schultz and Tsien, 1992). The obstacle was presumably the slow intracellular hydrolysis of the butyrates. The synthesis of Ins(1,4,5)P_3/AM without butyrates (Figure 10) avoided this problem, and the compound was active in lower extracellular doses (C. Schultz and R. Y. Tsien, unpublished data).

The synthesis of a membrane-permeant derivative of *myo*-inositol-3,4,5,6-tetrakisphosphate, 1,2-di-O-butyryl-Ins(3,4,5,6)P_4/AM (Figure 10), allowed the direct proof of the hypothesis that Ins(3,4,5,6)P_4 is involved in uncoupling the Cl^- secretion of T_{84} cells from the intracellular Ca^{2+} signaling pathway (Kachintorn *et al.*, 1993; Vajanaphanich *et al.*, 1994). A competition assay specifically designed to measure intracellular levels of Ins(3,4,5,6)P_4 revealed that incubation of T_{84} cells with 1,2-di-O-butyryl-Ins(3,4,5,6)P_4/AM indeed elevated intracellular Ins(3,4,5,6)P_4 levels. This suggested that at least for a fraction of the molecules added, all ten carboxylic acid esters were cleaved inside the cells.

The use of AM-esters is certainly limited with respect to medical applications, mainly because the compound gives rise to intracellular levels of formal-

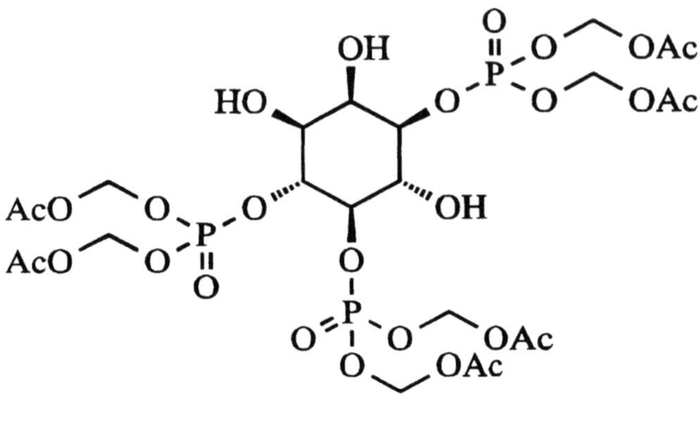

Ins(1,4,5)P_3/AM

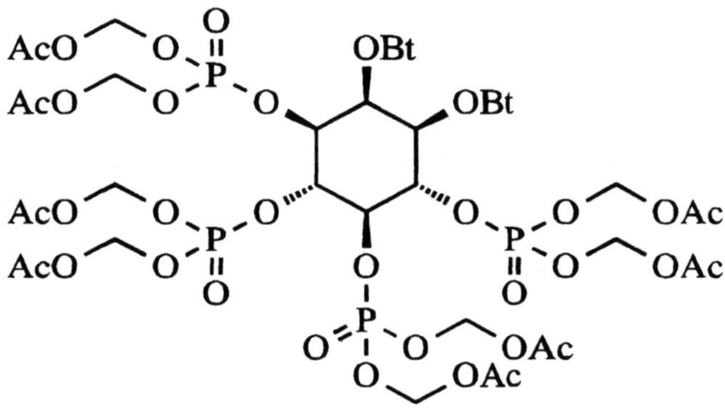

bt_2Ins(3,4,5,6)P_4/AM

FIGURE 10. Structures of membrane-permeant derivatives of Ins(1,4,5)P_3 and Ins(3,4,5,6)P_4.

dehyde (Figure 9). Therefore, AM-esters should be considered only a starting point in the development of membrane-permeant derivatives of intracellular phosphate-containing messengers and their use as prodrugs.

3. SEPARATION

The separation of *myo*-inositol phosphates is one of the most important prerequisites to investigate the turnover of *myo*-inositol phosphates. The techniques developed for the sophisticated analysis of the complex mixtures of inositol phosphates inside cells is predominantly based on liquid chromatography, which may also be transferred to a preparative scale. The purification of the naturally occurring isomers and separation of phytic acid lysates provide an important source for inositol phosphates without the need for chemical procedures.

This section is limited to the most widely used methods. Very detailed descriptions of the common biochemical techniques for the investigation of the inositol phosphate turnover are presented in *Methods in Inositide Research* edited by Irvine (1990). The purification of inositol phosphates usually started with isolation from biological material by various extraction methods (see below). Subsequently, negatively charged molecules other than inositol phosphates (e.g., ATP) were removed by treatment of the extracts with charcoal (Mayr, 1988; Meek, 1986). The separation of the major groups of inositol phosphates was achieved by anion-exchange chromatography. Finally, it was possible to separate different isomers of, e.g., $InsP_3$ by more refined methods.

In a few cases other methods such as thin-layer chromatography (TLC) (Scholz *et al.*, 1992; Hokin-Neaverson and Sadeghian, 1976), the time-consuming paper chromatography (Grado and Ballou, 1961), electrophoresis (Sekar *et al.*, 1987; Tate, 1968), and more recently, supercritical fluid chromatography (Chester *et al.*, 1989) have been employed in the analysis of inositol phosphates but must be considered subordinated methodologies.

3.1. Extraction

The reagents chosen for the extraction of inositol phosphates from biological material usually also terminate enzymatic activity. In general, acidic reagents such as perchloric acid or trichloroacetic acid were used for the extraction procedure. After removal of the denaturated proteins by centrifugation, the extracts were neutralized. Perchlorate was removed by precipitation as its potassium salt. Trichloroacetic acid could be extracted with water-saturated diethylether. Extractions were also performed under neutral conditions with chloroform/methanol solutions (Hawkins *et al.*, 1987). It should be kept in mind that all these conditions may cause some artifacts. In particular, the acidic conditions make it difficult to observe the acid-labile cyclic phosphates. For example, the hydrolysis of Ins(1:2-cyclic 4,5)P_3 led to Ins(1,4,5)P_3 (major product) and Ins(2,4,5)P_3 (minor product) (Wong *et al.*, 1988). Although phosphate monoesters are moderately stable under acidic conditions, migration of the phosphates may occur. Therefore, the extracts should be neutralized as rapidly as possible.

The extraction method chosen should be checked carefully with regard to recovery rates of the phosphates of interest. Nonspecific binding of some inositol phosphates by filter material or molecules from the extract itself was observed. This phenomenon appeared to depend on the kind of tissue used and even on the kind of agonist the cells were stimulated with (Wreggett *et al.*, 1987).

3.2. Anion-Exchange Chromatography

Anion-exchange chromatography was shown to be a simple and efficient method to separate complex mixtures of inositol phosphates into inositol monophosphates, bisphosphates, trisphosphates, etc. The soluble fraction of cell or tissue extracts was brought on a small column or even a Pasteur pipette filled with the anion-exchange resin. A washing procedure starting with water removed the uncharged molecules such as sugars and *myo*-inositol first, followed by stepwise increasing salt concentrations to elute the phosphates. Mostly ammonium formate buffers were used. A carefully prepared column allowed separation of each group of inositol phosphates from InsP up to $InsP_6$ (Christensen, 1990). This simple kind of anion-exchange chromatography was often used as a crude preseparation to prepare samples for more accurate analysis (Wreggett and Irvine, 1987).

Usually samples collected from anion-exchange columns need to be desalted for further investigations such as mass measurements or the final characterization by NMR-spectroscopy. Ammonium formate buffers could be removed easily by repeated lyophilization (Scholz *et al.*, 1992; Turk *et al.*, 1986). Another procedure involved the use of exchange resins in the chloride form. Elution of the phosphates with lithium chloride led to the lithium salts of the phosphates. Lithium chloride is soluble in ethanol and could be removed by extraction (Shayman *et al.*, 1987). To desalt inositol phosphates in phosphate buffers, the samples were applied to an anion-exchange column, and the inorganic phosphate was eluted with triethylammonium formate (0.2 M). A tenfold increase of the formate concentration eluted the inositol phosphates (Stephens *et al.*, 1988), which were lyophilized to remove the triethylammonium formate. A combination of anion- and cation-exchange processes was also used to change elution buffers to the more easily removable formic acid (Bird *et al.*, 1989).

3.3. High-Performance Liquid Chromatography

Anion-exchange resins for HPLC offer more refined applications for the separation of inositol phosphates. Both isocratic and gradient experiments were performed for this purpose. Isocratic flow conditions were useful for the separation of the inositol phosphate isomers $Ins(1,3,4)P_3$ and $Ins(1,4,5)P_3$ (Wreggett and Irvine, 1987). For the separation of more complex mixtures, a gradient flow technique was necessary; for example, the separation of several $InsP_3$ and $InsP_4$ isomers in one chromatographic process was possible (Kachintorn *et al.*, 1993; Mayr, 1988). The efficiency of this method was demonstrated by Goldschmidt (1990), who identified 11 of the 12 racemic $InsP_3$ isomers and all of the nine racemic $InsP_2$ isomers from the chemical hydrolysis of phytic acid.

In general, quarternary ammonium salts such as tetrabutylammonium hydrogen sulfate or dihydrogen phosphate and N-methylimipramine (Shayman and Barcellon, 1990; Sulpice *et al.*, 1989; Shayman and BeMent, 1988) were used for ion-pair chromatography on HPLC. The eluent was applied to reversed-phase HPLC columns and contained acetonitrile as the organic component. The order of elution obtained by this method was about the same as that observed by anion-exchange chromatography. The principle relied on the fact that the more charges an inositol phosphate carried, the more lipophilic counterions it bound, hence leading to longer retention times.

A similar HPLC system involved a micellar mobile phase with an octyldimethylsilyl stationary phase. Gradient elution separated glycerol phosphate, three isomeric inositol monophosphates, glycerolphosphoinositol phosphate, three inositol bisphosphates, glycerolphosphoinositol bisphosphate, three inositol trisphosphates, and Ins(1,3,4,5)P_4 (Brando *et al.*, 1990).

3.4. Gas Chromatography

Analysis of inositol phosphates by gas chromatography required the derivatization of the hydroxyl groups and the phosphates. This derivatization relied on either pertrimethylsilylation (Sherman *et al.*, 1986) or methylation of the phosphates after selective cleavage of the trimethylsilyl groups from the phosphates (Goldman *et al.*, 1990; Hsu *et al.*, 1990). However, this method was successful only for inositol monophosphates. Leavitt and Sherman (1982b) demonstrated the separation of modified InsP isomers as well as InsP enantiomers (Leavitt and Sherman, 1982a). Application of this approach to higher phosphorylated inositol phosphates was limited by adsorption phenomenons (Sherman *et al.*, 1986). Gas chromatography was used mainly for mass determination of inositol phosphates after their separation by liquid chromatography. The single isomers had to be completely dephosphorylated, and the resulting *myo*-inositol was modified. Recently, a method for the measurement of inositol polyphosphates, namely Ins(1,2,6)P_3, was described in which the modifcation preserved the original structural information. For this purpose, Ins(1,2,6)P_3 was permethylated with diazomethane, dephosphorylated, and converted to the methylated 1,2,6-tris(heptafluorobutyryl) derivative. Application of this method to biological samples seemed to be limited by phosphate migration and/or dephosphorylation during the modification process (Goldman *et al.*, 1990).

3.5. Detection

The difficulty in detecting inositol phosphates is based on the lack of a chromophore that would allow the use of a spectrophotometer as a detector. Although internal standards such as AMP, ADP, and ATP were employed to follow the progress of separations or the stability of a gradient (Foster *et al.*

1994; Irvine *et al.*, 1985), such techniques as radiolabeling or postcolumn derivatization represent common methods to detect inositol phosphates.

3.5.1. Radiolabeling

For radiolabeling experiments, cells were usually preincubated with tritiated *myo*-inositol (Prestwich and Bolton, 1991; Zilberman *et al.*, 1987) or inorganic [^{32}P]phosphate (Stephens and Downes, 1990; Sulpice *et al.*, 1989; Ishii *et al.*, 1986). After chromatographic separation, radioactivity was measured by taking small fractions (0.2–2 ml), adding scintillator liquid, and subsequent analysis with a liquid scintillation spectrometer (Zilberman *et al.*, 1987; Irvine *et al.*, 1985). More recently, this process was coupled directly to the chromatographic separation in on-line experiments (Foster *et al.*, 1994; Kachintorn *et al.*, 1993). It should be mentioned that radiolabeling is not the appropriate method for mass measurement of the inositol phosphates because basal levels of unlabeled inositol phosphates are undetectable.

3.5.2. Detection by Postcolumn Reactions

Because specific binding assays (Challiss and Nahorski, 1993) provide information for only a single inositol phosphate isomer, more general methods were developed for the identification of inositol phosphates eluting from chromatography columns. Enzymatic assays required the dephosphorylation of the inositol phosphates after separation by HPLC. The resulting *myo*-inositol was converted to *myo*/*scyllo*-inosose by the action of the NAD^+-dependent inositol dehydrogenase. Changes in the amounts of NADH were monitored fluorometrically (Shayman *et al.*, 1987) or by a bioluminescence approach (Prestwich and Bolton, 1991).

Meek and Nicoletti (1986) integrated the dephosphorylation into the HPLC process. After separation of the inositol phosphates on an anion-exchange column, the sample was passed through a column with immobilized alkaline phosphatase in an on-line experiment.

A very sophisticated detection system relied on the complexation of transition metals by polyanions. This so-called metal-dye detection avoided the hydrolysis of the phosphates (Mayr, 1988). The inositol phosphates were separated on an anion-exchange HPLC column using an HCl gradient. The eluent contained yttrium ions, which bound with high affinity to both polyanions such as inositol phosphates and a cation-specific dye 4-(2-pyridylazo)resorcinol (PAR). The PAR was mixed with the eluent for the postcolumn reaction. Changes in the absorbance of the PAR–yttrium complex were monitored at 546 nm.

Because of the high salt concentrations required for the elution, especially of higher phosphorylated inositols, a conductivity detector could be used only with reasonable sensitivity by means of chemically suppressed conductivity (Smith and MacQuarrie, 1988). The inositol phosphates were separated on an anion-exchange column by HPLC followed by a suppressor column that lowered the conductivity prior to detection.

4. IDENTIFICATION

The identification of unknown inositol phosphates, as well as the final proof of a synthetic structure, relies predominantly on two techniques: nuclear magnetic resonance spectroscopy and classic chemical analysis. Other methods such as mass spectroscopy and sometimes the analysis of X-ray structural data are also employed.

4.1. Nuclear Magnetic Resonance Spectroscopy

Nuclear magnetic resonance (NMR) data for several inositol phosphates including isomeric mono-, bis-, tris-, tetrakisphosphates, and cyclic InsP have been published (Cerdan *et al.*, 1986; Radenberg *et al.*, 1989). Because of the low sensitivity of NMR, its application for this purpose is limited to *in vitro* experiments. Since several signals are obtained for each single inositol phosphate, the correct identification requires almost complete separation (Radenberg *et al.*, 1989). Information about the structure can be extracted from the chemical shift values of the signals in combination with the multiplicity of their reosonances (Mayr and Dietrich, 1987). The combination of NMR experiments detecting different nuclei, e.g., ^{13}C, led to further information (Lindon *et al.*, 1986). Selective decoupling as well as two-dimensional NMR experiments allowed a precise determination of the structure (Johansson *et al.*, 1990; Radenberg *et al.*, 1989; Lindon *et al.*, 1987; Cerdan *et al.*, 1986). However, the results obtained from NMR experiments gave no information about enantiomeric inositol phosphates (Lindon *et al.*, 1987).

4.2. Chemical Analysis

The determination of the absolute configuration of inositol phosphates requires their conversion to known molecules in a definite way. The classical chemical procedure started with the selective cleavage of the inositol phosphate ring structure by periodate oxidation between vicinal hydroxyl groups. After reduction of the resulting aldehyde functions, the phosphates were hydrolyzed, and the open-chain polyols could be identified by comparison. In contrast to other identification methods, this procedure allows the identification of optical isomers (Grado and Ballou, 1961), which is essential for studying the biochemical pathways of inositol phosphates (Stephens and Downes, 1990) or the analysis of unknown inositol phosphates, e.g., from phytic acid lysates.

4.3. Discovery of Novel Inositol Phosphates

The HPLC measurement of extracts from the slime mold *Dictyostelium discoideum* showed that some peaks eluted even later than $InsP_6$ from an anion-exchange column. The corresponding compounds could be isolated (Mayr *et al.*, 1992). Their analysis by MS and NMR spectroscopy revealed the presence of pyrophosphate groups. The final characterization suggested the pyrophosphates to be 1-diphosphoinositol pentakisphosphate (D- or L-$PPInsP_5$) and bis-(1,4)-diphosphoinositol tetrakisphosphate (D- or L-bis$PPInsP_4$) (Stephens *et al.*, 1993). Though their biological function remains obscure at the moment, it was hinted that the inositol pyrophosphates might represent a new form of phosphate donor.

In the hyperthermophilic archaebacteria *Pyrococcus woesi*, unusually high potassium concentrations in the molar range are present. An important counterion was recently shown to be di-*myo*-inositol-1,1′-phosphate (DIP) (Scholz *et al.*, 1992). This compound was isolated by means of anion-exchange chromatography, preparative thin-layer chromatography, and gel filtration. Alkaline hydrolysis gave inositol and inositol monophosphate as analyzed by TLC. In accordance with the mass spectra, a phosphorus diester structure was determined by ^{1}H-NMR. To challenge the stereochemical configuration of DIP, all three possible stereoisomers (LL, DD, and DL) were prepared synthetically (van Leeuwen *et al.*, 1994). Comparison of the specific rotation values revealed that the naturally occurring DIP had the LL configuration.

ACKNOWLEDGMENTS. We are grateful to Bernd Jastorff, Agnes Roemer, Andrzej Skladanowski, and Werner Tegge for carefully reading the manuscript.

ABBREVIATIONS USED IN THIS CHAPTER

3-kinase:	*myo*-inositol-1,4,5-trisphosphate 3-kinase
5-phosphatase:	*myo*-inositol polyphosphate 5-phosphatase
AM:	acetoxymethyl
bisPP-$InsP_4$:	bis-(1,4)-diphosphoinositol tetrakisphosphate
bt:	butyryl
bu:	butyl
$[Ca^{2+}]_i$:	intracellular free calcium concentration
cyclic AMP:	cyclic adenosine monophosphate
cyclic GMP:	cyclic guanosine monophosphate

cyclic InsP:	cyclic *myo*-inositol monophosphate
cyclic $InsP_3$:	cyclic *myo*-inositol trisphosphate
DIP:	di-*myo*-inositol-1,1′-phosphate
GC:	gas chromatography
Gro:	glycerol
HPLC:	high performance liquid chromatography
IC_{50}:	concentration of unlabeled competitor that displaces half of specific binding
IMPase:	*myo*-inositol monophosphatase
Ins:	*myo*-inositol
InsP, $InsP_2$, $InsP_3$, $InsP_4$, $InsP_5$, and $InsP_6$:	inositol mono-, bis-, tris-, tetrakis-, pentakis-, and hexakisphosphates, respectively, with assignment of phosphate locants where appropriate [e.g., Ins(1,4,5)P_3]
Ins(1,4,5)P_3S_3:	*myo*-inositol-1,4,5-trisphosphorothioate
$InsP_6$:	*myo*-inositol hexakisphosphate
MS:	mass spectroscopy
NMR:	nuclear magnetic resonance
PAR:	4-(2-pyridylazo)resorcinol
PP-$InsP_5$:	1-diphosphoinositol pentakisphosphate
RBL:	rat basophilic leukemia cells
Ref-52:	rat embryo fibroblasts
T_{84}:	human colon epithelial cell line
TLC:	thin-layer chromatography
TMS:	trimethylsilyl
UV:	ultraviolet

5. REFERENCES

Aguilo, A., Martin-Lomas, M., and Penades, S., 1992, The regioselective synthesis of enantiomerically pure *myo*-inositol derivatives: Efficient synthesis of *myo*-inositol 1,4,5-trisphosphate, *Tetrahedron Lett.* **33:**401–404.

Andersch, P., and Schneider, M. P., 1993, Enzyme assisted synthesis of enantiomerically pure *myo*-inositol derivatives—chiral building blocks for inositol polyphosphates, *Tetrahedron: Asymmetry* **4:**2135–2138.

Anejo, R., and Parra, A., 1994, Facile optical resolution of DL-1,4,5,6-tetra-O-benzyl *myo*-inositol: Key synthons for the phosphoinositides, *Tetrahedron Lett.* **35:**525–526.

Angyal, S. J., and Tate, M. E., 1961, Cyclitols. Part X. *myo*-Inositol phosphates, *J. Chem. Soc.* **1961:**4122–4128.

Attwood, P. V., Ducep, J. B., and Chanal, M. C., 1988, Purification and properties of *myo*-inositol 1-phosphatase from bovine brain, *Biochem. J.* **253:**387–394.

Baker, R., Kulagowski, J. J., Billington, D. C., Leeson, P. D., Lennon, I. C., and Liverton, N. J., 1989, Synthesis of 2- and 6-deoxyinositol 1-phosphate and the role of the adjacent hydroxy groups in the mechanism of inositol monophosphatase. *J. Chem. Soc. Chem. Commun.* **1989:**1383–1385.

Baker, R., Leeson, P. D., Liverton, N. J., and Kulagowski, J. J., 1990, Identification of (1*S*)-phosphoryloxy-(2*R*,4*S*)-dihydroxycyclohexane as a potent inhibitor of inositol monophosphatase, *J. Chem. Soc. Chem. Commun.* **1990:**462–464.

Baker, R., Carrick, C., Leeson, P. D., Lennon, I. C., and Liverton, N. J., 1991, Design and synthesis of 6α-substituted 2β,4α-dihydroxy-1β-phosphoryloxycyclohexanes, potent inhibitors of inositol monophosphatase, *J. Chem. Soc. Chem. Commun.* **1991:**298–300.

Balla, T., Guillemette, G., Baukal, A. J., and Catt, K. J., 1987, Metabolism of inositol 1,3,4-trisphosphate to a new tetrakisphosphate isomer in angiotensin-stimulated adrenal glomerulosa cells, *J. Biol. Chem.* **262:**9952–9955.

Balla, T., Sim, S. S., Iida, T., Choi, K. Y., Catt, K. J., and Rhee, S. G., 1991, Agonist-induced calcium signalling is impaired in fibroblasts overproducing inositol 1,3,4,5-tetrakisphosphate, *J. Biol. Chem.* **266:**24719–24726.

Ballou, C. E., and Pizer, L. I., 1960, The absolute configuration of the *myo*-inositol 1-phosphates and a confirmation of the bornesitol configurations, *J. Am. Chem. Soc.* **82:**3333–3335.

Batty, I. R., Nahorski, S. R., and Irvine, R. F., 1985, Rapid formation of inositol 1,3,4,5-tetrakisphosphate following muscarinic receptor stimulation of rat cerebral cortical slices, *Biochem. J.* **232:**211–215.

Baudin, G., Glänzer, B. I., Swaminathan, K. S., and Vasella, A., 1988, A synthesis of 1D- and 1L-*myo*-inositol 1,3,4,5-tetrakisphosphate, *Helv. Chim. Acta* **71:**1367–1378.

Berridge, M. J., 1981, Phosphatidylinositol hydrolysis: A multifunctional transducing mechanism, *Mol. Cell. Endocrinol.* **24:**115–140.

Berridge, M. J., 1983, Rapid accumulation of inositol trisphosphate reveals that agonists hydrolyse polyphosphoinositides instead of phosphatidylinositol, *Biochem. J.* **212:**849–858.

Berridge, M. J., 1993, Inositol trisphosphate and calcium signalling, *Nature* **361:**315–325.

Berridge, M. J., and Irvine, R. F., 1989, Inositol phosphates and cell signalling, *Nature* **341:**197–205.

Billington, D. C., 1989, Recent developments in the synthesis of *myo*-inositol phosphates, *Chem. Soc. Rev.* **18:**83–122.

Billington, D. C., 1993, *The Inositol Phosphates: Chemical Synthesis and Biological Significance*, Verlag Chemie, Weinheim.

Bird, I. M., Sadler, I. H., Williams, B. C., and Walker, S. W., 1989, The preparation of *myo*-inositol 1,4-bisphosphate and D-*myo*-inositol 1,4,5-trisphosphate in milligram quantition from a readily available starting material, *Mol. Cell. Endocrinol.* **66:**215–229.

Brando, C., Hoffman, T., and Bonvini, E., 1990, High performance liquid chromatographic separation of inositol phosphate isomers employing a reversed phase column and a micellar mobile phase, *J. Chromatogr.* (*Biomed. Appl.*) **529:**65–80.

Bundgaard, H., 1993, Preparation of bioreversibly protected inositol phosphates as drugs, *CA* **119:**9102t.

Carless, H.A.J., and Busia, K., 1990, Total synthesis of *chiro*-inositol 2,3,5-trisphosphate: A *myo*-inositol 1,4,5-trisphosphate analogue from benzene via photo-oxidation, *Tetrahedron Lett.* **31:**1617–1620.

Cerdan, S., Hansen, C. A., Johanson, R., Inubushi, T., and Williamson, J. R., 1986, Nuclear magnetic resonance spectroscopic analysis of *myo*-inositol phosphates including inositol 1,3,4,5-tetrakisphosphate, *J. Biol. Chem.* **261:**14676–14680.

Challiss, R.A.J., and Nahorski, S. R., 1993, Measurement of inositol 1,4,5-trisphosphate, inositol 1,3,4,5-tetrakisphosphate, and phosphatidylinositol 4,5-bisphosphate in brain, *Methods Neurosci.* **18:**224–244.

Challiss, R.A.J., and Willcocks, A. L., Mulloy, B., Potter, B.V.L., and Nahorski, S. R., 1991, Characterization of inositol 1,4,5-trisphosphate- and inositol 1,3,4,5-tetrakisphosphate-binding sites in rat cerebellum, *Biochem. J.* **274:**861–867. (1)

Changya, L., Gallacher, D. V., Irvine, R. F., Potter, B.V.L., and Petersen, O. H., 1989, Inositol 1,3,4,5-tetrakisphosphate is essential for sustained activation of the Ca^{2+}-dependent K^+ current in single internally perfused mouse lacrimal acinar cells, *J. Membr. Biol.* **109:**85–93.

Chester, T. L., Pinkston, J. D., Innis, D. P., and Bowling, D. J., 1989, Separation, detection and identification of inositol trisphosphate and phytic acid derivatives by supercritical fluid chromatography and SFC-mass spectrometry, *J. Microcolumn Sep.* **1:**182–189.

Christensen, S., 1990, Serial separation of inositol phosphates including pentakis- and hexakisphosphates on small anion exchange columns, *J. Chromatogr.* (*Biomed. Appl.*) **553:**201–206.

Chung, S.-K., and Moon, S.-H., 1994, Synthesis and biological activities of (4,6-di-O-phosphonato-β-D-mannopyranosyl)-methylphosphonate as an analogue of 1L-*myo*-inositol 1,4,5-trisphosphate, *Carbohydrate Res.* **260:**39–50. (2)

Cooke, A. M., Gigg, R., and Potter, B.V.L., 1987a, Synthesis of DL-*myo*-inositol 1,4,5-trisphosphate, *Biochem. Soc. Trans.* **15:**904–906.

Cooke, A. M., Gigg, R., and Potter, B.V.L., 1987b, *myo*-Inositol 1,4,5-trisphosphorothioate: A novel analogue of a biological second messenger, *J. Chem. Soc. Chem. Commun.* **1987:**1525–1526.

Cooke, A. M., Nahorski, S. R., and Potter, B.V.L., 1989, *myo*-Inositol 1,4,5-trisphosphorothioate is a potent competitive inhibitor of human erythrocyte 5-phosphates, *FEBS Lett.* **242:**373–377. (3)

Cullen, P. J., Irvine, R. F., and Dawson, A. P., 1990, Synergistic control of Ca^{2+} mobilization in permeabilized mouse L1210 lymphoma cells by inositol 2,4,5-trisphosphate and inositol 1,3,4,5-tetrakisphosphate, *Biochem. J.* **271:**549–553.

DeLisle, S., Radenberg, T., Wintermantel, M. R., Tietz, C., Parys, J. B., Pittet, D., Welsh, M. J., and Mayr, G. W., 1994, Second messenger specificity of the inositol trisphosphate receptor: Reappraisal based on novel inositol phosphates, *Am. J. Physiol.* **266:**C429–C436.

Downes, C. P., and MacPhee, C. H., 1990, *myo*-Inositol metabolites as cellular signals, *Eur. J. Biochem.* **193:**1–18.

Dreef, C. E., Douwes, M., Elie, C.J.J., van der Marel, G. A., and van Boom, J. H., 1991a, Application of the bifuntional phosphonylating agent bis[6-(trifluoromethyl)benzotriazol-1-yl] methylphosphonate towards the preparation of isosteric D-*myo*-inositol phospholipid and phosphate analogues, *Synthesis* **1991:**443–447.

Dreef, C. E., Schiebler, W., van der Marel, G. A., and van Boom, J. H., 1991b, Synthesis of 5-phosphonate analogues of *myo*-inositol 1,4,5-trisphosphate: Possible intracellular calcium antagonists, *Tetrahedron Lett.* **32:**6021–6024. (4)

Drummond, A. H., 1987, Lithium and inositol lipid-linked signalling mechanisms, *Trends Pharmacol. Sci.* **8:**129–133.

Eckstein, F., and Gindl, H., 1968, Uridin-2′,3′-O,O-cyclothiophosphat, *Chem. Ber.* **101:**1670–1673.

Ely, J. A., Hunyady, L., Baukal, A. J., and Catt, K. J., 1990, Inositol 1,3,4,5-tetrakisphosphate stimulates calcium release from bovine adrenal microsomes by a mechanism independent of the inositol 1,4,5-trisphosphate receptor, *Biochem. J.* **268:**333–338.

Erneux, C., and Takazawa, K., 1991, Intracellular control of inositol phosphates by their metabolizing enzymes, *Trends Pharmacol. Sci.* **12:**174–176.

Estevez, V. A., and Prestwich, G. D., 1991a, Synthesis of enantiomerically pure, P-1 tethered inositol tetrakisphosphate affinity labels via a Ferrier rearrangement, *J. Am. Chem. Soc.* **113:**9885–9887.

Estevez, V. A., and Prestwich, G. D., 1991b, Affinity probes for Ins(1,3,4,5)P_4 receptors, *Tetrahedron Lett.* **32:**1623–1626.

Falck, J. R., and Abdali, A., 1993, Total synthesis of D-*myo*-inositol 3,4,5-trisphosphate and 1,3,4,5-tetrakisphosphate, *Bioorg. Med. Chem. Lett.* **3:**717–720.

Falck, J. R., and Yadagiri, P., 1989, Enantiospecific synthesis of D-*myo*-inositol 1,4,5-trisphosphate from (−)-quinic acid, *J. Org. Chem.* **54:**5851–5852.

Falck, J. R., Abdali, A., and Wittenberger, S. J., 1990, Total synthesis of the 5-methylenephosphonate analogue of D-*myo*-inositol 1,4,5-trisphosphate, *J. Chem. Soc. Chem. Commun.* **1990:** 953–955.

Ferris, C. D., and Snyder, S. H., 1992, IP_3 receptors: Ligand-activated calcium channels in multiple forms, in *Advances in Second Messenger and Phosphoprotein Research,* Vol. 26 (J. W. Putney, ed.), pp. 95–107, Raven Press, New York.

Foster, P. S., Hogan, S. P., Hansbro, P. M., O'Brien, R., Potter, B.V.L., Ozaki, S., and Denborough, M. A., 1994, The metabolism of D-*myo*-inositol trisphosphate and D-*myo*-inositol 1,3,4,5-tetrakisphosphate by porcine skeletal muscle, *Eur. J. Biochem.* **222:**955–964.

Freed, J. J., Farquhar, D., and Hampton, A., 1989, Evidence for acyloxymethyl esters of pyrimidine 5′-deoxyribonucleotides as extracellular sources of active 5′-deoxyribonucleotides in cultured cells, *Biochem. Pharmacol.* **38:**3193–3198.

Gawler, D. J., Potter, B.V.L., Gigg, R., and Nahorski, S. R., 1991, Interactions between inositol tris- and tetrakisphosphates, *Biochem. J.* **276:**163–167. (5)

Géro, S. G., 1966, The preparation of 1-O-tosyl-(−)-inositol from quebrachitol, *Tetrahedron Lett.* **6:**591–595.

Géro, S. G., Mercier, D., and Barrett, J. E., 1972, 1L-*myo*-Inositol 1-phosphate, *Methods Carbohydr. Chem.* **6:**403–408.

Gigg, R., and Warren, C. D., 1969, Derivatvies of DL-2,3,4,5,6-penta-O-benzyl *myo*-inositol, *J. Chem. Soc.* **1969:**2367–2371.

Gill, D. L., Ghosh, T. K., and Mullaney, J. M., 1989, Calcium signalling mechanisms in endoplasmatic reticulum activated by inositol 1,4,5-trisphosphate and GTP, *Cell Calcium,* **10:**363–374.

Goldman, H. D., Hsu, F.-F., and Sherman, W. R., 1990, Studies on the permethylation/dephosphorylation of inositol polyphosphates: An approach to a more sensitive assay, *Biomed. Environ. Mass Spectrom.* **19:**771–776.

Goldschmidt, B., 1990, Preparative separation of *myo*-inositol bis- and trisphosphate isomers by anion exchange chromatography, *Carbohydrate Res.* **208:**105–110.

Grado, C., and Ballou, C. E., 1961, *myo*-Inositol phosphates obtained by alkaline hydrolysis of beef brain phosphoinositide, *J. Biol. Chem.* **236:**54–60.

Guédat, P., Spiess, B., and Schlewer, G., 1994, Synthesis of (+/−)-1,2-dideoxy-1,2-diamino *myo*-inositol, *Tetrahedron Lett.* **35:**7375–7378.

Guidot, J. P., and Le Gall, T., 1993a, Synthesis of a sulfur-containing analogue of *myo*-inositol, D,L-1-deoxy-1-mercapto-*myo*-inositol, via an intramolecular sulfur-delivery reaction, *J. Org. Chem.* **58:**5271–5273.

Guidot, J. P., and Le Gall, T., 1993b, Highly regioselective functionalization of a *cis*-1,2-diol in the *myo*-inositol series: Application to the synthesis of 2-deoxy-2-mercapto-*myo*-inositol, *Tetrahedron Lett.* **34:**4647–4650.

Hansbro, P. M., Foster, P. S., Liu, C., Potter, B.V.L., and Denborough, M. A., 1994, Kinetic analysis of novel inhibitors of inositol polyphosphate metabolism, *Biochem. Biophys. Res. Commun.* **200:**8–15. (6)

Hashii, M., Hirata, M., Ozaki, S., Nozawa, Y., and Higashida, H., 1994, Ca^{2+} influx evoked by inositol 3,4,5,6-tetrakisphosphate in *ras*-transformed NIH/3T3 fibroblasts, *FEBS Lett.* **340:**276–280.

Hawkins, P. T., Berrie, C. P., Morris, A. J., and Downes, C. P., 1987, Inositol 1:2-cyclic 4,5-trisphosphate is not a product of muscarinic receptor stimulated phosphatidylinositol 4,5-bisphosphate hydrolysis in rat parotid glands, *Biochem. J.* **243:**211–218.

Henne, V., Mayr, G. W., Grabowski, B., Koppitz, B., and Söling, H.-D., 1988, Semisynthetic derivatives of inositol 1,4,5-trisphosphate substituted at the 1-phosphate group, *Eur. J. Biochem.* **174:**95–101.

Hirata, M., Watanabe, Y., Ishimatsu, T., Ikebe, T., Kimura, Y., Ymaguchi, K., Ozaki, S., and Koga, T., 1989, Synthetic inositol trisphosphate analogs and their effects on phosphatase, kinase, and the release of Ca^{2+}, *J. Biol. Chem.* **264:**20303–20308.

Hirata, M., Watanabe, Y., Yoshida, M., Koga, T., and Ozaki, S., 1993, Roles for hydroxyl groups of D-*myo*-inositol 1,4,5-trisphosphate in the recognition by its receptor and metabolic enzymes, *J. Biol. Chem.* **268:**19260–19266.

Hokin-Neaverson, M., and Sadeghian, K., 1976, Separation of [^{3}H]-inositol monophosphates and [^{3}H]-inositol on silica gel glass-fiber sheets, *J. Chromatogr.* **120:**502–505.

Hsu, F.-F., Goldman, D., and Sherman, W. R., 1990, Thermospray liquid chromatographic/mass spectrometric studies with inositol phosphates, *Biomed. Environ. Mass Spectrom.* **19:**597–600.

Irvine, R. F. (ed.), 1990, *Methods in Inositide Research,* Raven Press, New York.

Irvine, R. F., 1992, Is inositol tetrakisphosphate the second messenger that controls Ca^{2+} entry into cells, in *Advances in Second Messenger and Phosphoprotein Research,* Vol. 26 (J. W. Putney, ed.), pp. 161–185, Raven Press, New York.

Irvine, R. F., and Moor, R. M., 1986, Micro-injection of inositol 1,3,4,5-tetrakisphosphate activates sea urchin eggs by a mechanism dependent on external Ca^{2+}, *Biochem. J.* **240:**917–920.

Irvine, R. F., Änggard, E. E., Letcher, A. J., and Downes, C. P., 1985, Metabolism of inositol 1,4,5-trisphosphate and inositol 1,3,4-trisphosphate in rat parotid glands, *Biochem. J.* **229:**505–511.

Irvine, R. F., Letcher, A. J., Heslop, J. P., and Berridge, M. J., 1986, The inositol tris/tetrakisphosphate pathway—demonstration of Ins(1,4,5)P_3 3-kinase activity in animal tissues, *Nature* **320:**631–634.

Ishii, H., Connolly, T. M., Bross, T. E., and Majerus, P. W., 1986, Inositol cyclic trisphosphate [inositol 1,2-(cyclic)-4,5-trisphosphate] is formed upon thrombin stimulation of human platelets, *Proc. Natl. Acad. Sci. USA* **83:**6397–6401.

IUPAC, 1989, Numbering of atoms in *myo*-inositol, *Biochem. J.* **258:**1–2.

Jansen, A.B.A., and Russell, T. J., 1965, Some novel penicillin derivatives, *J. Chem. Soc.* **1965:**2127–2132.

Johansson, C., Koerdel, J., and Drakenberg, T., 1990, Analysis of *myo*-inositol phosphates by 2D proton NMR spectroscopy, *Carbohydrate Res.* **207:**177–183.

Johnson, S. C., Tagliaferri, F., and Baker, D. C., 1993, Synthesis of the 3-deoxy-3-C-(phosphonomethyl) analogue of 1D-*myo*-inositol 3-(dihydrogenphosphate), *Carbohydrate Res.* **250:**315–321.

Kachintorn, U., Vajanaphanich, M., Barrett, K. E., and Traynor-Kaplan, A. E., 1993, Elevation of inositol tetrakisphosphate parallels inhibition of Ca^{2+}-dependent Cl^- secretion in T_{84} cells, *Am. J. Physiol.* **264:**C671–C676.

Kiely, D. E., Abruscato, G. J., and Baburao, V., 1974, A synthesis of *myo*-inositol 1-phosphate, *Carbohydrate Res.* **34:**307–313.

Kozikowski, A. P., Fauq, A. H., Powis, G., Kurian, P., and Crews, F. T., 1992, Expedient synthetic routes to [^{3}H]-D-3-azido-deoxy-*myo*-inositol and D-3-azido-3-deoxy-*myo*-inositol 2,4,5-trisphosphate, *J. Chem. Soc. Chem. Commun.* **1992:**362–364. (10)

Kozikowski, A. P., Ognyanov, V. I., Fauq, A. H., Nahorski, S. R., and Wilcox, R. A., 1993, Synthesis of 1D-3-deoxy-, 1D-2,3-dideoxy-, and 1D-2,3,6-trideoxy *myo*-inositol 1,4,5-trisphosphate from quebrachitol, their binding affinities, and calcium release activity, *J. Am. Chem. Soc.* **115:**4429–4434. (11)

Kozikowski, A. P., Fauq, A. H., Wilcox, R. A., Challiss, R.A.J., and Nahorski, S. R., 1994a, The novel Ins(1,4,5)P_3 analogue 3-amino-3-deoxy-Ins(1,4,5)P_3: A pH-dependent Ins(1,4,5)P_3 receptor partial agonist in SH-SY5Y neuroblastoma cells, *J. Med. Chem.* **37:**868–872. (12)

Kozikowski, A. P., Ognyanov, V. I., Fauq, A. H., Wilcox, R. A., and Nahorski, S. R., 1994b, 1D-*myo*-inositol 1,4,5-trisphosphate and 1D-*myo*-inositol 1,3,4,5-tetrakisphosphate analogues modified at C-3; synthesis of 1D-3-C-(trifluoromethyl)-*myo*-inositol 1,4,5-trisphosphate and 1L-*chiro*-inositol 1,2,3,5-tetrakisphosphate from L-quebrachitol, *J. Chem. Soc. Chem. Commun.* **1994:**599–600.

Kulagowski, J. J., 1989, The synthesis of (+/−)-*myo*-inositol 1-phosphonate, *Tetrahedron Lett.* **30:**3869–3872.

Kulagowski, J. J., Baker, R., and Fletcher, S. R., 1991, Inhibitors of *myo*-inositol monophosphatase containing methylenebisphosphonic acid as a replacement for a phosphate group, *J. Chem. Soc. Chem. Commun.* **1991:**1649–1650.

Kurihara, N., Shibata, H., Saeki, H., and Nakajima, M., 1967, Synthesen von Inosit-monophosphaten, *Liebigs Ann. Chem.* **701:**225–232.

Lampe, D., and Potter, B.V.L., 1990, Synthesis of *myo*-inositol 1-phosphorothioate 4,5-bisphosphate: Preparation of a fluorescently labelled *myo*-inositol 1,4,5-trisphosphate analogue, *J. Chem. Soc. Chem. Commun.* **1990:**1500–1502.

Lampe, D., and Potter, B.V.L., 1993, Synthesis of 2-fluoro-2-deoxy *myo*-inositol 1,4,5-trisphosphate and *scyllo*-inositol 1,2,4-trisphosphate, novel analogues of the second messenger *myo*-inositol 1,4,5-trisphosphate, *Tetrahedron Lett.* **34:**2365–2368.

Leavitt, A. L., and Sherman, W. R., 1982a, Resolution of DL-*myo*-inositol 1-phosphate and other sugar enantiomers by gas chromatography, *Methods Enzymol.* **89:**3–9.

Leavitt, A. L., and Sherman, W. R., 1982b, Determination of inositol phosphates by gas chromatography, *Methods Enzymol.* **89:**9–18.

Leeson, P. D., James, K., Lennon, I. C., Liverton, N. J., Aspley, S., and Jackson, R. G., 1993, Adenosine-2′-monophosphate derivatives: Structural requirements as substrates for inositol monophosphatase, *Bioorg. Med. Chem. Lett.* **3:**1925–1930.

Leeuwen, S. H. van, van der Marel, G. A., Hensel, R., and van Boom, J. H., 1994, Synthesis of LL-di-*myo*-inositol-1,1′-phosphate: A novel inositol phosphate from *Pyrococcus woesi*, *Recl. Trav. Chim. Pays-Bas* **113:**335–336.

Ley, S. V., Parra, M., Redgrave, A. J., and Sternfeld, F., 1990, Microbial oxidation in synthesis: Preparation of *myo*-inositol phosphates and related cyclitol derivatives from benzene, *Tetrahedron* **46:**4995–5026.

Lindon, J. C., Baker, D. J., Farrant, R. D., and Williams, J. M., 1986, ^{1}H, ^{13}C and ^{31}P n.m.r. spectra and molecular conformation of *myo*-inositol 1,4,5-trisphosphate, *Biochem. J.* **233:**275–277.

Lindon, J. C., Baker, D. J., Williams, J. M., and Irvine, R. F., 1987, Conformation of the identities of inositol 1,3,4-trisphosphate and inositol 1,3,4,5-tetrakisphosphate by the use of one-dimensional and two-dimensional n.m.r. spectroscopy, *Biochem. J.* **244:**591–595.

Ling, L., and Ozaki, S., 1994, A chemoenzymatic synthesis of D-*myo*-inositol 1,4,5-trisphosphate, *Carbohydrate Res.* **256:**49–58.

Ling, L., Watanabe, Y., Akiyama, T., and Ozaki, S., 1992, A new efficient method for resolution of *myo*-inositol derivatives by enzyme catalyzed regio- and enantioselective esterifications in organic solvent, *Tetrahedron Lett.* **33:**1911–1914.

Liu, Y.-C., and Chen, C.-S., 1989, An efficient synthesis of optically active D-*myo*-inositol 1,4,5-trisphosphate, *Tetrahedron Lett.* **30:**1617–1620.

Maranto, A. R., 1994, Primary structure, ligand binding, and localization of the human type 3 inositol 1,4,5-trisphosphate receptor expressed in intestinal epithelium, *J. Biol. Chem.* **269:** 1222–1230.

Marecek, J. F., and Prestwich, G. D., 1989, Fluorinated analogs of Ins(1,4,5)P_3, *Tetrahedron Lett.* **30:**5401–5404.

Mayr, G. W., 1988, A novel metal-dye detection system permits picomolar range h.p.l.c. analysis of inositol polyphosphates from non-radioactively labelled cell or tissue specimens, *Biochem. J.* **254:**585–591.

Mayr, G. W., and Dietrich, W., 1987, The only tetrakisphosphate detectable in avian erythrocytes is the isomer lacking phosphate at position 3: A NMR study, *FEBS Lett.* **213:**278–282.

Mayr, G. W., Radenberg, T., Thiel, U., Vogel, G., and Stephens, L. R., 1992, Phosphoinositol diphosphates: Non-enzymic formation *in vitro* and occurrence *in vivo* in the cellular slime mold *Dictyostelium, Carbohydrate Res.* **234:**247–262.

Meek, J. L., 1986, Inositol bis-, tris-, and tetrakis(phosphate)s: Analysis in tissues by HPLC, *Proc. Natl. Acad. Sci. USA* **83:**4162–4166.

Meek, J. L., and Nicoletti, F., 1986, Detection of inositol trisphosphate and other organic phosphates by high performance liquid chromatography using an enzyme-loaded post column reactor, *J. Chromatogr.* **351:**303–311.

Menitti, F. S., Oliver, K. G., Nogimori, K., Obie, J. F., Shears, S. B., and Putney, J. W., 1990, Origins of *myo*-inositol tetrakisphosphates in agonist-stimulated rat pancreatoma cells, *J. Biol. Chem.* **265:**11167–11176.

Menitti, F. S., Oliver, K. G., Putney, J. W., and Shears, S. B., 1993a, Inositol phosphates and cell signalling: New views of $InsP_5$ and $InsP_6$, *Trends Biochem. Sci.* **18:**53–56.

Menitti, F. S., Miller, R. N., Putney, J. W., and Shears, S. B., 1993b, Turnover of inositol polyphosphate pyrophosphates in pancreatoma cells, *J. Biol. Chem.* **268:**3850–3856.

Mercier, D., and Géro, S. D., 1968, The synthesis of (1*S*,2*R*,4*R*,5*S*,6*R*)-1,2,4,5,6-penta-O-benzoyl *myo*-inositol, *Tetrahedron Lett.* **31:**3459–3462.

Mercier, D., Barnett, J.E.G., and Géro, S. D., 1969, Synthesis of optically active derivatives of *myo*-inositol: Preparation of 1L-*myo*-inositol 1-phosphate, *Tetrahedron* **25:**5681–5687.

Metschies, T., 1990, *Analoga des* myo-*Inositol-1-phosphats als Werkzeuge zur Aufklärung essentieller Wechselwirkungen der Inositmonophosphatase*, Dissertation, Universität Bremen.

Mikoshiba, K., 1993, Inositol 1,4,5-trisphosphate receptor, *Trends Pharmacol. Sci.* **14:**86–89.

Mills, S. J., Safrany, S. T., Wilcox, R. A., Nahorski, S. R., and Potter, B.V.L., 1993, Synthesis of *myo*-Inositol 1,2,4,5-tetrakisphosphate, a Ca^{2+}-mobilising tetrakisphosphate with a potency similar to *myo*-inositol 1,4,5-trisphosphate, *Bioorg. Med. Chem. Lett.* **3:**1505–1510. (13)

Nahorski, S. R., and Potter, B.V.L., 1989, Molecular recognition of inositol polyphosphates by intracellular receptors and metabolic enzymes, *Trends Pharmacol. Sci.* **10:**139–144.

Nahorski, S. R., Ragan, C. I., and Chaliss, R.A.J., 1991, Lithium and the phosphoinositide cycle: An example of uncompetitive inhibition and its pharmacological consequences, *Trends Pharmacol. Sci.* **12:**297–303.

Noble, N. J., Cooke, A. M., and Potter, B.V.L., 1992a, Synthesis of (+/−)-*myo*-inositol 1,4,5-trisphosphate and the novel analogue (+/−)-*myo*-inositol 1,4-bisphosphate 5-phosphorothioate, *Carbohydrate Res.* **234:**177–187.

Noble, N. J., Dubreuil, D., and Potter, B.V.L., 1992b, Total synthesis of *myo*-inositol 1-phosphate-4,5-pyrophosphate, a novel second messenger analogue, via *myo*-inositol 1-phosphate-4,5-bisphosphorothioate, *Bioorg. Med. Chem. Lett.* **2:**471–476. (14)

Oliver, K. G., Putney, J. W., Obie, J. F., and Shears, S. B., 1992, The interconversion of inositol 1,3,4,5,6-pentakisphosphate and inositol tetrakisphosphates in AR4-2J cells, *J. Biol. Chem.* **267:**21528–21534.

Ozaki, S., Watanabe, Y., Ogasawara, T., Kondo, Y., Shiotani, N., Nishii, H., and Matsuki, T., 1986, Total synthesis of optically active *myo*-inositol 1,4,5-trisphosphate, *Tetrahedron Lett.* **27:**3157–3160.

Ozaki, S., Watanabe, Y., Ogasawara, T., Hirata, M., and Kanematsu, T., 1992, Synthesis and biological properties of 2-substituted *myo*-inositol 1,4,5-trisphosphate analogues directed toward affinity chromatography and photoaffinity labelling, *Carbohydrate Res.* **234:**189–206.

Ozaki, S., Ling, L., Ogasawara, T., Watanabe, Y., and Hirata, M., 1994, A convenient chemoenzymatic synthesis of D- and L-*myo*-inositol 1,4,5,6-tetrakisphosphate, *Carbohydrate Res.* **259:**307–310.

Parthasarathy, R., and Eisenberg, F., 1986, The inositol phospholipids: A stereochemical view of biological activity, *Biochem. J.* **235:**313–322.

Paulsen, H., and Röben, W., 1983, Synthese von 1L-1-desoxy-1-C-hydroxymethyl-*myo*-inosit und 1L-1-desoxy-1-C-hydroxymethyl-*chiro*-inosit, *Liebigs Ann. Chem.* **1983:**1073–1077.

Pietrusiewicz, K. M., Salamonezyk, G. M., Bruzik, K. S., and Wieczork, W., 1992, The synthesis of homochiral inositol phosphates from *myo*-inositol, *Tetrahedron* **48:**5523–5542.

Polokoff, M. A., Bencen, G. H., Vacca, J. P., deSolms, S. J., Young, S. D., and Huff, J. R., 1988, Metabolism of synthetic inositol trisphosphate analogs, *J. Biol. Chem.* **263:**11922–11927. (15)

Potter, B.V.L., 1990, Recent advances in the chemistry and biochemistry of inositol phosphates of biological interest, *Nat. Prod. Rep.* **7:**1–24.

Potter, B.V.L., and Nahorski, S. R., 1992, Synthesis and biology of inositol polyphosphate analogues, *Biochem. Soc. Trans.* **20:**434–442.

Prestwich, G. D., Marecek, J. F., Mourey, R. J., Theibert, A. B., Ferris, C. D., Danoff, S. K., and Snyder, S. H., 1991, Tethered IP_3: Synthesis and biochemical applications of the 1-O-(3-aminopropyl) ester of inositol 1,4,5-trisphosphate, *J. Am. Chem. Soc.* **113:**1822–1825. (16)

Prestwich, S. A., and Bolton, T. B., 1991, Measurement of picomole amounts of any inositol phosphate isomer separable by h.p.l.c. by means of a bioluminescence assay, *Biochem. J.* **274:**663–672.

Radenberg, T., Scholz, P., Bergmann, G., and Mayr, G. W., 1989, The quantitative spectrum of inositol phosphate metabolites in avian erythrocytes, analysed by proton n.m.r. and h.p.l.c. with direct isomer detection, *Biochem. J.* **264:**323–333.

Reese, C. B., and Ward, J. G., 1987, Synthesis of D-*myo*-inositol 1,4,5-trisphosphate, *Tetrahedron Lett.* **28:**2309–2312.

Russell, M.G.N., Baker, R., and Billington, D. C., 1992, The synthesis of 2-O-alkyl-*myo*-inositol 1-phosphates as competitive inhibitors of inositol monophosphatase, *Carbohydrate Res.* **234:**263–286.

Safrany, S. T., Wojcikiewicz, R.J.H., Strupish, J., McBain, J., Cooke, A. M., Potter, B.V.L., and Nahorski, S. R., 1991a, Synthetic phosphorothioate-containing analogues of inositol 1,4,5-trisphosphate mobilize intracellular Ca^{2+} stores and interact differentially with inositol 1,4,5-trisphosphate 5-phosphatase and 3-kinase, *Mol. Pharmacol.* **39:**754–761.

Safrany, S. T., Wojcikiewicz, R.J.H., Strupish, J., Nahorski, S. R., Dubreuil, D., Cleophax, J., Géro, S. D., and Potter, B.V.L., 1991b, Interaction of synthetic D-6-deoxy-*myo*-inositol 1,4,5-trisphosphate with the Ca^{2+}-releasing D-*myo*-inositol 1,4,5-trisphosphate receptor, and the metabolic enzymes 5-phosphatase and 3-kinase, *FEBS Lett.* **278:**252–256. (18)

Safrany, S. T., Wilcox, R. A., Liu, C., Potter, B.V.L., and Nahorski, S. R., 1992a, 3-Position modification of *myo*-inositol 1,4,5-trisphosphate: Consequences for intracellular Ca^{2+} mobilisation and enzyme recognition, *Eur. J. Pharmacol.* **226:**265–272. (19)

Safrany, S. T., Sawyer, D. A., Nahorski, S. R., and Potter, B.V.L., 1992b, Synthetic D- and L-enantiomers of 2,2-difluoro-2-deoxy *myo*-inositol 1,4,5-trisphosphate interact differently with *myo*-inositol 1,4,5-trisphosphate binding proteins: Identification of a potent small molecule 3-kinase inhibitor, *Chirality* **4:**415–422. (20)

Safrany, S. T., Wilcox, R. A., Liu, C., Potter, B.V.L., and Nahorski, S. R., 1993, Identification of partial agonists with low intrinsic activity at the inositol-1,4,5-trisphosphate receptor, *Mol. Pharmacol.* **43:**499–503. (21)

Salamonczyk, G. M., and Pietrusiewicz, K. M., 1991, Expedient synthesis of D-*myo*-inositol 1,4,5-trisphosphate and D-*myo*-inositol 1,4-bisphosphate, *Tetrahedron Lett.* **32:**6167–6170.

Sastry, J. K., Nehete, P. N., Khan, S., Nowak, B. J., Plunkett, W., Arlinghaus, R. B., and Farquhar, D., 1992, Membrane-permeable dideoxyuridine 5′-monophosphate analogue inhibits human immunodeficiency virus infection. *Mol. Pharmacol.* **41:**441–445.

Schäfer, R., Nehls-Sahabandu, M., Grabowsky, B., Dehlinger-Kremer, M., Schulz, I., and Mayr,

G. W., 1990, Synthesis and application of photoaffinity analogues of inositol 1,4,5-trisphosphate selectively substituted at the 1-phosphate group, *Biochem. J.* **272:**817–825.

Scholz, S., Sonnenbichler, J., Schäfer, W., and Hensel, R., 1992, Di-*myo*-inositol-1,1′-phosphate: A new inositol phosphate isolated from *Pyrococcus woesi, FEBS Lett.* **306:**239–242.

Schultz, C., and Tsien, R. Y., 1992, Membrane-permeant derivatives of inositolpolyphosphates applied to REF-52 fibroblasts, *FASEB J.* **6**(5)**:**A1924.

Schultz, C., Metschies, T., Gerlach, B., Stadler, C., and Jastorff, B., 1990, Regioselective synthesis of inositol phosphate diesters via a cyclic phosphate triester approach, *Synlett* **1990:**163–165.

Schultz, C., Vajanaphanich, M., Barrett, K. E., Sammak, P. J., Harootunian, A. T., and Tsien, R. Y., 1993, Acetoxymethylesters of phosphates: Enhancement of the permeability and potency of cAMP, *J. Biol. Chem.* **268:**6316–6322.

Schultz, C., Vajanaphanich, M., Genieser, H.-G., Jastorff, B., Barrett, K. E., and Tsien, R. Y., 1994, Membrane-permeant derivatives of cAMP optimized for high potency, prolonged activity, or rapid reversibility, *Mol. Pharmacol.*, **46:**702–708.

Seewald, M. J., Aksoy, I. A., Powis, G., Fauq, A. H., and Kozikowski, A. P., 1990, Synthesis of D-3-deoxy-*myo*-inositol 1,4,5-trisphosphate and its effect on Ca^{2+} release in NIH 3T3 cells, *J. Chem. Soc. Chem. Commun.* **1990:**1502–1503.

Sekar, M., Dixon, J. F., and Hokin, L. E., 1987, The formation of inositol 1:2-cyclic 4,5-trisphosphate and 1,2-cyclic 4-bisphosphate on stimulation of mouse pancreatic minilobules with carbamyl-choline, *J. Biol. Chem.* **262:**340–344.

Sharp, A. H., Snyder, S. H., and Nigam, S. K., 1992, *myo*-Inositol 1,4,5-trisphosphate receptors, *J. Biol. Chem.* **267:**7444–7449.

Shayman, J. A., and Barcellon, F. S., 1990, Ion-pair chromatography of inositol polyphosphates with N-methylimipramine, *J. Chromatogr.* **528:**143–152.

Shayman, J. A., and BeMent, D. M., 1988, The separation of *myo*-inositol phosphates by ion-pair chromatography, *Biochem. Biophys. Res. Commun.* **151:**114–122.

Shayman, J. A., Morrison, A. R., and Lowry, O. H., 1987, Enzymatic fluorimetric assay for *myo*-inositol trisphosphate, *Anal. Biochem.* **162:**562–568.

Shears, S. B., 1992, Metabolism of inositol phosphates in *Advances in Second Messenger and Phosphoprotein Research,* Vol. 26 (J. W. Putney, ed.), pp. 63–92, Raven Press, New York.

Sherman, W. R., Ackermann, K. E., Berger, R. A., Gish, B. G., and Zinbo, M., 1986, Analysis of inositol mono- and polyphosphates by gas chromatography/mass spectrometry and fast atom bombardment, *Biomed. Environ. Mass Spectrom.* **13:**333–341.

Shvets, V. I., Klyashchitskii, B. A., Stepanov, A. E., and Evstigneeva, R. P., 1973, Resolution of asymmetrically substituted *myo*-inositols into optical antipodes, *Tetrahedron* **29:**331–340.

Sinha, N. D., Biernat, J., McManus, J., and Köster, H., 1984, Polymer support oligonucleotide synthesis XVIII: Use of β-cyanoethyl-N,N-dialkylamino/N-morpholino phosphoroamidite of deoxynucleosides for the synthesis of DNA fragments simplifying deprotection and isolation of the final product, *Nucleic Acids Res.* **12:**4539–4557.

Smith, R. E., and MacQuarrie, R. A., 1988, Determination of inositol phosphates and other biologically important anions by ion chromatography, *Anal. Biochem.* **170:**308–315.

Srivastva, D. N., and Farquhar, D., 1984, Bioreversible phosphate protective groups: Synthesis and stability of model acyloxymethyl phosphates, *Bioorg. Chem.* **12:**118–129.

Stepanov, A. E., and Shvets, V. I., 1979, Synthetic studies of phosphoinositides, *Chem. Phys. Lipids* **25:**247–263.

Stephens, L., and Downes, C. P., 1990, Product–precursor relationships amongst inositol polyphosphates, *Biochem. J.* **265:**435–452.

Stephens, L., Hawkins, P. T., Barker, C. J., and Downes, C. P., 1988, Synthesis of *myo*-inositol 1,3,4,5,6-pentakisphosphate from inositol phosphates generated by receptor activation, *Biochem. J.* **253:**721–733.

Stephens, L., Radenberg, T., Thiel, U., Vogel, G., Khoo, K.-H., Dell, A., Jackson, T. R., Hawkins, P. T., and Mayr, G. W., 1993, The detection, purification, structural characterization, and metabolism of diphosphoinositol pentakisphosphate(s) and bisdiphosphoinositol tetrakisphosphate(s), *J. Biol. Chem.* **268:**4009–4015.

Streb, H., Irvine, R. F., Berridge, M. J., and Schulz, I., 1983, Release of Ca^{2+} from a nonmitochondrial intracellular store in pancreatic acinar cells by inositol 1,4,5-trisphosphate, *Nature* **306:**67–69.

Sulpice, J.-C., Gascard, P., Journet, E., Rendu, F., Renard, D., Poggioli, J., and Giraud, F., 1989, The separation of [^{32}P]-inositol phosphates by ion-pair chromatography: Optimization of the method and biological applications, *Anal. Biochem.* **179:**90–97.

Takahashi, M., Tanzawa, K., and Takahashi, S., 1994, Adenophostins, newly discovered metabolites of *Penicillium brevicompactum,* act as potent agonists of the inositol 1,4,5-trisphosphate receptor, *J. Biol. Chem.* **269:**369–372. (22)

Takahashi, S., Kinoshita, T., and Takahashi, M., 1994, Adenophostins A and B: Potent agonists of inositol 1,4,5-trisphosphate receptor produced by *Pencillium brevicompactum;* Structure elucidation, *J. Antibiot.* **47:**95–100.

Tate, M. E., 1968, Separation of *myo*-inositol pentaphosphates by moving paper electrophoresis (MPE), *Anal. Biochem.* **23:**141–149.

Tegge, W., and Ballou, C. E., 1989, Chiral synthesis of D- and L-*myo*-inositol 1,4,5-trisphosphate, *Proc. Natl. Acad. Sci. USA* **86:**94–98.

Tegge, W., and Ballou, C. E., 1992, Syntheses of D-*myo*-inositol 1,4,5-trisphosphate affinity ligands, *Carbohydrate Res.* **230:**63–77.

Tegge, W., Denis, G. V., and Ballou, C. E., 1991, Synthesis and Ca^{2+}-release activity of D- and L-*chiro*-inositol 1,3,4-trisphosphate, *Carbohydrate Res.* **217:**107–116. (23)

Tsien, R. W., and Tsien, R. Y., 1990, Calcium channels, stores, and oscillations, *Annu. Rev. Cell Biol.* **6:**715–760.

Tsien, R. Y., 1981, A non-disruptive technique for loading calcium buffers and indicators into the cells, *Nature* **290:**527–528.

Turk, J., Wolf, B. A., and McDaniel, M. L., 1986, Quantitation of *myo*-inositol as its hexakis(trifluoroacetyl) derivative with negative ion chemical ionization mass spectrometry, *Biomed. Environ. Mass Spectrom.* **13:**237–244.

Vajanaphanich, M., Schultz, C., Rudolf, M. T., Wasserman, M., Enyedi, P., Craxton, A., Shears, S. B., Tsien, R. Y., Barrett, K. E., and Traynor-Kaplan, A. E., 1994, Long-term uncoupling of chloride secretion from intracellular calcium levels by Ins(3,4,5,6)P_4, *Nature* **371:**711–714.

Walker, J. W., Feeney, J., and Trentham, D. R., 1989, Photolabile precursors of inositol phosphates: Preparation and properties of 1-(2-nitrophenyl)ethyl esters of *myo*-inositol 1,4,5-trisphosphate, *Biochemistry* **28:**3272–3280.

Watanabe, Y., Mitani, M., Morita, T., and Ozaki, S., 1989, Highly efficient protection by the tetraisopropyldisiloxane-1,3-diyl group in the synthesis of *myo*-inositol phosphates as inositol 1,3,4,6-tetrakisphosphate, *J. Chem. Soc. Chem. Commun.* **1989:**482–483.

Watanabe, Y., Shinohara, T., Fujimoto, T., and Ozaki, S., 1990, A short step and practical synthesis of *myo*-inositol 1,3,4,5-tetrakisphosphate, *Chem. Pharm. Bull.* **38:**562–563.

Watanabe, Y., Fujimoto, T., Shinohara, T., and Ozaki, S., 1991, A short step synthesis of optically active *myo*-inositol 1,3,4,5-tetrakisphosphate and *myo*-inositol 1,4,5-trisphosphate from 1,3,5-tri-O-benzoyl *myo*-inositol, *J. Chem. Soc. Chem. Commun.* **1991:**428–429.

Watanabe, Y., Ogasawara, T., Ozaki, S., and Hirata, M., 1994, Synthesis of *myo*-inositol 1,4,6-trisphosphate, an analogue of *myo*-inositol 1,4,5-trisphosphate, *Carbohydrate Res.* **258:**87–92.

Westerduin, P., Willems, H.A.M., and van Boeckel, C.A.A., 1990, Total synthesis of new 1,5-bissubstituted *myo*-inositol derivatives: Synthesis of D-*myo*-inositol 1,5-bisphosphate and of rac.

1,5-bissulphated and 1,5-bissulphamoylated isosteric analogues, *Tetrahedron Lett.* **31:**6915–6918.

Westerduin, P., Willems, H.A.M., and van Boeckel, C.A.A., 1992, Synthesis of analogues of *myo*-inositol 1,4,5-trisphosphate that contain sulfonamide, sulfate, methylphosphonate, and carboxymethyl groups, *Carbohydrate Res.* **234:**131–140.

Wilcox, R. A., Nahorski, S. R., Sawyer, D. A., Liu, C., and Potter, B.V.L., 1992, The role of the 2- and 3-hydroxyl groups of 1D-*myo*-inositol 1,4,5-trisphosphate in the mobilisation of calcium from permealized human 1321N1 astrocytoma cells, *Carbohydrate Res.* **234:**237–246. (24)

Wilcox, R. A., Challiss, R.A.J., Liu, C., Potter, B.V.L., and Nahorski, S. R., 1993, Inositol 1,3,4,5-tetrakisphosphate induces calcium mobilization via the inositol 1,4,5-trisphosphate receptor in SH-SY5Y neuroblastoma cells. *Mol. Pharmacol.* **44:**810–817. (25)

Wilcox, R. A., Safrany, S. T., Lampe, D., Mills, S. J., Nahorski, S. R., and Potter, B.V.L., 1994, Modification at C2 of *myo*-inositol 1,4,5-trisphosphate produces inositol trisphosphates and tetrakisphosphates with potent biological activities, *Eur. J. Biochem.* **223:**115–124. (26)

Willcocks, A. L., Potter, B.V.L., Cooke, A. M., and Nahorski, S. R., 1988, *myo*-Inositol 1,4,5-trisphosphorothioate binds to specific [^{3}H]-inositol 1,4,5-trisphosphate sites in rat cerebellum and is resistant to 5-phosphatase, *Eur. J. Pharmacol.* **155:**181–183. (27)

Willcocks, A. L., Strupish, J., Irvine, R. F., and Nahorski, S. R., 1989, Inositol 1:2-cyclic 4,5-trisphosphate is only a weak agonist at inositol 1,4,5-trisphosphate receptors. *Biochem. J.* **257:**197–200. (28)

Wong, N. S., Barker, C. J., Shears, S. B., Kirk, C. J., and Michell, R. H., 1988, Inositol 1:2-cyclic 4,5-trisphosphate is not a major product of inositol phospholipid metabolism in vasopressin stimulated WRK1 cells, *Biochem. J.* **252:**1–5.

Wreggett, K. A., and Irvine, R. F., 1987, A rapid separation method for inositol phosphates and their isomers, *Biochem. J.* **245:**655–660.

Wreggett, K. A., Howe, L. R., Moore, J. P., and Irvine, R. F., 1987, Extraction and recovery of inositol phosphates from tissues, *Biochem. J.* **245:**933–934.

Yu, K.-L., and Fraser-Reid, B., 1988, A novel reagent for the synthesis of *myo*-inositol phosphates: N,N-Diisopropyl dibenzyl phosphoramidite, *Tetrahedron Lett.* **29:**979–982.

Zhou, M., Hu, Y., Schultz, C., Kandel, E. R., and Hawkins, R. D., 1994, Role of guanylyl cyclase and cGMP-dependent protein kinse in long-term potentiation, *Nature* **368:**635–639.

Zilberman, Y., Howe, L. R., Moore, J. P., Hesketh, T. R., and Metcalfe, J. C., 1987, Calcium regulates inositol 1,3,4,5-tetrakisphosphate production in lysed thymocytes and in intact cells stimulized with concanvalin A, *EMBO J.* **6:**957–962.

Index

Abscisic acid, 296, 302, 318, 327, 328, 330
Abscisic acid-induced morphogenic response, 260
Abscisic acid signaling, 318
Abscisic acid-resistant gene, 302
Acetylcholine (ACh), 3, 5, 44, 48, 52
Acetylcholine esterase, 166, 172
Acquorin, 293, 324
Actin-binding protein, 98, 109
Actin cytoskeleton, 97, 98, 100
Actin filament turnover to
  lamellipodia, 97
  microspikes, 97
Actin satellite protein, 101
Actin stress fiber, 102
Action potential, 304
Adenine nucleotide, effects of, 70
Adenophostins A & B, 60, 383, 390, 394
Adenylate cyclase, 18, 346
Adipocyte, 174
ADP-ribosylation, 19
AIDS, 154
Alkaline phosphatase, 170, 177
Alzheimer's disease, 54
Analogues of inositol phosphates, 243, 244
Anion exchange chromatography, 398
Anthocyanins, 305
AP-1, 103, 133, 142
AP-2, 211, 268, 313, 215, 268, 301
AP-3, 212, 213, 215
Arabidopsis thaliana, 295, 319, 329
Arachidonate, 104
Arachidonic acid, 305
ATPase, 306, 351, 363
Atropine, 2, 6, 11
Attachment of GPI, 170
Autocrine regulation, 104
Autophosphorylation, 48, 68
Auxins, 318, 349
Avian sarcoma virus, 26, 131

Basophilic cytoplasm, 7
Biosynthesis of *myo*-inositol phosphates, 239, 288, 289
Blue–light-induced kinase, 302
Bombesin, 24, 47
Bradykinin, 25, 47
Brain-derived growth factor, 119
bZIP, 142

cycli-AMP-dependent protein kinase, 67, 267; *see also* protein kinase A
Caffeine, 70
Calcineurin, 128, 137, 320, 328, 329
Ca^{2+}, 146, 297
Ca^{2+}-ATPase, 202
Ca^{2+} calmodulin-activated protein phosphatase, 328
Ca^{2+} channel blockers, 321
  La^{3+}, 323
  nifedipine, 330
  Verapamil, 330
Ca^{2+} channels, 48, 294
  $InsP_3$-activated/abscisic-acid-activated/Ca^{2+}-activated, 295–297
  Voltage-dependent (VDCC), 297

Ca^{2+}-dependent Cl^- secretion, 202, 203, 204
Ca^{2+}-dependent PIP_2 hydrolysis, 11
Ca^{2+} in deflagellation, 322
Ca^{2+} gradient, 358
Ca^{2+} homeostasis, 304
Ca^{2+}-induced turgor change, 332
Ca^{2+} ionophore A 23187, 8, 10, 332
Ca^{2+} mobilization by
  Ins(3,4,5,6)P_4
  Ins(1,3,4,5)P_4, 17, 61
Ca^{2+} mobilization in plant response, 293, 319
Ca^{2+}-mobilizing neurotransmitters, 10
Ca^{2+} receptor, 294
Ca^{2+} release, 60, 61, 70
  all-or-none/steady-state, 81, 82
Calmodulin, 68, 126, 127, 128, 136, 294, 295, 296, 346
Calmodulin-dependent protein kinases, 68
Calmodulin-dependent protein phosphorylation, 295
Calmodulin-independent proteinb phosphorylation, 295
Carbohydrate transport/capsule synthesis, 152
Carboxy-terminal domain (CTD) kinase, 141
Casein kinase II, 137, 155
Caveolae, 176
cDNA encoding Ins(1)P synthase, 260
CDP-diacylglycerol, 43
CDP-diacylglyceral-inositol-3-phosphatidyl transferase, 261, 290
CEA, 177
Cell-derived cycle factor ($cdc^{2,25,45}$), 320
Cell proliferation, 104
Cellulose microfibrils, 326
cGMP, 68, 125, 135
Cerebellar cortex, 51
Cerebellar granule cells, 49
che genes, 149
Chemotaxis, 149, 150
che proteins, 149
Chiroinositol, 167
L-*chiro*-Ins(1,4,6)P_3, 374, 375
Chlamydomonas reinhardtii, 319, 322, 323, 334
chloro-inositols, 385
Cholecytoskinin–pancreozymin (CCK/PZ), 3
Cholera toxin, 19
Cholin esterase, 53
Cholinergic neurons, 50, 52
Cholinergic receptor, 47, 48
Circadian rhythm, 306
c-jun, 142
Clathrin, 176, 211, 301
Colony-stimulating factor, 145
CMP-PA cytidyltransferase, 6
c-myc, 142
Concanavalin A, 144
Control by luminal Ca^{2+}, 81
COS cells, 175
CREB, 136, 142
CTP-PA cytidyl transferase, 290
Cyclin, 140
Cyclin A,B, 145, 147
Cyclin-dependent protein kinase (CDK), 302
Cyclophilin, 328, 329
Cyclosporin A, 328, 329
Cyclitols, 228, 249
Cytokinin, 318
Cytoskeletal domain, 144
Cytoskeletal protein, 104
Cytoskeleton, 70, 71

DAG–PKC pathway of Ca^{2+} signaling, 23
Decay-accelerating factor (DAF), 169, 170, 176, 177
Deflagellation, 322
Dephosphorylation of $InsP_6$, 208, 291
Deoxy-inositol trisphosphate (1,4,5)P_3, 382–383, 391–392
Detection of inositol phosphates, 399–401
Diacylglycerol (DAG), 3, 11, 20, 21, 23, 44, 133–135, 171, 249, 297, 320, 345–347, 353, 354, 361, 363, 364
Dictyostelium, 272–274, 277
Dihexanoyl glycerol, 21
Diphosphoinositol polyphosphates, 209, 210, 211
Dioctanoyl glycerol, 21
Dionaea muscipula, 326
DNA-binding protein, 132
Dopamine, 47
Dunalliela salina, 319

egr-a, 104
Endocytosis, 212, 214
Endocytotic vesicles, 176
Endoplasmic reticulum (ER), 7, 8, 15, 16, 168, 170

Entamoeba histolytica, 210
envZ protein, 152
Epithelial growth factor (EGF), 103, 107
Epidermal growth factor receptor (EGFR), 117, 132, 134, 140, 154
Epithelial cell lines
  MDCK I and II, 175
  LLC-PK_1, 175
  Caco-2, 175
  SK-CO15, 175
Electrical signal, 304
Elongation factor EF1a, 175
Erythropoietin, 124
Ethanolamine, 53, 166
Ethylene, 304, 305
ETS-like transcriptional regulators, 132
Expression of $InsP_3R$, 300
Extensor cell, 326
  protoplast, 305
Extraction of inositol phosphates, 397

Fibroblast growth factor (FGF), 117, 134
Fibroblast growth factor receptor (FGFR), 117
Fibronectin type III, 117
Flagellor gene (fliG), 149, 150
Flexor cells, 326
  protoplast, 305
Fussicoccin, 349

GTPase activating protein (GAP), 99, 102, 120
Galactinol, 232
Gas chromatography of inositol phosphates, 403
GDPβS, 19
Gelsolin, 100, 107
Gibberellins, 318
Glucuronic acid, 232
Glutamate receptor, 48, 52, 150
Glutamine synthetase, 150
Glycero-3-phosphate transacylase, 5, 7
Glycosyl phospholipid, 165
Glycosyl phosphoinositides, 165
Glycosylphosphatidylinositol (GPI), 165, 166, 167, 168, 169, 171
GPI in
  anchoring, 169
  mimetic insulin action, 172, 173
GPI-anchored proteins, 177
GPI-specific phospholipases, 178
G-protein, 18, 19, 46, 298, 299, 300, 306, 319, 323, 325, 346, 356
Golgi apparatus, 7, 261
grb 2-SOS1 complex, 123, 130
  GTPase, 120, 129
  GTP YS, 18, 19, 20, 325, 331, 356, 362
Guanine nucleotide exchange factors (GEFs), 100, 129
Guanylate cyclase, 125, 136, 137
Guard cell, 320

Helix–loop–helix domain, 141
Hematopoiesis, 141
Hexokinase, 167
Hexosamine, 288
High-performance liquid chromatography (HPLC) of inositol phosphates, 398
5-HT, 9, 28, 350, 359, 360
Human growth hormone (HGH), 170
Human mammary carcinomas, 154
Hydrogen peroxide, 325
Hydraulic movement, 326, 333
Hydraulic signal, 304
Hypersensitive response (HR), 305, 318, 350, 359, 360, 362, 364

Identification of inositol phosphates, 371, 401
Immunoelectron microscopy, 76
Immunoglobulin repeat, 81
Inactivation of $InsP_3$ receptor, 85
Indoleacetic acid, 349; *see also* Auxins
Indole esters, 232
INO-1 gene, 234, 259
Inositol auxotroph, 259, 260
Inositol bisphosphate, 236
Inositol cyclic phosphate, 12, 16
Inositol glycan antilipolytic lipogenic, 173
Inositol methyl phosphonate, 389
Inositol phosphates, 272, 284, 372
  analogues, 384, 386, 388
  formation, 9
  receptor, 267, 372
  receptor isoforms, 300
Inositol 1-phosphate kinase, 260, 261
  phosphatase, 259, 260, 269
Inositol phospholipid turnover, 346, 350, 358, 359
Inositol polyphosphate 5-phosphatase, 269, 270

Inositol trisphosphate ($InsP_3$), 13, 64, 236, 345, 353
as second messenger, 14
phytase (complex) action, 303, 306
heterogeneity, 71, 82
inactivation, 82
subtypes, 16
type I, 60, 72, 80
type II, 60, 72, 80
type III, 60, 72, 80
type IV, 60, 72, 80
type V, 60, 72, 80
Ins(1,4)P_2 (5) PS, 389
Ins(1,3,4)P_3, 13, 353
Ins(1,3,4)P_3 5-kinase, 193
Ins(1,3,4)P_3 6-kinase, 200, 259, 260
Ins(1,4,5)P_3, 13, 45, 268, 319, 323, 371, 378
Ins(1,4,5)P_3 degradation, 45
Ins(1,4,5)P_3 3-kinase, 45, 380
Ins(2,4,5)P_3, 303, 353
$InsP_3R$ isoforms, 78
Ins(1,3,4,5)P_4, 13, 49, 373
Ins(1,4,5,6)P_4 3-kinase, 195, 269
Ins(1,3,4,6)P_4 5-kinase, 193
Ins(1,3,4,5)P_4 3-phosphatase, 195
Ins(3,4,5,8)P_4, 372–374, 385, 386
$InsP_5$, 17, 187, 206, 216, 238, 289
Ins(1,3,4,5)P_5, formation, 17
Ins(1,3,4,5,6)P_5 2-kinase, 205
Ins(1,3,4,5,6)P_5 3-phosphatase, 195
$InsP_6$, 17, 187, 206, 216, 238
$InsP_6$-ADP-phosphotransferase, 289, 290, 293
Ins-1,2-cyclic phosphates, 244, 266
Insulin, 120, 171–174, 177, 178, 264
receptor, 120, 121
receptor substrate, 120
Interaction with cytoskeleton, 70, 71
Interferon, 139
Interferon-stimulated response elements, 139
Isocitrate dehydrogenase, 153

JAK, 122, 124, 139
Jasmonic acid (methyl jasmonate), 305

K^+-channel, 328, 354, 358
K^+ efflux, 304, 321
Ketamine, 28
Kentaxserin, 28
5-keto-glucose-6-phosphate, 232, 234
Kinases
Ca^{2+}/calmodulin-dependent protein, 125, 128
histidine, 149
Janus, 124
insulin-stimulated protein, 126
MAP, 125, 131
mek, 131
phosphatidyl inositol-3-, 124
ribosomal protein S_6, 116, 132
serine/threonine, 117, 124, 125, 131, 133
Src, 123
tyrosine, 117, 120, 123, 124
ZAP-70, 120, 123, 124

Lamprothamnium succinetum, 320
Leaf movement, 326, 331, 357
Leukotrienes, 104
Light singal, 349
LIM domain, 109
Lipid-based signaling, 305
Lipoprotein lipase, 177
Lithium-mediated NR gene expression, 360
Lithium, effect on PI metabolism, 13, 49, 235
.Lowe's oculocerabrorenal syndrome (OCRL), 270, 271
LPI/LPIP/LP1P_2, 362, 363
Lymphokines, 145
Lysophosphatidyl choline, 51
Lysophosphatidyl inositol, 4

Macrophage colony-stimulating factor, 119
Mannitol, 249
Mannose, 53
MAP kinase, 103, 116, 125, 131, 132
MARCKS proteins, 22, 135, 136
Mast cells, 18
Maturation-promoting factor (MPF), 145
Medicago sativa, 325
mek, 131, 133
mekk, 131, 132, 133
Membrane ruffles, 108
Memory formation, 51, 52
Metabolism
of $InsP_5$, 205, 288, 372
of $IndP_6$, 205, 288, 372
Mitogenes in PI turnover, 25
Motility, 119, 152

Muscarinic cholinergic agonists, 18
Muscarinic receptor, 48, 52
myo-Inositol-1-phosphate
  dehydrogenase, 293
  synthase, 231, 234, 259, 288
myo-Inositol phosphate
  genetics, 257
  phosphatase (MIPP), 195, 208, 209, 213; *see also* Phytases

N-acetyl glucosamine, 53
NAD^+, 234, 293
NDF, 118
NEL-GF, 118
Neoinositol, 232
Neomycin, 333, 362
Neuroblastoma cells, 49
Neurofibronin, 154
Neurokinins A&B, 48
Neuromodulin, 135
Neurophils, 177
Neuotrophin, 119
NGF, 19, 172
Nicotinic receptor, 48
Nifedipine, 330
Nitrate reductase, 355, 360
Nitella translucens, 321
Nitrogen assimilation, 150
Nitric oxide (NO), 136
  synthase (NOS), 128, 136, 137
NMDA receptor, 28
NMR spectroscopy of InsP, 401, 402
Nomenclature of *myo*-inositol, 228, 375
Nonreceptor tyrosine kinases, 122
NTPase, 306; *see also* JAK
ntr gene system, 150

Olfactory signals, 155
ompC, ompF, ompR, 151, 152
Osmotic stress, 151
Outer membrane porins, 151

PAL, 304
Paracrine regulation, 104
Paroxysmal nocturnal hemoglobinemea, 169
Patch clamp, 328
Pathogen responses, 304, 305, 324
Permeant inositol phosphates, 394
Pertussis toxin, 48
Pfr-mediated signal transduction, 351
pH (Pleckstrin homology) domain, 100, 130, 197
pH receptor, 324
Phorbol estser, 21, 177, 354
  receptor, 21
Phosphatidylcholine, 3, 4, 348, 355
Phosphatidylethanolamine, 3, 355
Phosphatidylinositol (PI), 2, 3, 4, 258, 261, 290
  glycans, 52, 53, 232
  inositol phosphatidyl transferase, 290
  transfer proteins, 261, 262, 265
Phosphatidylinositol(4,5)bisphosphate (PIP_2), 43, 44, 49, 347
Phosphatidylinositol 3-kinase, 9, 26, 52, 116, 121, 262, 277, 290, 334
Phosphoinositides, 1, 43, 44, 49, 288, 290, 297, 317, 318, 334, 348
Phospholamban, 127
Phospholipase A-, 2, 43, 103
Phospholipase C, 19, 45, 46, 53, 235, 264, 265, 266, 299, 346, 348, 351, 358
Phospholipase γ, 99, 116, 120, 134
Phospholipase C (GPI-specific), 165, 166, 167, 171, 177, 178
Phospholipase D (GPI-specific), 51, 53, 178
Phosphomyoinositol kinase, 292
Phosphorylation of protein, transcription factor, 115, 117, 118
Phosphoserine, 155
Phosphotransduction in *Drosophila*, 285
Phosphotyrosine, 120, 121, 123
  on $PLC_{\gamma 1}$, 105
Phototransduction, 350
Phytoalexins, 304, 325
Phytochromes, 318, 331, 349, 350, 352, 356, 358, 361
Phytophthora nicotianae, 325
Phytases, 241, 242, 271, 272, 292
Phytic acid (PA), 238–240, 242, 258, 260, 272, 274, 287, 288, 376, 398, 402
  as antioxidant, 258, 260, 272–275
  as anticolon cancer agent, 258, 260, 272–275
  as chelator, 258, 260, 272–275
  mutants, 273, 275, 276
pin gene, 304
D-Pinitol, 249, 376
PI-4 kinase, 107
PIP kinase, 351

Plasma membrane, 52, 53
Polyene antibiotics, 304
Polyphosphoinositides, 347, 349, 356
  metabolism, 10
Polysaccharides, 306
Porins, 151
Potassium channel, 328, 354, 358
P21, 109
P ras, 109
Prenylation, 271
Priming specific factor, 261
Profilin, 100, 105, 106
Prostaglandins, 104
Protein kinase
  A, 67, 68, 125
  C, 20, 45, 50, 51, 132, 135, 190, 204, 205, 295, 320, 346, 354
  G, 67, 68, 125
  substrates, 135
Protein phosphatases, 302, 303, 329
Protein phosphorylation, 115
Proteinase inhibitors, 304
Protein serine/threonine kinase, 122, 124, 125, 127, 130, 131, 133, 145, 154
Protein tyrosine phosphatase, 99, 120, 144, 302, 303
Proton ATPase, 334
Protoplast culture, 356, 361
Pulvinus, 326
Pyrophosphorylated phytic acid, 248
Pyruvate kinase, 173

Quinic acid, 376
L-Quebrachetol, 376

Radiolabelling of InP, 400
raf, 130
  ANP, 116, 129, 130, 132, 135
  Cd-interleukins (IL), 116, 129, 130, 132, 135
  T cell erythropoietin, 116, 129, 130, 132, 135
ras, 102
ras-GEF, 130
ras-transformed cell line, 103
Receptor-linked PI hydrolysis, 44
  growth factor, 47
Receptors
  cytokine, 116
  EGF, 116
  PDGF, 107
  tyrosine kinase, 116, 132, 302
retinal degeneration gene B (*rdgB*), 282
  Ca^{2+} binding, 262, 265, 276
  in olfaction, 262, 265, 276
  ptd transfer protein, 262, 265, 276
  in retinal degeneration, 262, 265, 276
Rhodopsin, 214
rho protein: *see* G-protein
Ribulose-5-P, 291, 293
  mRNA, 50
RSV-transformed fibroblast, 28
RTK, 98, 103
Ryanodine receptor, 63, 72

Saponin-permeabilized cells, 49, 384
Satellite proteins, 99
Scylloinositol, 232
Second messengers, 14–15, 21, 135, 146, 153, 155, 171, 174, 190, 243, 263, 321, 324, 372
  cAMP, 345
  DAG, 345
  $Ins(1,4,5)P_3$, 14
Separation of inositol phosphates, 371, 396
Serum response
  element, 142
  factor, 142
Serine/threonine phosphatase, 173
SH_2 domain in
  Ptd Ins 3-kinase, 121
  tyrosine kinase, 263, 264
SH_3 domain, 100, 106, 120
Sphingolipids, 232
Sphingomyelin, 176
Signal transduction, 249, 262, 264, 266, 271, 276, 277, 317, 346
Signaling by inositol hexakisphosphate, 298, 303
Signaling by wound, 304
Site-directed mutagenesis, 155
Son of sevenless (sos), 102
src Homology 2 (SH2), 120
STD, 154
Stomata, 327
  closure, 328
  guard cells, 305, 329
  pores, 328
Substance P receptor, 214

SV40, 147
Synaptic transmission, 51, 215
Synaptosomes, 5
Synaptotagmin, 215
Synthesis of,
  $Ins(1,3,4,6)P_4$, 191
  $Ins(1,4,5,6)P_4$, 193, 194
  $Ins(3,4,5,6)P_4$, 193, 198, 199, 289
  $Ins(1,3,4,5,6)_5$, 188, 289
  $InsP_6$, 207, 289
  $InsP_7/InsP_8$, 210, 246, 257, 274
Systemic wound response, 304
Synthetic inositol phosphates, 377, 378

TGF-β-signaling, 133
Throbin, 7, 12, 25
Thy-1, 166, 168, 169, 176, 177
T/pp6oc src complex, 107
Transdųcers, 129, 148
Transmembrane integrins, 129
Transpeptidase, 169
Transformimng growth factor β (TGF-β), 133
3-5,6-trideoxy Ins(1)P, 387, 388
Trifluoromethylsulfonic acid, 385
Troponin C, 127
Troponin I, 127

UDP-galactose (UDP-GlcNac), 169

Vanadate, 15
Vascular endothelial cell growth factor, 117, 119
Virulence, 152
v-fos, 154
v-jun, 154
v-raf, 131, 154
v-ras, 131, 154
v-src, 131, 154
Vasodilator-stimulated protein (VASP), 137
Vasopressin, 13, 18
Verapamil, 330
Verticillium albatrum, 325
Virulence, 152
Voltage-dependent Ca^{2+} channels, 297
von Recklinghausen neurofibromatosis, 154
VSV G proteins, 175

Wound-inducible genes, 305

Xenopus oocytes, 14, 15, 29, 63, 72

Yttrium ions, 400

MIX
Papier aus verantwortungsvollen Quellen
Paper from responsible sources
FSC® C105338
FSC
www.fsc.org

If you have any concerns about our products,
you can contact us on
ProductSafety@springernature.com

In case Publisher is established outside the EU,
the EU authorized representative is:
Springer Nature Customer Service Center GmbH
Europaplatz 3, 69115 Heidelberg, Germany

Printed by Libri Plureos GmbH
in Hamburg, Germany